Philosophy of Nonlinear Control Systems

Advances
in
Science
and
Technology
in
the USSR

Technology Series

ADVANCES IN
SCIENCE AND TECHNOLOGY
IN THE USSR

Philosophy of Nonlinear Control Systems

Edited by B. NAUMOV

Mir Publishers
Moscow

CRC Press
Boca Raton Ann Arbor Boston

Library of Congress Cataloging-in-Publication Data

Philosophy of nonlinear control systems / edited by B. Naumov.
p. cm. - - (Advances in science and technology in the USSR. Technology series)
Translated from the Russian by M. Edelev.
Includes bibliographical references.
ISBN 0-8493-7127-9
1. Nonlinear control theory. I. Naumov, B. N. (Boris Nikolaevich) II. Series.
QA402.35.P45 1990
629.8′312—dc20 90-5769
CIP

This edition published by CRC Press, Inc., 2000 Corporate Blvd. N.W. , Boca Raton, Florida.

International Standard Book Number 0-8493-7127-9

Library of Congress Card Number 90-5769
Printed in the United States

CONTENTS

PREFACE

High accuracy and performance required of modern automatic systems can be achieved with the use of control systems using high gain coefficients, relay elements, and physical elements with saturation or dead zone. With a sine input the output of such nonlinear systems is no longer sinusoidal. Existing theories of nonlinear systems are by no means exhaustive, as the processes in such systems often avoid both analytical description and numerical analysis.

Development of closed-loop nonlinear systems has to do with a number of physical phenomena. These include above all self-excited oscillations that disturb normal operation and may cause breadkdown of system elements and controlled plants. Nonlinear system behaviour may reveal intermittent ranges of stability and self-excited oscillations with parameters depending on the magnitude of system input and element gains. The influence of self-excited oscillations can be diminished by harnessing the effect of entrainment that leads to periodic motion with higher frequency and much lower amplitude. Damping of self-excited oscillations can also occur as a result of intense random disturbances.

In some cases the form of the periodic output is controlled by higher harmonics, so that the system analysis by the amplitude of the fundamental component may lead to erroneous conclusions about the system stability. In the vicinity of the resonant frequency the amplitude and phase of the output can change jumpwise and have an undesired effect on the plant. This is but a short list of specific features of the nonlinear systems considered in this book.

The nonlinearities in closed-loop systems prevent the use of classical methods of system analysis and design. The most popular techniques of stability analysis instead are the second (or direct) method of Lyapunov, the harmonic balance method first developed by Krylov and Bogoljubov, and the absolute stability criterion due to Popov. The physically appealing techniques of frequency-domain analysis have also become widely accepted in control engineering. However, the application of this analysis to complicated systems is limited by a considerable amount of computations involved, therefore the progress in this area depends heavily on the advance in numerical methods and computer hardware.

The major purpose of this multiauthor volume has been to highlight the advances of Soviet science in the development of control for multiloop nonlinear systems. The emphasis is on the problems of system design with the sufficient criteria of conditional and absolute stability. A vector-matrix technique has been developed for selection of compensating facilities that ensure the performance specifications and accuracy for both time-invariant and dynamic systems. The compensating elements are realized in the form of piecewise-linear components and in the form of microprocessor and computer programs. The last two cases draw on the methods of logical control and multidimensional z-transform.

One of the chapters in this volume is devoted to the application of the path integral method to the design of systems subjected to random disturbances. Breakdown of control is investigated and the probability of rare events is estimated to forecast the likely emergency conditions and failures and to find a way to cope with such situations. A method of statistical nodes has been devised to build mathematical models of random processes. It is helpful in solving the problems of analysis and synthesis of complex nonlinear dynamic systems.

The performance and cost of computer-assisted development projects depend on the degree of utilization of both the computer capabilities and the intangible experience of the designer. For many a practical problem it is convenient to join the construction and selection of effective system solution so that they be handled on the basis of multiple objective optimization. In place of traditional optimization techniques the authors use an optimization procedure that is based on a multistage man-machine decision making discarding sequentially the areas of parameter space devoid of the optimal solution. The procedure results in the construction of a Pareto-optimal set of nonlinear-system versions of which the more suitable can be selected.

Although the majority of techniques discussed in this volume are widely known, some of them are under development and should find their application in the nearest future. Nevertheless we hope that these discussions may suggest the reader a way to solving similar problems.

Moscow 1988 B. N. Naumov

1 Introduction

B. N. Naumov

The present stage of control system development is associated with the study of multiloop nonlinear systems. Unfortunately the classical methods of single-loop nonlinear system design cannot be extended to the new class of systems owing to a number of physical phenomena that either have not occurred in or have had only a weak effect on the processes in the simple systems. Although the control problems in complex nonlinear systems have been extensively reported, the pace of advance in the relevant techniques of design is rather slow. This is associated with the fact that such systems may experience a few limit cycles whose amplitudes and frequencies vary in a wide range due to both control or disturbance inputs and the effect of higher harmonics.

The behavior of autonomous nonlinear systems is also significantly affected by the initial conditions. A perturbation of initial conditions leads to a change of the stable state, i.e. the output is no longer single-valued. A sine input to the nonlinear system produces low-frequency subharmonic oscillations of various multiplicity, jumpwise resonance phenomena in amplitude and frequency, entrainment of frequency, and a rapid change of frequency. This list can be continued at length, i.e. the class of such systems is characteristic of a wide variety of intrinsic processes.

The situation complicates further if there is a need for the incorporation of random disturbances, since at low intensity of noise the system may experience "spontaneous" generation of self-excited oscillations, alteration of oscillatory modes, and a transition from one stable state to another. However, when the intensity of the random signal increases, the limit cycles diffuse out and completely disappear, and the subharmonics and jumpwise resonance phenomena cease to exist. Practical adaptation of such artificial disturbances for damping of self-excited oscillations is retarded by severely penalized system accuracy. Therefore, the designer is presented with a multicriterial optimization problem very much alike to other cases.

The problems of analysis and design of high-order nonlinear systems are as a rule solved by such approximate methods as harmonic and statistical linearization, modeling, and multicriterial synthesis assisted by a computer. Computer simulation does not always lead to correct results even if the stability of the system being designed has been ascertained. To prove that the conclusions derived are correct, one needs accurate analytical techniques such as the direct Lyapunov method and the absolute stability criterion due to Popov. These techniques require a simplified mathematical model adequately describing the processes concerned for all operating modes. On the basis of the analytical methods the designer can rapidly select the best version and parameters of the linear and nonlinear compensating elements that suppress self-excited oscillations and other undesirable operating modes.

One of the more versatile practical approximations is the method of harmonic linearization that has been developed on the basis of asymptotic expansions of solutions to nonlinear equations in a series over a complete system of orthogonal functions. This method developed by N. M. Krylov and N. N. Bogoljubov [1.1, 2] has found wide acceptance. It is used to evaluate the conditions leading to self-excited oscillations, the parameters and stability of such oscillations in analysis of oscillations and phenomena associated with a jumpwise resonance.

The most developed version of the method is the one assuming a sine signal applied to the input of any nonlinear element in the loop. This assumption simplifies appreciably the derivation of harmonic linearization coefficients with respect to the fundamental component. The concept of describing function introduced with the aid of these coefficients has proved useful in solving the problems of analysis and design of nonlinear systems.

This method helps revealing the effects of higher harmonics on the fundamental component and obtaining a generalized describing function that defines a more accurate bound of stability areas. It also enables the parameters of self-excited oscillations to be determined with solely the higher harmonics (in the fundamental component such oscillations may be absent) and the applicability conditions of the harmonic linearization method to be delineated. A recent impetus to the development of this direction has been due to the evaluation of performance indices, accuracy and synthesis of compensating elements [1.3].

With an eye to the direct method of Laypunov, V. M. Popov of Romania has formulated a frequency criterion defining a sufficient condition of asymptotic stability for single-loop systems with various types of single-valued non-linearities [1.4]. In the following these findings have been generalized over the dynamic systems with multi-valued and nonstationary nonlinearities [1.5]. Asymptotic stability and a number of constraints imposed on the position of zeros and poles of the linear system guarantee absolute stability to the entire system.

An original absolute-stability definition has been developed by B. N. Naumov [1.6]; the method is based on the construction of a Nyquist locus and application of a Nichols chart. Other developments of this author worthy of mention include a method of synthesis of cascade and parallel correcting elements that provide the specified degree of absolute stability, and the evaluation of asymptotic behavior of the control action and plant input [1.7]. The two last techniques have been generalized to the case of nonlinear systems [1.8].

Stochastic studies of control systems as a rule draw on the idea of statistical linearization devised by I. E. Kazakov [1.9, 10]. The method consists essentially in the evaluation of the best stochastic approximation of a nonlinear lag-free functional transformation by a linearized dependence. In its simple version this search reduces to a requirement that the initial and fitting functions should possess sufficiently close first and second moments. Another analytical approach to stochastic systems is based on the concurrent canonic developments of the random input and stochastic nonlinearity. A substantial contribution to the theory of statistical linearization of stochastic transformations more general than stochastic nonlinearities has been made by V. S. Pugachev with his method of stochastic system normalization [1.11]. This approach has markedly simplified analysis of nonlinear stochastic systems and revealed the principal regularities of such systems. An extensive coverage of the methods of statistical linearization can be found in the review paper by Sinitsyn [1.12].

If the input of a nonlinear system is corrupted by a noise bounded in magnitude, the stability analysis of the system may be carried out with the absolute stability criterion. Such a criterion has been generalized also to the case of noise unbounded in magnitude but obeying a normal distribution law [1.6].

An outstanding role in the studies of dynamic processes in nonlinear systems belongs to A. M. Lyapunov. His rigorous definitions of dynamic system stability and instability [1.13] formed a basis of the two famous methods of solving stability problem. In the first method, the linearized equations of system dynamics suggest in most cases whether the equilibrium position of the nonlinear system is stable or unstable. This method establishes the exact applicability limits of nonlinear system stability analysis with linearized equations.

According to the second, or direct method, stability and instability of a nonlinear system are defined by the sign of the Lyapunov function and its time derivative. This approach enables the analyst to handle the stability problem without having to solve the nonlinear equations of motion, thus the analysis of stability, performance and accuracy becomes much neater [1.14].

P. L. Chebyshev contributed a great deal into the stability theory of dynamic systems described by high-order differential equations. He formulated and solved the problem of direct-action controller with Coulomb friction [1.15]. This method led to the development of a high accuracy control system for angular velocity of a steam engine, and paved the way for a new design of the Watt governor.

The need for high-speed controllers with relay elements in the control loop called for design techniques based on the theory of bifurcations of Poincaré [1.16]. A development of this theory for nonlinear systems has been suggested by A. A. Andronov in the form of an image point method enabling an easy location of self-excited oscillation modes [1.17]. This method was used to study the conditions of sliding modes and complicated forms of self-excited oscillations.

Considerable growth of the order of the differential equations describing the behavior of new large control systems motivated the development of relevant approximated methods of system design. N. M. Krylov and N. N. Bogoljubov suggested the principle of harmonic balance that formed a basis for harmonic analysis of nonlinear systems. This approach has been extended to multiple frequency forced oscillations with large and small frequency differences [1.3]. The graphical [1.18] and analytical [1.19] methods have been developed for evaluation of the linearization coefficients of complex nonlinearities. An exact method has been developed for Fourier-series evaluation of periodic modes for various forms of piecewise linear single-valued characteristics. As a result, the nonlinear problem solving boils down to solution of transcendental equations describing system motion within one period [1.20].

Other Soviet developments in the field include the separation of motions in short-period and slow components [1.21], small parameter technique [1.22], and an analytical approach to controller design [1.23]. L. S. Pontryagin has devised a mathematical theory of optimal processes and formulated the maximum principle valid for both linear and nonlinear control systems [1.24].

One could hardly overestimate the advance of Soviet science in the studies of Markov processes [1.25], optimal nonlinear filtering, and statistical identification of plant parameters. A. N. Kolmogorov [1.26] was the first to derive the equation for transition probability density in nonlinear systems. V. S. Pugachev [1.11] used this equation to obtain the equation for the characteristic function. A considerable contribution has been made in the field of stochastic differential equations [1.27], systems of such equations [1.28], and statistically equivalent decompositions [1.29]. A method of equivalent perturbations [1.30] has been devised and a method of functional series in the central moments of higher orders [1.31] has been developed. These findings have been widely reported in the Soviet Union and elsewhere.

This introduction by no means pretends to an exhaustive coverage of the work of Soviet researchers in the field of nonlinear control systems. The focus of this volume is on the techniques and developments that might not be well known to readers outside the Soviet Union as well as on some new promising directions of research.

References

1.1 N. M. Krylov and N. N. Bogoljubov: *Novye metody nelineinoi mekhaniki* (New methods of nonlinear mechanics). Gostekhizdat, Moscow 1934.

1.2 N. M. Krylov and N. N. Bogoljubov: *Vvedeniye v nelineinuyu mekhaniku* (Introduction to nonlinear mechanics). Izdatelstvo AN Ukr. SSR, Kiev 1937.

1.3 E. P. Popov: *Prikladnaya teoriya protsessov upravleniya v nelineinykh sistemakh* (Applied theory of control in nonlinear systems). Nauka, Moscow 1973.

1.4 V. M. Popov: *Studii si Cercetari de Energetica* **9**(4), 119-135 (1959).

1.5 B. N. Naumov: *Teoriya nelineinykh avtomaticheskikh sistem* (Theory of nonlinear systems). Nauka, Moscow 1972.

1.6 B. N. Naumov: *Avtomatika i Telemekhanika* No. 4, 591-600 (1965).

1.7 B. N. Naumov: in *Tochnye metody issledovaniya nelineinykh system avtomaticheskogo upravleniya* (Exact methods of nonlinear control system analysis). Mashinostroyeniye, Moscow 1971.

1.8 V. A. Yakubovich: *Avtomatika i Telemekhanika* No. 6 (1967).

1.9 I. E. Kazakov: *Avtomatika i Telemekhanika* **17**(5), 385-409 (1956).

1.10 I. E. Kazakov: *Statisticheskiye metody proyektirovaniya sistem upravleniya* (Statistical methods of control system design). Mashinostroyeniye, Moscow 1969.

1.11 V. S. Pugachev: *Teoriya sluchainykh funktsii i ee primeneniye k zadacham avtomaticheskogo upravleniya* (Theory of stochastic functions and its application to problems of automatic control). Fizmatgiz, Moscow 1962.

1.12 I. N. Sinitsyn: *Avtomatika i Telemekhanika* No. 5, 74-94 (1974).

1.13 A. M. Lyapunov: *General problem of stability of motion.* D.Sc. thesis. Kharkov 1892.

1.14 N. G. Chetayev: *Ustoichivost dvizheniya* (Stability of motion). Gostekhizdat, Moscow-Leningrad 1946.

1.15 P. L. Chebyshev: On the centrifugal equation. *Collection of works.* Vol. 2. St. Petersburg 1907.

1.16 A. Poincaré: *Les methodes novelles de la mécanique céleste.* Vol. 1. Gauthiez-Vellars 1892.

1.17 A. A. Andronov and S. E. Khaikin: *Teoriya kolebanii* (Oscillation theory). ONTI, Moscow 1937.

1.18 Ya. Z. Tsypkin: *Avtomatika i Telemekhanika* No. 4, 1956.

1.19 E. P. Popov and I. P. Paltov: *Approximate methods for the investigation of non-linear control systems.* Akademische Verlag Gesellschaft, Leipzig 1963.

1.20 Ya. Z. Tsypkin: *Teoriya releinykh sistem avtomaticheskogo regulirovaniya* (Theory of relay control systems). Gostekhizdat, Moscow 1955.

1.21 E. I. Gerashchenko: in *Metody sinteza nelineinykh sistem avtomaticheskogo upravleniya* (Methods of nonlinear control system design). Mashinostroyeniye, Moscow 1970.

1.22 B. V. Bulgakov: *Kolebaniya* (Oscillations). Gostekhizdat, Moscow-Leningrad 1949.

1.23 A. M. Letov: *Stability in Nonlinear Control Systems.* Princeton University Press, Princeton, New Jersey 1961.

1.24 L. S. Pontryagin, V. G. Boltyanskii, R. V. Gamkrelidze, and E. P. Mishchenko: *Matematicheskaya teoriya optimalnykh protsessov* (Mathematical theory of optimal control), Nauka, Moscow 1969.

1.25 V. I. Tikhonov and M. A. Mironov: *Markovskiye protsessy* (Markov processes). Sovetskoye Radio, Moscow 1977.

1.26 A. N. Kolmogorov: *Math. Ann.* 104, 415-458 (1931).

1.27 R. L. Stratonovich: *Izbrannye voprosy teorii fluktuatsii v radiotekhnike* (Application of selected topics of fluctuation theory to radio engineering). Sov. Radio, Moscow, 1961.

1.28 V. S. Pugachev and I. N. Sinitsyn: *Stokhasticheskiye differentsialnye sistemy* (Stochastic differential systems), Nauka, Moscow 1985.

1.29 I. E. Kazakov and S. V. Malchikov: *Analiz stokhasticheskikh sistem v prostranstve sostoyanii* (Analysis of stochastic systems in state space). Nauka, Moscow 1973.

1.30 I. E. Kazakov and B. G. Dostupov: *Statisticheskaya dinamika nelineinykh avtomaticheskikh sistem* (Statistical dynamics of nonlinear automatic systems). Fizmatgiz, Moscow 1962.

1.31 V. I. Demukh: *Avtomatika i Telemekhanika* **26**(6), 417-424 (1965).

2 Application of Lyapunov's Second Method to Analysis and Design of Control Systems

B. N. Naumov and Yu. I. Topcheyev

Nonlinear problems of control theory had for some years attracted attention of such outstanding researchers as A. I. Wischnegradski, N. E. Zhukovsky, A. Poincaré, P. Chebyshev and A. Stodola. They formulated a number of important concepts describing the effect of nonlinearities on the behavior of conservative dynamic systems. However, their efforts had not resulted in a general theory of stability. This goal was achieved in 1888-92 by the outstanding Russian scientist A. M. Lyapunov [2.1, 2]. The results of his fundamental studies had not been used for 60 years and it were only the publications of N. G. Chetayev [2.3, 4], I. G. Malkin [2.5, 6], and G. V. Kamenkov [2.7] that highlighted the ideas of Lyapunov that paved the way for their wide application in this country and abroad [2.8, 2.9, 2.10, 2.11].

The Lyapunov theory was first used to advantage in the problems of analysis [2.3, 7, 10, 11] and later was extended to the problems of synthesis [2.12]. This resulted in a uniform procedure for design of nonlinear time-invariant and time-varying dynamic systems. The methods thus developed have been reduced to a form of algorithms or computer programs that simplify application in studies of complicated dynamic systems described by high-order differential equations.

2.1 Generation of Lyapunov Functions for Continuous, Time-Invariant and Time-Varying Systems

At the beginning we consider the general concepts of the Lyapunov stability theory in the context of time-invariant nonlinear systems. We assume that a system of the given class described by n-th order differential equations is associated with a continuous function $V(\mathbf{x})$ possessing continuous partial derivatives of the first order in some domain Γ including the origin. In this domain the function $V(\mathbf{x})$ is positive definite and its time derivative

$$\dot{V}(\mathbf{x}) = \sum_{i=1}^{n} \frac{\partial V(\mathbf{x})}{\partial x_i} \frac{\mathrm{d}x_i}{\mathrm{d}t} = [\nabla V(\mathbf{x})]^T \dot{\mathbf{x}} \tag{2.1}$$

along the trajectory $\mathbf{x}(t)$ of the system $\dot{\mathbf{x}} = f(\mathbf{x})$ is negative, i.e. $\dot{V}(\mathbf{x}) \leqslant 0$. The gradient $\nabla V(\mathbf{x})$ is an n-tuple with the i-th component $\partial V(\mathbf{x})/\partial x_i$. The function defined in this manner is a Lyapunov function.

It is important to note that no general method has been developed as yet to determine Lyapunov functions for nonlinear systems. These functions assume a rather simple form for the following cases.

1. The function has a matrix-product form

$$V(\mathbf{x}) = \sum_{i,j=1}^{n} p_{ij} x_i x_j, \qquad p_{ij} = p_{ji} \tag{2.2}$$

or

$$V(\mathbf{x}) = \mathbf{x}^T \mathbf{P} \mathbf{x} \tag{2.3}$$

where $\mathbf{P}$ is a symmetric $n \times n$ matrix.

2. The function $V(\mathbf{x})$ is a homogeneous polynomial of degree k

$$V(\alpha x_1, \quad \alpha x_2, \ \ldots, \quad \alpha x_n) = \alpha^k V(x_1, x_2, \ \ldots, x_n) \tag{2.4}$$

3. The function can be represented as the sum of two components

$$V(\mathbf{x}) = V_1(\mathbf{x}) + V_2(\mathbf{x}) \tag{2.5}$$

with

$$V_1(\mathbf{x}) = \sum_{j=1}^{n} \left| \sum_{i=1}^{n-j+1} k_{ji} x_i \right|; \qquad V_2(\mathbf{x}) = \sum_{j=1}^{n} |k_j x_j| \tag{2.6}$$

suitable for analysis of piecewise-linear systems [2.9, 12]. Let

$$\sum_{i=1}^{n-j+1} k_{ji} x_i = g_j \tag{2.7}$$

$$|g_j| = g_j \operatorname{sgn} g_j, \quad j = 1, \ \ldots, n \tag{2.8}$$

and find the derivative

$$\frac{\mathrm{d}}{\mathrm{d}t} |g_j| = \frac{\mathrm{d}}{\mathrm{d}t} (g_j \operatorname{sgn} g_j) \tag{2.9}$$

of this piecewise-linear function

$$\frac{\mathrm{d}}{\mathrm{d}t} |g_j| = \left(\sum_{i=1}^{n-j+1} k_{ji} \dot{x}_i \right) \operatorname{sgn} \left(\sum_{i=1}^{n-j+1} k_{ji} x_i \right), \quad j = 1, \ \ldots, n \tag{2.10}$$

Then the derivative of the Lyapunov function will be

$$\dot{V}(\mathbf{x}) = \sum_{i=1}^{n} (\dot{g}_j \operatorname{sgn} g_j + k_j x_j \operatorname{sgn} x_j) \tag{2.11}$$

with the coefficients defined by the equation

$$\dot{V}(\mathbf{x}) = -kV(\mathbf{x}) \tag{2.12}$$

Among other methods of determination of Lyapunov functions for nonlinear problems we should like to note the works of Letov [2.13], Zubov [2.9], G. Szege, and Schultz [2.11]. This chapter covers only those works that have found wide acceptance in practice.

The stability conditions stipulated by the second method of Lyapunov for nonlinear, dynamic, time-invariant systems are sufficient only, therefore, if $V(\mathbf{x})$ avoids evaluation, this by no means implies that the system is unstable. Two stability conditions of nonlinear stationary systems are most common.

First stability condition. If for a system described by the nonlinear differential equations

$$\dot{\mathbf{x}} = f(\mathbf{x}) \tag{2.13}$$

with $\mathbf{x}(0) = \mathbf{0}$ there exists a Lyapunov function $V(\mathbf{x})$ in some area Γ including the origin, then the origin of such a system is stable.

Second stability condition. If in an area Γ including the origin for a system (2.13) a Lyapunov function $V(\mathbf{x})$ can be prescribed such that its derivative $\dot{V}(\mathbf{x})$ is negative definite, then the unperturbed solution to this system is asymptotically stable.

We illustrate these definitions with two examples.

Example 2.1. For a system of nonlinear equations

$$\begin{aligned} \dot{x}_1 &= -3x_2 - x_1^3 \\ \dot{x}_2 &= -2x_2 + x_1^3 \end{aligned} \tag{2.14}$$

the validity of the first condition can be verified by setting the Lyapunov function in the form

$$V(x_1, x_2) = \frac{1}{2}x_1^4 + 3x_2^2 \tag{2.15}$$

Example 2.2. The validity of the second condition for the system of nonlinear equations

$$\begin{aligned} \dot{x}_1 &= x_2 \\ \dot{x}_2 &= -x_2 + |x_1|x_2 - x_1 \end{aligned} \tag{2.16}$$

can be demonstrated by writing for $x_1 \geqslant 0$ a Lyapunov function in the form

$$V(x_1, x_2) = 2x_1^2 - |x_1|x_1^2 + \frac{1}{4}x_1^4 + 2x_1x_2 - |x_1|x_1x_2 + x_2^2 > 0 \tag{2.17}$$

then

$$\dot{V}(x_1, x_2) = 4x_1\dot{x}_1 - 3x_1^2\dot{x}_1 + x_1^3\dot{x}_1 + 2\dot{x}_1x_2 + 2x_1\dot{x}_2 - 2x_1x_2\dot{x}_1 - x_1^2\dot{x}_2 + 2x_2\dot{x}_2 \tag{2.18}$$

Substituting (2.16) into (2.18) yields

$$\dot{V}(x_1, x_2) = -x_1^2(2 - |x_1|) \tag{2.19}$$

The system (2.16) is seen to be asymptotically stable for $|x_1| < 2$, unstable for $|x_1| > 2$, and experiences self-excited oscillations at $|x_1| = 2$.

Now, we formulate three instability conditions for nonlinear systems due to Lyapunov.

First instability condition. If for a system of differential equations (2.13) there exists a continuous function $V(\mathbf{x})$ possessing continuous partial derivatives in an area Γ including the origin where $V(\mathbf{0}) = 0$, the derivative $\dot{V}(\mathbf{x})$ is positive definite along the trajectory $\mathbf{x}(t)$, and $V(\mathbf{x})$ is nonnegative in some vicinity of the origin, then the unperturbed solution is unstable.

Second instability condition. If for a system of differential equations (2.13) a function $V_1(\mathbf{x})$ exists in an area about the origin such that $V_1(\mathbf{0}) = 0$ and

$$\dot{V}_1 = \Lambda V_1 + W(\mathbf{x}) \tag{2.20}$$

along the trajectory $\mathbf{x}(t)$, where $\Lambda > 0$ and $W(\mathbf{x}) \geqslant 0$ in Γ, and $V_1(\mathbf{x})$ is nonnegative in an arbitrary neighborhood of the origin, then the origin is unstable.

Third instability condition. If Γ is some area about the origin containing Γ_1 that includes the origin and there exists a function $V(\mathbf{x})$ continuously differentiable with respect to $\mathbf{x}$ in Γ, positive definite along with $\dot{V}(\mathbf{x})$ in Γ_1 and vanishing on the boundary of Γ_1, then the origin of the system is unstable.

The last definition of instability was given by Chetayev [2.3] and therefore is commonly referred to as the Lyapunov-Chetayev instability condition.

The examples that follow are intended to illustrate these instability conditions.

Example 2.3. For the system of equations

$$\begin{aligned} \dot{x}_1 &= -3x_2 + x_1^3 \\ \dot{x}_2 &= 2x_2 + x_1^3 \end{aligned} \tag{2.21}$$

we construct the Lyapunov function in the form

$$V(x_1, x_2) = \frac{1}{2}x_1^4 + 3x_2^2 \tag{2.22}$$

which vanishes at the origin, $V(0, 0) = 0$.

Constructing the derivative

$$\dot{V}(x_1, x_2) = 2x_1^3\dot{x}_1 + 6x_2\dot{x}_2 \tag{2.23}$$

and substituting equations (2.21) for the derivatives in this expression, we see that for all x_1 and x_2

$$\dot{V}(x_1, x_2) = 2x_1^6 + 12x_2^2 > 0 \tag{2.24}$$

Thus, equations (2.22) and (2.24) indicate that the first instability condition of Lyapunov is satisfied, hence the nonlinear system in question is unstable.

Example 2.4. For the system

$$\begin{aligned} \dot{x}_1 &= x_1(1 + x_2^2) \\ \dot{x}_2 &= x_2 \end{aligned} \tag{2.25}$$

we take the Lyapunov function in the form

$$V(x_1, x_2) = x_1^2 + 2x_2^2 \tag{2.26}$$

with the derivative

$$\dot{V}(x_1, x_2) = 2x_1\dot{x}_1 + 4x_2\dot{x}_2 \tag{2.27}$$

Substituting (2.25) in (2.27) yields

$$\dot{V}(x_1, x_2) = 2(x_1^2 + 2x_2^2) + 2x_1^2x_2^2 \tag{2.28}$$

From this expression it follows that $\Lambda = 2$, and $W(x_1, x_2) = 2x_1^2x_2^2 > 0$, i.e. the second instability condition is satisfied, hence the nonlinear system (2.25) is unstable at the origin in Lyapunov's sense.

Example 2.5. For the system of differential equations

$$\begin{aligned} \dot{x}_1 &= ax_1^2 + bx_2^3 \\ \dot{x}_2 &= -cx_2 + dx_1^3 \end{aligned} \tag{2.29}$$

with positive a, b, c, and d, we construct the Lyapunov function in the form

$$V(x_1, x_2) = x_1 - x_2^2/2 \tag{2.30}$$

Its time derivative

$$\dot{V}(x_1, x_2) = \dot{x}_1 - x_2\dot{x}_2 \tag{2.31}$$

after substitution from (2.29) becomes

$$\dot{V}(x_1, x_2) = ax_1^2 + bx_2^3 + cx_2^2 - dx_2x_1^3 \tag{2.32}$$

Observing that $V(0, 0) = 0$, and inside the Γ region, i.e. at points close to the origin, where the sign of (2.32) is governed by the positive sum $ax_1^2 + cx_2^2$, the derivative $\dot{V}(x_1, x_2)$ is positive, we see that the third condition of instability is satisfied and the origin is unstable.

Let us look now at the stability of time-varying nonlinear system. This system is described by the differential equations

$$\dot{\mathbf{x}} = f(\mathbf{x}, t), \quad f(\mathbf{0}, t) = \mathbf{0} \tag{2.33}$$

We assume that for this system there exists a function $V(\mathbf{x}, t)$ that is positive definite in region Γ encompassing the origin, i.e. $V(\mathbf{0}, t) = 0$ and such that $V(\mathbf{x}, t) = \varphi(\|\mathbf{x}\|)$, where φ is a monotonously increasing function and $\varphi(0) = 0$. In view of (2.33) for all t the time derivative of V is

$$\begin{aligned}\dot{V}(\mathbf{x}, t) &= \sum_{i=1}^{n} \frac{\partial V(\mathbf{x}, t)}{\partial x_i} f_i(\mathbf{x}, t) + \frac{\partial V(\mathbf{x}, t)}{\partial t} \\ &= [\nabla V(\mathbf{x}, t)]^T f(\mathbf{x}, t) + \frac{\partial V(\mathbf{x}, t)}{\partial t}\end{aligned} \tag{2.34}$$

Now, we are in a position to formulate three stability conditions for time-varying nonlinear systems.

First stability condition. If the scalar function $V(\mathbf{x}, t)$ generated for the system (2.33) is positive definite in some region Γ including the origin, has continuous partial derivatives in $\mathbf{x}$ and t, and its derivative $\dot{V}(\mathbf{x}, t)$ is negative in this region, then the origin of this system is stable.

Second stability condition. If there exists a positive, nondecreasing function $\theta(z)$ in one variable such that in addition to being positive, $V(\mathbf{x}, t) > 0$ (first stability condition), it has a derivative for which in some vicinity Γ of the origin the condition

$$\dot{V}(\mathbf{x}, t) \leqslant -\theta(\|\mathbf{x}\|) \tag{2.35}$$

holds, then the origin of this system is asymptotically stable.

Third stability condition. If there exists a continuous, strictly increasing function $\varrho(z)$ in one variable such that $\varrho(0) = 0$ and in addition to the condition $V(\mathbf{x}, t) > 0$ for $\mathbf{x} \neq \mathbf{0}$ and the inequality (2.35) the condition

$$V(\mathbf{x}, t) \leqslant \varrho(\|\mathbf{x}\|) \tag{2.36}$$

holds in Γ for all t, then the origin of this system is uniformly asymptotically stable.

This list may be complemented by a few more stability conditions in practical use. For uniform asymptotic stability, the following conditions must hold: $V(\mathbf{0}, t) = 0$, $V(\mathbf{x}, t) > 0$ for $\mathbf{x} \neq 0$, and the derivative $\dot{V}(\mathbf{x}, t)$ must satisfy the condition (2.36). It is worth noting that the stability condition of nonlinear time-varying systems is defined by an inequality involving the integral $\int_0^{x_1} \xi F(\xi)d\xi$ that avoids analytical evaluation in some problems and the analyst has to resort to numerical techniques and a computer. This approach therefore complicates solution of system synthesis problems.

The Schultz method [2.11] may be recommended for construction of Lyapunov functions. This method is suitable for both time-varying and time-invariant systems. In brief it is as follows. Let for the system (2.13) there exist a region Γ where the vector ∇V with the elements $\nabla V_i = \partial V/\partial x_i$ is such that

$$\frac{\partial \nabla V_i}{\partial x_j} = \frac{\partial \nabla V_j}{\partial x_i}$$

and ∇V does not vanish at any point of Γ except the origin:

$$\begin{aligned} \frac{d}{dt} V(\mathbf{x}) &= [\nabla V(\mathbf{x})]^T \dot{\mathbf{x}} < 0 \qquad \text{for } \mathbf{x} \neq 0 \\ &= 0 \qquad \text{for } \mathbf{x} = 0 \end{aligned}$$

and $dV(\mathbf{x})/dt$ is not identically zero in Γ, then the scalar function

$$V(\mathbf{x}) = \int_0^{\mathbf{x}} [\nabla V(\mathbf{x})]^T dt$$

possesses the following properties: $V(\mathbf{x})$ is positive definite and one of the surfaces $V = k$ bounds the region Γ. In this case the time-invariant nonlinear system is asymptotically stable in Γ. The integration is executed by the rule

$$\begin{aligned} V(\mathbf{x}) = &\int_0^{x_1} \nabla V_1(\xi_1, 0, \ldots, 0)d\xi_1 \\ &+ \int_0^{x_2} \nabla V_2(x_1, \xi_2, 0, \ldots, 0)d\xi_2 + \ldots \\ &+ \int_0^{x_n} \nabla V_n(x_1, x_2, \ldots, x_{n-1}, \xi_n)d\xi_n \end{aligned} \tag{2.37}$$

The Schultz method results in a uniformly valid scalar function $V(\mathbf{x})$. The subsequent analysis is performed with the aid of the relevant stability or instability systems.

2.2 Stability Analysis of Dynamic Systems by the Direct Method of Lyapunov

The direct method of Lyapunov offers the advantage of accurate investigation of asymptotic stability in the large and even total stability. This method is suitable for both time-invariant and time-varying systems. The Lyapunov function depends on the state coordinates and time for time-varying systems only. This method may also be used for comparison of control system performance. Below we give some examples of stability analysis for time-invariant and time-varying nonlinear systems [2.14].

Example 2.6. Reducing the differential equation

$$\dddot{x} + b\ddot{x} + c\dot{x} + F(x)x = 0$$

to the system

$$\begin{aligned} \dot{x}_1 &= x_2 \\ \dot{x}_2 &= x_3 \\ \dot{x}_3 &= -bx_3 - cx_2 - F(x_1)x_1 \end{aligned} \tag{2.38}$$

we analyze the effect of the type of nonlinearity on the system stability. The block diagram of the respective system is shown in Fig. 2.1*a*. The nonlinearity $F(x_1)$ is represented as a single element to simplify its analysis, say, by the method of harmonic linearization.

Writing the gradient of Lyapunov function in the form

$$\nabla V = \begin{bmatrix} \alpha_{11}x_1 + \alpha_{12}x_2 + \alpha_{13}x_3 \\ \alpha_{21}x_1 + \alpha_{22}x_2 + \alpha_{23}x_3 \\ \alpha_{31}x_1 + \alpha_{32}x_2 + \alpha_{33}x_3 \end{bmatrix} \tag{2.39}$$

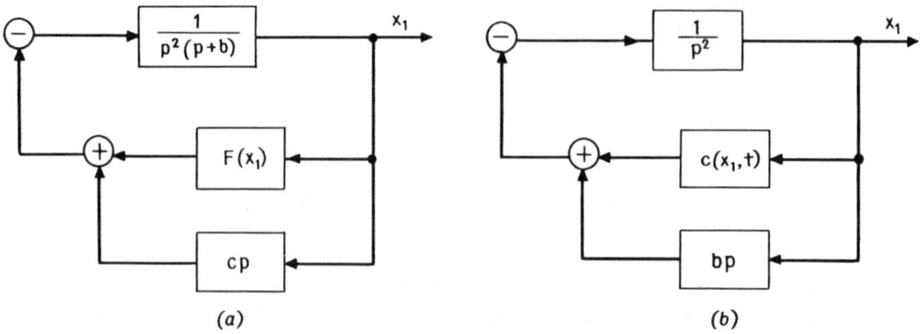

Fig. 2.1

and making use of (2.1) and (2.38), we obtain the time derivative in the form

$$\frac{dV}{dt} = -F(x_1)x_1^2\alpha_{31} + [\alpha_{11} - c\alpha_{31} - F(x_1)\alpha_{32}]x_1x_2 + [\alpha_{21} - b\alpha_{31} - F(x_1)\alpha_{33}]x_1x_3 + [\alpha_{12} - c\alpha_{32}]x_2^2 + [\alpha_{13} + \alpha_{22} - b\alpha_{32} - c\alpha_{33}]x_2x_3 + [\alpha_{23} - b\alpha_{33}]x_3^2 \tag{2.40}$$

To obtain the inequality $dV/dt < 0$ we put the expressions in brackets equal to zero, namely,

$$\begin{aligned} \alpha_{11} &= c\alpha_{31} + F(x_1)\alpha_{32} \\ \alpha_{21} &= b\alpha_{31} + F(x_1)\alpha_{33} \\ \alpha_{23} + \alpha_{22} &= b\alpha_{32} + c\alpha_{33} \end{aligned} \tag{2.41}$$

and let $\alpha_{13} = \alpha_{31} = 0$, $\alpha_{33} = 2$, then

$$\begin{aligned} \alpha_{11} &= F(x_1)\alpha_{32} \\ \alpha_{21} &= 2F(x_1) \\ \alpha_{22} &= 2c + b\alpha_{32} \end{aligned} \tag{2.42}$$

Examination of equation (2.40) under the above assumptions indicates that its x_1^2 term will vanish. Let also

$$\alpha_{23} = \alpha_{32} = 2b \tag{2.43}$$

The coefficient α_{12} is not a constant and will be determined subject to the condition

$$\frac{\partial \nabla V_1}{\partial x_2} = \frac{\partial \nabla V_2}{\partial x_1} \tag{2.44}$$

where the derivatives ∇V_1 and ∇V_2 are obtained from equations (2.39), (2.42) and (2.43) as

$$\begin{bmatrix} \nabla V_1 \\ \nabla V_2 \end{bmatrix} = \begin{bmatrix} 2bF(x_1)x_1 + \alpha_{12}x_2 \\ 2F(x_1)x_1 + 2(b^2 + c)x_2 + 2bx_3 \end{bmatrix} \tag{2.45}$$

By virtue of (2.44) and (2.45) we get

$$\frac{\partial \nabla V_1}{\partial x_2} = \alpha_{12}$$

$$\frac{\partial \nabla V_2}{\partial x_1} = 2F(x_1) + 2x_1 \frac{\partial F(x_1)}{\partial x_1}$$

whence

$$\alpha_{12} = 2F(x_1) + 2x_1 \frac{\partial F(x_1)}{\partial x_1} \tag{2.46}$$

Thus, we can write the gradient of Lyapunov function (2.39) in terms of the coefficients (2.42), (2.43) and (2.46) as

$$\nabla V = \begin{bmatrix} 2bx_1F(x_1) + \left(2F(x_1) + 2x_1 \dfrac{\partial F(x_1)}{\partial x_1}\right) x_2 \\ 2x_1F(x_1) + 2(b^2 + c)x_2 + 2bx_3 \\ 2bx_2 + 2x_3 \end{bmatrix} \tag{2.47}$$

With the expressions (2.39) and (2.47) we find

$$V(x_1, x_2, x_3) = 2b \int_0^{x_1} \xi F(\xi)\mathrm{d}\xi + 2x_1x_2F(x_1) + (b^2 + c)x_2^2 + 2bx_2x_3 + x_3^2 \tag{2.48}$$

from which

$$\frac{\mathrm{d}V}{\mathrm{d}t} = -2[bc - F(x_1)]x_2^2 + 2x_2^2 \frac{\partial F(x_1)}{\partial x_1} \tag{2.49}$$

For nonsymmetric nonlinearities $F(x_1)$ we introduce simplifications as follows:

$$\left.\begin{aligned} \alpha_{12} &= c\alpha_{32} \\ \alpha_{11} &= c\alpha_{31} + F(x_1)\alpha_{32} \end{aligned}\right\} \tag{2.50}$$

At $\alpha_{12} = \alpha_{21}$, we get

$$\alpha_{21} = b\alpha_{31} = c\alpha_{32} \tag{2.51}$$

The remaining α_{ij}, except α_{11}, are found in the same manner. Observing (2.50-51), we find

$$\frac{\mathrm{d}V}{\mathrm{d}t} = -\frac{c}{b} F(x_1)\alpha_{32}x_1^2 - 2x_1x_3F(x_1) - (2b - \alpha_{23})x_3^2 \tag{2.52}$$

which is seen to be negative definite provided that

$$b > 0,\ c > 0,\ F(x_1) > 0 \quad \text{and} \quad [bc - F(x_1)] > 0 \tag{2.53}$$

It follows that asymptotic stability at the origin will be guaranteed if the conditions (2.53) are met.

For the parameters b and c Fig. 2.2 shows the regions of asymptotic stability confined between the coordinate axes and the dashed hyperbolas for unsymmetric single-valued nonlinearities $F(x_1)$. For the nonlinearities represented by solid lines, the origin is asymptotically stable.

Example 2.7. Let us investigate the stability of the nonlinear time-varying system represented by the block diagram of Fig. 2.1*b*. The system

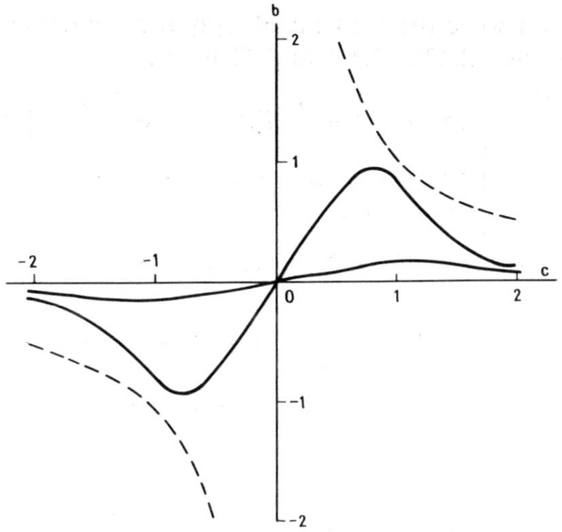

Fig. 2.2

is described by the second-order differential equation

$$\ddot{x} + b\dot{x} + c(x, t)x = 0 \tag{2.54}$$

which rewrites in state variables as the system

$$\left.\begin{aligned} \dot{x}_1 &= x_2 \\ \dot{x}_2 &= -bx_2 - c(x_1, t)x_1 \end{aligned}\right\} \tag{2.55}$$

Determine the gradient of Lyapunov function in a form similar to equation (2.39), namely,

$$\nabla V = \begin{bmatrix} \alpha_{11}x_1 + \alpha_{12}x_2 \\ \alpha_{21}x_1 + \alpha_{22}x_2 \end{bmatrix} \tag{2.56}$$

Assume that $\alpha_{22} = 2$, then

$$\frac{dV}{dt} = [\alpha_{11} - b\alpha_{21} - 2c(x_1, t)]x_1x_2 + [\alpha_{12} - 2b]x_2^2 - c(x, t)\alpha_{21}x_1^2 + \frac{\partial V}{\partial t} \tag{2.57}$$

Let

$$\alpha_{11} = b\alpha_{21} + 2c(x_1, t) \tag{2.58}$$

then the gradient of Lyapunov function takes the form

$$\nabla V = \begin{bmatrix} b\alpha_{21}x_1 + 2c(x_1, t)x_1 + \alpha_{12}x_2 \\ \alpha_{21}x_2 + 2x_2 \end{bmatrix} \tag{2.59}$$

By virtue of equation (2.37) we construct the Lyapunov function as

$$V(x_1, x_2) = \frac{b}{2}\alpha_{21}x_1^2 + 2\int_0^{x_1} \xi c(\xi, t)\mathrm{d}\xi + \alpha_{21}x_1x_2 + x_2^2 \tag{2.60}$$

whence

$$\frac{\mathrm{d}V}{\mathrm{d}t} = -(2b - \alpha_{21})x_2^2 - c(x_1, t)\alpha_{21}x_1^2 + 2\int_0^{x_1} \xi \frac{\partial}{\partial t} c(\xi, t)\,\mathrm{d}\xi \tag{2.61}$$

Represent

$$\alpha_{21} = 2(b - \gamma)$$

then

$$V(x_1, x_2) = b(b - \gamma)x_1^2 + 2(b - \gamma)x_1x_2 + x_2^2 + 2\int_0^{x_1} \xi c(\xi, t)\,\mathrm{d}\xi \tag{2.62}$$

and

$$\frac{\mathrm{d}V}{\mathrm{d}t} = -2(b - \gamma)c(x_1, t)x_1^2 - 2\gamma x_2^2 + 2\int_0^{x_1} \xi \frac{\partial c(\xi, t)}{\partial t}\,\mathrm{d}\xi \tag{2.63}$$

If the integral in (2.63) is positive and V is above the positive definite function

$$W_1(\mathbf{x}) = \frac{1}{2}[b(b - \gamma)x_1^2 + 2(b - \gamma)x_1x_2 + x_2^2] \tag{2.64}$$

then to secure $\mathrm{d}V/\mathrm{d}t$ negative we represent $c(x_1, t)$ as the sum

$$c(x_1, t) = c_0 + c_1(x_1, t) \tag{2.65}$$

where c_0 is a small constant. Now,

$$-\frac{\mathrm{d}V}{\mathrm{d}t} = 2c_0(b - \gamma)x_1^2 + 2\gamma x_2^2 + 2(b - \gamma)x_1^2c_1(x_1, t) - 2\int_0^{x_1} \xi \frac{\partial c_1(\xi, t)}{\partial t}\,\mathrm{d}\xi \tag{2.66}$$

If the inequality

$$(b - \gamma)c_1(x_1, t)x_1^2 > \int_0^{x_1} \frac{\partial c_1(\xi, t)}{\partial t}\,\xi\,\mathrm{d}\xi \tag{2.67}$$

holds for all x_1 and t, then $-\mathrm{d}V/\mathrm{d}t$ is higher than the time-invariant positive-definite function

$$W_2(\mathbf{x}) = \frac{1}{2}\,[2c_0(b - \gamma)x_1^2 + \gamma x_2^2]$$

consequently, $\mathrm{d}V/\mathrm{d}t < 0$.

Note that inequality (2.67) holds for certain functions $c(x_1, t)$ only. For example, if $c(x_1, t) = (c_0 + c_1 t - c_2 t^3)x_1^2$ with c_0, c_1, c_2, and c all positive, then $V(x_1, x_2)$ is positive definite and bounded in x_1, and $\mathrm{d}V/\mathrm{d}t$ is negative, which guarantees asymptotic stability of the system at the origin.

For $c(x_1, t) = (c_0 + c_1 e^t)x_1^2$ with positive c_0 and c_1, $V(x_1, x_2)$ is not bounded in x_1 and the nonlinear time-varying system is unstable.

2.3 Performance Criteria of Nonlinear Systems

Stability of nonlinear systems is a necessary but not sufficient condition to qualify the system performance as adequate. Stable systems are characteristic of transient damping in handling setpoints or disturbances. However the fact of system stability does not define the time of damping, maximum deviation of controlled variable (or overshoot), and the number of oscillations until the asymptotic approximation to equilibrium begins. Performance criteria of transients should be given as values of these parameters or bounds on such parameters in the form of inequalitites.

Let us consider the general approach to assessing the performance of system transients with the aid of Lyapunov functions. We shall assume that for the given initial conditions the Lyapunov function $V(\mathbf{x}, t_0)$ is prescribed. Then, by forming the derivative $\mathrm{d}V(\mathbf{x}, t)/\mathrm{d}t$, we may estimate the variation of $V(\mathbf{x}(t), t)$. Since in most practical problems the curves $V = \text{const}$ are closed loops, estimation of the distance from the origin as a function of time is straightforward. Clearly this method comes in handy only when the derivative is determined as a function of time without integration of system equations [2.13].

To assess the performance of nonlinear system, we resort to the relation involving only the Lyapunov function and its derivative

$$\dot{V}(\mathbf{x}, t) = \left[\frac{\dot{V}(\mathbf{x}, t)}{V(\mathbf{x}, t)}\right] V(\mathbf{x}, t) \tag{2.68}$$

If in some vicinity of the origin the magnitude of the ratio in the brackets does not exceed some constant K, then

$$\dot{V}(\mathbf{x}, t) \leqslant -KV(\mathbf{x}, t) \tag{2.69}$$

whence

$$V(\mathbf{x}(t),\ t) \leqslant V(\mathbf{x}_0,\ t_0) \exp\left[-K(t - t_0)\right] \tag{2.70}$$

where K is the degree of stability for the system deviated from the equilibrium position.

Let $V(\mathbf{x},\ t)$ constructed for the system of nonlinear equations satisfy all three instability conditions presented above for time-varying systems, i.e. $\|\mathbf{x}\| \to \infty$ as $\varrho\ (\|\mathbf{x}\|)$ goes to infinity. Then the equilibrium state at the origin is asymptotically stable. In this case the degree of stability in the neighborhood Γ can be found from the second and third stability conditions (2.35) and (2.36) with

$$K < \theta(\|\mathbf{x}\|)/\varrho(\|\mathbf{x}\|) \tag{2.71}$$

In some problems, to increase the accuracy of the bound, one may choose degrees of stability K_1 and K_2 corresponding to the adjacent levels $V_1(\mathbf{x},\ t)$ and $V_2(\mathbf{x},\ t)$, viz.,

$$\left.\begin{aligned} V_1(\mathbf{x}(t),\ t) &\leqslant V_1(\mathbf{x}_0,\ t_0) \exp\left[-K_1(t - t_0)\right] \\ V_2(\mathbf{x}(t),\ t) &\leqslant V_2(\mathbf{x}_0,\ t_0) \exp\left[-K_2(t - t_0)\right] \end{aligned}\right\} \tag{2.72}$$

For every time we assume that

$$V_1(\mathbf{x},\ t) > V_2(\mathbf{x},\ t)$$

Then in agreement with (2.69)

$$\frac{\mathrm{d}V(\mathbf{x},\ t)}{V(\mathbf{x},\ t)} \leqslant -K\,\mathrm{d}t \tag{2.73}$$

Now, the bound for t_K is obtained by integration of (2.73) at the given initial conditions

$$t_K \leqslant \frac{1}{K} \ln \frac{V(\mathbf{x},\ t_0)}{V(\mathbf{x},\ t_K)} \tag{2.74}$$

This expression is valid for both monotonous and oscillatory processes. Therefore, having specified a system speed in terms of t_K, we assume that the transient is over if the coordinates of the state vector remain within the domain

$$\|\mathbf{x}(t_K)\| \leqslant \varepsilon \tag{2.75}$$

where ε is the error in measuring transient time.

Example 2.8. Let the control system be described by the equation

$$\dot{\mathbf{x}} = \mathbf{A}(t)\mathbf{x} + \mathbf{B}(t)u \tag{2.76}$$

where

$$\mathbf{A}(t) = \begin{bmatrix} 0 & 1 \\ -a_1(t) & -a_2(t) \end{bmatrix} \quad \text{and} \quad \mathbf{B}(t) = \begin{bmatrix} 0 \\ b(t) \end{bmatrix}$$

Then the Lyapunov function (2.5) subject to the condition (2.6) may be written as

$$\begin{aligned} V(\mathbf{x},\ t_0) &= |K_{1,1}\mathbf{x}(t_0) + K_{1,2}\dot{\mathbf{x}}(t_0)| \\ &\quad + K_1|\mathbf{x}(t_0)| + K_2|\dot{\mathbf{x}}(t_0)| \end{aligned} \tag{2.77}$$

$$\begin{aligned} V(\mathbf{x},\ t_K) &= (K_{1,1} + K_1 + K_{2,1})\varepsilon \\ \mathbf{x}(t_K) &= \varepsilon \\ \dot{\mathbf{x}}(t_K) &= 0 \end{aligned}$$

Substituting (2.77) in (2.74) yields

$$t_K \leqslant \frac{1}{K} \ln \frac{|K_{1,1}\mathbf{x}(t_0) + K_{1,2}\dot{\mathbf{x}}(t_0)| + K_1|\mathbf{x}(t_0)| + K_2|\dot{\mathbf{x}}(t_0)|}{(K_{1,1} + K_1 + K_{2,1})\varepsilon} \tag{2.78}$$

By selecting the coefficients $K_{j,i}$ one may satisfy the requirements imposed on the transient process. It is worth noting that the bound (2.78) holds only when the system (2.76) has no singular solutions, i.e. experiences no sliding modes for all times t (see Sect. 2.5).

For $K_2 > K_1$ and $t_{K1} < t_{K2}$, the Nemytsky theorem [2.13] may be invoked to define the conditions for a criterion on the maximum overshoot. For example, if $t_{K1} < t_{K2}$, then for peak overshoots $\sigma_{1\max} < \sigma_{2\max}$ for the same ε.

2.4 Evaluation of Control Laws

Increasing gain coefficients in the loop on an attempt to achieve a better reproduction of control signals can give rise to unduly unstable processes. Resorting to the methods of synthesis developed for large-gain systems, one may arrive at control laws precluding unstable modes. In the general case this problem is solved by dynamic programming but subject to the condition that the optimal laws are satisfied by the Lyapunov function for the closed-loop system [2.13]. Using both methods in conjunction enables synthesis problem solving by applying the well-based Lyapunov theory to the perturbed and unperturbed solutions.

Let the equations of unperturbed motion in a time-varying system have the form

$$\left.\begin{aligned} \dot{\mathbf{x}} &= \mathbf{A}(t)\mathbf{x} + \mathbf{B}(t)u \\ \mathbf{y} &= \mathbf{C}(t)\mathbf{x} \end{aligned}\right\} \tag{2.79}$$

where

$$\mathbf{A}(t) = \begin{bmatrix} 0 & 1 & 0 & \dots & 0 \\ 0 & 0 & 1 & \dots & 0 \\ \vdots & \vdots & \vdots & \vdots & \vdots \\ -a_1(t) & -a_2(t) & -a_3(t) & \dots & -a_n(t) \end{bmatrix};$$

$$\mathbf{B}(t) = \begin{bmatrix} 0 \\ 0 \\ \vdots \\ b(t) \end{bmatrix} \qquad \mathbf{C}(t) = \begin{bmatrix} 1 & 0 & 0 & \dots & 0 \\ 0 & 1 & 0 & \dots & 0 \\ \vdots & \vdots & \vdots & \vdots & \vdots \\ 0 & 0 & 0 & \dots & 1 \end{bmatrix}$$

We represent the functional invoked for selection of an optimal control law in the general form

$$J = \int_0^\infty Q(\mathbf{x}, u, t)\mathrm{d}t \tag{2.80}$$

This functional will be minimized so that in the vicinity of the origin the motion will be asymptotically stable with respect to initial perturbations x_0 in the domain $\|\mathbf{x}_0\| < \Gamma_0 < \Gamma$. Below we formulate this statement as the sufficient optimality condition of Letov.

Let $V(\mathbf{x}, t)$ and u_0 satisfying the system of equations (2.79) be defined in the region $\|\mathbf{x}\| < \Gamma$, $V(\mathbf{x}, t)$ be positive definite and close to zero at the origin with

$$\sup_{\|\mathbf{x}\| < \Gamma_0} V < \inf_{\|\mathbf{x}\| = \Gamma} V$$

and the derivative

$$\dot{V}|_{u_0} = -Q(\mathbf{x}, u_0, t)$$

satisfy the equation

$$\dot{V}|_{u_0} + Q(\mathbf{x}, u_0, t) = \min_u [\dot{V}|_u + F(\mathbf{x}, u, t)] \tag{2.81}$$

then $u_0(t)$ is the optimal control. Here $V(\mathbf{x}, t)$ is a Lyapunov function and equation (2.81) is the Bellman equation. The aforementioned conditions, therefore, define the optimal stable motion in the sense of Bellman-Lyapunov.

Optimal laws are normally implemented by analog means in the form of correcting elements, and in systems computer control in the form of respective programs. In time-varying systems with linear plants, optimal correcting facilities may be linear or nonlinear depending on the integrand Q. For example, if Q is a square function of control, and system coefficients vary linearly with time, then the control will be a linear function also dependent on time. In this case the correcting element will be a phase-lead linear filter with variable parameters. The use of such a filter will, however, substantially widen the passband for the input, thus increasing the effect of noise which will reduce the system accuracy.

In time-varying systems with linear plants and non-quadratic Q, control will involve relay elements. Depending on the amplitude of input signal or time variation of the parameters such systems may experience regions of unstable states of the plant. To eliminate these deficiencies new time-varying closed-loop systems have been equipped with pseudo-linear correcting elements providing phase-lead action without amplitude distortions. However, choice of type and parameters for such facilities has been mostly supported by heuristic techniques. As an alternative approach the following paragraphs outline the synthesis of piecewise linear laws by the Bellman-Lyapunov method.

Rewrite the Lyapunov function (2.5), making use of (2.6), (2.7), as

$$V(\mathbf{x}) = \sum_{j=1}^{n} (|g_j| + |K_j \mathbf{x}_j|) \tag{2.82}$$

and form its derivative

$$\dot{V}(\mathbf{x}) = \sum_{j=1}^{n} [\dot{g}_j \operatorname{sgn} g_j + K_j \dot{x}_j \operatorname{sgn} x_j] \tag{2.83}$$

where

$$\dot{g}_j = \sum_{j=1}^{n-j+1} K_{ji} \dot{x}_i, \quad j = 1, \ldots, n$$

The derivative of (2.5) is defined for all $\mathbf{x}$ provided that $g_j \neq 0$ and $x_j \neq 0$. Substituting (2.79) in (2.5) gives

$$\dot{V}(\mathbf{x}) = \sum_{j=2}^{n} \dot{g}_j \operatorname{sgn} g_j + \left(\sum_{j=1}^{n-1} K_{1,i} \dot{x}_{i+1} \right) \operatorname{sgn} g_1 + \sum_{j=1}^{n-1} K_j \dot{x}_j \operatorname{sgn} x_j$$
$$+ \left(\sum_{j=1}^{n} a_j x_j + b_u \right) (K_{1n} \operatorname{sgn} g_1 + K_n \operatorname{sgn} x_n) \tag{2.84}$$

From the Lyapunov stability condition it follows that the derivative of (2.5) must have the form (2.12). The relation (2.12) will hold true with a suitable choice of control law. To find this law we substitute (2.84) and (2.82) in

(2.12), viz.,

$$\sum_{j=2}^{n} \dot{g}_j \operatorname{sgn} g_j + \sum_{j=1}^{n} [K_j \dot{x}_j \operatorname{sgn} x_j + (a_j x_j + bu)(K_{1,n} \operatorname{sgn} g_1 + K_n \operatorname{sgn} x_n)] + \left(\sum_{i=1}^{n-1} K_{1,i} x_{i+1}\right) \operatorname{sgn} g_1 = -K \sum_{j=1}^{n} (|g_j| + |K_j x_j|) \tag{2.85}$$

This expression may be transformed to a formula helpful in deriving the desired control law. We rewrite it in the form

$$\begin{aligned} u = &-[b(K_{1,n} \operatorname{sgn} g_1 + K_n \operatorname{sgn} x_n)]^{-1} \\ &\times \left\{ \left[\sum_{j=1}^{n} (KK_{1,j} + K_{1,j-1} + a_j K_1) x_j \right] \operatorname{sgn} g_1 \right. \\ &+ \left[\sum_{j=1}^{n} (a_j x_j + K x_n) \right] K_n \operatorname{sgn} x_n + \sum_{j=2}^{n} (\dot{g}_j + K g_i) \operatorname{sgn} g_j \\ &\left. + \sum_{i=1}^{n-1} K_j (x_{j+1} + K x_j) \operatorname{sgn} x_j \right\} \end{aligned} \tag{2.86}$$

In order to reduce (2.86) to a form convenient for practical realization of the control law we multiply its numerator and denominator by $(K_{1,n} \operatorname{sgn} g_1 - K_n \operatorname{sgn} x_n)$, then

$$\begin{aligned} u = &-[b(K_{1,n}^2 - K_n^2)]^{-1} \times \left\{ \sum_{j=1}^{n} d_{j,1} x_j - \sum_{j=1}^{n-1} d_{j,2} w_j \operatorname{sgn} g_1 \operatorname{sgn} x_n \right. \\ &+ \sum_{j=2}^{n} K_{1,n} g_j \operatorname{sgn} g_j \operatorname{sgn} g_1 - \sum_{i=2}^{n} K_n g_j \operatorname{sgn} g_j \operatorname{sgn} x_n \\ &\left. + \sum_{j=1}^{n-1} d_{j,3} w_j \operatorname{sgn} x_j \operatorname{sgn} g_1 - \sum_{j=1}^{n-1} d_{j,4} w_j \operatorname{sgn} x_j \operatorname{sgn} x_n \right\} \end{aligned} \tag{2.87}$$

where

$$\begin{aligned} q_j &= \dot{g}_j + K g_j \\ d_{j,1} &= a_j(K_{1,n}^2 - K_n^2) + (K_{1,j} K + K_{1,j-1}) K_{1,n} \\ d_{j,2} &= K_n K_{1,j} \\ d_{j,3} &= K_{1,n} K_j \\ d_{j,4} &= K_n K_j \\ w_j &= x_j + K \dot{x}_j \end{aligned}$$

Equation (2.87) is helpful in deriving any suitable network for the particular piecewise linear control law. An appropriate procedure is essentially as follows: the hyperplanes $g_j = 0$, for $j = 1, \ldots, n$, divide the phase space into $2n - 1$ regions with invariable structure of control $u_j \in U$ in each of them. The switching hyperplanes in the phase space are governed by the relations between the respective laws. For example, the controls for u_j, $j = 1, \ldots, n$, are as follows:

$$u_1: \quad \operatorname{sgn} x_1 = \operatorname{sgn} x_j, \quad \operatorname{sgn} x_1 = \operatorname{sgn} g_j$$

$$u_2: \quad \operatorname{sgn} x_1 \neq \operatorname{sgn} x_2, \quad \operatorname{sgn} x_1 = \operatorname{sgn} x_j = \operatorname{sgn} g_1$$

$$\vdots$$

$$u_n: \quad \operatorname{sgn} x_1 \neq \operatorname{sgn} x_2, \quad \operatorname{sgn} x_1 \neq \operatorname{sgn} g_j$$

We note that a required structure of control law may then be given in each of the regions. Then the method of undetermined coefficients is invoked to construct a system of constraints of n-th order which yields the coefficients of (2.87). When the phase coordinate of the system crosses from one domain to another, the structure of control law can alter because the terms proportionate to the desired phase coordinate are introduced in the formula of type (2.87) or dropped from it. Therefore, the total system of constraints has the order $n(2n - 1)$ represented as a system of linear algebraic equations of the type

$$\mathbf{A}_1 \mathbf{z} = \mathbf{B}_1 \tag{2.88}$$

where $\mathbf{z} = (d_{1,j}, d_{2,j}, d_{3,j}, d_{4,j}, q_j)$ is the vector of unknown coefficients; $\mathbf{B}_1$ is the vector of the desired parameters; and $\mathbf{A}_1$ is a matrix.

To close this topic we note that structural schemes of piecewise linear control laws may be formed by choosing the appropriate coefficients in (2.88). To illustrate this statement consider the original functional scheme of the control system described by equations (2.79). The respective block diagram is shown in Fig. 2.3. The coefficients of matrix $\mathbf{K}$ of the correcting

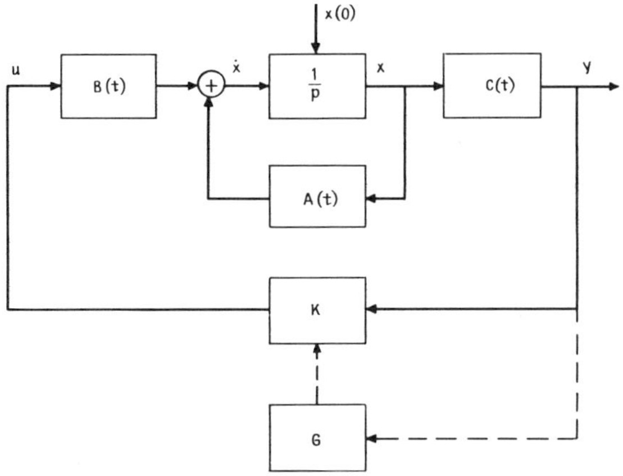

Fig. 2.3

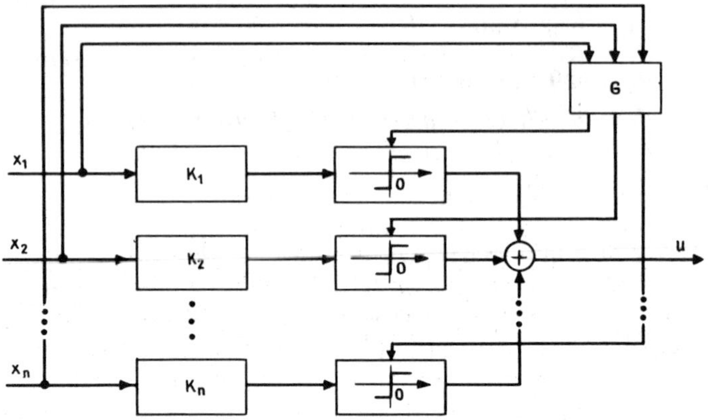

Fig. 2.4

facility vary in steps as functions of control commands generated by a generator G (Fig. 2.4).

To simplify the calculations we illustrate the construction of a pseudo-linear correcting facility for a control system described by a second-order differential equation.

Example 2.9. Let the control system be described by equation (2.76), then the Lyapunov function and its derivative will be

$$V(x) = |K_{1,1}x_1 + K_{1,2}x_2| + |K_1' x_1| + |K_2 x_2| \tag{2.89}$$

$$\dot{V}(x) = \dot{g}_1 \operatorname{sgn} g_1 + K_1' \dot{x}_1 \operatorname{sgn} x_1 + K_2 \dot{x}_2 \operatorname{sgn} x_2 \tag{2.90}$$

where $K_1' = K_1 + K_{2,1}$, $g_1 = K_{1,1}x_1 + K_{1,2}x_2$ for $x_1 \neq 0$, $x_2 \neq 0$, and $\dot{g}_1 \neq 0$. Substitute these expressions in (2.12) to get

$$K_{1,1}x_2 \operatorname{sgn} g_1 + (b(t)u + a_1(t)x_1 - a_2(t)x_2) \times (K_{1,2} \operatorname{sgn} g_1 + K_2 \operatorname{sgn} x_2) + K_1 x_2 \operatorname{sgn} x_2 + K[|K_{1,1}x_1 + K_{1,2}x_2| + |K_1' x_1| + |K_2 x_2|] = 0 \tag{2.91}$$

From (2.91) the control law will be as follows:

$$u = -[b(t)(K_{1,2} \operatorname{sgn} g_1 + K_2 \operatorname{sgn} x_2)]^{-1} \times \{[(K_{1,1}K - a_1(t)K_{1,2})x_1 + (KK_{1,2} + K_{1,1} - K_{1,2}a_2(t))x_2] \operatorname{sgn} g_1 + [(K - a_2(t))K_2 x_2 - a_1(t)K_2 x_1] \operatorname{sgn} x_2 + (KK_1 x_2 + a_2(t)x_2) \operatorname{sgn} x_1\} \tag{2.92}$$

Multiplying and dividing this epxression by $(K_{1,2} \operatorname{sgn} g_1 - K_2 \operatorname{sgn} x_2)$ gives

$$u = -[b(t)(K_{1,2}^2 - K_2^2)]^{-1} \times \{d_1 x_1 + d_2 x_2 + (d_3 x_1 + d_4 x_2) \operatorname{sgn} g_1 \operatorname{sgn} x_2 + (d_5 x_1 + d_6 x_2) \operatorname{sgn} g_1 \operatorname{sgn} x_1 - (d_7 x_1 + d_8 x_2 \operatorname{sgn} x_1 \operatorname{sgn} x_2\} \tag{2.93}$$

where

$$
\begin{aligned}
d_1 &= K_{1,2}(K_{1,1}K - K_{1,2}a_1(t)) - K_2^2 a_2(t) \\
d_2 &= K_{1,1}K_{1,2} + KK_{1,2}^2 + a_2(t)K_{1,2}^2 - K_2^2 K - a_2(t)K_2^2 \\
d_3 &= -K_2 K_{1,1} K \\
d_4 &= K_{1,2}K_2(K - a_2(t)) + K_{1,2}K_2 a_2(t) - K_2(K_{1,1} + KK_{1,2}) \\
d_5 &= K_1 K_{1,2} K; \quad d_6 = K_1 K_{1,2}; \quad d_7 = K_1 K_2 K \\
d_8 &= K_1 K_2
\end{aligned}
$$

Let us consider a few cases dealing with the generation of control laws. By letting, say, $d_i = 0$ in (2.93) for $i = 3, \ldots, 8$, we arrive at a linear control law. For $d_i = 0$ $(i = 1, 2)$ and $d_i \to \infty$ $(i = 3, \ldots, 8)$ with $|x_1| \leqslant x_{1\max}$ and $|x_2| \leqslant x_{2\max}$ we have an on-off control law. For finite d_i $(i = 1, \ldots, 8)$, one may form various piecewise linear laws.

For example, in the region $\operatorname{sgn} x_1 = \operatorname{sgn} x_2 = \operatorname{sgn} g_1$, the control is proportional to the deviation, $u_1 = K_1 x_1$; in the region $\operatorname{sgn} x_1 \neq \operatorname{sgn} x_2$, $\operatorname{sgn} x_1 = \operatorname{sgn} g_1$, the control is proportional to the deviation and its first derivative, $u_2 = K_1 x_1 + K_2 x_2$; and in the region $\operatorname{sgn} x_1 \neq \operatorname{sgn} x_2$, $\operatorname{sgn} x_1 \neq \operatorname{sgn} g_1$, the control is absent, $u_3 = 0$. Summarizing, the control may be written as

$$
u = \begin{cases} \tilde{K}_1 x_1 & \text{for} \quad \operatorname{sgn} x_1 = \operatorname{sgn} x_2 \\ \tilde{K}_1 x_1 + \tilde{K}_2 x_2 & \text{for} \quad |\tilde{K}_1 x_1| > |\tilde{K}_2 x_2|, \quad \operatorname{sgn} x_1 \neq \operatorname{sgn} x_2 \\ 0 & \text{for} \quad |\tilde{K}_1 x_1| \leqslant |\tilde{K}_2 x_2|, \quad \operatorname{sng} x_1 \neq \operatorname{sgn} x_2 \end{cases} \tag{2.94}
$$

This control law has a considerable edge on the linear law as it reduces input transient, maximum overshoot, and the deviation of controlled variable in handling a disturbance.

Now we turn to selection of coefficients d_i. With the aid of the method of undetermined coefficients we construct a system of linear algebraic equations of type (2.88) for each region. Then the respective matrix and vectors will be

$$
A_1 = \begin{bmatrix} 1 & 0 & 1 & 0 & 1 & 0 & -1 & 0 \\ 0 & 1 & 0 & 1 & 0 & 1 & 0 & -1 \\ 1 & 0 & -1 & 0 & 1 & 0 & 1 & 0 \\ 0 & 1 & 0 & -1 & 0 & 1 & 0 & 1 \\ 1 & 0 & 1 & 0 & -1 & 0 & 1 & 0 \\ 0 & 1 & 0 & 1 & 0 & -1 & 0 & 1 \end{bmatrix};
$$

$$
B_1 = \begin{bmatrix} \tilde{K}_1 \\ 0 \\ \tilde{K}_1 \\ \tilde{K}_2 \\ 0 \\ 0 \end{bmatrix}; \qquad Z = \begin{bmatrix} d_1 \\ d_2 \\ d_3 \\ d_4 \\ d_5 \\ d_6 \end{bmatrix} \tag{2.95}
$$

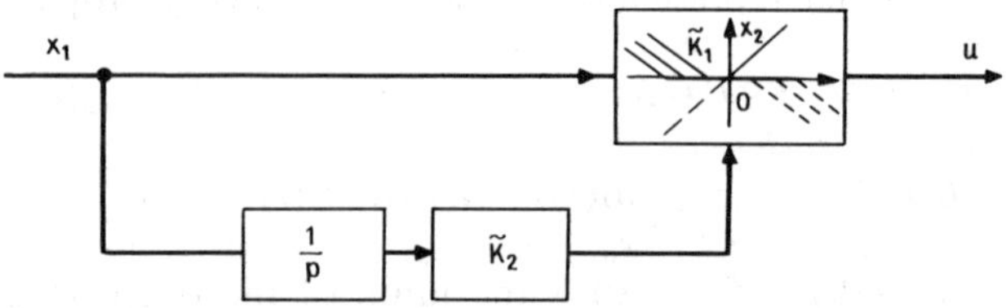

Fig. 2.5

Solving the system (2.88) subject to (2.95) yields

$$\begin{aligned} d_1 &= \tilde{K}_1/2 - d_3 \quad & d_2 &= \tilde{K}_2/2 - d_8 \\ d_3 &= d_7 & d_4 &= -\tilde{K}_2/2 + d_8 \\ d_5 &= \tilde{K}_1/2 + d_3 & d_6 &= d_8 \end{aligned} \tag{2.96}$$

For simplification let $d_3 = d_8 = 1$, then (2.93) may be rewritten subject to (2.94), (2.95), and (2.96) as

$$u = \left(\frac{\tilde{K}_1}{2} - 1\right)x_1 + \left(\frac{\tilde{K}_2}{2} - 1\right)x_2 + \left[x_1 + \left(1 - \frac{\tilde{K}_2}{2}\right)x_2\right] \operatorname{sgn} g_1 \operatorname{sgn} x_2$$
$$+ \left[\left(\frac{\tilde{K}_1}{2} + 1\right)x_1 + x_2\right] \operatorname{sgn} g_1 \operatorname{sgn} x_1 - (x_1 + x_2) \operatorname{sgn} x_1 \operatorname{sgn} x_2 \tag{2.97}$$

The implementation of this control law is illustrated in Fig. 2.5. The blocks of the generating and correcting facilities are seen to be united in a single element. This approach markedly simplifies the manufacture of the correcting facility made of analog components. In what follows we shall refer to the joint unit as a nonlinear correcting unit. This constructional procedure may be employed to form any other structures of piecewise linear control laws.

At this point we wish to demonstrate that the control law thus derived and realized in the form of a nonlinear correcting unit may be related to pseudo-linear laws. For this purpose we resort to the method of harmonic linearization of the pseudolinear correcting unit. Any periodic, piecewise continuous function $f(t)$ satisfying the Dirichlet conditions, having n discontinuities and $n - 1$ continuity intervals may be written as

$$f(t) = \sum_{i=1}^{n-1} A_i \exp\left[-\alpha(t - t_i)\right] \tag{2.98}$$

where α is a real positive or complex number, t_i is the left bound of interval i, and A_i is an integer.

This exponential representation has the following properties:

$$f'(t) = -\alpha f(t)$$
$$f(t)e^{\beta t}\,|_0^T = -\sum_{i=1}^{n} \varkappa_i e^{\beta t_i} \tag{2.99}$$

Here, $\varkappa_i = f(t_i + 0) - f(t_i - 0)$ is the jump of the function at t_i, β is an integer, $f(t_i - 0)$ is the value on the left of t_i, and $f(t_i + 0)$ the value on the right of t_i. Then the Fourier coefficients for $f(t)$ can be determined as

$$F_n = \frac{2}{T}\int_0^T f(t)\exp(-jn't)\,dt \tag{2.100}$$

where $n' = (2\pi/T)n$, $n = 0, 1$, etc.

Making use of the properties (2.99) and integrating (2.100), we get

$$F_n = \frac{2}{(\alpha + jn')T}\sum_{i=1}^{n} \varkappa_i \exp(-jn't_i) \tag{2.101}$$

From this expression it follows that all harmonics are defined by jumps of the function at discontinuities. An inhomogeneous function $f(t)$ can be represented as a sum of homogeneous functions for which the n-th order harmonic of the inhomogeneous function equals the sum of the n-th order harmonics of n homogeneous functions. If $f(t)$ is constant in each interval, then $\alpha = 0$ in (2.101).

Let us consider the case of $\alpha = -j\nu$, then F_n is a sine or cosine function and all n harmonics, except the first one, are defined by the formula

$$F_n = \frac{2}{T(n' - \nu)}\sum_{i=1}^{n} \varkappa_i \exp(-jn't_i) = \frac{2}{T}\,\frac{\sum_{i=1}^{n} (\varkappa_i + jn'\varkappa_i')\exp(-jn't_i)}{\nu^2 - n'^2} \tag{2.102}$$

For the first harmonic

$$F_1 = \frac{2}{T}\left[\frac{1}{(\alpha + j)}\sum_{i=1}^{n} \varkappa_i \exp(-j\psi_i) + (\psi_i - \psi_{i-1})\right] \tag{2.103}$$

where $\psi_i = \omega(t_i)$, $i = 1, \ldots, n - 1$.

For higher harmonics of $n > 1$ employ formula (2.102). Then

$$F_n = \frac{2A}{j\pi(n-1)}\left\{-\frac{\tilde{K}_1}{2j}[\exp(j\psi_1)\exp(-jn\psi_1) - \exp(j\psi_1(1+n))]\right.$$
$$-\frac{\tilde{K}_2}{2}\omega[\exp(j\psi_1(1-n)) + \exp(-j\psi_1(1+n))] + \frac{\tilde{K}_1}{2j}[\exp(j(\pi-\psi_1)$$
$$\times(1-n)) - \exp(j(\pi-\psi_1)(1+n))] + \frac{\tilde{K}_2}{2}\omega[\exp(j(\pi-\psi_1)(1-n))$$
$$\left.-\exp(-j(\pi-\psi_1)(1+n)]\right\} + \frac{2A}{j\pi n}\left\{\tilde{K}_1\sin\frac{\pi}{2}\times\exp\left(-j\frac{\pi}{2}n\right)\right.$$
$$-\tilde{K}_1\sin(\pi-\psi_1)\exp(-jn(\pi-\psi_1))$$
$$\left.-\tilde{K}_2\omega\cos(\pi-\psi_1)\exp(jn(\pi-\psi_1))\right\} \tag{2.104}$$

where $\psi_1 = \arctan(\tilde{K}_2/\tilde{K}_1)\omega$.

To determine the coefficients of harmonic linearization, we partition the control signal u leaving the nonlinearity (Fig. 2.5) in two components u_1 and u_2 as illustrated in Fig. 2.6. Then, taking the fundamental compo-

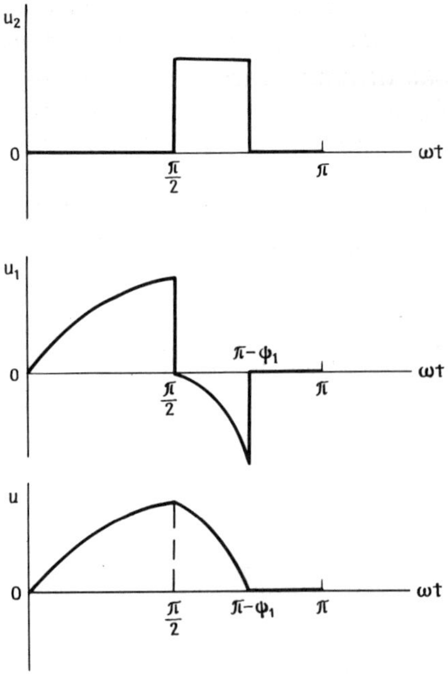

Fig. 2.6

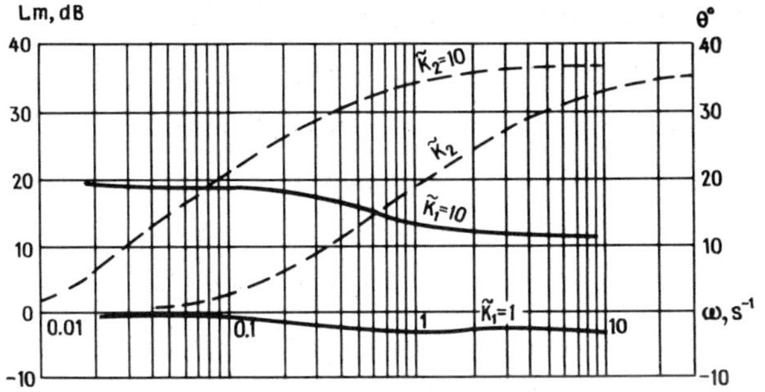

Fig. 2.7

nent only, we obtain two coefficients of harmonic linearization

$$\left.\begin{aligned} a_1(\omega) &= \tilde{K}_1\left(1 - \frac{1}{\pi}\arctan\frac{\tilde{K}_2}{\tilde{K}_1}\,\omega\right) \\ b_1(\omega) &= \tilde{K}_2\omega\left(\frac{1}{2} - \frac{1}{\pi}\arctan\frac{\tilde{K}_2}{\tilde{K}_1}\,\omega\right) \end{aligned}\right\} \tag{2.105}$$

If we take into account the higher harmonics, then equation (2.104) yields the coefficients $a_2(\omega)$, $b_2(\omega)$, . . . , $a_n(\omega)$, and $b_n(\omega)$.

From (2.105) it follows that the coefficients of harmonic linearization depend on the frequency, rather than on the amplitude, of the input signal. This proves that the nonlinear correcting unit synthesized in this way possesses the property of pseudolinearity. The Bode diagrams constructed for this unit with the aid of equations (2.105) for two values of $\tilde{K}_1$ and $\tilde{K}_2$ are presented in Fig. 2.7. The selected pseudo-linear correcting facility is seen to increase the phase lead up to 36° at small amplitude distortions.

Example 2.10. Consider the application of the control law (2.94) to the control system, for an actuator equipped plant, described by the equations

$$\begin{aligned} \dot{\mathbf{x}} &= \mathbf{A}(t)\mathbf{x} + \mathbf{B}(t)\mathbf{u} \\ \mathbf{y} &= \mathbf{C}\mathbf{x} \\ \dot{\mathbf{u}} &= \mathbf{R}\mathbf{u} + \mathbf{S}\mathbf{g} \end{aligned} \tag{2.106}$$

where

$$\mathbf{A} = \begin{bmatrix} 0 & 1 & 0 \\ 0 & -a_1(t) & -a_2(t) \\ 0 & 1 & -a_4(t) \end{bmatrix}; \quad \mathbf{B} = \begin{bmatrix} 0 \\ -a_3(t) \\ -a_5(t) \end{bmatrix}; \quad \mathbf{x} = \begin{bmatrix} x_1 \\ x_2 \\ x_3 \end{bmatrix}$$

$$\mathbf{R} = \begin{bmatrix} a_8 & a_9 \\ 0 & a_6 \end{bmatrix}; \quad \mathbf{S} = \begin{bmatrix} 0 \\ a_7 \end{bmatrix}; \quad \mathbf{u}^T = \begin{bmatrix} u_1 \\ u_2 \end{bmatrix}$$

We rewrite the control law for this system as

$$
\begin{aligned}
g &= \tilde{K}_1 \Delta x_1 && \text{for } \operatorname{sgn} x_2 = \operatorname{sgn} \Delta x_1 \\
&= \tilde{K}_2 \Delta x_1 + \tilde{K}_2 x_2 && \text{for } \operatorname{sgn} x_2 \neq \operatorname{sgn} \Delta x_1 \\
& && |\tilde{K}_1 \Delta x_1| > |K_2 x_2| \\
&= 0 && \text{for } \operatorname{sgn} x_2 \neq \operatorname{sgn} \Delta x_1 \\
& && |\tilde{K}_1 \Delta x_1| < |\tilde{K}_2 x_2|
\end{aligned}
\tag{2.107}
$$

Here, $\Delta x_1 = x_1 - x_{1,sp}$, and x_{1sp} is the setpoint. If we apply the method of "frozen" coefficients, depending on operating modes the system may then be investigated in the frequency domain. The respective transfer functions in terms of the Laplace variable are as follows:

$$
\begin{aligned}
W_0(s) &= \frac{X_2(s)}{U_1(s)} = \frac{K_1(T_1 s + 1)}{s(T_0^2 s^2 + 2\xi_0 T_0 s + 1)} \\
W_1(s) &= \frac{U_2(s)}{G(s)} = \frac{K_2}{T_2 s + 1} \\
W_2(s) &= \frac{U_1(s)}{U_2(s)} = \frac{K_3}{T_3 s + 1} \\
W_3(s) &= \frac{X_1(s)}{X_2(s)} = \frac{1}{s}
\end{aligned}
\tag{2.108}
$$

where for some constant time interval of system operation

$$
K_1 = \frac{a_3 a_4 - a_2 a_5}{a_2 + a_1 a_4}; \qquad \xi_0 = \frac{a_1 + a_2}{2\sqrt{a_2 + a_1 a_4}}
$$

$$
T_0 = \frac{a_3}{a_3 a_4 - a_2 a_5}; \qquad T_2 = \frac{1}{\sqrt{a_2 + a_1 a_4}}
$$

$$
K_2 = \frac{a_7}{a_6}; \quad K_3 = \frac{a_9}{a_8}; \quad T_3 = \frac{1}{a_6}; \quad T_4 = \frac{1}{a_8}
$$

The numerical values of the coefficients in (2.106) for various operating modes are summarized below.

Mode	a_1	a_2	a_3	a_4	a_5	
1	0.25	0.2	0.5	0.5	0.1	The remaining a's equal 10
2	0.3	0.3	1.5	0.4	0.05	
3	0.1	0.4	0.6	0.3	0.02	
4	0.06	0.05	0.3	0.15	0.003	

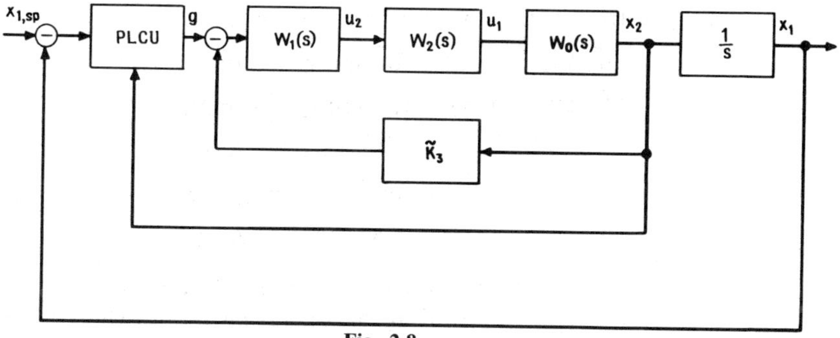

Fig. 2.8

The transfer functions (2.108) have evolved in the block diagram shown in Fig. 2.8 for the control law

$$u_3 = g - \tilde{K}_3 x_2 \tag{2.109}$$

Figure 2.9 shows log-log frequency responses for four system operating modes ($s = j\omega$) with the parameters of the pseudo-linear correcting unit (PLCU) as follows: $\tilde{K}_1 = 5$, $\tilde{K}_2 = 2.5$, and $\tilde{K}_3 = 1.5$. For comparison, the dashed lines represent the response of the system with a linear correcting facility. The incorporation of PLCU is seen to increase the stability margins in phase $\theta(\omega) \geqslant 60°$ and the magnitude $H_m \geqslant -10$ dB, retaining a rather fast response to the setpoint. In addition, the stability and performance indices of the PLCU system are less sensitive to variation in the system duty as respective H and θ Bode plots become closer (see Fig. 2.9b).

Complicated systems of automatic control should be equipped with multichannel correcting facilities with nonlinearity of the form $F(x, p)$ and linear filters $W_1(p), W_2(p), \ldots, W_n(p)$, and $W_f(p)$; $p = \mathrm{d}/\mathrm{d}t$. The general form of a multichannel PLCU

$$\mathscr{F}(x, p) = \frac{K}{b(t)K_{1,n}}\left(\sum_{j=2}^{n} |W_j(p)x_1|\right) + \frac{1}{b(t)K_{1,n}}[pW_j(p)x_1]\,\mathrm{sgn}\,[W_j(p)]x_1$$

$$W_j(p) = \sum_{i=1}^{n-j+1} K_{j,i}p^{i-1}, \quad j = 2, 3, \ldots, n \tag{2.110}$$

The block diagram of this correcting unit is presented in Fig. 2.10. A simple double-loop PLCU may be described by

$$u = |W_a(p)x_1|\,\mathrm{sgn}\,[W_f(p)x_1] \tag{2.111}$$

where the transfer functions of the filters will be $W_a(s) = 1/(T_{2,k}s + 1)$ and $W_f(s) = T_{k2}(T_{k1}s + 1)/[T_{k1}(T_{k2}s + 1)]$. The block diagram constructed for (2.111) is presented in Fig. 2.11.

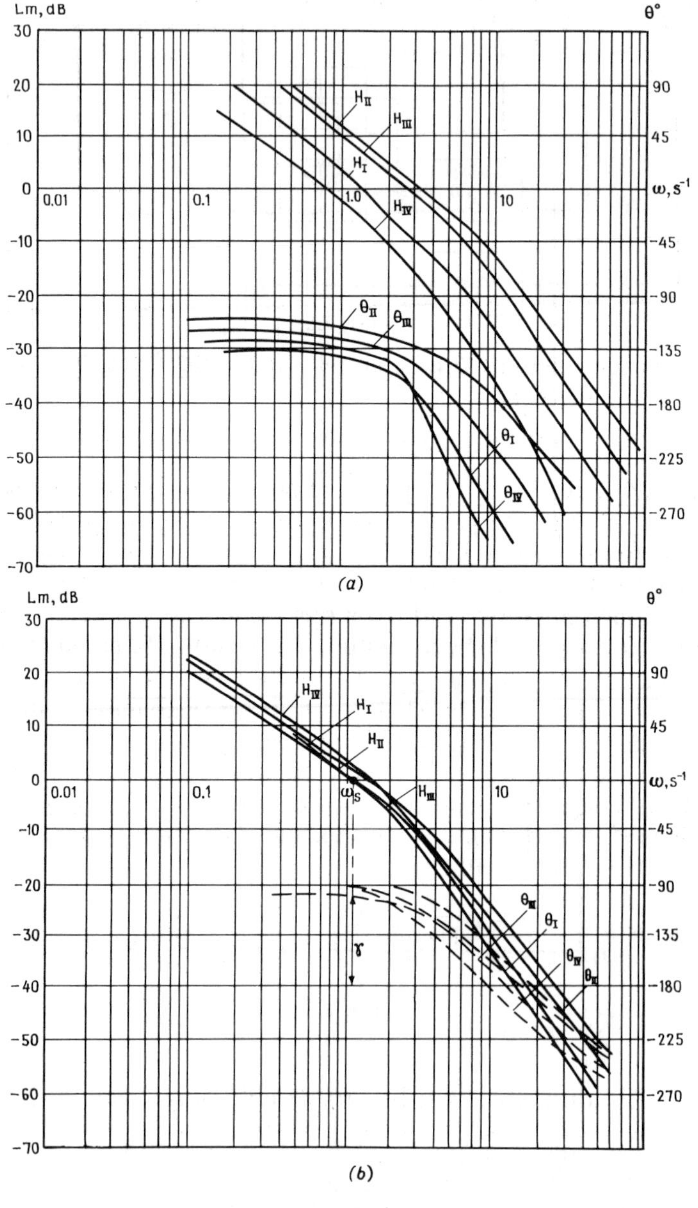
Lm, dB
θ°
30
20
10
0
-10
-20
-30
-40
-50
-60
-70
90
45
-45
-90
-135
-180
-225
-270
ω, s⁻¹
0.01
0.1
1.0
10
H_II
H_III
H_I
H_IV
θ_II
θ_III
θ_I
θ_IV
(a)
ω_s
γ
(b)

Fig. 2.9

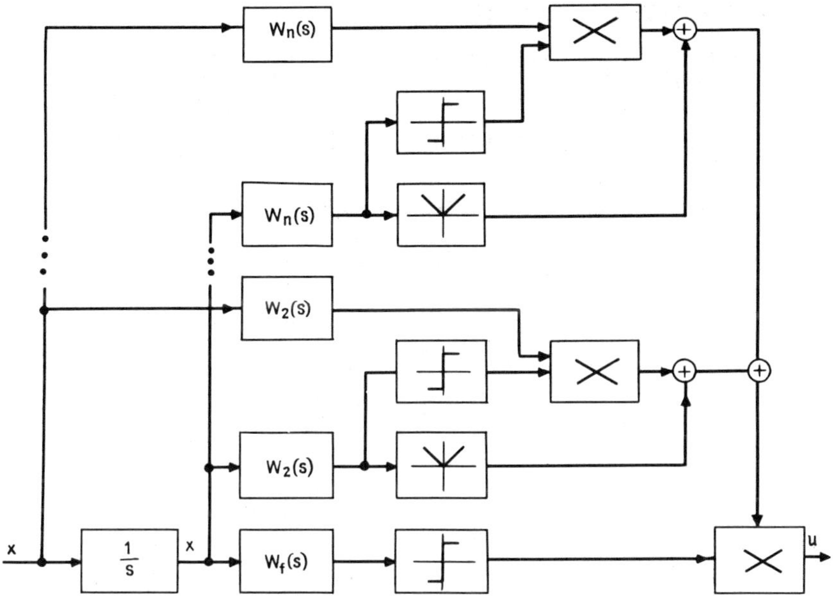

Fig. 2.10

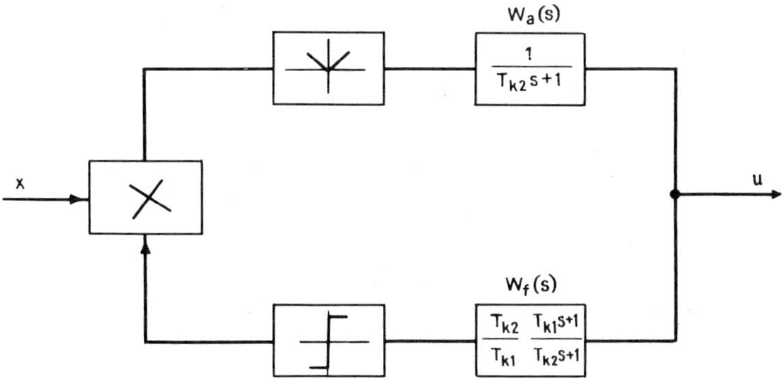

Fig. 2.11

2.5 Evaluating Conditions Precluding Sliding Modes

For any switching hyperplane g_j ($j = 2, 3, \ldots, n$) conditions can be found that preclude sliding modes in control systems with nonlinear correcting unit. Let

$$g = \sum_{i=1}^{n} K_{1,i} x_i \tag{2.112}$$

then (2.112) may be written as

$$
\begin{aligned}
&\dot g \operatorname{sgn} g + \left[\frac{K_{2,n-1}}{K_{1,n}} g - \frac{K_{2,n-1}K_{1,1}}{K_{1,n}} x_1 + \left(K_{2,1} - \frac{K_{2,n-1}K_{1,2}}{K_{1,n}}\right) x_2 + \dots \right.\\
&\quad \left. + \left(K_{2,n-2} - \frac{K_{2,n-1}K_{1,n-1}}{K_{1,n}}\right) x_{n-1}\right] \operatorname{sgn} g_2 + \dot g_3 \operatorname{sgn} g_3 + \dots\\
&\quad + \dot g_{n-1} \operatorname{sgn} g_{n-1} + (K_1 + K_{n+1}) x_2 \operatorname{sgn} x_1 + K_2 x_3 \operatorname{sgn} x_2 + \dots\\
&\quad + K_{n-2} x_{n-1} \operatorname{sgn} x_{n-2} + \left(\frac{K_{n-1}}{K_{1,n}^2} g - \frac{K_{1,1}K_{n-1}}{K_{1,n}} x_1 - \dots \right.\\
&\quad \left. - x_{n-1} \frac{K_{n-1}K_{1,n-1}}{K_{1,n}}\right) \operatorname{sgn} x_{n-1} + \frac{K_n}{K_{1,n}} \operatorname{sgn}\left[\frac{K_n}{K_{1,n}}(g - K_{1,1}x_1 - \dots \right.\\
&\quad \left. - K_{1,n-1}x_{n-1})\right] + \frac{K_{1,n-1}}{K_{1,n}}\left[-g + \frac{K_{1,1}}{K_{1,n}} x_1 + \left(\frac{K_{1,2}}{K_{1,n}} \right.\right.\\
&\quad \left.\left. - K_{1,1}\right) x_2 + \dots + \left(\frac{K_{1,n-2}}{K_{1,n}} - K_{1,n-2}\right) x_{n-1}\right]\\
&\quad \times \operatorname{sgn}\left[\frac{K_n}{K_{1,n}}(g - g_{1,1}x_1 - \dots - K_{1,n-1}x_{n-1})\right] + K(|g|\\
&\quad + \sum_{j=2}^{n-1} |g_j| + \left|\frac{K_n}{K_{1,n}}(g - g_{1,1}x_1 - \dots - K_{1,n-1}x_{n-1})\right|\Bigg) = 0
\end{aligned}
\tag{2.113}
$$

which at $g = 0$ may be rewritten for the derivative as

$$
\begin{aligned}
&\dot g\left\{\operatorname{sgn} g - \frac{K_n}{K_{1,n}} \operatorname{sgn}\left[\frac{K_n}{K_{1,n}}\left(\sum_{i=1}^{n-1} K_{1,i}x_i\right)\right]\right\} = \left\{\left[\sum_{i=1}^{n-1}\left(KK_{2,i} + K_{2,i-1}\right.\right.\right.\\
&\quad \left.\left. - \frac{K_{2,n-1}}{K_{1,n}}\right) x_i\right] \operatorname{sgn} g_2 + K\left(\sum_{j=3}^{n-1} |g_j| + \sum_{i=1}^{n-2} K_i |x_i|\right)\\
&\quad + \sum_{i=1}^{n-2} K_i x_{i+1} \operatorname{sgn} x_i + \sum_{j=3}^{n-1} \dot g_j \operatorname{sgn} g_j - \frac{1}{K_{1,n}}\left[\sum_{i=1}^{n-1}\left(\frac{K_{1,n-1}}{K_{1,n}} K_{1,i}\right.\right.\\
&\quad \left. - K_{1,n-1}K_{1,i-1} - KK_nK_{1,i}\right) \operatorname{sgn}\left(\sum_{i=1}^{n} K_{1,i}x_i\right)\\
&\quad \left.\left. - \left(\sum_{i=1}^{n-1} \frac{K_{1,i}K_{n-1}}{K_{1,n}} x_i\right) \operatorname{sgn} x_{n-1}\right]\right\}
\end{aligned}
\tag{2.114}
$$

We impose the following constraints on the coefficients $K_{i,j}$ in (2.114)

$$KK_{2,i} + K_{2,i-1} = \frac{K_{2,n-1}K_{1,i}}{K_{1,n}}$$

$$\frac{K_{1,i}K_{1,n-1}}{K_{1,n}} = K_{1,i-1} + KK_nK_{1,i}; \quad i = 1, \ldots, n-1 \tag{2.115}$$

$$K_{1,n} > K_n; \quad KK_i > K_{i-1}; \quad KK_{j,i} > K_{j,i-1}; \quad i = 1, \ldots, n;$$

$$j = 4, \ldots, n$$

$$KK_{3,i} > K_{3,i-1} + \frac{K_{n-1}}{K_n} K_{1,i}; \quad i = 1, \ldots, n$$

The sign of g in (2.114) will be found by the relation

$$\dot{g} = -K\left(\sum_{i=1}^{n-2} K_i |x_i| + \sum_{j=3}^{n-1} |g_j| \right) \operatorname{sgn} g \tag{2.116}$$

from which it follows that at all times

$$g\dot{g} < 0 \tag{2.117}$$

This inequality is a necessary and sufficient condition for a sliding mode to exist along the hyperplane g_1. Thus, if the constraints (2.115) are satisfied for the system $\dot{\mathbf{x}} = \mathbf{A}(t)\mathbf{x}$, it experiences a motion along the hyperplane g_1.

The same type of reasoning may be applied to obtain constraints that ensure motion along the hyperplanes $g_j = 0$, $x_i = 0$, $j = 2, \ldots, n$ $i = 1, \ldots, n$ in the system $\dot{\mathbf{x}} = \mathbf{A}(t)\mathbf{x}$. Analysis of these constraints can be carried out to select coefficients of the Lyapunov function (2.82) such that no motions along hyperplanes occur in this system.

Example 2.11. Determine conditions eliminating sliding mode for the system $\dot{\mathbf{x}} = \mathbf{A}(t)\mathbf{x}$ with frozen coefficients for different modes of operation as shown in the table on p. 42. Let the Lyapunov function be in the form (2.59). Find conditions for motion along the trajectory

$$g_1 = K_{1,1}x_1 + K_2x_2 = 0 \tag{2.118}$$

In this case (2.113) may be rewritten as

$$\dot{g}_1 \operatorname{sgn} g_1 + \left[\frac{K_1'}{K_{1,2}} (g_1 - K_{1,1}x_1) \operatorname{sgn} x_1 + \frac{K_2}{K_{1,2}} (K_1 - K_{1,1}x_1) \right]$$
$$\times \operatorname{sgn} \left[\frac{1}{K_{1,2}} (g - g_{1,1}x_1) \right] = -K\left(|g_1| + K_1' |x_1| + \frac{K_2}{K_{1,2}} |g_1| - K_{1,1} |x_1| \right) \tag{2.119}$$

where $K_1' = K_1 + K_{2,1}$.

Let here $g = 0$, then

$$\dot{g}_1\left[\operatorname{sgn} g_1 - \frac{K_2}{K_{1,2}} \operatorname{sgn} \frac{K_{1,1}}{K_{1,2}} x_1\right]$$
$$= -|x_1|\left[K\left(K_1' + \frac{K_2}{K_{1,2}}\right) - \frac{1}{K_{1,2}}\left(K_1' + \frac{K_2K_{1,1}}{K_{1,2}}\right)\right] \tag{2.120}$$

This expression suggests that the condition (2.117) is satisfied only subject to the following constraints

$$K_{1,2} > K_2$$
$$K\left(K_1' + \frac{K_2}{K_{1,2}}\right) > \frac{1}{K_{1,2}}\left(K_1' + \frac{K_2K_{1,1}}{K_{1,2}}\right) \tag{2.121}$$

Let us evaluate constraints providing motion along the trajectory $x_2 = 0$. With the substitution $x_2 = g$ (2.119) rewrites

$$(K_{1,1}g + K_{1,2}\dot{g}) \operatorname{sgn}(K_{1,1}x_1 + K_{1,2}g) + K_1'\dot{g} \operatorname{sgn} g + K_2\dot{x}_2 \operatorname{sgn} x_2$$
$$= -K(|K_{1,1}g + K_{1,2}\dot{g}| + K_1'|g| + K_2|\dot{g}|) \tag{2.122}$$

whence for $g = 0$ we obtain

$$\dot{g}[(K_{1,2} + K_2 + KK_{1,2}) \operatorname{sgn} x_2 + K_1' \operatorname{sgn} g] = -K_2' \operatorname{sgn} x_2 \tag{2.123}$$

From (2.123) it follows that the condition (2.117) is met provided that

$$\operatorname{sgn} x_1 = \operatorname{sgn} x_2 = \operatorname{sgn} \dot{x}_2$$

However, satisfying these equations removes the system described by the equations $\dot{\mathbf{x}} = \mathbf{A}(t)\mathbf{x}$ from the equilibrium state and disturbs the condition (2.12). Therefore, this system will have no singular solution if the coefficients $K_{i,j}$ satisfy the inequalities

$$K_{1,2} > K_2$$
$$\frac{1}{K_{1,2}}\left(K_1' + \frac{K_2K_{1,1}}{K_{1,2}}\right) > K\left(K_1' + \frac{K_2}{K_{1,2}}\right) \tag{2.124}$$

This example illustrates derivation of the constraints on the coefficients $K_{i,j}$, that preclude sliding modes in the control system.

Concluding this chapter we note that the use of pseudo-linear correcting facilities in automatic control systems not only provides operation without sliding modes, but also improves the system performance in handling transients. The required system performance is achieved without having to retune the coefficients of the correcting facilities, i.e. independently of process parameter variation or alteration of the forcing function and disturbance.

References

2.1 A. M. Lyapunov: *Lectures on analytical mechanics*. Kharkov 1895.
2.2 A. M. Lyapunov: *General problem of stability of motion*. Kharkov 1892.
2.3 N. G. Chetayev: *Dokl. Akad. Nauk SSSR* **1**(9), (1934).
2.4 N. G. Chetayev: *Ustoichivost dvizheniya* (Stability of motion). Gostekhizdat, Moscow 1946.
2.5 I. G. Malkin: *Izv. Kazanskogo Fiz-Math. Obshchestva* **5**, (1931).
2.6 I. G. Malkin: *Stability Theory of Motion*. Oldenburg, Munich 1959.
2.7 G. V. Kamenkov: *Sbornik Trudov Kazanskogo Aviatsionnogo Instituta*, No. 9 (1939).
2.8 Ye. A. Barbashin: *Vvedeniye v teoriyu ustoichivosti* (Introduction to stability theory). Nauka, Moscow 1967.
2.9 B. I. Zubov: *Lektsii po teorii upravleniya* (Lectures on control theory). Nauka, Moscow 1975.
2.10 A. I. Lurye: *Nekotorye nelineinye zadachi teorii avtomaticheskogo regulirovaniya* (Some nonlinear problems of control theory). Gostekhizdat, Moscow 1959.
2.11 D. G. Schultz: The generation of Lyapunov functions, in: *Advances in Control Systems* **2** (ed. by C. T. Leondes), Academic Press, N. Y. 1965, pp. 1-64.
2.12 A. A. Kozlov and B. F. Shumilov: In *Sbornik MIFI*. Energoatomizdat, Moscow 1982.
2.13 A. M. Letov: *Stability in Nonlinear Control Systems*. Princeton University Press, Princeton, N.J. 1961.
2.14 Yu. I. Topcheyev and A. P. Tsyplyakov: *Zadachnik po teorii avtomaticheskogo regulirovaniya* (Problems in control theory). Mashinostroyeniye, Moscow 1977.

3 Nonlinear Control Systems: Frequency Analysis of Absolute Stability and Performance

B. N. Naumov, A. I. Barkin, and I. N. Sinitsyn

3.1 General

The analytical techniques based on frequency response characteristics of system components are fundamental to modern system design. Efficiency and simplicity of the harmonic analysis have tempted researchers to extend this approach to substantially nonlinear systems. The prime application was in the analysis of nonlinear oscillations (harmonic balance method). In 1942 Bulgakov [3.1] formulated the stability problem for a nonlinear dynamic system composed of a linear inertial element and a lag-free element with a nonlinear steady-state response $\varphi(\sigma)$ that must be bounded within a given angle. Stability was associated with the absence of self-excited oscillations. However the first general results were obtained with Lyapunov functions. This direction was pioneered by the paper of Lurie and Postnikov [3.2] published in 1944. The classical problem of absolute stability formulated in this publication was solved by Popov [3.3] of Romania. This solution formed a base for further development of frequency methods and paved the way for wide acceptance of the frequency criteria of stability and performance of nonlinear systems. This chapter covers some achievements in this area of control theory.

3.2 Basic Models

Modern control theory studies various systems and methods of their control by means of mathematical methods and models. To construct a mathematical model of a system one should first of all decide which quantities could characterize the system performance and establish relations between these quantities to describe how the system operates.

In the linear control theory, the mathematical models are predominantly systems of ordinary linear differential or integral equations. This situation owes its existence to the fact that the prime objects of study were linear systems with constant parameters. All methods of description are equivalent for such systems and a particular choice is a matter of convenience.

With nonlinear systems the situation is quite different. Here the structure of a particular mathematical model may be of decisive importance in solving the problem at hand. By way of example, a simple linear transformation from the set of differential equations to one integral equation results in the efficiently verifiable frequency conditions of absolute stability.

Any system of automatic control contains a plant (or process) to be controlled and a controller. In many a situation the process is linear and time invariant (stationary) whereas the controller is substantially nonlinear. Therefore, it is customary that such control systems be broken down into a linear part and one or a few nonlinear elements. The most popular model is built as the differential equations in R^n

$$\dot{x} = Ax + b\xi + rg, \quad x(0) = x_0 \tag{3.1}$$

$$\xi = \varphi(\sigma, t) \tag{3.2}$$

$$\sigma = c^{\mathrm{T}}x \tag{3.3}$$

where $x(t)$ is the state vector; the scalar functions $\xi(t)$ and $g(t)$ are referred to as feedback and system input, respectively, and b, c and r are constant column n-vectors. The coefficient matrix A is assumed to be a constant Hurwitz matrix.

The important particular cases of the system (3.1-3) are those of the autonomous system when the system input is absent, i.e. $g(t)$ is identically zero, and of the time-invariant, or stationary, system when $\varphi(\sigma, t)$ does not depend explicitly on time.

For some systems it is convenient to use the description of the dynamic system by a single integral equation

$$\sigma(t) = f(t) - \int_0^t w(t - \lambda)\xi(\lambda)d\lambda + s(t) \tag{3.4}$$

where

$$\begin{aligned} f(t) &= \mathscr{L}^{-1}\{c^{\mathrm{T}}(sI - A)^{-1}x_0\} \\ s(t) &= \mathscr{L}^{-1}\{c^{\mathrm{T}}(sI - A)^{-1}rG(s)\} \\ G(s) &= \mathscr{L}\{g(t)\} \\ w(t) &= \mathscr{L}^{-1}\{W(s)\} \\ W(s) &= c^{\mathrm{T}}(A - sI)^{-1}b \end{aligned} \tag{3.5}$$

and $\mathscr{L}$ and $\mathscr{L}^{-1}$ stand for the direct and inverse Laplace transforms.

With the definition of variables according to (3.5) equation (3.4) is equivalent to (3.1) and (3.3). This equation, however, may be used to describe a more complex system. In particular, the description (3.4) is valid for some systems with distributed parameters [3.4, 5] and time-lag [3.6]. The functions $w(t)$ and $W(s) = c^{\mathrm{T}}(A - sI)^{-1}b$ are the impulse response

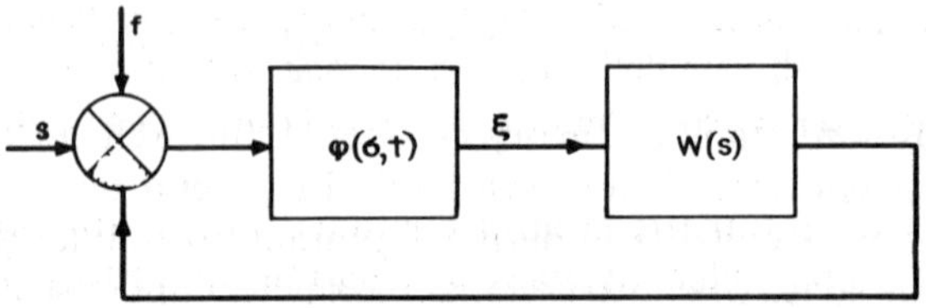

Fig. 3.1

and the impulse transfer function of the linear part. Figure 3.1 shows the block diagram of the system.

Let us turn to the description of nonlinearity $\varphi(\sigma, t)$. We shall assume that the function $\varphi(\sigma, t)$ satisfies the conditions

$$\varphi(0, t) \equiv 0, \quad 0 \leqslant \varphi(\sigma, t)\sigma \leqslant k\sigma^2 \tag{3.6}$$

In the σ, ξ plane the curve $\xi = \varphi(\sigma, t)$ is confined in the sector $[0, k]$ formed by the straight line $\xi = k\sigma$ and the σ axis. It may have common points with the sector boundaries or coincide with one of them (Fig. 3.2). If the nonlinearity concerned is time-invariant, i.e. $\varphi(\sigma, t) \equiv \varphi(\sigma)$, then it is commonly assumed that $\varphi(\sigma)$ is single-valued and continuous.

The class of nonlinearities thus covered is rather wide. It includes such widely used nonlinearities as dead band, an ideal relay, etc. Clearly, the nonstationary linear characteristics $\xi(t) = u(t)\sigma$, $0 \leqslant u(t) \leqslant k$ also belong to this class.

Although in a finite-dimensional case the differential and integral models are equivalent, the methodology of their application is quite different. The differential model is basic to the Lyapunov method, whereas the integral description is fundamental to the method of the *a priori* integral criteria of Popov. More complicated systems will be treated in Sec. 3.13.

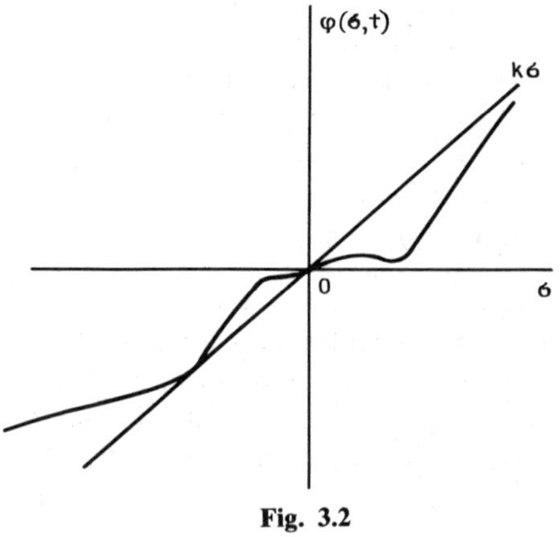

Fig. 3.2

3.3 Absolute Stability of Control Systems

Absolute stability theory is one of the well advanced fields of nonlinear system analysis and synthesis. In general terms, the subject matter of this theory consists in deriving guaranteed bounds for domains of stability in the space of constructional parameters under scarce information on the nonlinearities. The theory is being developed in several directions determined by different approaches and applications. This chapter will focus on the principal direction associated with the analysis of finite-dimensional models that consist of a linear part and one nonlinear element. The linear part is described by ordinary differential equations and the nonlinear element by quadratic relations.

In this chapter, stability will be understood in the sense of Lyapunov, that is, based on the concepts of perturbed and unperturbed (stationary) motions. Accordingly, a certain motion of the system (3.1-3) to be studied for stability will be referred to as the stationary solution. It is associated with the particular solution $x(t) = \tilde{x}(t)$ of the system subject to the initial condition $x(0) = x_0$. If we perturb the initial conditions by letting $x(0) = x_0 + r_0$, then we obtain a new motion corresponding to these new initial conditions. The difference between the stationary and perturbed motions will be denoted by $y(t) = x(t) - \tilde{x}(t)$.

Definition 3.1. An unperturbed motion is stable in the sense of Lyapunov if for any scalar $\varepsilon > 0$ there exists $\delta > 0$ such that $\|r_0\| < \delta$ implies $\|x\| < \varepsilon$ for all $t > t_0$.

Definition 3.2. An unperturbed motion is called asymptotically stable if it is stable as defined by Lyapunov and for a given t_0 there exists a positive scalar $\Delta \leqslant \delta$ such that if $\|r_0\| < \Delta$, then $x(r_0, t) \to 0$ as $t \to \infty$.

Lurie and Postnikov [3.2] were the first to discuss the stability of an unperturbed motion of (3.1-3) at $g(t) \equiv 0$.

Definition 3.3. For $g(t) \equiv 0$, the system (3.1-3) is absolutely stable in the sector $[0, k]$ if for any $\varphi(\sigma, t)$ satisfying (3.6), the system origin is asymptotically stable in the large. This implies that the origin is asymptotically stable in the sense of Lyapunov (Definition 3.2) and the domain of attraction of the origin $x = 0$ is the entire phase space.

It is instructive to note that this definition of absolute stability relates to the class of systems having identical linear parts but different nonlinear elements satisfying (3.6).

Lurie and Postnikov [3.2] investigated the stability of a time-invariant autonomous system with a Lyapunov function of the form

$$V(x) = x^{\mathrm{T}} L x + q \int_0^{c^T x} \varphi(\sigma) \mathrm{d}\sigma \tag{3.7}$$

where $L = L^T > 0$ is a positive definite matrix, and q is a real number. If the nonlinearity is time-varying, then we let $q = 0$.

Differentiating $V(x)$ with respect to time and making use of (3.1-3), we obtain

$$\dot{V}(x) = x^T(A^TL + LA)x + 2x^TLb\xi + q\dot{\sigma}\xi \tag{3.8}$$

By designating $S(x, \xi) = -\dot{V}(x) - \xi(\sigma - k^{-1}\xi)$ we obtain

$$S(x, \xi) = \gamma^2\xi^2 - 2\xi\gamma x^Tu - x^T(A^TL + LA)x \tag{3.9}$$

where

$$\gamma = (k^{-1} - qc^Tb)^{1/2}$$

$$\gamma u = Lb + \frac{1}{2}c + \frac{1}{2}qA^Tc \tag{3.10}$$

The derivative $\dot{V}(x)$ will be negative definite if the quadratic form $S(x, \xi)$ in $n + 1$ variable is positive definite. Following Aizerman and Gantmacher [3.7] we shall refer to this method of construction of Lyapunov functions as the S-procedure.

Theorem 3.1 (Yakubovich [3.8]). Let X be Euclidean space, $\mathscr{F}(x)$ and $\mathscr{G}(x)$ be arbitrary quadratic forms and there exist a vector x_0 such that $\mathscr{G}(x_0) > 0$. The following statements are equivalent

(i) $\mathscr{F}(x) \geqslant 0$ in the set where $\mathscr{G}(x) \geqslant 0$,

(ii) there exists $\tau \geqslant 0$ such that $\mathscr{F}(x) - \tau\mathscr{G}(x) \geqslant 0$ for all x and X.

In (3.2) the role of $\mathscr{F}(x)$ is played by $-\dot{V}(x)$ and the role of $\mathscr{G}(x)$ is played by the form $-\xi(\sigma - k^{-1}\xi)$. The implication of Theorem 3.1 is that the S-procedure, though seemingly artificial, does not result in a loss of some set in the region of stability.

From (3.8-10) it is apparent that the form $S(x, \xi)$ is positive definite provided that

$$A^TL + LA = -uu^T - \varepsilon P$$

$$Lb + \frac{1}{2}c + \frac{1}{2}qA^Tc = \gamma u \tag{3.11}$$

$$\gamma = (k^{-1} - qc^Tb)^{1/2}$$

where $P = P^T > 0$ and $\varepsilon > 0$. In the Russian literature on the subject the system (3.11) is known as sublimited equations of Lurie, while the limited equations result at $\varepsilon = 0$. The major result of Lurie and Postnikov [3.2] may be formulated as follows.

Theorem 3.2. For the system (3.1-3) to be absolutely stable in the sector $[0, k]$ it is sufficient that for some $\varepsilon > 0$ there exists a solution to (3.11) in the form of a positive definite matrix $L = L^T$, a real vector u, and a real number γ.

A direct proof of existence of a feasible solution to (3.11) is a cumbersome algebraic problem for which no analytical solution is available in the general form.

3.4 Frequency Criteria

Criteria of absolute stability of nonlinear systems in the frequency domain have been derived by Popov [3.3, 9] on the base of *a priori* integral criteria. This approach consists essentially in deriving the bounds for coordinates of the system under study in a finite time interval $[0, T]$. Investigating the behavior of such bounds by letting T go to infinity lends a possibility to reveal the character of motions in the system.

Popov's investigation [3.3] was based on the integral model (3.4) and the integral criterion

$$\varrho(T) = \int_0^T \xi(\sigma - k^{-1}\xi + q\dot{\sigma})\mathrm{d}t \tag{3.12}$$

which is also referred to as Popov's functional.

Theorem 3.3 (Popov [3.3]). For the autonomous time-invariant system (3.1-3) to be absolutely stable in the sector $[0, k]$ it is sufficient that there exists a real number q such that for all $\omega \geqslant 0$

$$P(q, k, \omega) \equiv \mathrm{Re}\,[(1 + qj\omega)W(j\omega)] + k^{-1} > 0 \tag{3.13}$$

where $j = \sqrt{-1}$.

To prove this theorem one has to establish the following properties of Popov's functional: $\varrho(T) \geqslant 0$, $\mathrm{d}\varrho(T)/\mathrm{d}T \geqslant 0$, and $\varrho(T) < \infty$.

The criterion (3.13) is valid for a time-varying nonlinearity at $q = 0$. The generalization of this criterion over the case of nonlinearity belonging to the sector $[r, k]$ is, however, more popular.

Theorem 3.4 (Circle criteria [3.3, 10-12]). In order that the autonomous system (3.1-3) with a time-varying nonlinearity be absolutely stable in the sector $[r, k]$ it is sufficient that the linear system with the transfer function $W(s)/(1 + rW(s))$ be stable and that for all $\omega \geqslant 0$

$$\left[\mathrm{Re}\,W(j\omega) + \frac{1}{2}\left(\frac{1}{r} + \frac{1}{k}\right)\right]^2 + [\mathrm{Im}\,W(j\omega)]^2 > \frac{1}{4}\left(\frac{1}{r} - \frac{1}{k}\right)^2 \tag{3.14}$$

At $r = 0$ this condition reduces to

$$P(0, k, \omega) \equiv \mathrm{Re}\,W(j\omega) + k^{-1} > 0 \tag{3.15}$$

The criterion (3.14) derives its name from the geometric interpretation, namely, the Nyquist locus of $W(j\omega)$ must not enter or encircle the circle

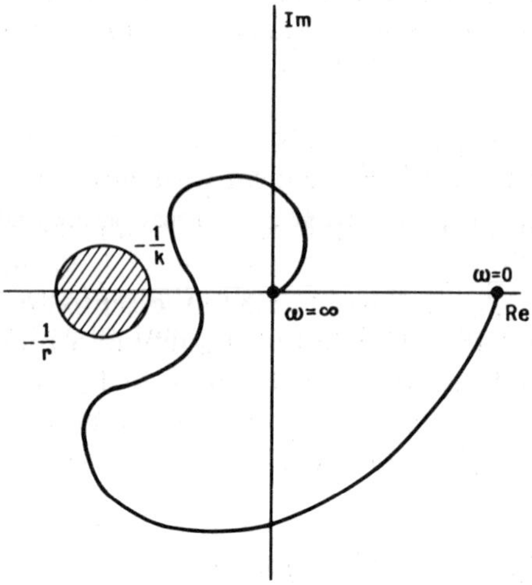

Fig. 3.3

with center $-(r^{-1}+k^{-1})/2+j0$ and radius $(r^{-1}-k^{-1})/2$ (Fig. 3.3). In what follows we shall refer to the condition (3.14) as a circle criterion, too.

Since the pioneer publication of Popov [3.3] there has been a great deal of effort in generalizing the primary formulation of the absolute stability problem. The most general criterion is that due to Yakubovich [3.13]; it embraces the basic criteria of absolute stability in the frequency domain available at the moment.

Consider the system with nonlinearity in vector form

$$\begin{aligned}\dot{x} &= Ax + b\xi \\ \sigma &= c^{\mathrm{T}}x\end{aligned} \tag{3.16}$$

where x is a column n-vector function, and $\xi(t)$ and $\sigma(t)$ are n-vector and m-vector functions, respectively. The functions $\xi(t)$ and $\sigma(t)$ are assumed to satisfy the l local constraints

$$F_i(\xi,\ \sigma,\ \dot{\sigma}) \geqslant 0, \quad i = 1, \ldots, l$$

and s integral constraints

$$\int_0^t F_j(\xi(t),\ \sigma(t),\ \dot{\sigma}(t))\mathrm{d}t \geqslant -\gamma_0, \quad j = 1, \ldots, s$$

where F_i and F_j are quadratic forms in three variables and the constant γ_0 depends on the initial conditions.

To formulate the quadratic criterion we write the form

$$F(\xi, \sigma, \dot{\sigma}) = \sum_{j=1}^{l+s} \tau_j F_j(\xi, \sigma, \dot{\sigma}) \tag{3.17}$$

and introduce a transformation of this form as follows:

(1) considering ξ, σ and $\dot{\sigma}$ as independent variables, we extend the form, retaining its Hermitian property, to the complex values of the variables denoted by $\tilde{\xi}$, $\tilde{\sigma}$, and $\tilde{\dot{\sigma}}$;

(2) substitute $\tilde{\sigma} = -W(s)\tilde{\xi}$ and $\tilde{\dot{\sigma}} = -sW(s)\tilde{\xi}$ in $F(\tilde{\xi}, \tilde{\sigma}, \tilde{\dot{\sigma}})$, where $W(s)$ is the matrix transfer function of the linear part of the system (3.16)

$$W(s) = c^{\mathrm{T}}(A - sI)^{-1}b \tag{3.18}$$

After this transformation we obtain the form $\tilde{F}(s, \tilde{\xi})$ in one vector variable $\tilde{\xi}$.

Theorem 3.5 (Yakubovich [3.14]). Let the matrix A be Hurwitz and $F(0, \sigma, \dot{\sigma}) \geqslant 0$ for all x. For the system (3.16) satisfying the inequality $F(\xi, \sigma, \dot{\sigma}) \geqslant 0$ to be absolutely stable it is sufficient to have the form $\tilde{F}(j\omega, \tilde{\xi})$ negative definite for any $\omega \geqslant 0$.

A general relation of Lyapunov functions with frequency domain techniques has been established by Yakubovich [3.10] and Kalman [3.15] in the form of the frequency theorem.

Given a Hurwitz $n \times n$ matrix A, $n \times m$ matrix B, and a bilinear form

$$\mathscr{G}(x, \xi) = x^{\mathrm{T}}Gx + 2x^{\mathrm{T}}g_1\xi + \xi^{\mathrm{T}}\Gamma\xi \tag{3.19}$$

in vectors x in R^n and ξ in R^m with G and Γ being symmetric $n \times n$ and $m \times m$ matrices, and g_1 and $n \times m$ matrix. It is required to find conditions for existence of a symmetric matrix H that satisfies the inequality

$$2x^{\mathrm{T}}H(Ax + B\xi) - \mathscr{G}(x, \xi) \leqslant 0 \tag{3.20}$$

for all x in R^n and ξ in R^m.

Theorem 3.6 (Yakubovich [3.10]). For a symmetric matrix H satisfying (3.20) to exist, it is necessary and sufficient that

$$\mathscr{G}[(j\omega I - A)^{-1}B\tilde{\xi}, \tilde{\xi}] > 0 \tag{3.21}$$

where $\mathscr{G}[\tilde{x}, \tilde{z}]$ is the extension of (3.19) to the Hermitian form.

Provided that the condition (3.21) is met, there exists an $n \times n$ matrix $H > 0$ and an $n \times m$ matrix h such that

(1) the identity

$$2x^{\mathrm{T}}H(Ax + B\xi) - \mathscr{G}(x, \xi) = -(\xi - h^{\mathrm{T}}x)^{\mathrm{T}}\Gamma(\xi - h^{\mathrm{T}}x) \tag{3.22}$$

holds, where $\Gamma = \Gamma^{\mathrm{T}} > 0$ is a matrix in (3.19); and

(2) $A = A + Bh$ is a Hurwitz matrix.

By virtue of (3.22), equating the matrices in the quadratic and bilinear forms leads to the generalized Lurie equations

$$\begin{aligned} A^{\mathrm{T}}H + HA - G &= -h\Gamma h^{\mathrm{T}} \\ Hb - g_1 &= h\Gamma \end{aligned} \tag{3.23}$$

The frequency theorem is a convenient analytic tool for many problems as it affords easy transfer from Lyapunov functions to frequency domain criteria and vice versa.

A few publications have thus far addressed the development of a more rigorous circle criterion. The more general results have been reported by Pyatnitskii [3.16, 17] for autonomous systems with a time-varying nonlinearity. They have been derived with the variational approach. This approach is essentially a search for a least stable reference system among those constituting the class of systems under consideration. If this system, referred to as limiting, appears to be stable, then the other systems of this class are stable, too.

The limiting system is described by the equations

$$\begin{aligned} \dot{x} &= Ax + \frac{k}{2}\, bc^{\mathrm{T}}x[1 - \operatorname{sgn}(c^{\mathrm{T}}x)(z^{\mathrm{T}}b)] \\ \dot{z} &= -\mu z - A^{\mathrm{T}}z - \frac{k}{2}\, cz^{\mathrm{T}}b[1 - \operatorname{sgn}(c^{\mathrm{T}}x)(z^{\mathrm{T}}b)] \end{aligned} \tag{3.24}$$

where z is an n-vector, and μ is a positive number satisfying the inequality

$$\mu > \mu^* = \frac{n(n+1)}{2} \max\left\{\max_{i,j} |a_{ij}|,\ \max_{s,m} |kb_s c_m|\right\}$$

$$A = \{a_{ij}\}, \quad B = \{b_s\}, \quad c = \{c_m\} \tag{3.25}$$

Theorem 3.7 (Pyatnitskii [3.17]). For the autonomous system (3.1-3) to be absolutely stable in the sector $[0, k]$ it is necessary and sufficient that the trivial solution $x = 0$, $z = 0$ of the system (3.24) be asymptotically stable in the small (in the sense of Lyapunov) for $\|x_0\|\ \|z_0\| > 0$, where $\|x_0\| = (x_0^{\mathrm{T}}x_0)^{1/2}$ and $\|z_0\| = (z_0^{\mathrm{T}}z_0)^{1/2}$.

Pyatnitskii [3.17] points out that for large μ asymptotic stability takes place in z variables. In addition, it suffices to investigate only the initial conditions of the form $\|x_0\| = 1$, $\|z_0\| = 1$. Unfortunately, beyond direct modeling there exist no other efficient methods for the investigation of stability of the limiting system (3.24).

For second-order systems, readily verifiable necessary and sufficient conditions of absolute stability have been obtained by Levin [3.18]. A new substantial finding makes the subject of the following theorem.

Theorem 3.8 (Leonov [3.19]). Let there be given the system

$$\begin{aligned} \dot{x} &= Ax + b\varphi(\sigma, t) \\ \sigma &= c^{\mathrm{T}}x \\ r\sigma^2 &\leqslant \varphi(\sigma, t)\sigma \leqslant \mu\sigma^2, \text{ for all } \sigma, t \end{aligned} \tag{3.26}$$

whose transfer function $c^{\mathrm{T}}(A - sI)^{-1}b = W(s)$ is nonsingular, has one zero pole and $n - 1$ poles λ_i with negative real parts, and $\lim sW(s) = 0$ as $s \to \infty$. Let $W_1(s) = sW(s)$ and $\lambda_0 = \min_i \{-\mathrm{Re}\,\lambda_i\}$. If for some scalar $T > (2\lambda_0)^{-1}$

$$\lim_{\omega \to \infty} \left\{ \omega^2 \,\mathrm{Re} \left[\left(j\omega + \frac{1}{2T} \right) W_1 \left(j\omega - \frac{1}{2T} \right) \right] \right\} > 0 \tag{3.27}$$

$$\mu < \frac{1.3\mathrm{Re}\,[(j\omega + 1/2T)W_1(j\omega - 1/2T)]}{T^2 |(j\omega + 1/2T)W_1(j\omega - 1/2T)|} \tag{3.28}$$

then the system (3.26) is absolutely stable.

Examples can be provided where this criterion is stronger than the circle criterion. However, situations with the circle criterion having edge on Leonov's are also known.

The aforementioned examples of improvement of the circle criterion relate but to systems with a time-varying nonlinearity. For systems with time-invariant nonlinearity, an improvement of Popov's criterion has been achieved by Barkin and Zelentsovsky [3.20].

3.5 Absolute Stability of Processes

In nonlinear systems, the stability of an equilibrium state does not guarantee stability for forced processes. Let us consider the model of time-invariant system (3.1-3) with bounded system input $g(t)$. The response $x(t)$ of the system to $g(t)$ will be called an unperturbed motion.

The system (3.1-3) with perturbed initial conditions provides the equations of perturbed motion

$$\dot{y} = Ay + b\varphi(\sigma) + g(t) \tag{3.29}$$

$$y(0) = y_0, \; y_0 \neq x_0 \tag{3.30}$$

$$\sigma = c^{\mathrm{T}}y \tag{3.31}$$

$$r\sigma^2 \leqslant \varphi(\sigma)\sigma \leqslant k\sigma^2 \tag{3.32}$$

Definition 3.4. The processes in the system (3.29-32) will be referred to as absolutely stable if for any function $\varphi(\sigma)$ satisfying (3.32) for any scalar $\varepsilon > 0$ there exists $\delta > 0$ such that the condition $\|x_0 - y_0\| < \delta$ implies that

$$\|x(t) - y(t)\| < \varepsilon, \; t \geqslant 0 \tag{3.33}$$

Definition 3.5. The processes in the system (3.29-32) will be called absolutely asymptotically stable if they satisfy Definition 3.4 and $x(t) - y(t) \to 0$ as $t \to \infty$.

The criterion of absolute asymptotic stability was simultaneously developed by Naumov and Tsypkin [3.21] and Yakubovich [3.22].

Theorem 3.9. The processes in the system (3.29-32) are absolutely asymptotically stable if

(1) the linear system with the transfer function $W(s)/[1 + rW(s)]$ is stable,

(2) the nonlinearity $\varphi(\sigma)$ possesses the property

$$r \leqslant \frac{\varphi(\sigma_1) - \varphi(\sigma_2)}{\sigma_1 - \sigma_2} \leqslant k, \text{ and} \tag{3.34}$$

(3) for all $\omega \geqslant 0$ the frequency condition

$$\left[\operatorname{Re} W(j\omega) + \frac{1}{2}\left(\frac{1}{r} + \frac{1}{k}\right)\right]^2 + [\operatorname{Im} W(j\omega)]^2 \geqslant \frac{1}{4}\left(\frac{1}{r} - \frac{1}{k}\right)^2 \tag{3.35}$$

Proof. Denote $z = y - x$ and write the variational equations

$$\dot{z} = Az + b\tilde{\varphi}(z, t), \qquad z(0) = y_0 - x_0 \tag{3.36}$$

where

$$\tilde{\varphi}(z, t) = \varphi[c^{\mathrm{T}}(x + z)] - \varphi(c^{\mathrm{T}}x) \tag{3.37}$$

By virtue of (3.33) we have

$$r \leqslant \tilde{\varphi}(c^{\mathrm{T}}z, t)/c^{\mathrm{T}}z \leqslant k \tag{3.38}$$

Thus the problem of absolute stability of the processes is reduced to investigation of absolute stability of the equilibrium state $z = 0$ of the time-varying system (3.36-38). The solution to the last problem is given by Theorem 3.4.

Remark 3.1. The condition (3.34) will obviously be satisfied if there exists $d\varphi/d\sigma$ such that $r \leqslant d\varphi/d\sigma \leqslant k$.

Practical application of the criterion (3.35) may be facilitated as follows [3.23, 24]. If we introduce the parameter $\alpha = k/r > 0$, then the condition (3.35) rewrites as

$$\left[k \operatorname{Re} W(j\omega) + \frac{1}{2}(\alpha + 1)\right]^2 + k[\operatorname{Im} W(j\omega)]^2 > \frac{1}{4}(\alpha - 1)^2$$

This is the equation of the circles of radius $R = (\alpha - 1)/2$ passing through the point $(-1, j0)$ and placed on the left from the straight line $k \operatorname{Re} W(j\omega) = -1$ (Fig. 3.4). Each circle corresponds to its own value of $\alpha = k/r \geqslant 1$. At $\alpha = 1$, the circle shrinks to a point $(-1, j0)$ and the sufficient condition of absolute stability coincides with the necessary and suffi-

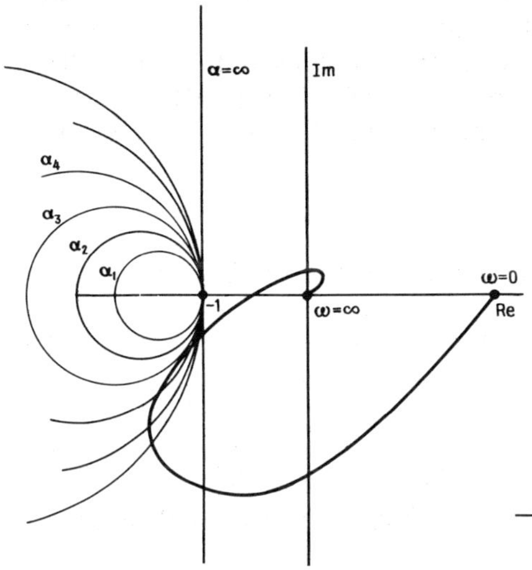

Fig. 3.4

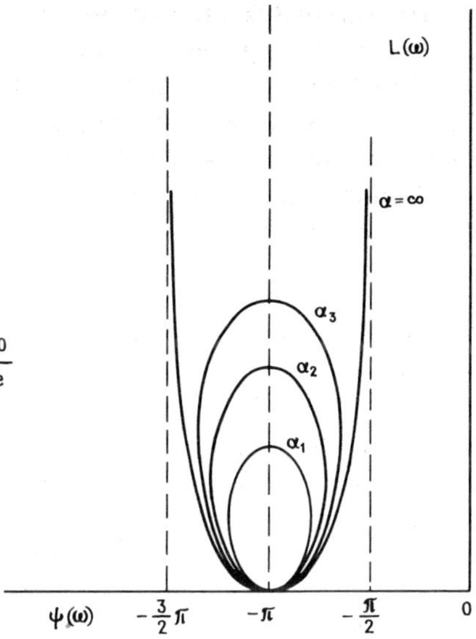

Fig. 3.5

cient condition of stability of the linear system with the transfer function $W(s)/[1 + rW(s)]$. At $\alpha = \infty$, the α circle becomes a vertical straight line passing through the point $(-1, j0)$.

Now we are in a position to give geometrical interpretation to the absolute stability criterion of the processes in a nonlinear system. For absolute stability of the processes in a control system with one nonlinear element it is sufficient that the derivative of the nonlinear element response belong to the interval $[r, k]$ and that the Nyquist plot of the open-loop transfer function $kW(j\omega)$ do not enter the respective α circle.

We can also formulate the frequency criterion of absolute stability in nonlinear systems in terms of logarithmic plots. Let the abscissa be $\psi(\omega) = \arg W(j\omega)$ and the ordinate be $L(\omega) = 20 \log |kW(j\omega)|$. The boundary curves [3.25] corresponding to α circles plotted to these coordinates are shown in Fig. 3.5.

For absolute stability of processes in a system with one nonlinear element it is sufficient that the derivative of the nonlinear element response satisfy the condition

$$r \leqslant \mathrm{d}\varphi/\mathrm{d}\sigma \leqslant k$$

and that the logarithmic plot of the open-loop system

$$L(\omega) = Q[\psi(\omega)]$$

satisfy the Nyquist criterion and do not cross the respective boundary curve. It is worth noting that the boundary α curves plotted in Fig. 3.5 coincide with Nichols charts [3.26], that are used to determine the magnitude of the closed-loop frequency response of a linear system from the Bode plot of the open-loop transfer function. Each Nichols chart reveals its highest value of magnitude B_m. In stability analysis B_m should be replaced with α by the following rule

$$\alpha - 1 = 1/|1 - B_m|$$

The portion of a Nichols chart that corresponds to the boundary curves is defined by the positive values of B_m in the range from unity to infinity.

Unfortunately the frequency criterion (3.35) is very rough in many cases and results in underestimated bounds for stability regions and overestimated performance indices. Because the criterion (3.35) coincides formally with the circle criterion for the time-varying system, it could be improved if there would be a possibility to reinforce the circle criterion. Below we wish to demonstrate that this could be done by invoking a wider class of Lyapunov functions.

3.6 The Power Transform Method

To demonstrate the idea of this approach we begin with a simple illustrative example. Let us consider the second order nonlinear system

$$\begin{aligned} \dot{x}_1 &= x_2 \\ \dot{x}_2 &= -a_1 x_1 - a_2 x_2 - b\varphi(x_1, t) \end{aligned} \tag{3.39a}$$

Introduce the new variables $y_1 = x_1^2$, $y_2 = x_1 x_2$, and $y_3 = x_2^2$. Differentiating these variables carries the set (3.39a) to the form

$$\begin{aligned} \dot{y}_1 &= 2y_2 \\ \dot{y}_2 &= -a_1 y_1 - a_2 y_2 + y_3 + \psi_1(x_1, t) \\ \dot{y}_3 &= -2a_1 y_1 - 2a_2 y_2 + \psi_2(x_1, x_2, t) \end{aligned} \tag{3.39b}$$

where $\psi_1(x_1, t) = \varphi(x_1, t)x_1$, and $\psi_2(x_1, x_2, t) = \varphi(x_1, t)x_2$.

The presence of nonlinearity $\varphi(x_1, t)$ in the initial set results in both sets (3.39a) and (3.39b) being coupled. In addition, the dimension of the state vector and the number of nonlinear elements increase. However, the transformed system (3.39b) simplifies drastically if $\varphi(x_1, t) = u(t)x_1(t)$, as in this case $\psi_1(x_1, t) = u(t)y_1$ and $\psi_2(x_1, x_2, t) = u(t)y_2$, that is the nonlinear transformation of the state vector maps a linear system onto a linear system with a similar structure.

Application of the power transformation to analysis of linear (generally time-varying) systems has been developed by Brockett [3.27].

Let us consider an n-variable control system whose phase coordinates form the state vector x. Following Brockett's theory [3.27] we introduce a vector $x^{[p]}$ composed of linear-independent products of p (positive integer) elements of the vector x.

$$\sqrt{\binom{p}{p_1}\binom{p-p_1}{p_2}\cdots\binom{p-p_1-\cdots-p_{n-1}}{p_n}}\, x_1^{p_1}x_2^{p_2}\cdots x_n^{p_n}$$

$$\sum_{i=1}^{n} p_i = p, \quad p_i \geqslant 0 \tag{3.40}$$

The elements of $x^{[p]}$ will be ordered lexigraphically. The dimension of $x^{[p]}$ is

$$m = \binom{n+p-1}{p} = \frac{(n+p-1)!}{p!(n-1)!} \tag{3.41}$$

Choosing the coefficients in (3.40), we can secure the condition

$$(x^{\mathrm{T}}x)^p = x^{[p]^{\mathrm{T}}}x^{[p]^{\mathrm{T}}} \tag{3.42}$$

and the general equality

$$(x^{\mathrm{T}}y)^p = x^{[p]^{\mathrm{T}}}y^{[p]^{\mathrm{T}}} \tag{3.43}$$

Let us look at how this power transformation applies to linear equations. Application to the equation

$$z = Ax \tag{3.44}$$

yields

$$z^{[p]} = A^{[p]}x^{[p]} \tag{3.45}$$

where $A^{[p]}$ is an $m \times m$ matrix. By way of example, for $m = p = 2$

$$A = \begin{bmatrix} a_{11}a_{12} \\ a_{21}a_{22} \end{bmatrix}, \quad A^{[2]} = \begin{bmatrix} a_{11}^2 & \sqrt{2}a_{11}a_{12} & a_{12}^2 \\ \sqrt{2}a_{11}a_{21} & a_{11}a_{22} + a_{12}a_{21} & \sqrt{2}a_{12}a_{22} \\ a_{21}^2 & \sqrt{2}a_{21}a_{22} & a_{22}^2 \end{bmatrix}$$

It is worth noting that the elements of $A^{[p]}$ are nonlinear functions of the elements of A. The properties of $A^{[p]}$ transformation are as follows [3.27].

Theorem 3.10. Let A and B be $n \times n$ matrices, then

(1) $I_n^{[p]} = I_m$;

(2) $(AB)^{[p]} = A^{[p]}B^{[p]}$;

(3) $(A^q)^{[p]} = (A^{[p]})^q$, where q is an integer and A^q existent;

(4) $(A^{\mathrm{T}})^{[p]} = (A^{[p]})^T$.

The eigenvalues of $A^{[p]}$ are m products $\lambda_i\lambda_j \ldots \lambda_k$ of p cofactors each, variable in various sets of indices. It follows that

(i) the eigenvalues of $A^{[p]}$ lie within a unit circle if $|\lambda_i| < 1$, $i = 1, 2, \ldots, n$;

(ii) if $A^{\mathrm{T}} = A > 0$, then $A^{[p]} > 0$.

Let the linear differential equation

$$\dot{x} = A(t)x \tag{3.46}$$

be given in R^n. Then [3.27]

$$\frac{\mathrm{d}}{\mathrm{d}t} x^{[p]} = A_{[p]} x^{[p]} \tag{3.47}$$

where $A_{[p]}$ is an $m \times m$ matrix whose elements are linear functions of the elements of $A(t)$.

By way of example, for $p = 2$ the image corresponding to the matrix

$$A = \begin{bmatrix} a_{11} & a_{12} \\ a_{21} & a_{22} \end{bmatrix}$$

will be

$$A_{[2]} = \begin{bmatrix} 2a_{11} & \sqrt{2}a_{12} & 0 \\ \sqrt{2}a_{21} & a_{11} + a_{22} & \sqrt{2}a_{12} \\ 0 & \sqrt{2}a_{21} & 2a_{22} \end{bmatrix}$$

It is worth noting the correspondence between the eigenvalues of A and $A_{[p]}$. The eigenvalues of the $m \times m$ matrix $A_{[p]}$ are m sums $\lambda_i + \lambda_j + \ldots + \lambda_k$ of p addends each taken over various sets of indices. This specifically implies that if A is a Hurwitz matrix, then $A_{[p]}$ is also Hurwitzean.

Now that we introduced the definitions and some properties of power transformations of a state vector $x^{[p]}$ and the associated upper-index $A^{[p]}$ and lower-index $A_{[p]}$ transformations of square matrices, we should recognize that these are insufficient to study nonlinear systems with feedback or nonautonomous systems. The missing link is the lower-index transformation $b_{[p]}$ of a constant vector b. This transformation is introduced below [3.28].

Definition 3.6. Consider the differential equation

$$\dot{x} = b\xi(t) \tag{3.48}$$

given in R^n, where $\xi(t)$ is a scalar function. Applying the power transformation, we obtain

$$\frac{\mathrm{d}}{\mathrm{d}t} x^{[p]} = b_{[p]}\xi(t)x^{[p-1]} \tag{3.49}$$

where $b_{[p]}$ is an $m \times l$ matrix which will be taken as the desired transformation of b. It is natural to assume also that

$$(b^{\mathrm{T}})_{[p]} = b^{\mathrm{T}}_{[p]} \tag{3.50}$$

By way of illustration,

$$\begin{bmatrix} b_1 \\ b_2 \\ b_3 \end{bmatrix} = \begin{bmatrix} 2b_1 & 0 & 0 \\ \sqrt{2}b_2 & \sqrt{2}b_1 & 0 \\ \sqrt{2}b_3 & 0 & \sqrt{2}b_1 \\ 0 & 2b_2 & 0 \\ 0 & \sqrt{2}b_3 & \sqrt{2} \\ 0 & 0 & 2b_3 \end{bmatrix}$$

Making use of the definitions of the matrices $A^{[p]}$, $A_{[p]}$, $b_{[p]}$ by equations (3.45), (3.47), (3.50) and the property (3.43) of the power transformation, one can obtain the algebraic relation between the upper-index and lower-index transformations of vectors and matrices displayed in Table 3.1. These relations form a mathematical basis for construction of new models of control systems.

Table 3.1. Power Transformation Algebra

(i) *Square matrix transformation*

(1) $I_n^{[p]} = I_m$

(2) $(AB)^{[p]} = A^{[p]}B^{[p]}$

(3) $(A^q)^{[p]} = (A^{[p]})^q$

(4) $(A^{\mathrm{T}})^{[p]} = A^{[p]T}$

(5) $(I_n)_{[p]} = pI_m$

(6) $(A + B)_{[p]} = A_{[p]} + B_{[p]}$

(7) $(A^{\mathrm{T}})_{[p]} = A^{\mathrm{T}}_{[p]}$

(8) $(BAB^{-1})_{[p]} = B^{[p]}A_{[p]}(B^{[p]})^{-1}$

(ii) *Vector transformations (a and b are column vectors of the same dimension)*

(9) $pa^{\mathrm{T}}ba^{[p-1]} = b^{\mathrm{T}}_{[p]}a^{[p]}$

(10) $p(ab^{\mathrm{T}})_{[p]} = a_{[p]}b^{\mathrm{T}}_{[p]}$

(11) $(a + b)_{[p]} = a_{[p]} + b_{[p]}$

(12) $(ab^{\mathrm{T}})^{[p]} = a^{[p]}b^{[p]\mathrm{T}}$

(13) $(a^{\mathrm{T}}b)^{[p]} = a^{[p]\mathrm{T}}b^{[p]}$

(14) $pa^{[p]} = a_{[p]}a^{[p-1]}$

(15) $a^{[p]} = \frac{1}{p!} a_{[p]}a_{p-1} \ldots a_{[2]}a$

(16) $a_{[p]}b_{[p-1]} = b_{[p]}a_{[p-1]}$

(17) $\frac{1}{p} a^{\mathrm{T}}_{[p]}b_{[p]} = \frac{1}{p-1} b_{p-1}a^{\mathrm{T}}_{p-1} + (a^{\mathrm{T}}b)I_l$

(iii) *Vector-matrix relations*

(18) $(Ab)_{[p]}A^{[p-1]} = A^{[p]}b_{[p]}$

(19) $(Ab)_{[p]}b^{[p-1]} = A_{[p]}b^{[p]}$

(20) $b^{\mathrm{T}}_{[p]}A_{[p]} = (b^{\mathrm{T}}A)_{[p]} + A_{[p-1]}b^{\mathrm{T}}_{[p]}$

3.7 Model Hierarchy of Nonlinear Control

To construct a new model of a nonlinear system we need a general expression for the derivative of x.

Theorem 3.11. Let $x(t)$ be an n vector function having a derivative $\dot{x}(t)$, then

$$\frac{\mathrm{d}}{\mathrm{d}t} x^{[p]}(t) = (\dot{x}(t))_{[p]}x^{[p-1]}(t) \tag{3.51}$$

Proof. By definition

$$x(t + h) = x(t) + h\dot{x}(t) + O(h^2)$$

By applying the upper-index transformation $(\cdot)^{[p]}$ we obtain

$$x^{[p]}(t + h) = (x(t) + h\dot{x}(t))^{[p]} + O(h^2)$$

According to item (15) of Table 3.1,

$$(x(t) + h\dot{x}(t))^{[p]} = \frac{1}{p!} \prod_{r=p}^{1} (x(t) + h\dot{x}(t)_{[r]} = \frac{1}{p!} \prod_{r=p}^{1} x_{[r]}(t)$$

$$+ h \frac{1}{p!} \{x_{[p]}x_{[p-1]}, \ldots, x_{[2]}\dot{x} + x_{[p]} \ldots \dot{x}_{[2]}x + \ldots$$

$$+ (\dot{x})_{[p]}x_{[p-1]} \ldots x_{[2]}x\} + O(h^2)$$

The braces of this expression confine r terms of the form

$$x_{[p]}x_{[p-1]} \ldots x_{[r]}(\dot{x})_{[r-1]}x_{[r-2]} \ldots x$$

Using item (16) of Table 3.1, we obtain that all the terms in the braces are identical and equal to

$$(\dot{x})_{[p]}x_{[p-1]}x_{[p-2]}\ldots x_{[2]}x$$

A repeated application of property (15) leads to

$$[x(t) + h\dot{x}(t)]^{[p]} = x^{[p]} + h(\dot{x})_{[p]}x^{[p-1]}$$

whence

$$x^{[p]}(t+h) - x^{[p]}(t) = h(\dot{x})_{[p]}x^{[p-1]} + O(h^2)$$

The identity (3.51) is attained by going to the limit as $h \to 0$. QED.

Corollary. Let $x(t)$ satisfy the differential equation

$$\dot{x} = F(x, t) \tag{3.52}$$

then

$$\frac{\mathrm{d}}{\mathrm{d}t}x^{[p]} = (F(x, t))_{[p]}x^{[p-1]} \tag{3.53}$$

This formula is helpful in deriving power transforms of nonlinear differential equations. Successful application of (3.53) to (3.1) at $g(t) \equiv 0$ yields

$$\frac{\mathrm{d}}{\mathrm{d}t}x(t) = Ax(t) + b\xi(t) \tag{3.54}$$

$$\frac{\mathrm{d}}{\mathrm{d}t}x^{[2]}(t) = (Ax(t) + b\xi(t))_{[2]}x(t) = A_{[2]}x^{[2]}(t) + b_{[2]}x(t)\xi(t) \tag{3.55}$$

$$\frac{\mathrm{d}}{\mathrm{d}t}x^{[p]}(t) = A_{[p]}x^{[p]}(t) + b_{[p]}x^{[p-1]}(t)\xi(t) \tag{3.56}$$

Each subsequent equation of this system is a consequence of (3.54). For the case of the linear system with variable feedback, i.e. for

$$\xi(t) = u(t)c^{\mathrm{T}}x \tag{3.57}$$

equation (3.56) may be made independent explicitly of $x^{[p-1]}$. In agreement with property (9) we have

$$z = c^{\mathrm{T}}xx^{[p-1]} = \frac{1}{p}c^{\mathrm{T}}_{[p]}x^{[p]} \tag{3.58}$$

Therefore, (3.56) is reduced to the form

$$\frac{\mathrm{d}}{\mathrm{d}t}x^{[p]} = \left(A_{[p]} + \frac{1}{p}u(t)b_{[p]}c^{\mathrm{T}}_{[p]}\right)x^{[p]} \tag{3.59}$$

or by (10) of Table 3.1

$$\frac{\mathrm{d}}{\mathrm{d}t}x^{[p]} = (A + u(t)bc^{\mathrm{T}})_{[p]}x^{[p]} \tag{3.60}$$

In addition to this equation, one may also investigate the equation

$$\frac{d}{dt} y = (A + u(t)bc^{T})_{[p]} y \tag{3.61}$$

in R^m.

Under consistent initial conditions, i.e. at $y(0) = x_0^{[p]}$, the solutions to (3.60) and (3.61) coincide. Mismatched initial conditions produce different solutions of these equations. However, all solutions to (3.61) may be investigated for all initial conditions $y(0)$ in R^m. In this case the solutions to (3.60) constitute a subset of the set of solutions to (3.61).

In the general case, the system (3.54-56) may be decoupled by substituting $y \in R^m$ and $\eta \in R^l$ for $x^{[p]}$ and $x^{[p-1]}$, respectively. This, however, will break the connection to the initial problem. This hazard can be mitigated to a considerable degree by allowing for nonlinear, specifically quadratic, relations between the elements of $x^{[p]}$ and $x^{[p-1]}$. A compact description of such relations involving the least number of free parameters is not a trivial problem. It is desirable that this description be adapted to absolute stability theory and enabled improvement of existent results.

We suggest as a basic constraint

$$\eta^{T} S c_{[p]}^{T} y = 0 \tag{3.62}$$

where S is an arbitrary skew-symmetric $l \times l$ matrix. This product is identically zero if $y = x^{[p]}$ and $\eta = \xi x^{[p-1]}$.

Definition 3.7. The system

$$\left.\begin{aligned} \dot{y} &= A_{[p]} y + b_{[p]} \eta \\ \eta^{T} S c_{[p]}^{T} y &= 0 \end{aligned}\right\} \tag{3.63}$$

will be referred to as the p-level model for the equation $\dot{x} = Ax + b\xi$.

The following step is a description of the nonlinear feedback in the new coordinates. It is achieved by multiplying the inequality $\xi(c^{T}x - k^{-1}\xi) \geqslant 0$ by an arbitrary positive definite quadratic form

$$x^{[p-1]T} R x^{[p-1]}, \quad R = R^{T} > 0$$

thus producing the inequality

$$\eta^{T} R \left(\frac{1}{p} c_p^{T} y - k^{-1} \eta \right) \geqslant 0 \tag{3.64}$$

Combining (3.64) and (3.63), we obtain a model of level p for the original system.

Definition 3.8. The model of level p for the autonomous system (3.1-3, 6) is the set of equations

$$\dot{y} = A_{[p]}y + b_{[p]}\eta$$
$$\eta^{\mathrm{T}} Q\left(\frac{1}{p} c_{[p]}^{\mathrm{T}} y - k^{-1}\eta\right) \geqslant 0 \tag{3.65}$$

in R^m, where $Q = R + S$ is an asymmetric positive definite matrix.

Definition 3.9. A family of p-level models ordered with respect to p will be called a hierarchic system of models.

3.8 Improved Criterion of Absolute Stability

We shall decide whether or not the initial system (3.1-6) is stable by its p-level model (3.65). If this model is absolutely stable, then $y^{\mathrm{T}}y \to 0$ as $t \to \infty$. Specifically, $x^{[p]\mathrm{T}}x^{[p]} \to 0$ as $t \to \infty$. Clearly, all the stability properties intrinsic to the model (3.65) (stability in the sense of Lyapunov, exponential stability, etc.) carry over onto the initial system.

Theorem 3.12 (Barkin and Zelentsovsky [3.29]). A sufficient condition for the absolute stability of the autonomous time-varying system (3.1-3, 6) is that A be a Hurwitz matrix and for all $\omega \geqslant 0$

$$\mathrm{Re}\,[Q(\Phi(j\omega) + k^{-1}I_l)] > 0 \tag{3.66}$$

where

$$\Phi(j\omega) = \frac{1}{p} c_{[p]}^{\mathrm{T}}(A_{[p]} - j\omega I_m)^{-1} b_{[p]} \tag{3.67}$$

Proof. We resort to the quadratic criterion of Yakubovich (Theorem 3.5) and apply it to the linear system (3.65) with one local bound. The frequency condition (3.66) is sufficient for

$$\tilde{F}[j\omega, \eta] = \tilde{\eta}^* Q(-\Phi(j\omega) - k^{-1}I_l)\eta < 0$$

for all $\eta \in C^n$ and $\omega \geqslant 0$. QED.

Remark. The condition (3.66) implies positive definiteness of the Hermite matrix

$$\Gamma(\omega) = Q\Phi(j\omega) + \Phi^*(j\omega)Q^{\mathrm{T}} + k^{-1}(Q + Q^{\mathrm{T}}) \tag{3.68}$$

This matrix is positive definite if [3.14]

$$\Gamma(\omega) > 0 \tag{3.69}$$

and

$$\det \Gamma(\omega) \neq 0 \tag{3.70}$$

The first of these conditions holds automatically as $\Phi(j\omega) \to 0$ as ω is allowed to go to infinity.

It is worth noting that for practical computations all the coefficients in (3.40) may be set equal to unity. Below we give a few examples of applicability of the new absolute stability criterion [3.30, 31].

Example 3.1. Consider a second-order system with time-varying nonlinearity modeled by the system of equations

$$\begin{aligned} \dot{x}_1 &= a_{11}x_1 + a_{12}x_2 + b_1\xi \\ \dot{x}_2 &= a_{21}x_1 + a_{22}x_2 + b_2\xi \\ \xi(t) &= \varphi(\sigma_1, t) \\ \sigma_1 &= c_1x_1 + c_2x_2 \end{aligned}$$

The new variables $\sigma_1 = c_1x_1 + c_2x_2$ and $\sigma_2 = x_2$ carry the system to the form

$$\begin{aligned} \dot{\sigma}_1 &= p_{11}\sigma_1 + p_{12}\sigma_2 + d_1\varphi(\sigma_1, t) \\ \dot{\sigma}_2 &= p_{21}\sigma_1 + p_{22}\sigma_2 + d_2\varphi(\sigma_1, t) \end{aligned}$$

Application of the transformation of the second degree leads to the second-level model

$$\begin{aligned} \dot{y}_1 &= 2p_{11}y_1 + 2p_{12}y_2 + 2d_2\eta_1 \\ \dot{y}_2 &= p_{21}y_1 + (p_{11} + p_{22})y_2 + p_{13}y_3 + d_2\eta_1 + d_1\eta_2 \\ \dot{y}_3 &= 2p_{21}y_2 + 2p_{22}y_3 + 2d_2\eta_2 \end{aligned}$$

where $y_1 = \sigma_1^2$, $y_2 = \sigma_1\sigma_2$, $y_3 = \sigma_2^2$, $\eta_1 = \xi x_1$, and $\eta_2 = \xi x_2$.

The instability condition takes the form

$$\det \Gamma(\omega) = [q_{11}(\operatorname{Re}\varphi_{11} + k^{-1}) + q_{12}\operatorname{Re}\varphi_{21}][\operatorname{Re}\varphi_{22} + k^{-1} + q_{21}\operatorname{Re}\varphi_{12}] - \frac{1}{4}\,|q_{11}\varphi_{12} + \overset{*}{\varphi}_{21} + q_{12}(\varphi_{22} + k^{-1}) + q_{21}(\overset{*}{\varphi}_{11} + k^{-1})|^2 > 0$$

where φ_{mn} are elements of the matrix $\Phi(j\omega)$ and the asterisk implies the complex conjugate. Assuming $p_{11} = d_1 = 0$, $p_{12} = 1$, $d_2 = -1$, $p_{21} = -a_0$, and $p_{22} = -a_1$ yields

$$\begin{aligned} \varphi_{11} &= 2(s + 2a_1)/\Delta \qquad & \varphi_{12} &= 4/\Delta \\ \varphi_{21} &= s(s + 2a_1)/\Delta \qquad & \varphi_{22} &= 2s/\Delta \end{aligned}$$

where $\Delta = s^3 + 3a_1s^2 + (2a_1^2 + 4a_0)s + 4a_0a_1$.

In the trial example we let $a_1 = 2$ and $a_0 = 1$. The computer aided solution of the nonlinear programming problem gave for $q_{11} = 0.514$, $q_{12} = 1.193$, and $q_{21} = -0.206$ the maximum value for k equal to 5.47. For these parameters $\det \Gamma(\omega)$ is plotted in Fig. 3.6. For comparison, we point out that Popov's criterion gives $k \leqslant 3.82$ in this case.

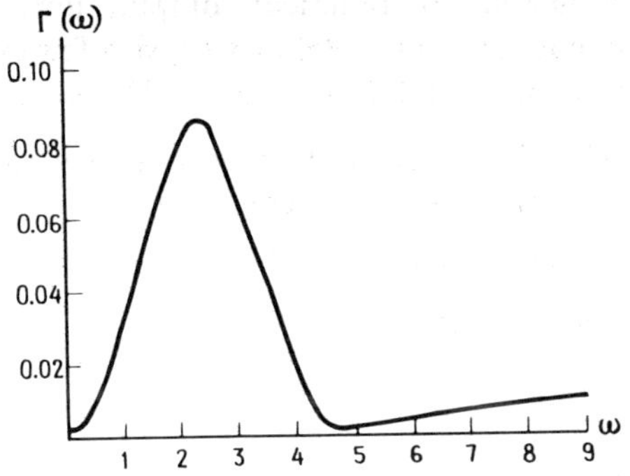

Fig. 3.6

Example 3.2. Consider the third-level model for the system of the previous example. The model will be obtained as the power-3 transform. For the reference system we have (the coefficients in (3.40) are all unities)

$$A_{[3]} = \begin{bmatrix} 3p_{11} & 3p_{12} & 0 & 0 \\ p_{21} & 2p_{11} + p_{22} & 2p_{12} & 0 \\ 0 & 2p_{21} & p_{11} + 2p_{22} & p_{12} \\ 0 & 0 & 3p_{21} & 3p_{22} \end{bmatrix},$$

$$b_{[3]} = \begin{bmatrix} 3b_1 & 0 & 0 \\ b_2 & 2b_1 & 0 \\ 0 & 2b_2 & b_1 \\ 0 & 0 & 3b_2 \end{bmatrix}$$

The stability with respect to k was computed for $a_{11} = 0$, $a_{12} = 1$, $b_1 = 0$, $a_{21} = -2$, $a_{22} = -1$, and $b_2 = -1$. The matrix Q has the form

$$Q = \begin{bmatrix} -1.0 & 4.075 & 1.575 \\ -0.6508 & 4.628 & 2.867 \\ 0.4068 & 0.5683 & 1.246 \end{bmatrix}$$

The maximum value of k thus determined equals 5.98.

Example 3.3. Consider the third-order system

$$\dot{x}_1 = x_2$$
$$\dot{x}_2 = x_2$$
$$\dot{x}_3 = -20x_1 - 10x_2 - 10x_3 - 10\varphi(x_1, t)$$

Construct the second-level model in the new variables $y_1 = x_1^3$, $y_2 = x_1x_2$, $y_3 = x_1x_3$, $y_4 = x_2^2$, $y_5 = x_2x_3$, and $y_6 = x_3^2$. The transformed system takes on the form

$$\begin{aligned}
\dot{y}_1 &= 2y_2 \\
\dot{y}_2 &= y_3 + y_4 \\
\dot{y}_3 &= -20y_1 - 10y_2 - 10y_3 + y_5 - 10\eta_1 \\
\dot{y}_4 &= 2y_5 \\
\dot{y}_5 &= -20y_2 - 10y_4 - 10y_5 + y_6 - 10\eta_2 \\
\dot{y}_6 &= -40y_3 - 20y_5 - 20y_6 + 20\eta_3
\end{aligned}$$

where $\eta_1 = x_1\varphi(x_1, t)$, $\eta_2 = x_2\varphi(x_1, t)$ and $\eta_3 = x_3\varphi(x_1, t)$.

According to Popov's criterion, the system is stable for $k < 2.43$. Making use of the criterion (3.66) and solving the respective nonlinear programming problem, extends a stability domain up to $k = 3.0$.

These illustrative examples indicate that the criterion (3.66) yields a greater region of absolute stability in the space of parameters than the circle criterion (3.15). The purpose of the following theorem is to establish relation between these criteria in the general form.

Theorem 3.13 (Barkin [3.28]). Let the frequency condition of the circle criterion (3.15) hold. Then there exists a symmetric positive definite matrix Q for which the criterion (3.66) holds.

The assertion of this theorem tells essentially that all the results derivable from the circle criterion can be derived also from the condition (3.66), i.e. the new criterion is at least not inferior to the circle cirterion. The following lemma is of prime significance for the proof of the theorem.

Lemma 3.1. Let a real and positive definite matrix $L = L^{\mathrm{T}}$ and a real vector u satisfy the system of Lurie equations [3.2]:

$$A^{\mathrm{T}}L + LA = -uu^{\mathrm{T}} - \varepsilon P \tag{3.71}$$

$$Lb + \frac{1}{2}c = \gamma u \tag{3.72}$$

where A is a Hurwitz matrix, b and c are real n-vectors, $P = P^{\mathrm{T}}$ is a real positive definite matrix, and ε and γ are positive numbers. Then

$$A_{[p]}^{\mathrm{T}}L^{[p]} + L^{[p]}A_{[p]} = -\frac{1}{p}u_{[p]}L^{[p-1]}u_{[p]}^{\mathrm{T}} - \varepsilon F \tag{3.73}$$

$$F = F^{\mathrm{T}} = L^{[p]}(L^{-1}P)_{[p]} \tag{3.74}$$

$$L^{[p]}b_{[p]} + \frac{1}{2}c_{[p]}L^{[p-1]} = \gamma u_{[p]}L^{[p-1]} \tag{3.75}$$

Proof of the lemma. Assume that the conditions of the lemma are satisfied. We premultiply (3.71) by L^{-1} and apply the lower-index transformation $(\cdot)_{[p]}$ to both sides of the identity, viz.

$$(L^{-1}AL)_{[p]} + A_{[p]} = -(L^{-1}uu^{\mathrm{T}})_{[p]} - \varepsilon(L^{-1}P)_{[p]}$$

By property (8) of Table 3.1 we have

$$A^{\mathrm{T}}_{[p]}L^{[p]} + L^{[p]}A_{[p]} = -L^{[p]}(L^{-1}uu^{\mathrm{T}})_{[p]} - \varepsilon F$$

where $F = L^{[p]}(L^{-1}P)_{[p]}$. The same property may be used to establish that $F = F^{\mathrm{T}} > 0$. By virtue of property (10) we have

$$(L^{-1}uu^{\mathrm{T}})_{[p]} = \frac{1}{p}(L^{-1}u)_{[p]}u^{\mathrm{T}}_{[p]}$$

and by property (18) we get

$$(L^{-1}u)_{[p]} = (L^{[p]})^{-1}u_{[p]}L^{[p-1]}$$

The two last relations indicate that

$$L^{[p]}(L^{-1}uu^{\mathrm{T}})_{[p]} = \frac{1}{p}u_{[p]}L^{[p-1]}u^{\mathrm{T}}_{[p]}$$

which proves (3.73). By applying $(\cdot)_{[p]}$ to (3.72) and observing property (18) we arrive at (3.75).

Proof of the theorem. Assume that the frequency condition (3.15) is satisfied. Then, by the frequency theorem of Yakubovich-Kalman (Theorem 3.6), equations (3.71-72) hold for sufficiently small ε. By the lemma we have that (3.73-75) hold true. Consider the quadratic form

$$\begin{aligned} V = {} & y^{\mathrm{T}}(A^{\mathrm{T}}_{[p]}L^{[p]} + L^{[p]}A_{[p]})y + 2y^{\mathrm{T}}L^{[p]}b_{[p]}\eta \\ & + p\eta^{\mathrm{T}}L^{[p-1]}\left(\frac{1}{p}c^{\mathrm{T}}_{[p]}y - k^{-1}\eta\right) \end{aligned} \tag{3.76}$$

Using (3.73-75), we obtain

$$V = -\left(y^{\mathrm{T}}u_{[p]}\frac{1}{\sqrt{p}} - \sqrt{\frac{p}{k}}\,\eta^{\mathrm{T}}\right)L^{[p-1]}\left(u^{\mathrm{T}}_{[p]}y\frac{1}{\sqrt{p}} - \sqrt{\frac{p}{k}}\,\eta\right) - \varepsilon y^{\mathrm{T}}Fy \tag{3.77}$$

From the properties of the power transformation it follows that the matrix $F = F^{\mathrm{T}}$ is positive definite. Then $V < 0$ for all y in R^m, η in R^l, and $|y| + |\eta| \neq 0$. Therefore from the frequency theorem

$$\eta^{*}L^{[p-1]}\left(\frac{1}{p}c^{\mathrm{T}}_{[p]}y - k^{-1}\eta\right) > 0 \tag{3.78}$$

for $y = (j\omega I_m - A)^{-1}b_{[p]}\eta$. From (3.78) we have

$$\operatorname{Re}\left\{L^{[p-1]}\left[\frac{1}{p}c^{\mathrm{T}}_{[p]}(A_{[p]} - j\omega I_m)^{-1}b_{[p]} + k^{-1}I_l\right]\right\} > 0 \tag{3.79}$$

Thus, we obtained that if the circle criterion (3.15) is satisfied, then the inequality (3.66) holds true with $Q = L^{[p-1]}$.

This theorem suggests that the absolute stability region derived in the parameter space of an autonomous system (3.1-3, 6) by the circle criterion is embedded in the absolute stability region derived by the criterion (3.66). Because examples are known when (3.66) does expand the stability domain as compared to (3.15), we are in a position to say that the new criterion is "stronger" than the circle criterion.

In the examples displayed above, the largest stability domain was obtained for unsymmetric matrix Q_1. Now we wish to demonstrate that the circle criterion cannot be improved by confining to the use of only symmetric matrices in (3.66).

Theorem 3.14. For the transfer matrix

$$\Phi(pj\omega) = \frac{1}{p} c_{[p]}^{\mathrm{T}} (A_{[p]} - pj\omega I_m)^{-1} b_{[p]}$$

one of the eigenvalues is the frequency response $W(j\omega) = c^{\mathrm{T}}(A - j\omega I)^{-1}b$, and the respective right eigenvector is

$$d(j\omega) = [(j\omega I - A)^{[p-1]}]^{-1} b^{[p-1]} \tag{3.80}$$

Proof. Consider the stationary motion of the system

$$\dot{x} = Ax + be^{j\omega t} \tag{3.81}$$

The stationary solution is

$$x(t) = (j\omega I - A)^{-1} be^{j\omega t} \tag{3.82}$$

Taking the upper-index transform of (3.82), we obtain

$$\begin{aligned} x^{[p-1]}(t) &= [(j\omega I - A)^{[p-1]}]^{-1} b^{[p-1]} e^{j(p-1)\omega t} \\ &= \exp[j(p-1)\omega t] d(j\omega) \end{aligned} \tag{3.83}$$

$$z(t) = \frac{1}{p} c_{[p]}^{\mathrm{T}} x^{[p]} = c^{\mathrm{T}} x x^{[p-1]} = -\exp(jp\omega t) W(j\omega) d(j\omega) \tag{3.84}$$

On the other hand,

$$\eta(t) = \xi x^{[p-1]}(t) = e^{jp\omega t} d(j\omega) \tag{3.85}$$

$$z(t) = -\Phi(pj\omega) d(j\omega) e^{jp\omega t} \tag{3.86}$$

Comparing (3.84) and (3.86), we arrive at the assertion of the theorem.

Let now $Q = Q^{\mathrm{T}}$ in the condition (3.66). The condition (3.66) is equivalent to the inequality

$$Q\Phi(pj\omega) + \Phi^{\mathrm{T}}(-pj\omega)Q^{\mathrm{T}} + k^{-1}(Q + Q^{\mathrm{T}}) > 0 \tag{3.87}$$

Premultiplying (3.87) by $d^*(j\omega)$ and postmultiplying by $d(j\omega)$, we have

$$d^*(j\omega)Qd(j\omega)[W(j\omega) + W(-j\omega) + 2k^{-1}] > 0 \qquad (3.88)$$

Because $Q = Q^T > 0$, the circle criterion follows from the last inequality. Hence the circle criterion cannot be improved for $Q = Q^T$.

3.9 Stability of Nonautonomous Systems

The three preceding sections discussed autonomous systems where the additive system input, i.e. $g(t)$, was identically zero. Now we wish to consider the stability of the system in which $g(t) \neq 0$. For this purpose we introduce the concept of stability in the functional space $L_{2p}(0, \infty)$.

Definition 3.10. A vector function $x(t)$ belongs to $L_{2p}(0, \infty)$ if

$$\|x\|_{2p} = \left[\int_0^\infty (x^T x)^p \mathrm{d}t\right]^{1/2p} < \infty \qquad (3.89)$$

Definition 3.11. The system (3.1-3, 6) is said to be stable in L_{2p} if for any $\varepsilon > 0$ there exist $\delta_1 > 0$ and $\delta_2 > 0$ such that $\|x\|_{2p} < \varepsilon$ is satisfied as soon as $x_0^T x < \delta_1$ and $\|g\|_{2p} < \delta_2$.

Theorem 3.15. (Popov [3.3]). Given a positive definite matrix Q such that the condition (3.66) is satisfied for all $\omega \geqslant 0$. Then the system (3.1-3, 6) is stable in $L_{2p}(0, \infty)$.

Remark. If $|g(t)| < L < \infty$, then the system (3.1-3, 6) being stable in $L_{2p}(0, \infty)$ implies that it is asymptotically stable. To prove this statement let

$$\int_0^\infty (x^T x)^p \mathrm{d}t < \infty, \qquad \int_0^\infty \xi^{2p}(t)\mathrm{d}t < \infty$$

and $V^p = (x^T x)^p$, then

$$\frac{\mathrm{d}}{\mathrm{d}t} V^p = 2p(x^T x)^{p-1}(x^T Ax + x^T b\xi + x^T rg)$$

On integration we obtain

$$[x^T(t)x(t)]^p - (x_0^T x_0)^p = 2p \int_0^t (x^T x)^{p-1}(x^T Ax + x^T b\xi + x^T rg)\mathrm{d}t \qquad (3.90)$$

Let us estimate the terms on the right-hand side of (3.90). If μ is the largest eigenvalue of the matrix $(A + A^T)/2$, then

$$\int_0^t (x^T x)^{p-1} x^T Ax \,\mathrm{d}t \leqslant \mu \int_0^t (x^T x)^p \,\mathrm{d}t$$

$$\int_0^t (x^T x)^{p-1} x' b\xi \,\mathrm{d}t \leqslant (b^T b)^{\frac{1}{2}} \left(\int_0^t (x^T x)^p \,\mathrm{d}t\right)^{\frac{2p-1}{2p}} \left(\int_0^t |\xi|^{2p} \,\mathrm{d}t\right)^{\frac{1}{2p}}$$

$$\int_0^t (x^{\mathrm{T}}x)^{p-1}x^{\mathrm{T}}rg\,\mathrm{d}t \leqslant (r^{\mathrm{T}}r)^{\frac{1}{2}}\left(\int_0^t (x^{\mathrm{T}}x)^p\,\mathrm{d}t\right)^{\frac{2p-1}{2p}}\left(\int_0^t g^{2p}(t)\,\mathrm{d}t\right)^{\frac{1}{2p}}$$

These criteria suggest that

$$(x^{\mathrm{T}}x)^p \leqslant (x_0^{\mathrm{T}}x_0)^p + C_1\int_0^t (x^{\mathrm{T}}x)^p\,\mathrm{d}t + C_2\left[\int_0^t (x^{\mathrm{T}}x)^p\,\mathrm{d}t\right]^{\frac{2p-1}{2p}} < C_3 < \infty$$

$$(x^{\mathrm{T}}x) < \mu_1 = \mathrm{const}$$

Clearly $\dot{V} = 2x\dot{x} = x^{\mathrm{T}}Ax + x^{\mathrm{T}}b\xi + x^{\mathrm{T}}rg < \mu_2 < \infty$ if $|g(t)| < L$. To complete the proof we refer to Barbalat's lemma [3.3].

In view of the above remark Theorem 3.12 is a particular case of Theorem 3.15 at $g(t) \equiv 0$. The proof of Theorem 3.15 can be found in the paper of Barkin [3.30].

3.10 Improved Criterion of Process Stability

Like in the theorem of Naumov-Tsypkin-Yakubovich (see Sect. 3.5) we assume that the equations of unperturbed motion are given by the system (3.1-3, 6). Perturbations will be supplied by the variation of the initial conditions and the magnitude-bounded input $\nu(t)$ that belongs to $L_{2p}(0, \infty)$, $p = 1, 2$, etc.

The equations of perturbed motion are given by

$$\dot{y} = Ay + b\varphi(\sigma) + g(t) + \nu(t) \tag{3.91}$$

$$y(0) = y_0 \neq x_0, \quad \sigma = c^{\mathrm{T}}y(t) \tag{3.92}$$

$$r\sigma^2 \leqslant \varphi(\sigma)\sigma \leqslant k\sigma^2 \tag{3.93}$$

The norm of the vector $x(t)$ in $L_{2p}(0, \infty)$ is

$$\|x\|_{2p} = \left[\int_0^\infty (x^{\mathrm{T}}x)^p\,\mathrm{d}t\right]^{1/2p}$$

Definition 3.12. The processes in the system (3.91-93) are said to be absolutely asymptotically stable if for any $\varepsilon > 0$ there exist $\delta_1 > 0$ and $\delta_2 > 0$ such that

$$\|x - y\|_{2p} < \varepsilon$$

as soon as

$$|x_0 - y_0| < \delta_1, \qquad \|\nu\|_{2p} < \delta_2$$

and $x(t) - y(t) \to 0$ as $t \to \infty$.

Theorem 3.16. The processes in the system (3.91-93) are absolutely asymptotically stable if

(1) $\tilde{A} = A + kbc^{\mathrm{T}}$ is a Hurwitz matrix,

(2) the nonlinearity possesses the property

$$r \leqslant \frac{\varphi(\sigma_1) - \varphi(\sigma_2)}{\sigma_1 - \sigma_2} \leqslant k$$

(3) the frequency condition

$$\mathrm{Re}\left\{Q\left[\frac{1}{p}\, c_{[p]}^{\mathrm{T}}(\tilde{A}_{[p]} - j\omega I)^{-1} b_{[p]} + (k - r)^{-1} I_l\right]\right\} > 0 \quad (3.94)$$

holds for all $\omega \geqslant 0$.

Proof. We write, like in Sect. 3.5, the variational equations in $z = y - x$

$$\begin{aligned} &\dot{z} = Az + b\tilde{\varphi}(c^{\mathrm{T}}z,\ t) + \nu(t) \\ &r \leqslant \frac{\tilde{\varphi}(c^{\mathrm{T}}z,\ t)}{c^{\mathrm{T}}z} \leqslant k \end{aligned} \quad (3.95)$$

The substitution

$$\tilde{\varphi}_1(c^{\mathrm{T}}z,\ t) = \tilde{\varphi}(c^{\mathrm{T}}z,\ t) - kc^{\mathrm{T}}z$$

carries the system (3.95) to the form

$$\left.\begin{aligned} &\dot{z} = \tilde{A}z + b\tilde{\varphi}_1(c^{\mathrm{T}}z,\ t) + \nu(t) \\ &0 \leqslant \tilde{\varphi}(c^{\mathrm{T}}z,\ t)c^{\mathrm{T}}z \leqslant (k - r)(c^{\mathrm{T}}z)^2 \end{aligned}\right\} \quad (3.96)$$

Thus the analysis of process stability reduces to the investigation of the absolute stability of the equilibrium position of the time-varying system (3.96), in which $\nu(t)$ belongs to $L_{2p}(0,\ \infty)$.

3.11 Performance Indices of Absolutely Stable Systems

The transient behavior of a nonlinear system may be indirectly ascertained with the aid of degree of stability. A nonlinear system exhibits a given degree of stability δ if the deviation of the process from the equilibrium position $\xi(t)$ satisfies the condition [3.5]

$$|\xi(t)| < Me^{-\delta t} \quad (3.97)$$

where M is a constant determined by the system parameters and initial conditions.

Investigate the conditions that ensure for the time-varying system

$$\begin{aligned} &\dot{x} = Ax + b\xi \\ &\xi = \varphi(\sigma,\ t) \\ &\sigma = c^{\mathrm{T}}x \\ &\xi(\sigma - k^{-1}\xi) \geqslant 0 \end{aligned} \quad (3.98)$$

a degree of stability at least equal to the given degree of stability δ. For this purpose we introduce the vector

$$y = e^{\delta t}x \tag{3.99}$$

From (3.98) and (3.99) it follows that

$$\dot{y} = (A + \delta I)y + b\tilde{\xi} \tag{3.100}$$

where

$$\tilde{\xi} = e^{\delta t}\xi, \quad \tilde{\sigma} = c^{\mathrm{T}}y \tag{3.101}$$

$$\tilde{\xi}(\tilde{\sigma} - k^{-1}\tilde{\xi}) \geqslant 0 \tag{3.102}$$

As can be seen the equations of motion in variables y, $\tilde{\xi}$, $\tilde{\sigma}$ are similar to the system (3.98) with the substitution of the matrix $A + \delta I$ for A.

Let

(1) $A + \delta I$ be a Hurwitz matrix,

(2) the inequality

$$\operatorname{Re} W(j\omega - \delta) + k^{-1} > 0 \tag{3.103}$$

hold for any $\omega \geqslant 0$.

Then the system (3.100-102) is absolutely stable which implies that the modulus $|y|$ is bounded, namely,

$$|y| < M \tag{3.104}$$

and, hence, it follows that $|x| < M\exp(-\delta t)$.

Application of logarithmic plots for analysis of nonlinear system degree of stability has been treated by Naumov [3.5, 24].

The integral quadratic criterion for the nonlinearity input was first obtained by Popov [3.9]. Let us consider the system (3.98) in which the nonlinearity is assumed time invariant, i.e. $\varphi(\sigma, t) \equiv \varphi(\sigma)$, single-valued and continuous. Assume also that the frequency condition

$$P(q, k, \omega) \equiv \operatorname{Re}\{W(j\omega)(1 + qj\omega)\} + k^{-1} > 0 \tag{3.105}$$

is satisfied. We note that since (3.105) is a strong inequality there always exists $\mu > k$ for which

$$P(q, \mu, \omega) > 0 \tag{3.106}$$

By virtue of (3.105) the time-invariant system is stable and there exists the integral

$$\varrho = \int_0^\infty \xi(\sigma - \mu^{-1}\xi + q\dot{\sigma})\,\mathrm{d}t = \int_0^\infty \xi(\sigma - \mu^{-1}\xi)\,\mathrm{d}t - q\psi(\sigma_0),$$

$$\sigma_0 = \sigma(0) \tag{3.107}$$

where

$$\psi(\sigma) = \int_0^{\sigma} \varphi(\lambda)\, d\lambda$$

By virtue of (3.98)

$$J_{\xi} = \int_0^{\infty} \xi(\sigma - \mu^{-1}\xi)\, dt = \int_0^{\infty} \xi^2(\sigma/\xi - \mu^{-1})\, dt \geqslant (k^{-1} - \mu^{-1}) \int_0^{\infty} \xi^2\, dt \tag{3.108}$$

From (3.107-108) we have

$$\int_0^{\infty} \xi^2\, dt \leqslant \frac{\mu k}{\mu - k}\,[\varrho + q\psi(\sigma_0)]$$

To estimate ϱ we resort to Parseval's theorem

$$\varrho = \frac{1}{2\pi} \int_{-\infty}^{\infty} \tilde{\xi}(-j\omega)[\tilde{\sigma}(j\omega) - \mu^{-1}\tilde{\xi}(j\omega) + q\tilde{\dot{\sigma}}(j\omega)]\, d\omega \tag{3.109}$$

where $\tilde{\xi}(j\omega)$, $\tilde{\sigma}(j\omega)$, and $\tilde{\dot{\sigma}}(j\omega)$ are the Laplace transforms of $\xi(t)$, $\sigma(t)$, and $\dot{\sigma}(t)$, respectively.

Since

$$\begin{aligned} \tilde{\sigma}(j\omega) &= -W(j\omega)\tilde{\xi}(j\omega) + F(j\omega) \\ \tilde{\dot{\sigma}}(j\omega) &= j\omega\tilde{\sigma}(j\omega) - \sigma(0) \end{aligned} \tag{3.110}$$

$$F(j\omega) = c^{\mathrm{T}}(j\omega I - A)^{-1}x_0 \tag{3.111}$$

then

$$\varrho = \frac{1}{2\pi} \int_{-\infty}^{\infty} \{ -|\tilde{\xi}(j\omega)|^2 P(q, \mu, \omega) + \tilde{\xi}(-j\omega)R(j\omega) \}\, d\omega \tag{3.112}$$

where

$$R(j\omega) = (1 + qj\omega)F(j\omega) - q\sigma(0)$$

Forming a full square in (3.112)

$$\varrho = -\frac{1}{2\pi} \int_{-\infty}^{\infty} \left| \tilde{\xi}(j\omega)\sqrt{P(q, \mu, \omega)} - \frac{R(j\omega)}{2\sqrt{P(q, \mu, \omega)}} \right|^2 d\omega + \frac{1}{8\pi} \int_{-\infty}^{\infty} \frac{|R(j\omega)|^2}{P(q, \mu, \omega)}\, d\omega$$

we obtain

$$J_\xi = \int_0^\infty \xi^2\,\mathrm{d}t \leqslant \frac{\mu k}{\mu - k}\left[q\int_0^{\sigma(0)} \varphi(\sigma)\,\mathrm{d}\sigma + \frac{1}{8\pi}\int_{-\infty}^{\infty} \frac{|R(j\omega)|^2}{P(q,\,\mu,\,\omega)}\,\mathrm{d}\omega\right] \tag{3.113}$$

For the time-varying system we let $q = 0$ and obtain

$$J_\xi \leqslant \frac{\mu k}{\mu - k}\,\frac{1}{8\pi}\int_{-\infty}^{\infty} \frac{|F(j\omega)|^2}{P(q,\,\mu,\,\omega)}\,\mathrm{d}\omega \tag{3.114}$$

The estimates (3.113, 114) can be computed with reference to the familiar representations of the integrals on the right-hand sides of these expressions [3.32].

Unfortunately, these criteria are rather rough. A simple way to improve them is as follows [3.32, 33]. Expression (3.112) may be rewritten as

$$\varrho = \frac{1}{2\pi j}\int_{-j\infty}^{j\infty} \{-\xi(-s)\xi(s)\Phi(s) + \xi(-s)R(s)\}\,\mathrm{d}s \tag{3.115}$$

where

$$\Phi(s) = \frac{1}{2}\,[W(s)(1 + qs) + W(-s)(1 - qs)] + \mu^{-1} \tag{3.116}$$

Because $\Phi(j\omega) = P(q,\,\mu,\,\omega) > 0$, $\Phi(s)$ allows the factorization

$$\Phi(s) = Q(s)Q(-s) \tag{3.117}$$

where $Q(s)$ is a rational function whose zeros and poles lie in the left half-plane $\operatorname{Re} s < 0$. Using (3.117) in (3.115) yields

$$\varrho = \frac{1}{2\pi j}\int_{-j\infty}^{j\infty} \{-\xi(-s)Q(-s)[\xi(s)Q(s) + R(s)/Q(-s)]\,\mathrm{d}s \tag{3.118}$$

The residue $R(s)/Q(-s)$ may be represented as

$$R(s)/Q(-s) = U(s) + V(-s) \tag{3.119}$$

where $U(s)$ has its poles in the left half-plane $\operatorname{Re} s < 0$, whereas the poles of $V(-s)$ lie in the right half-plane only.

Substituting (3.119) into (3.118) yields

$$\varrho = \frac{1}{2\pi j}\int_{-j\infty}^{j\infty} \{-\xi(-s)Q(-s)\xi(s)Q(s) + \xi(-s)Q(-s)U(s) + \xi(-s)Q(-s)V(-s)\}\,\mathrm{d}s \tag{3.120}$$

We recognize, however, that

$$\frac{1}{2\pi j}\int_{-\infty}^{\infty}\xi(-s)Q(-s)V(-s)\,\mathrm{d}s = 0 \tag{3.121}$$

for the integrand is analytical in the left half-plane and falls off at infinity as s^{-2}. Forming a full square in the rest of (3.120), we obtain

$$\varrho = \frac{1}{2\pi}\int_{-\infty}^{\infty} - \left|\xi(j\omega)Q(j\omega) - \frac{1}{2}U(j\omega)\right|^2 \mathrm{d}\omega + \frac{1}{8\pi}\int_{-\infty}^{\infty} |U(j\omega)|^2\,\mathrm{d}\omega$$

Whence we finally arrive at the new bound

$$J_\xi \leqslant \frac{\mu k}{\mu - k}\left[q\int_0^{\sigma(0)}\varphi(\sigma)\,\mathrm{d}\sigma + \frac{1}{8\pi}\int_{-\infty}^{\infty}|U(j\omega)|^2\,\mathrm{d}\omega\right] \tag{3.122}$$

Barkin [3.32] refers to the criterion (3.122) as optimal because it is always more accurate than Popov's criterion (3.113) and cannot be improved in the context of the stability criterion (3.105). This paper presents similar results for criteria of the general form

$$\int_0^{\infty}[x^{\mathrm{T}}Nx + \alpha\xi^2]\,\mathrm{d}t$$

3.12 Definition of Stability Region in the Large

To illustrate the application of the optimal integral criterion we consider the problem of defining the region of stability in the large.

Let the frequency condition (3.13) be satisfied. The system with a time-invariant nonlinearity is stable in the large if the inequality $0 \leqslant \varphi(\sigma)\sigma \leqslant k\sigma^2$ is satisfied.

Let, however, the function $\varphi(\sigma)$ be such that its plot lies within the sector $[0, k]$ only if

$$-\sigma' < \sigma < \sigma'' \tag{3.123}$$

where both σ' and σ'' are positive (Fig. 3.7). We wish to determine the set of initial conditions for which asymptotic stability of the stationary solution of (3.1-3) prevails [3.34].

Let at this point $q > 0$ in (3.123). We revisit the function

$$\varrho(T) = \int_0^T \xi(\sigma - k^{-1}\xi)\,\mathrm{d}t + q\int_0^T \xi\dot{\sigma}\,\mathrm{d}t = J_\xi + q[\psi(\sigma(T)) - \psi(\sigma(0))] \tag{3.124}$$

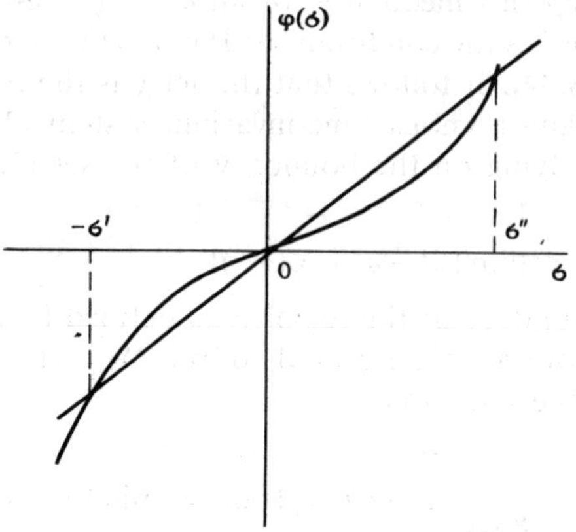

Fig. 3.7

where

$$J_\xi = \int_0^T \xi(\sigma - k^{-1}\xi)\,dt, \qquad \psi(\sigma) = \int_0^\sigma \varphi(\lambda)\,d\lambda \tag{3.125}$$

We rewrite (3.124) in the form

$$q\psi(\sigma(T)) = \varrho(T) - J_\xi + q\psi(\sigma(0)) \tag{3.126}$$

Assume that we have a bound for $\varrho(T)$ independent of T but dependent on the initial conditions

$$\varrho(T) < M(x_0) \tag{3.127}$$

Denote by χ the set of initial states x_0 satisfying the conditions

$$-\sigma' < c^T x_0 < \sigma'' \tag{3.128}$$

$$q\psi[c^T x_0] + M(x_0) < q \min[\psi(-\sigma'), \psi(\sigma'')] \tag{3.129}$$

Taken in the range $[-\sigma', \sigma'']$ the function $\varphi(\sigma)$ lies within the sector $[0, k]$, therefore $\psi(-\sigma') > 0$ and $\psi(\sigma'') > 0$. The function $\psi(\sigma)$ is continuous with $\psi(0) = 0$. The coordinates of the points close to the origin satisfy the inequalities (3.128-129), therefore the set χ is not empty.

If σ is varied from $-\sigma'$ to zero, $\psi(\sigma)$ falls off monotonously from $\psi(-\sigma')$ to $\psi(0) = 0$. When σ varies from 0 to σ'', $\psi(\sigma)$ monotonously increases from 0 to $\psi(\sigma'') > 0$. The following assertion holds [3.34]: if $x_0 \in \chi$, then (3.124) is valid for all $t \geqslant 0$.

Let us now alter the function $\varphi(\sigma)$ outside the interval $[-\sigma', \sigma'']$ so that it satisfy the inequality $0 \leqslant \varphi(\sigma)\sigma \leqslant k\sigma^2$. Then we obtain a new function $\tilde{\varphi}(\sigma)$ satisfying this inequality for all σ. At $\xi = \tilde{\varphi}(\sigma)$ the system under consideration satisfies the conditions of Popov's theorem and, hence, is stable in the large [3.34]. It follows that the set χ is the region of asymptotic stability for the autonomous time-invariant system (3.1-3).

The points x_0 lying on the boundary of the set (3.128) do not satisfy the inequality

$$\psi(\sigma_0) < \min[\psi(-\sigma'), \psi(\sigma'')] \tag{3.130}$$

Therefore, the boundary of the region χ is defined by (3.129). This region is wider for a more accurate bound $M(x_0)$. We determine the boundary of the set from the equation

$$\psi(\sigma_0) + \frac{1}{8\pi q}\int_{-\infty}^{\infty} |U(j\omega)|^2 \, d\omega = \min[\psi(-\sigma'), \psi(\sigma'')] \tag{3.131}$$

where $U(s)$ is defined by the representation (3.119).

The function $f(t)$ may be thought of also as an external system input if it is continuous and belongs to the intersection of $L_1(0, \infty)$ and $L_2(0, \infty)$.

The inequality (3.131) imposes a constraint on the initial conditions which ensure that the system remains asymptotically stable. If $\varphi(\sigma)$ is an odd function, i.e. $\varphi(\sigma) = -\varphi(-\sigma)$, then $\sigma' = \sigma''$, and the boundary of the region is described by the equation

$$\psi(\sigma_0) + \frac{1}{8\pi q}\int_{-\infty}^{\infty} |U(j\omega)|^2 \, d\omega = \psi(\sigma') \tag{3.132}$$

The region χ is seen to be symmetric about the origin.

When constructing the region χ for a particular nonlinear system, the choice of k, q, σ', σ'' should be such that maximizes χ. This region increases with σ' and σ'' and decreases as k increases. However, the possibilities to increase σ' and σ'' are limited by the inequality $0 \leqslant \varphi(\sigma)\sigma \leqslant k\sigma^2$.

Example 3.4. Consider the system of differential equations [3.34]

$$\begin{aligned} &\dot{x}_1 + x_2 \\ &\dot{x}_2 = a_{22}x_2 + a_{23}x_3 + \xi \\ &\dot{x}_3 = a_{32}x_2 + a_{33}x_3 \end{aligned} \tag{3.133}$$

The feedback is of the on-off type

$$\xi = y_0 \operatorname{sgn}(c_1x_1 + c_2x_2) \tag{3.134}$$

The system parameters are as follows: $a_{22} = -1$, $a_{23} = -1$, $a_{32} = 2$, $a_{33} = -1$, $y_0 = -1$, and $c_1 = c_2 = 1$.

The characteristic equation of (3.133) solved at $\xi = 0$ has one zero, one negative and one positive root. The system (3.133-134) is not stable in the large, therefore it is required to construct the domain of attraction χ.

Let us transform the system (3.133) so that to render the system intrinsically stable, i.e. stable at $\xi \equiv 0$. Then we shall be able to apply the method of construction of stability region outlined in the preceding section. We add and subtract $x_1 + x_2$ on the right-hand side of the second equation and denote $\xi_1 = -\xi - x_1 - x_2$, then

$$\begin{aligned} \dot{x}_1 &= x_2 \\ \dot{x}_2 &= -x_1 - 2x_2 + x_3 - \xi_1 \\ \dot{x}_3 &= 2x_2 - x_3 \\ \xi_1 &= \varphi(\sigma) = \operatorname{sgn}\sigma - \sigma, \qquad \sigma = x_1 + x_2 \end{aligned} \tag{3.135}$$

The characteristic equation of (3.135) has the roots $\lambda_1 = -2.769$ and $\lambda_{2,3} = -0.115 \pm 0.590j$. Hence, the system (3.135) is intrinsically stable. The nonlinear function $\varphi(\sigma)$ is discontinuous at $\sigma = 0$; consequently, $k = \infty$. Following Formalskii [3.34] we define ξ_1 at $\sigma = 0$ as follows:

$$\begin{aligned} \xi_1 &= -1 && \text{for } x_3 \leqslant -1 \\ &= x_3 && \text{for } |x_3| < 1 \\ &= 1 && \text{for } x_3 > 1 \end{aligned}$$

It follows that $x(t) \equiv 0$ is a solution of (3.135). Since $\varphi(\sigma)$ is an odd function, $\sigma' = \sigma'' = 1$. From (3.135) we have

$$W(p) = \frac{(p+1)^2}{p^3 + 3p^2 + p + 1}$$

Denote

$$X(\omega) = \operatorname{Re} W(j\omega) = \frac{(1-\omega^2)^2}{(1-3\omega^2)^2 + \omega^2(1-\omega^2)^2}$$

$$Y(\omega) = \omega \operatorname{Im} W(j\omega) = \frac{\omega^2(\omega^4 + 4\omega^2 - 1)}{(1-3\omega^2)^2 + \omega^2(1-\omega^2)^2}$$

In Fig. 3.8 the plot of $X(\omega) + jY(\omega)$ lies on the right of the straight line with slope coefficient $1/q = 1$ passing through the origin ($k = \infty$). Consequently, the nonlinear system in question satisfies the Popov criterion of absolute stability (3.13).

Construct the boundary of the region χ by (3.132). We have

$$\psi(\sigma) = |\sigma| - \frac{1}{2}\sigma^2 \quad \text{and} \quad \psi(\sigma') = \psi(1) = \frac{1}{2}$$

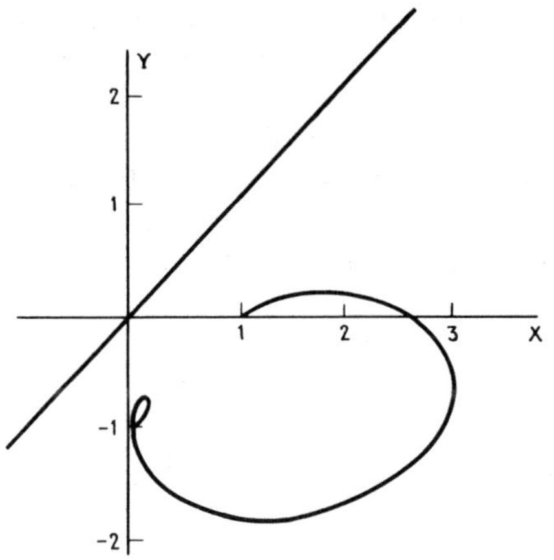

Fig. 3.8

At $q = 1$,

$$\Phi(p) = \frac{1}{2}\frac{(1 + p)^3(-p^3 + 3p^2 - p + 1) + (1 - p)^3(p^3 + 3p^2 + p + 1)}{G(p)G(-p)}$$

$$G(p) = p^3 + 3p^2 + p + 1$$

$$R(p) = \frac{-2x_1(0) + 2px_2(0) + (1 - p)^2x_3(0)}{G(p)}$$

We factor $\Phi(p)$ as follows:

$$\Phi(p) = N(p)N(-p)/G(p)G(-p), \quad N(p) = p^3 + c_2p^2 + c_1p + 1$$

where $c_1 = 1.6650$ and $c_2 = 2.8862$.

According to the definition (3.119)

$$U(p) = \frac{l_0 + l_1p + l_2p^2}{G(p)}, \quad V(-p) = \frac{s_0 + s_1p + s_2p^2}{N(-p)}$$

and the quantities l_0, l_1, and l_2 will be defined from the equation

$$\frac{R(p)G(-p)}{N(-p)} = \frac{l_0 + l_1p + l_2p^2}{G(p)} + \frac{s_0 + s_1p + s_2p^2}{N(-p)}$$

by equating the coefficients at like powers of p.

With the known formulas [3.32] we obtain

$$J_\xi = \frac{1}{8\pi}\int_{-\infty}^{\infty} |U(j\omega)|^2 \, d\omega = \frac{(l_0 - l_2)^2 + l_1^2 + 2l_0^2}{16}$$

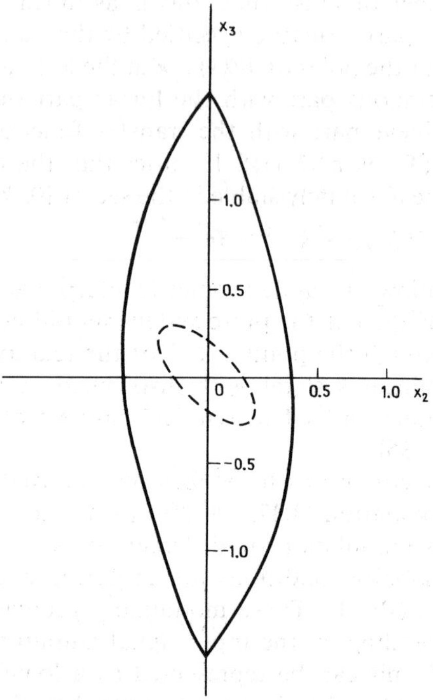

Fig. 3.9

Let $x_1(0) = 0$ and $x_3(0) = \lambda x_2(0)$. Assigning various λ, we construct the region boundary by the equation

$$|x_2(0)| + J_\xi = 0.5$$

The region χ is symmetric about the origin. Its section by the plane $x_1(0) = 0$ is shown in Fig. 3.9 by solid line. For comparison the dashed line shows the boundary of region χ computed by Formalskii [3.34]. Considerable expansion of the stability region in this example resulted from the application of the more accurate integral criterion.

3.13 Absolute Stability of Pulse Systems

Consider first a nonlinear control system with one pulse-function element that performs AM quantization of the nonlinear function $\varphi(\sigma)$ belonging to the sector $[0, k]$ (Fig. 3.10). The system nonlinearity is in two parts: the continuous nonlinear part $\varphi(\sigma)$, and the ideal pulse element converting a continuous signal $\xi(t) = \varphi(\sigma(t))$ into a sequence of delta-functions. The shaper of pulse functions is assumed linear and included into the continuous part structure specified by the transfer function $W(s)$. We shall assume that the poles of $W(s)$ are in the left half s-plane. Combining the linear continuous part with the linear part of the pulse element gives an impulse linear part with the transfer function $W^*(q)$.

Theorem 3.17 (Tsypkin [3.35]). In order that the time-invariant pulse system (Fig. 3.10) be absolutely stable in the sector $[0, k]$ it is sufficient that

$$\mathrm{Re}\,\{W^*(j\bar{\omega})\} + k^{-1} > 0 \tag{3.135}$$

This criterion allows a simple geometric interpretation, namely, the frequency response $W^*(j\bar{\omega})$ of the pulse system should lie on the right of the vertical passing through the point $-1/k$ of the real axis. This criterion is also valid for nonlinear systems with distributed parameters [3.4, 5].

A stronger formulation of Theorem 3.17 has been obtained by the power transfer method [3.36].

Global stability conditions have been also derived for pulse elements with frequency modulation [3.37, 38, 39], pulse-width modulation [3.25], and pulse-position modulation of different types.

Consider the stability conditions for nonlinear systems with pulse frequency modulators (PFM). These modulators generate pulse trains with pulse spacings depending on the input signal amplitude. A rather general scheme of a PFM unit can be represented as a low-frequency filter that is impressed by an input signal and a module that yields an output pulse each time the magnitude of the filter output attains a certain level after which it falls off to zero. The polarity of a PFM pulse coincides with that of the filter output at the instant of reset zeroing.

Dymkov [3.39] has demonstrated that the PFM unit can be represented as an equivalent closed-loop nonlinear system. Its feedforward path contains in a cascade connection an element of transfer function $W_f(s)$, that defines the type of PFM used, and a relay with hysteresis. The feedback loop contains a lag-free element with gain coefficient r/c, where r implies the level at which the $W_f(s)$ output signal reset begins, and $c = \lim sW_f(s)$ as $s \to \infty$. At $W_f(s) = k_f/s$, we have integral and at $W_f(s) = k_f/(T_f s + 1)$ additive (Σ) pulse width modulation. Now, the closed-loop PFM control

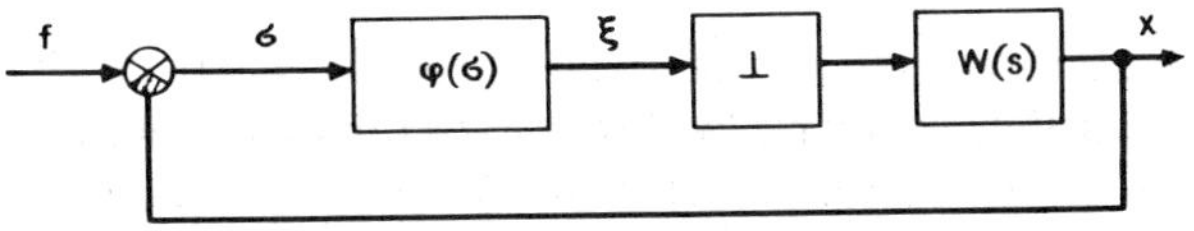

Fig. 3.10

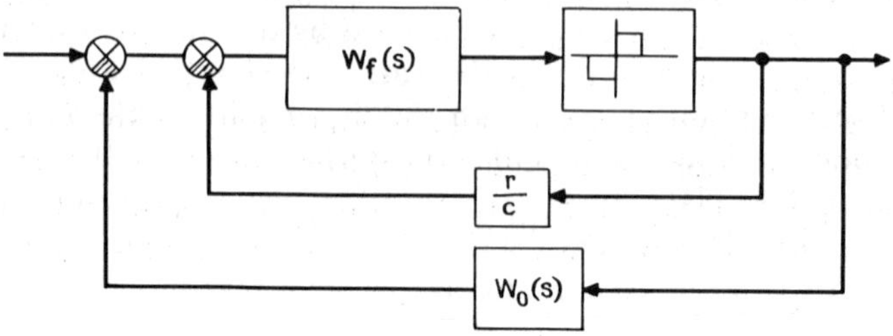

Fig. 3.11

system can be represented in the form shown in Fig. 3.11, where $W_0(s)$ is the transfer function of the linear plant to which, if necessary, the shaping unit may be related.

If we denote $W(s) = W_f(s)\,[W_0(s) + r/c]$, then the properties of the nonlinear function φ may be stated as follows:

(1) $\varphi(0) = 0$ for $\sigma = 0$,

(2) there exists a region of $\sigma \neq 0$ $(\sigma < |r|)$ such that

$$\varphi(\sigma) = 0,$$

(3) $0 \leqslant \varphi(\sigma)/\sigma \leqslant k < \infty$,

(4) the hysteretic loop obeys the following condition:

$$\int_0^t \varphi(\sigma,\ \varphi_0)_\tau \, d\sigma(\tau) \geqslant -\gamma_0, \ t \geqslant 0$$

$$\varphi_0 \in F(\sigma(0)), \qquad \gamma_0 \geqslant 0$$

The following findings are due to Dymkov [3.39].

Theorem 3.18. The PFM system is absolutely stable if

(i) the transfer function $W(s) = W_f(s)[W_0(s) + r/c]$, where $W_f(s) = k_f/s$ has all its poles in the left half-plane,

(ii) the inequality

$$\text{Re}\, W(j\omega) > 0, \tag{3.136}$$

holds for $0 \leqslant \omega < \infty$

If we define $\overline{W}_0(\omega)$ by

$$\begin{aligned} \text{Re}\, \overline{W}_0(\omega) &= \text{Re}\, W_0(j\omega) \\ \text{Im}\, \overline{W}_0(\omega) &= -\text{Im}\, W_0(j\omega)\, \text{Im}\, W_f(j\omega)/\text{Re}\, W_f(j\omega) \end{aligned} \tag{3.137}$$

then Theorem 3.18 may be interpreted in geometrical terms as follows.

The PFM system is absolutely stable if condition (i) is satisfied and for all ω satisfying $\text{Re}\, W_f(j\omega) > 0$ the frequency response $\overline{W}_0(\omega)$ lies on the right of the straight line of slope -1 passing through point $-r/c$ on the real axis, and for all ω satisfying Re $W_f(j\omega) < 0$ it lies on the left of this line.

For a Σ-PFM system when $W_f(s) = k_f/(T_f s + 1)$ (Re $W_f(j\omega) > 0$ for $0 \leqslant \omega < \infty$) condition (ii) assumes the form

$$\operatorname{Re} W_0(j\omega) + T_f\omega \operatorname{Im} W_0(j\omega) + rT_f/k_f > 0 \tag{3.138}$$

Therefore, if we introduce a modified frequency response $W^*(\omega)$ for which

$$\left.\begin{aligned} \operatorname{Re} W_0^*(\omega) &= \operatorname{Re} W_0(j\omega) \\ \operatorname{Im} W_0^*(\omega) &= \omega \operatorname{Im} W_0(j\omega) \end{aligned}\right\} \tag{3.139}$$

then the following theorem takes place.

Theorem 3.19. A Σ-PFM system is absolutely stable if condition (i) of Theorem 3.18 is satisfied and for $0 \leqslant \omega < \infty$ the modified $W_0^*(\omega)$ lies on the right of the straight line passing through the point $-rT_f/k_f$ on the real axis with the slope $-1/T_f$.

More general frequency conditions of stability have been obtained by Gelig [3.38, 40].

Based on the simple physical concepts the frequency methods have been widely accepted as engineering tools of analysis, synthesis and design of nonlinear control systems with both distributed and lumped parameters. The results outlined above have been obtained for the nonlinear systems with lumped parameters or systems reduced to these models, that, in addition, are assumed strictly deterministic.

Application of the frequency methods to substantially complicated nonlinear systems assumed a considerable amount of computing. Therefore the progress in computing has appreciably advanced the development of the frequency methods. The recent achievements and methods of numerical analysis, mathematical modeling, computing, and information science have brought about a new information technology based on the frequency concepts of investigation and design of complicated nonlinear control systems.

References

3.1 B. V. Bulgakov: *Dokl. Akad. Nauk SSSR.* **37**(39), 283-287 (1942).

3.2 A. I. Lurie and V. N. Postnikov: *Prikl. Mat. Mekh.* **8**(3) (1944).

3.3 V. M. Popov: *Avtomat. Telemekh.* No. 8, 961-979 (1961).

3.4 A. A. Voronov: *Osnovy teorii avtomaticheskogo upravleniya: Osobye lineinye i nelineinye sistemy* (Fundamentals of automatic control. Special linear and nonlinear systems). 2nd ed. Energoizdat, Moscow 1981.

3.5 B. N. Naumov: *Teoriya nelineinykh avtomaticheskikh sistem. Chastotnye metody* (Nonlinear system theory. Frequency methods) Nauka, Moscow 1972.

3.6 I. N. Sinitsyn: *Probl. Control and Information Theory.* **15**(4), 287-298 (1986).

3.7 M. A. Aizerman and F. R. Gantmacher: *Absolyutnaya ustoichivost reguliruemykh sistem* (Absolute stability of controlled systems). Izd. AN SSSR, Moscow 1963.

3.8 V. A. Yakubovich. *Vestn. LGU, Ser. Mat. Mekh. Astr.* No. 1, 62-77 (1971).

3.9 V. M. Popov: In *Proc. 1st Congress of IFAC. Vol. 1. Teoriya nepreryvnykh sistem* (Continuous system theory). Izd. AN SSSR, Moscow 1961.

3.10 V. A. Yakubovich: *Dokl. Akad. Nauk SSSR.* **143**(6), 1304-1307 (1962).

3.11 J. J. Bongiorno: *IEEE Trans. Automat. Control* **8**(2), 166-170 (1963).

3.12 J. J. Bongiorno: *Proc. IEEE* **52**(7), 832-841 (1964).

3.13 V. A. Yakubovich: *Avtomat. Telemekh.* No. 6, 5-30 (1967).

3.14 V. A. Yakubovich: In *Metody issledovaniya nelineinykh sistem avtomaticheskogo upravleniya* (Methods of nonlinear system analysis). Nauka, Moscow 1975, pp. 177-182.

3.15 R. E. Kalman: *Proc. Nat. Acad. Sci.* USA **49**(2), 201-205 (1963).

3.16 E. S. Pyatnitskii: *Avtomat. Telemekh.* No. 1, (1970).

3.17 E. S. Pyatnitskii: *Avtomat. Telemekh.* No. 3 (1970).

3.18 A. Yu. Levin: *Dokl. Akad. Nauk SSSR.* **141**(6), 1298-1301 (1961).

3.19 G. A. Leonov: *Avtomat. Telemekh.* No. 11, 21-26 (1980).

3.20 A. I. Barkin and A. L. Zelentsovsky: *Dokl. Akad. Nauk SSSR.* **276**(4), 809-812 (1984).

3.21 B. N. Naumov and Ya. Z. Tsypkin: *Avtomat. Telemekh.* No. 6, 852-867 (1964).

3.22 V. A. Yakubovich: *Avtomat. Telemekh.* No. 7, 577-590 (1964).

3.23 B. N. Naumov: *Avtomat. Telemekh.* No. 4 (1965).

3.24 B. N. Naumov: *Report to the 3rd Int. Congress of IFAC*, London 1966.

3.25 A. H. Gelig: *Avtomat. Telemekh.* No. 7, 33-43 (1968).

3.26 H. M. Jams, N. B. Nichols, and R. S. Filips (eds): *Servo-system Theory.*

3.27 R. W. Brockett: *Lie algebras and Lie groups in control theory*, in *Geometric Methods in Systems Theory* (Edited by R. Brockett and D. Mayne). Reidel Publ. Co., Dordrecht, Holland 1973.

3.28 A. I. Barkin: *Avtomat. Telemekh.* No. 1, (1984).

3.29 A. I. Barkin and A. L. Zelentsovsky: *Systems and Control Letters.* **3**(5), 303-310 (1983).

3.30 A. I. Barkin: *Avtomat. Telemekh.* No. 10, 64-69 (1983).

3.31 A. I. Barkin and A. L. Zelentsovsky: *Avtomat. Telemekh.* No. 7, 5-10 (1981).

3.32 A. I. Barkin: *Otsenki kachestva nelineinykh sistem regulirovaniya* (Performance indices of nonlinear control systems). Nauka, Moscow 1982.

3.33 A. I. Barkin: *Avtomat. Telemekh.* No. 7, 22-30 (1972).

3.34 A. M. Formalskii: *Upravlyaemost i ustoichivost sistem s ogranichennymi resursami* (Controllability and stability of systems with limited resources). Nauka, Moscow 1977.

3.35 Ya. Z. Tsypkin: in *Proc. 2nd Congress of IFAC*, Basel, 1963. *Diskretnye i samonastraivayushchiyesya sistemy* (Discrete and adaptive systems). Nauka, Moscow 1965, pp. 89-103.

3.36 A. I. Barkin and A. L. Zelentsovsky: *Int. J. Systems Science.* **14** (10), 1217-1228 (1983).

3.37 A. H. Gelig: *Avtomat. Telemekh.* No. 6, 75-82 (1967).

3.38 A. H. Gelig: *Avtomat. Telemekh.* No. 1, 188 (1973).

3.39 V. I. Dymkov: *Avtomat. Telemekh.* No. 10, 109-114 (1967).

3.40 A. H. Gelig: *Avtomat. Telemekh.* No. 11, (1971).

4 Evaluation of Regions of Stability, Instability, and Self-Excited Oscillations in Nonlinear Systems

Yu. I. Topcheyev

Even though both continuous and discrete systems of automatic control belong to nonlinear systems no common method of their analysis has been devised thus far. Practical investigations rely on various approximate techniques, which substantially complicates problem solving of nonlinear system analysis and synthesis since more than one method is normally involved. The analytical findings obtained in this way may be recognized as being true only when different methods produce identical numerical evidence for the same nonlinear problem. Clearly this approach to nonlinear system design is rather unwieldy. In attempts to simplify it, methods have been devised to separate the nonlinear system part from the linear part. This allows for the application of the harmonical linearization (harmonic balance) method and for the determination of the regions of stable and unstable states as functions of the system parameters. The linear parts are assumed to filter out all the harmonics, except the first, of the periodic outputs of the nonlinear elements[1].

The following consideration relies on the fundamental concepts of the method of harmonic balance, first developed by Krylov and Bogoljubov [4.1] and extended by many scientists [4.2-8]. These researchers have demonstrated that autonomous nonlinear systems experience operating modes with self-excited oscillations of parameter-dependent constant frequencies and amplitudes.

Self-excited oscillations are prone to disturb system operation. High amplitudes of such oscillations can lead to breakdown of the controller or have a mutilating effect on the plant. To avoid such phenomena system designers introduce various correcting elements (of cascade or parallel action) and select their parameters so as to suppress these oscillations or expand the stability regions.

---

[1] It is worthy of note that if this assumption is not satisfied completely, then this approach to nonlinear system analysis may be in error, since with a sinusoidal input the output of the nonlinear element is no longer sinusoidal. For example, in the presence of only the first harmonic, the nonlinear system exhibits regions of unstable states, and the action of the first and third harmonics leads to the formation of only a region of unstable states.

This chapter analyzes the effect of higher harmonics on the behavior of some types of nonlinear systems. For simplicity of the consideration it will be confined to only the first and third harmonics, though the approach outlined below allows for any number of harmonics [4.9].

4.1 Generalized Coefficients of Harmonic Linearization

We assume that the nonlinear element is a symmetric two-valued nonlinearity with the input impressed by the signal

$$x(t) = A \sin \omega t + A_3 \sin(3\omega t + \varphi_3) \tag{4.1}$$

where A is the amplitude of the first harmonic, A_3 is the amplitude of the third harmonic, and φ_3 is the phase shift of the third harmonic.

If we denote $\delta_3 = A_3/A$ and $\psi = \omega t$, then (4.1) can be rewritten as

$$x(t) = A[\sin \psi + \delta_3 \sin(3\omega t + \varphi_3)] \tag{4.2}$$

and the nonlinear-element output is

$$y(t) = \mathscr{F}\{A[\sin \psi + \delta_3 \sin(3\omega t + \varphi_3)]\} \tag{4.3}$$

If for the odd two-valued nonlinearity we confine ourselves to the first and third harmonics, then (4.3) may be represented as

$$\begin{aligned} y'(t) = A[a(A)\sin \psi + b(A)\cos \psi + a_3(A)\sin(3\psi + \varphi_3) \\ + b_3(A)\cos(3\psi + \varphi_3)] \end{aligned} \tag{4.4}$$

The error in reproducing the system output is

$$\delta(t) = y(t) - y'(t) \tag{4.5}$$

and the largest mean square deviation can be defined as

$$\begin{aligned} \bar{\delta}_m^2 = \frac{1}{2\pi}\int_0^{2\pi} \{ \mathscr{F}[A(\sin \psi + \delta_3 \sin(3\psi + \varphi_3)] - A[a(A)\sin \psi \\ + b(A)\cos \psi + a_3(A)\sin(3\psi + \varphi_3) + b_3(A)\cos(3\psi + \varphi_3)]\}^2 d\psi \end{aligned} \tag{4.6}$$

where $\delta(t)$ is a periodic function.

To determine the minimum mean square error we form the relations

$$\begin{aligned} \frac{\partial \bar{\delta}_m^2}{\partial A \partial a(A)} = 0, \qquad \frac{\partial \bar{\delta}_m^2}{\partial A \partial a_3(A)} = 0 \\ \frac{\partial \bar{\delta}_m^2}{\partial A \partial b(A)} = 0, \qquad \frac{\partial \bar{\delta}_m^2}{\partial A \partial b_3(A)} = 0 \end{aligned} \tag{4.7}$$

which by virtue of (4.6) lead to

$$\int_0^{2\pi} \{\mathscr{F}[A(\sin\psi + \delta_3 \sin(3\psi + \varphi_3))] - A[a(A)\sin\psi + b(A)\cos\psi + a_3(A)\sin(3\psi + \varphi_3) + b_3(A)\cos(3\psi + \varphi_3)]\}\sin\psi \, d\psi = 0 \tag{4.8}$$

$$\int_0^{2\pi} \{\mathscr{F}[A(\sin\psi + \delta_3 \sin(3\psi + \varphi_3))] - A[a(A)\sin\psi + b(A)\cos\psi + a_3(A)\sin(3\psi + \varphi_3) + b_3(A)\cos(3\psi + \varphi_3)]\}\sin(3\psi + \varphi_3) d\psi = 0 \tag{4.9}$$

$$\int_0^{2\pi} \{\mathscr{F}[A(\sin\psi + \delta_3 \sin(3\psi + \varphi_3)] - A[a(A)\sin\psi + b(A)\cos\psi + a_3(A)\sin(3\psi + \varphi_3) + b_3(A)\cos(3\psi + \varphi_3)]\}\cos\psi \, d\psi = 0 \tag{4.10}$$

$$\int_0^{2\pi} \{\mathscr{F}[A(\sin\psi + \delta_3 \sin(3\psi + \varphi_3))] - A[a(A)\sin\psi + b(A)\cos\psi + a_3(A)\sin(3\psi + \varphi_3) + b_3(A)\cos(3\psi + \varphi_3)]\}\cos(3\psi + \varphi_3) d\psi = 0 \tag{4.11}$$

After algebraic manipulations we get

$$a(A) = \frac{1}{\pi A} \int_0^{2\pi} \mathscr{F}[A(\sin\psi + \delta_3 \sin(3\psi + \varphi_3))]\sin\psi \, d\psi \tag{4.12}$$

$$b(A) = \frac{1}{\pi A} \int_0^{2\pi} \mathscr{F}[A(\sin\psi + \delta_3 \sin(3\psi + \varphi_3))]\cos\psi \, d\psi \tag{4.13}$$

$$a_3(A) = \frac{1}{\pi A} \int_0^{2\pi} \mathscr{F}[A(\sin\psi + \delta_3 \sin(3\psi + \varphi_3))]\sin(3\psi + \varphi_3) d\psi \tag{4.14}$$

$$b_3(A) = \frac{1}{\pi A} \int_0^{2\pi} \mathscr{F}[A(\sin\psi + \delta_3 \sin(3\psi + \varphi_3))]\cos(3\psi + \varphi_3) d\psi \tag{4.15}$$

Equations (4.12-15) suggest that the mean square error in replacing the exact function $y(t)$ with the approximate function $y'(t)$ will be a minimum when the coefficients of the harmonic linearization are the first and third terms of the Fourier series.

We shall assume that the integrands in (4.12-15) are generalized functions [4.9]. Expanding these functions in a Taylor series and taking the first terms of the expansion, we obtain generalized formulas for harmonic balance coefficients as follows:

$$a(A) = \frac{1}{\pi A}\int_0^{2\pi} \{\mathcal{F}[A \sin \psi) + \dot{\mathcal{F}}[A(\sin \psi + \varepsilon\delta_3 \sin(3\psi + \varphi_3))] \times A\delta_3 \sin(3\psi + \varphi_3)\}\sin \psi \, d\psi \tag{4.16}$$

$$b(A) = \frac{1}{\pi A}\int_0^{2\pi} \{\mathcal{F}(A \sin \psi) + \dot{\mathcal{F}}[A(\sin \psi + \varepsilon\delta_3 \sin(3\psi + \varphi_3))] \times A\delta_3 \sin(3\psi + \varphi_3)\}\cos \psi \, d\psi \tag{4.17}$$

$$a_3(A) = \frac{1}{\pi A}\int_0^{2\pi} \{\mathcal{F}(A \sin \psi) + \dot{\mathcal{F}}[A(\sin \psi + \varepsilon\delta_3 \sin(3\psi + \varphi_3))] \times A\delta_3 \sin(3\psi + \varphi_3)\}\sin(3\psi + \varphi_3) d\psi \tag{4.18}$$

$$b_3(A) = \frac{1}{\pi A}\int_0^{2\pi} \{\mathcal{F}(A \sin \psi) + \dot{\mathcal{F}}[A(\sin \psi + \varepsilon\delta_3 \sin(3\psi + \varphi_3))] \times A\delta_3 \sin(3\psi + \varphi_3)\}\cos(3\psi + \varphi_3) d\psi \tag{4.19}$$

where

$$\dot{\mathcal{F}}\{A(\sin \psi + \varepsilon\delta_3 \sin(3\psi + \varphi_3))\} = \frac{\partial \mathcal{F}\{A(\sin \psi + \varepsilon\delta_3 \sin(3\psi + \varphi_3))\}}{\partial(A \sin \psi)}$$

and ε is a parameter allowing for the effect of the remaining terms of the generalized function expansion in a Taylor series. For all forms of nonlinearity, $0 < \varepsilon < 1$.

Assume that $\varepsilon\delta_3$ is a negligibly small value, then (4.16-19) rewrites in the form

$$a(A) = \frac{1}{2\pi}\int_0^{2\pi} \mathcal{F}(A \sin \psi)\sin \psi \, d\psi + \frac{\delta_3 \sin \varphi_3}{\pi}\int_0^{2\pi} \dot{\mathcal{F}}(A \sin \psi)\cos 3\psi \sin \psi \, d\psi + \frac{\delta_3 \cos \varphi_3}{\pi}\int_0^{2\pi} \dot{\mathcal{F}}(A \sin \psi)\sin 3\psi \sin \psi \, d\psi \tag{4.20}$$

$$b(A) = \frac{1}{\pi A}\int_0^{2\pi} \mathscr{F}(A \sin \psi)\cos \psi \, d\psi$$

$$+ \frac{\delta_3 \cos \varphi_3}{\pi}\int_0^{2\pi} \dot{\mathscr{F}}(A \sin \psi)\sin 3\psi \cos \psi \, d\psi$$

$$- \frac{\delta_3 \sin \varphi_3}{\pi}\int_0^{2\pi} \dot{\mathscr{F}}(A \sin \psi)\cos 3\psi \cos \psi \, d\psi \tag{4.21}$$

$$a_3(A) = \frac{1}{\pi A}\int_0^{2\pi} \mathscr{F}(A \sin \psi)\sin(3\psi + \varphi_3) \, d\psi$$

$$+ \frac{\delta_3 \cos^2 \varphi_3}{\pi}\int_0^{2\pi} \dot{\mathscr{F}}(A \sin \psi)\sin^2 3\psi \, d\psi$$

$$+ \frac{2\delta_3 \sin \varphi_3 \cos \varphi_3}{\pi}\int_0^{2\pi} \dot{\mathscr{F}}(A \sin \psi)\sin 3\psi \cos 3\psi \, d\psi$$

$$+ \frac{\delta_3 \sin^2 \varphi_3}{\pi}\int_0^{2\pi} \dot{\mathscr{F}}(A \sin \psi)\cos^2 3\psi \, d\psi \tag{4.22}$$

$$b_3(A) = \frac{1}{\pi A}\int_0^{2\pi} \mathscr{F}(A \sin \psi)\cos(3\psi + \varphi_3) d\psi$$

$$+ \frac{\delta_3 \cos^2 \varphi_3}{\pi}\int_0^{2\pi} \dot{\mathscr{F}}(A \sin \psi)\sin 3\psi \cos 3\psi \, d\psi$$

$$- \frac{\delta_3 \sin \varphi_3 \cos \varphi_3}{\pi}\int_0^{2\pi} \dot{\mathscr{F}}(A \sin \psi)\sin^2 3\psi \, d\psi$$

$$+ \frac{\delta_3 \sin \varphi_3 \cos \varphi_3}{\pi}\int_0^{2\pi} \dot{\mathscr{F}}(A \sin \psi)\cos^2 3\psi \, d\psi$$

$$- \frac{\delta_3 \sin^2 \varphi_3}{\pi}\int_0^{2\pi} \dot{\mathscr{F}}(A \sin \psi)\sin 3\psi \cos 3\psi \, d\psi \tag{4.23}$$

Now we write (4.20-23) with the aid of the basic and additional coefficients of harmonic linearization

$$a(A) = a_1(A) + \delta_3 \cos \varphi_3 \Delta a_{13}(A) + \delta_3 \sin \varphi_3 \Delta' a_{13}(A) \tag{4.24}$$

$$b(A) = b_1(A) + \delta_3 \cos \varphi_3 \Delta' b_{13}(A) - \delta_3 \sin \varphi_3 \Delta b_{13}(A) \tag{4.25}$$

where

$$a_1(A) = \frac{1}{\pi A} \int_0^{2\pi} \mathscr{F}(A \sin \psi) \sin \psi \, d\psi \tag{4.26}$$

$$\Delta a_{13}(A) = \frac{1}{\pi} \int_0^{2\pi} \dot{\mathscr{F}}(A \sin \psi) \sin 3\varphi \sin \psi \, d\psi \tag{4.27}$$

$$\Delta' a_{13}(A) = \frac{1}{\pi} \int_0^{2\pi} \dot{\mathscr{F}}(A \sin \psi) \cos 3\psi \sin \psi \, d\psi \tag{4.28}$$

$$b_1(A) = \frac{1}{\pi A} \int_0^{2\pi} \mathscr{F}(A \sin \psi) \cos \psi \, d\psi \tag{4.29}$$

$$\Delta' b_{13}(A) = \frac{1}{\pi} \int_0^{2\pi} \dot{\mathscr{F}}(A \sin \psi) \sin 3\psi \cos \psi \, d\psi \tag{4.30}$$

$$\Delta b_{13}(A) = \frac{1}{\pi} \int_0^{2\pi} \dot{\mathscr{F}}(A \sin \psi) \cos 3\psi \cos \psi \, d\psi \tag{4.31}$$

Here $a_1(A)$ and $b_1(A)$ are the coefficients of harmonic linearization in the first harmonic, and $\Delta a_{13}(A)$, $\Delta' a_{13}(A)$, $\Delta' b_{13}(A)$, $\Delta b_{13}(A)$ are the additional coefficients of harmonic linearization taking into account the influence of the third harmonic on the first.

Rewrite formulas (4.22, 23) as follows:

$$a_3(A) = a_3'(A) + \delta_3 \cos^2 \varphi_3 \Delta a_{33}(A) + 2\delta_3 \sin \varphi_3 \cos \varphi_3 \Delta' a_{33}(A) + \delta_3 \sin^2 \varphi_3 \Delta b_{33}(A) \tag{4.32}$$

$$b_3(A) = b_3'(A) + \delta_3 \cos^2 \varphi_3 \Delta' a_{33}(A) + \delta_3 \sin \varphi_3 \cos \varphi_3 \Delta b_{33}(A) - \delta_3 \sin \varphi_3 \cos \varphi_3 \Delta a_{33}(A) - \delta_3 \sin^2 \varphi_3 \Delta' a_{33}(A) \tag{4.33}$$

where

$$a_3'(A) = \frac{1}{\pi A}\int_0^{2\pi} \mathscr{F}(A \sin \psi)\sin(3\psi + \varphi_3)d\psi \tag{4.34}$$

$$\Delta a_{33}(A) = \frac{1}{\pi}\int_0^{2\pi} \dot{\mathscr{F}}(A \sin \psi)\sin^2 3\psi \, d\psi \tag{4.35}$$

$$\Delta' a_{33}(A) = \frac{1}{\pi}\int_0^{2\pi} \dot{\mathscr{F}}(A \sin \psi)\sin 3\psi \cos 3\psi \, d\psi \tag{4.36}$$

$$b_3'(A) = \frac{1}{\pi A}\int_0^{2\pi} \mathscr{F}(A \sin \psi)\cos(3\psi + \varphi_3)d\psi \tag{4.37}$$

$$\Delta b_{33}(A) = \frac{1}{\pi}\int_0^{2\pi} \dot{\mathscr{F}}(A \sin \psi)\cos^2 3\psi \, d\psi \tag{4.38}$$

$a_3'(A)$ and $b_3'(A)$ are the basic coefficients of harmonic linearization in the third harmonic, and $\Delta a_{33}(A)$, $\Delta' a_{33}(A)$ and $\Delta b_{33}(A)$ are the additional coefficients of harmonic linearization in the third harmonic.

After completion of the generalized harmonic linearization, the characteristic of bivalued nonlinearity $\mathscr{F}(x)$ may be written as

$$\begin{aligned}\mathscr{F}(x) = {} & [a_1(A) + \delta_3 \cos \varphi_3 \Delta a_{13}(A) + \delta_3 \sin \varphi_3 \Delta' a_{13}(A)]x_1 \\ & + \frac{1}{\omega}[b_1(A) + \delta_3 \cos \varphi_3 \Delta' b_{13}(A) - \delta_3 \sin \varphi_3 \Delta b_{13}(A)]\dot{x}_1 \\ & + \frac{1}{\delta_3}\{[a_3'(A) + \delta_3 \cos^2 \varphi_3 \Delta a_{33}(A) + 2\delta_3 \sin \varphi_3 \cos \varphi_3 \Delta' a_{33}(A) \\ & + \delta_3 \sin^2 \varphi_3 \Delta b_{33}(A)]x_3 + \frac{1}{3\omega}[b_3'(A) - \delta_3 \cos^2 \varphi_3 \Delta' a_{33}(A) \\ & - \delta_3 \sin \varphi_3 \cos \varphi_3 \Delta b_{33}(A) - \delta_3 \sin^2 \varphi_3 \Delta a_{33}'(A)]\dot{x}_3\}\end{aligned} \tag{4.39}$$

where $x_1 = A \sin \psi$ and $x_3 = A_3 \sin(3\psi + \varphi_3)$.

For single- and two-valued nonlinearities most often encountered in practical problems Tables 4.1 and 4.2 summarize the formulas for calculation of the basic and additional coefficients of harmonic linearization.

For single-valued odd nonlinear functions, the formulas of generalized harmonic linearization take on the form

$$a(A) = a_1(A) + \delta_3 \cos \varphi_3 \Delta a_{13}(A) \tag{4.40}$$

Table 4.1 Basic and Additional Coefficients of Harmonic Linearization of Single-Valued Nonlinearities

Parameters of nonlinear functions and harmonic linearization coefficients	Expressions for harmonic linearization coefficients		
	characteristic of saturation zone	relay with deadband	characteristic with dead zone
$y = F(x)$ $k = \tan\beta$ $A \geqslant C$	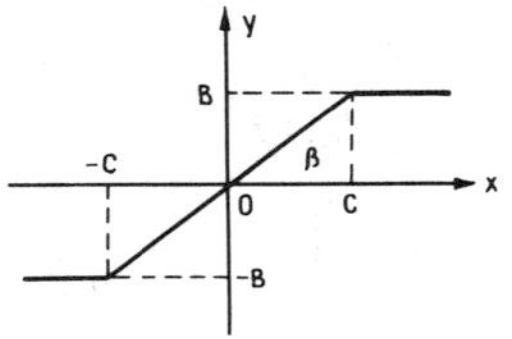	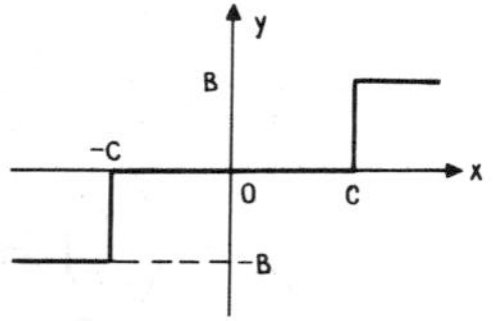	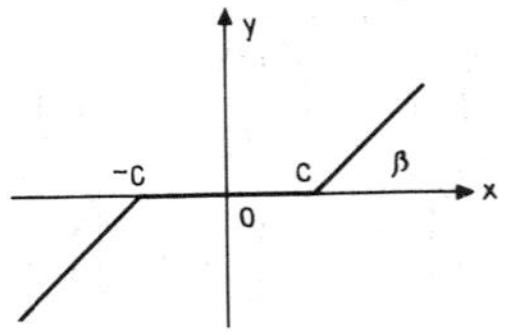
$a_1\left(\frac{C}{A}\right)$	$a_1\left(\frac{C}{A}\right) = \frac{2k}{\pi}\left(\arcsin\frac{C}{A} + \frac{C}{A}\sqrt{1-\left(\frac{C}{A}\right)^2}\right)$	$a_1\left(\frac{C}{A}\right) = \frac{4B}{\pi A}\sqrt{1-\left(\frac{C}{A}\right)^2}$	$a_1\left(\frac{C}{A}\right) = k - \frac{2k}{\pi}\left(\arcsin\frac{C}{A} + \frac{C}{A}\sqrt{1-\left(\frac{C}{A}\right)^2}\right)$
$a'_{33}\left(\frac{C}{A}\right)$	$a'_{33}\left(\frac{C}{A}\right) = \frac{4kC}{3\pi A}\left(1-\left(\frac{C}{A}\right)^2\right) \times \sqrt{1-\left(\frac{C}{A}\right)^2}$	$a'_{33}\left(\frac{C}{A}\right) = \frac{4B}{3\pi A} \times \left(1-4\left(\frac{C}{A}\right)^2\right)\sqrt{1-\left(\frac{C}{A}\right)^2}$	$a'_{33}\left(\frac{C}{A}\right) = -\frac{4kC}{3\pi A}\left(1-\left(\frac{C}{A}\right)^2\right) \times \frac{C}{A}\sqrt{1-\left(\frac{C}{A}\right)^2}$
$\Delta a_{13}\left(\frac{C}{A}\right)$	$\Delta a_{13}\left(\frac{C}{A}\right) = \frac{4k}{\pi}\left(\frac{C}{A}\right)^3 \times \sqrt{1-\left(\frac{C}{A}\right)^2}$	$\Delta a_{13}\left(\frac{C}{A}\right) = \frac{4BC^2\left[3-4\left(\frac{C}{A}\right)^2\right]}{\pi A^2}$	$\Delta a_{13}\left(\frac{C}{A}\right) = -\frac{4k}{\pi}\left(\frac{C}{A}\right)^3 \times \sqrt{1-\left(\frac{C}{A}\right)^2}$

$\Delta b_{13}\left(\frac{C}{A}\right)$	$\Delta b_{13}\left(\frac{C}{A}\right)=\frac{4k}{\pi}\left(\frac{C}{A}\right)\times\left[1-\left(\frac{C}{A}\right)^2\right]\times\sqrt{1-\left(\frac{C}{A}\right)^2}$	$\Delta b_{13}\left(\frac{C}{A}\right)=\frac{4B}{\pi}\times\left(1-4\left(\frac{C}{A}\right)^2\right)\times\left[1-\left(\frac{C}{A}\right)^2\right]$	$\Delta b_{13}\left(\frac{C}{A}\right)=-\frac{4k}{\pi}\times\left(1-\left(\frac{C}{A}\right)^2\right)\frac{C}{A}\sqrt{1-\left(\frac{C}{A}\right)^2}$
$\Delta a_{33}\left(\frac{C}{A}\right)$	$\Delta a_{33}\left(\frac{C}{A}\right)=\frac{2k}{\pi}[\arcsin\frac{C}{A}-\frac{C}{3A}\left(3-4\left(\frac{C}{A}\right)^2\right)\times\left(1-4\left(\frac{C}{A}\right)^2\right)\sqrt{1-\left(\frac{C}{A}\right)^2}]$	$\Delta a_{33}\left(\frac{C}{A}\right)=\frac{4BC^2}{\pi A^2}\times\left(3-4\left(\frac{C}{A}\right)^2\right)^2$	$\Delta a_{33}\left(\frac{C}{A}\right)=\frac{k}{3}-\frac{2k}{\pi}[\arcsin\frac{C}{A}-\left(3-16\left(\frac{C}{A}\right)^2+16\left(\frac{C}{A}\right)^4\right)\times\frac{C}{3A}\sqrt{1-\left(\frac{C}{A}\right)^2}]$
$\Delta b_{33}\left(\frac{C}{A}\right)$	$\Delta b_{33}\left(\frac{C}{A}\right)=\frac{2k}{\pi}[\arcsin\frac{C}{A}+\frac{C}{3A}\left(3-4\left(\frac{C}{A}\right)^2\right)\times\left(1-4\left(\frac{C}{A}\right)^2\right)\times\sqrt{1-\left(\frac{C}{A}\right)^2}]$	$\Delta b_{33}\left(\frac{C}{A}\right)=\frac{4B}{\pi}\times\left(1-4\left(\frac{C}{A}\right)^2\right)^2\times\left(1-\left(\frac{C}{A}\right)^2\right)$	$\Delta b_{33}\left(\frac{C}{A}\right)=\frac{k}{3}-\frac{2k}{\pi}[\arcsin\frac{C}{A}+\left(3-16\left(\frac{C}{A}\right)^2+16\left(\frac{C}{A}\right)^4\right)\times\frac{C}{3A}\sqrt{1-\left(\frac{C}{A}\right)^2}]$

Table 4.2 Basic and Additional Coefficients of Harmonic Linearization of Two-Valued Nonlinearities

Parameters of nonlinear functions and harmonic linearization coefficients	Type of nonlinear element		
$y = F(x)$ $K = \tan\beta$ $A \geqslant C$	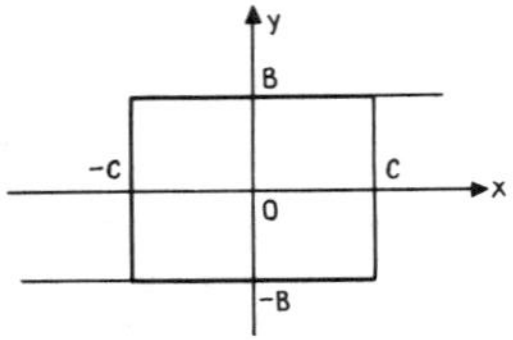	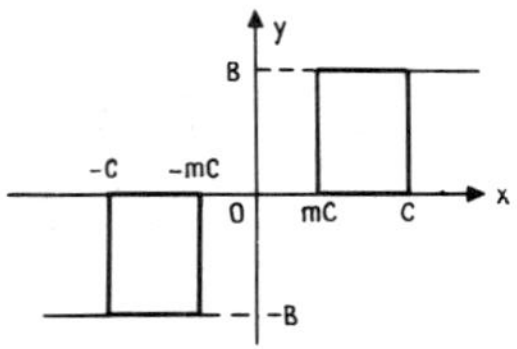	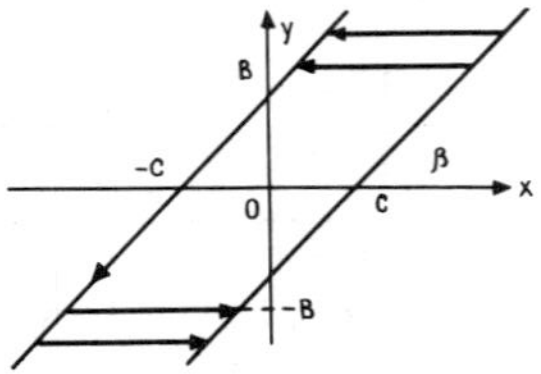
$a_1\left(\frac{C}{A}\right)$	$a_1\left(\frac{C}{A}\right) = \frac{4B}{\pi A}\sqrt{1 - \left(\frac{C}{A}\right)^2}$	$a_1\left(\frac{C}{A}\right) = \frac{2B}{\pi A}\left(\sqrt{1 - \left(\frac{C}{A}\right)^2} + \sqrt{1 - m^2\left(\frac{C}{A}\right)^2}\right)$	$a_1\left(\frac{C}{A}\right) = \frac{k}{\pi}\left[\frac{\pi}{2} + \arcsin\left(1 - 2\frac{C}{A}\right) + 2\left(1 - 2\frac{C}{A}\right)\sqrt{\frac{C}{A}\left(1 - \frac{C}{A}\right)}\right]$
$b_1\left(\frac{C}{A}\right)$	$b_1\left(\frac{C}{A}\right) = -\frac{4BC}{\pi A^2}$	$b_1\left(\frac{C}{A}\right) = -\frac{2BC}{\pi A^2}(1 - m)$	$b_1\left(\frac{C}{A}\right) = -\frac{4k}{\pi}\left(\frac{C}{A}\right)\left(1 - \frac{C}{A}\right)$

$a'_{33}\left(\frac{C}{A}\right)$	$a'_{33}\left(\frac{C}{A}\right) = -\frac{4B}{3\pi C}\left(\frac{C}{A}\right) \times \left[\left(-1 + 4\left(\frac{C}{A}\right)^2\right) \times \sqrt{1-\left(\frac{C}{A}\right)^2}\right]$	$a'_{33}\left(\frac{C}{A}\right) = -\frac{2B}{3\pi C}\left(\frac{C}{A}\right)\left[\left(-1 + 4\left(\frac{C}{A}\right)^2\right)\sqrt{1-\left(\frac{C}{A}\right)^2} + \left(-1 + 4m^2\left(\frac{C}{A}\right)^2\right)\sqrt{1-m^2\left(\frac{C}{A}\right)^2}\right]$	$a'_{33}\left(\frac{C}{A}\right) = \frac{8k}{\pi}\left(1-2\frac{C}{A}\right)^3 \times \sqrt{\left(1-\frac{C}{A}\right)\frac{C}{A}} - \frac{8k}{3\pi}\left(\frac{C}{A}\right)\left(-3 + 16\frac{C}{A} - 16\left(\frac{C}{A}\right)^2\right)\sqrt{\left(1-\frac{C}{A}\right)\frac{C}{A}} + \frac{8B}{3\pi C}\left(\frac{C}{A}\right)\left(-3 + 16\frac{C}{A} - 16\left(\frac{C}{A}\right)^2\right)\sqrt{\left(1-\frac{C}{A}\right)\frac{C}{A}}$
$b'_{33}\left(\frac{C}{A}\right)$	$b'_{33}\left(\frac{C}{A}\right) = -\frac{4B}{3\pi C}\left(\frac{C}{A}\right) \times \left(3 - 4\left(\frac{C}{A}\right)^2\right)\frac{C}{A}$	$b'_{33}\left(\frac{C}{A}\right) = \frac{2B}{3\pi C}\left(\frac{C}{A}\right)\left[-\left(3 - 4\left(\frac{C}{A}\right)^2\right)\frac{C}{A} + m\frac{C}{A}\left(3 - 4m^2\left(\frac{C}{A}\right)^2\right)\right]$	$b'_{33}\left(\frac{C}{A}\right) = \frac{8k}{\pi}\left(-3 + 8\frac{C}{A} - 8\left(\frac{C}{A}\right)^2\right)\left(1-\frac{C}{A}\right)\frac{C}{A} - \frac{4k}{3\pi}\left(\frac{C}{A}\right) \times \left[-1 - \left(1 - 16\frac{C}{A} + 16\left(\frac{C}{A}\right)^2\right)\left(1 - 2\frac{C}{A}\right)\right] + \frac{4B}{3\pi C} \times \left(\frac{C}{A}\right)\Bigl[-1 - \Bigl(1 - 16\frac{C}{A} + 16\left(\frac{C}{A}\right)^2\Bigr)\left(1 - 2\frac{C}{A}\right)$

Table 4.2 *(cont.)*

Parameters of nonlinear functions and harmonic linearization coefficients	Type of nonlinear element		
$\Delta a_{13}\left(\frac{C}{A}\right)$	$\Delta a_{13}\left(\frac{C}{A}\right) = \frac{4B}{\pi}\left(\frac{C}{A}\right)^2 \times \left[3 - 4\left(\frac{C}{A}\right)^2\right]$	$\Delta a_{13}\left(\frac{C}{A}\right) = \frac{2B}{\pi}\left(\frac{C}{A}\right)^2 \left\{3 - 4\left(\frac{C}{A}\right)^2 + m^2\left[3 - 4m^2\left(\frac{C}{A}\right)^2\right]\right\}$	$\Delta a_{13}\left(\frac{C}{A}\right) = \frac{8k}{\pi}\left(1 - 2\frac{C}{A}\right)^3 \times \sqrt{\left(1 - \frac{C}{A}\right)\frac{C}{A}}$
$\Delta' a_{13}\left(\frac{C}{A}\right)$	$\Delta' a_{13}\left(\frac{C}{A}\right) = \frac{4B}{\pi}\left(\frac{C}{A}\right)\left(1 - 4\left(\frac{C}{A}\right)^2\right)\sqrt{1 - \left(\frac{C}{A}\right)^2}$	$\Delta' a_{13}\left(\frac{C}{A}\right) = \frac{2B}{\pi}\left(\frac{C}{A}\right)^2 \times \left[\left(1 - 4\left(\frac{C}{A}\right)^2\right)\sqrt{1 - \left(\frac{C}{A}\right)^2} + m\sqrt{1 - m^2\left(\frac{C}{A}\right)^2}\left(1 - 4m^2\left(\frac{C}{A}\right)^2\right)\right]$	$\Delta' a_{13}\left(\frac{C}{A}\right)^2 = \frac{8k}{\pi}\left(-3 + 8\frac{C}{A} - 8\left(\frac{C}{A}\right)^2\right)\left(1 - \frac{C}{A}\right)\frac{C}{A}$
$\Delta' b_{13}\left(\frac{C}{A}\right)$	$\Delta' b_{13}\left(\frac{C}{A}\right) = \frac{4B}{\pi}\left(\frac{C}{A}\right) \times \left(1 - 4\left(\frac{C}{A}\right)^2\right) \times \sqrt{1 - \left(\frac{C}{A}\right)^2}$	$\Delta' b_{13}\left(\frac{C}{A}\right) = \frac{2B}{\pi c}\left(\frac{C}{A}\right)^2\left[\left(3 - 4\left(\frac{C}{A}\right)^2\right) + m\left(3 - 4m^2\left(\frac{C}{A}\right)^2\right)\right]$	$\Delta' b_{13}\left(\frac{C}{A}\right) = \frac{2k}{\pi} - \frac{2k}{\pi}\left(1 + 8\frac{C}{A} - 8\left(\frac{C}{A}\right)^2\right)\left(1 - 2\frac{C}{A}\right)^2$

$\Delta b_{13}\left(\frac{C}{A}\right)$	$\Delta b_{13}\left(\frac{C}{A}\right) = \frac{4B}{\pi C}\left(\frac{C}{A}\right)\left(1 - 4\left(\frac{C}{A}\right)^2\right)\sqrt{1 - \left(\frac{C}{A}\right)^2}$	$\Delta b_{13}\left(\frac{C}{A}\right) = \frac{2B}{\pi C}\left(\frac{C}{A}\right)\left\{\left(3 - 4\left(\frac{C}{A}\right)^2\right)\left[1 - \left(\frac{C}{A}\right)^2\right] + \left(3 - 4m^2\left(\frac{C}{A}\right)^2\right) \times \left[1 - m^2\left(\frac{C}{A}\right)^2\right]\right\}$	$\Delta b_{13}\left(\frac{C}{A}\right) = \frac{32k}{\pi}\left(1 - 3\frac{C}{A} + 2\left(\frac{C}{A}\right)^2\right)\frac{C}{A}\sqrt{\left(1 - \frac{C}{A}\right)\frac{C}{A}}$
$\Delta a_{33}\left(\frac{C}{A}\right)$	$\Delta a_{33}\left(\frac{C}{A}\right) = \frac{4B}{\pi C}\left(\frac{C}{A}\right)^2\left(3 - 4\left(\frac{C}{A}\right)^2\right)^2$	$\Delta a_{33}\left(\frac{C}{A}\right) = \frac{2B}{\pi C}\left(\frac{C}{A}\right)^2\left[\left(3 - 4\left(\frac{C}{A}\right)^2\right)^2 + m^2\left(3 - 4m^2 \times \left(\frac{C}{A}\right)^2\right)^2\right]$	$\Delta b_{33}\left(\frac{C}{A}\right) = \left[k + \frac{2k}{\pi}\arcsin\left(1 - 2\frac{C}{A}\right)\right] + \frac{4k}{3\pi}\left(-3 + 64\frac{C}{A} - 320\left(\frac{C}{A}\right)^2 + 512\left(\frac{C}{A}\right)^3 - 256\left(\frac{C}{A}\right)^4\right)\left(1 - 2\frac{C}{A}\right) \times \sqrt{\left(1 - \frac{C}{A}\right)\frac{C}{A}}$

Table 4.2 (*concl.*) 10

Parameters of nonlinear functions and harmonic linearization coefficients	Type of nonlinear element		
$\Delta b_{33}\left(\frac{C}{A}\right)$	$\Delta b_{33}\left(\frac{C}{A}\right) = \frac{4B}{\pi C}\left(\frac{C}{A}\right)\left(1 - 4\left(\frac{C}{A}\right)^2\right)\sqrt{1 - \left(\frac{C}{A}\right)^2}$	$\Delta b_{33}\left(\frac{C}{A}\right) = \frac{2B}{\pi C}\left(\frac{C}{A}\right)\left[\left(1 - 4\left(\frac{C}{A}\right)^2\right)\sqrt{1 - \left(\frac{C}{A}\right)^2} + \left(1 - 4m^2\left(\frac{C}{A}\right)^2\right) \times \sqrt{1 - m^2\left(\frac{C}{A}\right)^2}\right]$	$\Delta b_{33}\left(\frac{C}{A}\right) = \left[k + \frac{2k}{\pi}\arcsin\left(1 - 2\frac{C}{A}\right)\right] + \frac{4k}{3\pi}\left(3 - 64\frac{C}{A} + 320\left(\frac{C}{A}\right)^2 - 512\left(\frac{C}{A}\right)^3 + 256\left(\frac{C}{A}\right)^4\right)\left(1 - 2\frac{C}{A}\right) \times \sqrt{\left(1 - \frac{C}{A}\right)\frac{C}{A}}$
$\Delta' a_{33}\left(\frac{C}{A}\right)$	$\Delta' a_{33}\left(\frac{C}{A}\right) = \frac{4B}{\pi}\frac{C}{A}\left(3 - 4\left(\frac{C}{A}\right)^2\right)\left(1 - 4\left(\frac{C}{A}\right)^2\right) \times \sqrt{1 - \left(\frac{C}{A}\right)^2}$	$\Delta' a_{33}\left(\frac{C}{A}\right) = \frac{2B}{\pi}\left(\frac{C}{A}\right)\left[\left(3 - 4\left(\frac{C}{A}\right)^2\right)\left(1 - 4\left(\frac{C}{A}\right)^2\right) \times \sqrt{1 - \left(\frac{C}{A}\right)^2} + m\left(3 - 4\left(\frac{C}{A}\right)^2\right) \times \left(1 - 4\left(\frac{C}{A}\right)^2\right)\sqrt{1 - m^2\left(\frac{C}{A}\right)^2}\right]$	$\Delta' a_{33}\left(\frac{C}{A}\right) = \frac{8k}{3\pi}\left(3 - 16\frac{C}{A} + 16\left(\frac{C}{A}\right)^2\right)^2\left(1 - \frac{C}{A}\right)\frac{C}{A}$

$$b(A) = -\delta_3 \sin \varphi_3 \Delta b_{13}(A) \tag{4.41}$$

$$a_3(A) = a_3'(A) + \delta_3 \cos^2 \varphi_3 \Delta a_{33}(A) + \delta_3 \sin^2 \varphi_3 \Delta b_{33}(A) \tag{4.42}$$

$$b_3(A) = b_3'(A) + \delta_3 \sin \varphi_3 \cos \varphi_3 \Delta b_{33}(A) - \delta_3 \sin \varphi_3 \cos \varphi_3 \Delta a_{33}(A) \tag{4.43}$$

where the basic and additional coefficients of harmonic linearization are calculated by the expressions (4.26, 27, 31, 34-38).

The characteristics of single-valued nonlinearities may also be represented in terms of the generalized coefficients, namely,

$$\begin{aligned} F(x) = {} & [a_1(A) + \delta_3 \cos \varphi_3 \Delta a_{13}(A)]x_1 + [-\delta_3 \sin \varphi_3 \Delta b_{13}(A)]\dot{x}_1 \\ & + \frac{1}{\delta_3}\{[a_3'(A) + \delta_3 \cos^2 \varphi_3 \Delta a_{33}(A) + \delta_3 \sin^2 \varphi_3 \Delta b_{33}(A) \\ & + \frac{1}{3\omega}[b_3'(A) + \delta_3 \sin \varphi_3 \cos \varphi_3 \Delta b_{33}(A) \\ & - \delta_3 \sin \varphi_3 \cos \varphi_3 \Delta a_{33}(A)]]\dot{x}_3\} \end{aligned} \tag{4.44}$$

The appendices for single-valued (A1) and bivalued (A2) nonlinearities present the plots for the coefficients of harmonic linearization as functions of C/A. These graphs are convenient means in the location of stability regions and regions of self-excited oscillations of nonlinear systems.

4.2 Equivalent Bode Plots for Single- and Two-Valued Nonlinearities

To construct the equivalent Bode plots of nonlinearities with allowance for higher harmonic components we resort to the characteristic equations of closed-loop systems derivable in a first and subsequent approximations.

For two-valued nonlinearities, the first approximation has the form

$$1 + J(C/A)W(j\omega) = 0 \tag{4.45}$$

or

$$-1 = q(C/A)H(\omega)\exp\{j[\mu(C/A) + \theta(\omega)]\} \tag{4.46}$$

where $q(C/A)$ and $\mu(C/A)$ are the equivalent amplitude-ratio response and phase response of the nonlinearity, respectively; and $H(\omega)$ and $\theta(\omega)$ are the gain and phase responses of the linear part, respectively. The last expression can be rewritten as

$$\exp(-j\pi) = q(C/A)H(\omega)\exp\{j[\mu(C/A) + \theta(\omega)]\} \tag{4.47}$$

from which it is an easy matter to obtain the harmonic balance condition for the amplitudes and phases of the first approximation

$$20 \log H(\omega) = 20 \log \frac{1}{q(C/A)} \tag{4.48}$$

$$\theta(\omega) = -\pi - \mu(C/A) \tag{4.49}$$

which allows only for the coefficients of harmonic linearization $a(C/A)$ and $b(C/A)$. Then the equivalent values of the amplitudes and phases are

$$q(C/A) = \sqrt{a^2(C/A) + b^2(C/A)} \tag{4.50}$$

$$\mu(C/A) = \arctan[b(C/A)/a(C/A)] \tag{4.51}$$

Fot the first approximation, such dependences, allowing for the effect of the third harmonic on the first harmonic, can be obtained with equations (4.39, 34, 35, 32, and 33), *viz.*,

$$\begin{aligned} J_1(C/A) = {} & [a_1(C/A) + \delta_3 \cos \varphi_3 \Delta a_{13}(C/A) + \delta_3 \sin \varphi_3 \Delta' a_{13}(C/A)] \\ & + j[b_1(C/A) + \delta_3 \cos \varphi_3 \Delta' b_{13}(C/A) - \delta_3 \sin \varphi_3 \Delta b_{13}(C/A)] \end{aligned} \tag{4.52}$$

and

$$\begin{aligned} J_3(C/A) = {} & \frac{1}{\delta_3} \{[a_3'(C/A) + \delta_3 \cos^2 \varphi_3 \Delta a_{33}(C/A) \\ & + 2\delta_3 \sin \varphi_3 \cos \varphi_3 \Delta' a_{33}(C/A) + \delta_3 \sin^2 \varphi_3 \Delta b_{33}(C/A)] \\ & + j[b_3'(C/A) + \delta_3 \cos^2 \varphi_3 \Delta' a_{33}(C/A) \\ & - \delta_3 \sin \varphi_3 \cos \varphi_3 \Delta b_{33}(C/A) - \delta_3 \sin \varphi_3 \cos \varphi_3 \Delta a_{33}(C/A) \\ & - \delta_3 \sin^2 \varphi_3 \Delta' a_{33}(C/A)]\} \end{aligned} \tag{4.53}$$

By virtue of (4.52) we determine the harmonic balance conditions in the form of the second approximation when the effect of the third harmonic on the first is taken into account. In this case the balance conditions become

$$20 \log H(\omega) = 20 \log \frac{1}{q_1(C/A)} \tag{4.54}$$

$$\theta(\omega) = -\pi - \mu_1(C/A) \tag{4.55}$$

where

$$\begin{aligned} q_1(C/A) = {} & \{[a_1(C/A) + \delta_3 \cos \varphi_3 \Delta a_{13}(C/A) + \delta_3 \sin \varphi_3 \Delta' a_{13}(C/A)]^2 \\ & + [b_1(C/A) + \delta_3 \cos \varphi_3 \Delta' b_{13}(C/A) - \delta_3 \sin \varphi_3 \Delta b_{13}(C/A)]^2\}^{1/2} \end{aligned} \tag{4.56}$$

$$\mu_1(C/A) = \arctan \frac{b_1(C/A) + \delta_3 \cos \varphi_3 \Delta' b_{13}(C/A) - \delta_3 \sin \varphi_3 \Delta b_{13}(C/A)}{a_1(C/A) + \delta_3 \cos \varphi_3 \Delta a_{13}(C/A) + \delta_3 \sin \varphi_3 \Delta' a_{13}(C/A)} \tag{4.57}$$

Equations (4.56-57) include unknown values of the relative amplitude δ_3 and phase shift φ_3 which should be found from the characteristic equation for the third harmonic, namely,

$$1 + \delta_3^{-1} J_3(C/A) W(3j\omega) = 0 \tag{4.58}$$

where $J_3(C/A)$ is the describing function of the nonlinearity in the third harmonic,

$$J_3(C/A) = a_3'(C/A) + jb_3'(C/A) \tag{4.59}$$

Here, $a_3'(C/A)$ and $b_3'(C/A)$ are the harmonic linearization coefficients of the nonlinearity $\mathscr{F}(x)$ derived with no account of the additional coefficients due to third harmonic interference. If these coefficients are small, they may be neglected along with the effect of the additional coefficients $a_{33}'(C/A)$ and $b_{33}'(C/A)$. For most of nonlinear control systems, however, these coefficients are appreciable and have significant effect on the regions of stable states and self-excited oscillations and on the characteristics of forced oscillations.

To take into account the coefficients $a_{33}'(C/A)$ and $b_{33}'(C/A)$, we substitute (4.59) into (4.58) and obtain

$$\delta_3 = -W(3j\omega)[a_3'(C/A) + jb_3'(C/A)] \tag{4.60}$$

The basic coefficients of the harmonic linearization in the third harmonic may be written as

$$\begin{aligned} a_3'(C/A) &= a_{33}'(C/A)\cos\varphi_3 + b_{33}'(C/A)\sin\varphi_3 \\ b_3'(C/A) &= b_{33}'(C/A)\cos\varphi_3 - a_{33}'(C/A)\sin\varphi_3 \end{aligned} \tag{4.61}$$

where

$$\begin{aligned} a_{33}'(A) &= \frac{1}{\pi A}\int_0^{2\pi} \mathscr{F}(A\sin\psi)\sin 3\psi \, d\psi \\ b_{33}'(A) &= \frac{1}{\pi A}\int_0^{2\pi} \mathscr{F}(A\sin\psi)\cos 3\psi \, d\psi \end{aligned} \tag{4.62}$$

Substituting (4.61) into (4.60), we get

$$\delta_3 = -W(3j\omega)[a_{33}'(C/A) + jb_{33}'(C/A)(\cos\varphi_3 - j\sin\varphi_3)] \tag{4.63}$$

or

$$\delta_3 \exp(-j\varphi_3) = -W(3j\omega)[a_{33}'(C/A) + jb_{33}'(C/A)] \tag{4.64}$$

where

$$W(3j\omega) = H(3\omega)\exp[j\theta(3\omega)] \tag{4.65}$$

By virtue of (4.63-65) we may write

$$\delta_3 \exp(-j\varphi_3) = H(3\omega)q'_{33}(C/A)\exp\{j[\mu(C/A) + \theta(3\omega)]\} \quad (4.66)$$

Now, the equations of harmonic balance are written as

$$\delta_3 = H(3\omega)[a'^2_{33}(C/A) + b'^2_{33}(C/A)]^{1/2} \quad (4.67)$$

$$\varphi_3 = -\theta(3\omega) - \arctan[b'_{33}(C/A)/a'_{33}(C/A)] \quad (4.68)$$

For systems with symmetric single-valued odd nonlinearities of the type (4.40, 41), the equivalent amplitude-ratio responses may be computed with the expressions

$$q(C/A) = \{[a_1(C/A) + \delta_3 \cos \varphi_3 \Delta a_{13}(C/A)]^2 + [\delta_3 \sin \varphi_3 \Delta b_{13}(C/A)]^2\}^{1/2} \quad (4.69)$$

$$\mu_1(C/A) = -\arctan \frac{\delta_3 \sin \varphi_3 \Delta b_{13}(C/A)}{a_1(C/A) + \delta_3 \cos \varphi_3 \Delta a_{13}(C/A)} \quad (4.70)$$

It has been reported [4.2-8] that for the harmonic balance of systems with single-valued nonlinearities it is required in a first approximation that

$$20 \log H(\omega) = 20 \log \frac{1}{q(C/A)} \quad (4.71)$$

and

$$\theta(\omega) = -\pi \quad (4.72)$$

In a second approximation, the formulas (4.71) and (4.72) must be replaced with (4.54, 55), where the equivalent values of the amplitude ratio and phase responses are found with (4.69) and (4.70). The coefficient δ_3 and phase shift φ_3 are determined by (4.67, 68).

Comparing the conditions of harmonic balance (4.72) and (4.55) for systems with single-valued nonlinearities suggests that in a second approximation there occurs an additional phase shift due to the effect of the third harmonic.

The procedure thus outlined may be employed to take into account the effect of the fifth [4.9] and higher harmonics when the assumption that the linear elements of the system filter out all harmonics, except the first, of the periodic outputs of the nonlinear elements does not come true in practice.

A sufficient advantage of this approach, accounting for the effect of higher harmonics is associated with the possibility to use the graphical gauges $-20 \log[q_1(C/A)]$ and $-\pi - \mu_1(C/A)$, which substantially simplify the analysis of nonlinear systems.

Consider the procedure of evaluating the second and subsequent approximations for defining the reduced values of amplitudes and phases of system nonlinearities. The analytical steps are essentially as follows.

1. Find the frequency ω_{s1} and amplitude A_{s1} of self-excited oscillations with the use of graphical gauges or with the aid of a computer, allowing for the harmonic balance conditions for the first approximation.

2. Compute the basic coefficients of harmonic linearization in the third harmonic, $a'_{33}(C/A)$ and $b'_{33}(C/A)$.

3. Determine relative values of amplitude δ_3 and phase shift φ_3 with due account for $H(3\omega_{s1})$ and $\theta(3\omega_{s1})$ of the linear part.

4. Find the equivalent amplitude and phase characteristics, $q_1(C/A)$ and $\mu_1(C/A)$ respectively, for the second approximation.

5. Determine refined values of frequency ω_{s2} and amplitude A_{s2} of self-excited oscillations with the aid of curve gauges or a computer allowing for the harmonic balance conditions for the second approximation.

6. If the accuracy of the ω_{s2} and A_{s2} is insufficient, repeat steps 1 through 4 taking into account the coefficients $\Delta a_{33}(C/A)$, $\Delta' a_{33}(C/A)$ and $\Delta b_{33}(C/A)$. This will result in the third approximation for ω_{s3} and A_{s3} of self-excited oscillations.

The computations terminate if ω_{s2}, ω_{s3}, and A_{s2}, A_{s3} are sufficiently close.

If one has to resort to fifth and higher harmonics, evaluation of refined frequencies and amplitudes is carried out in the same order, however, the effect of the linearization coefficients in the fifth and higher harmonics on the third and first ones has to be taken into account and the parameters δ_3, δ_5, etc., and φ_3, φ_5, etc. have to be computed.

We make use of the outlined procedures to construct regions of stable and unstable states graphically, i.e. with the use of nonlinearity gauges.

4.3 The Curve Gauge Method

Before we embark on a detailed investigation of nonlinear systems taking into account higher harmonics, let us have an illustrative insight into how this method works for frequencies and amplitudes of self-excited oscillations in the first harmonic. Illustrative examples will be simple single-loop and two-loop systems with backlash and saturation.

Example 4.1. Figure 4.1*a* shows the block diagram of systems with the linear part

$$W(s) = \frac{K_l}{s(Ts + 1)} \tag{4.73}$$

Let $K_l = 40\ s^{-1}$ and $T = 0.1$ s. Then, substituting these parameters and

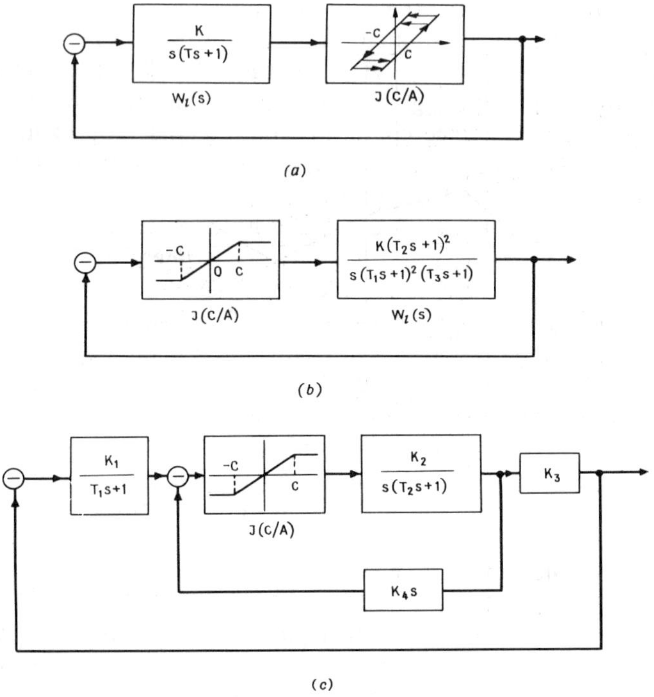

Fig. 4.1

$s = j\omega$ into the transfer function, we obtain the Bode plot for the amplitude $H(\omega)$ (solid line) and phase $\theta(\omega)$ (dashed line) shown in Fig. 4.2*a*.

We assume that the nonlinearity parameters (backlash) are $K_n = 1$ and $C = 0.004$ rad and compute, by expressions summarized in Table 4.2, $\text{Lm}[1/q(C/A)]$ in dB and $\mu(C/A)$ in degrees. The next step will be to construct a curve gauge (first approximation) with the characteristics $\text{Lm}(1/q)$ and $-180° - \mu°$ on transparent paper (Fig. 4.3*a*). It should not be overlooked that the equivalent phase will be negative. Clearly, the scales of the responses and curve gauges in both abscissa and ordinate are identical.

Apply the gauge with its back side[2] on the Bode plots (Fig. 4.2*a*) so that the axes ω and C/A be aligned. Then the curves $\text{Lm}(1/q)$ and

---

[2] When the gauge is applied with its back side, intercepts of $\text{Lm}(1/q)$ and $-180° - \mu$ with θ can be located more accurately. Figure 4.2 shows the third position of the gauge applied with its face side (dashed lines). The slopes of the gauge are seen to be close to those of the plots of the linear part of the system, thus reducing the accuracy in locating the periodic modes. Therefore, in the following we shall put the curve gauges with the back side to locate the respective intersection more reliably.

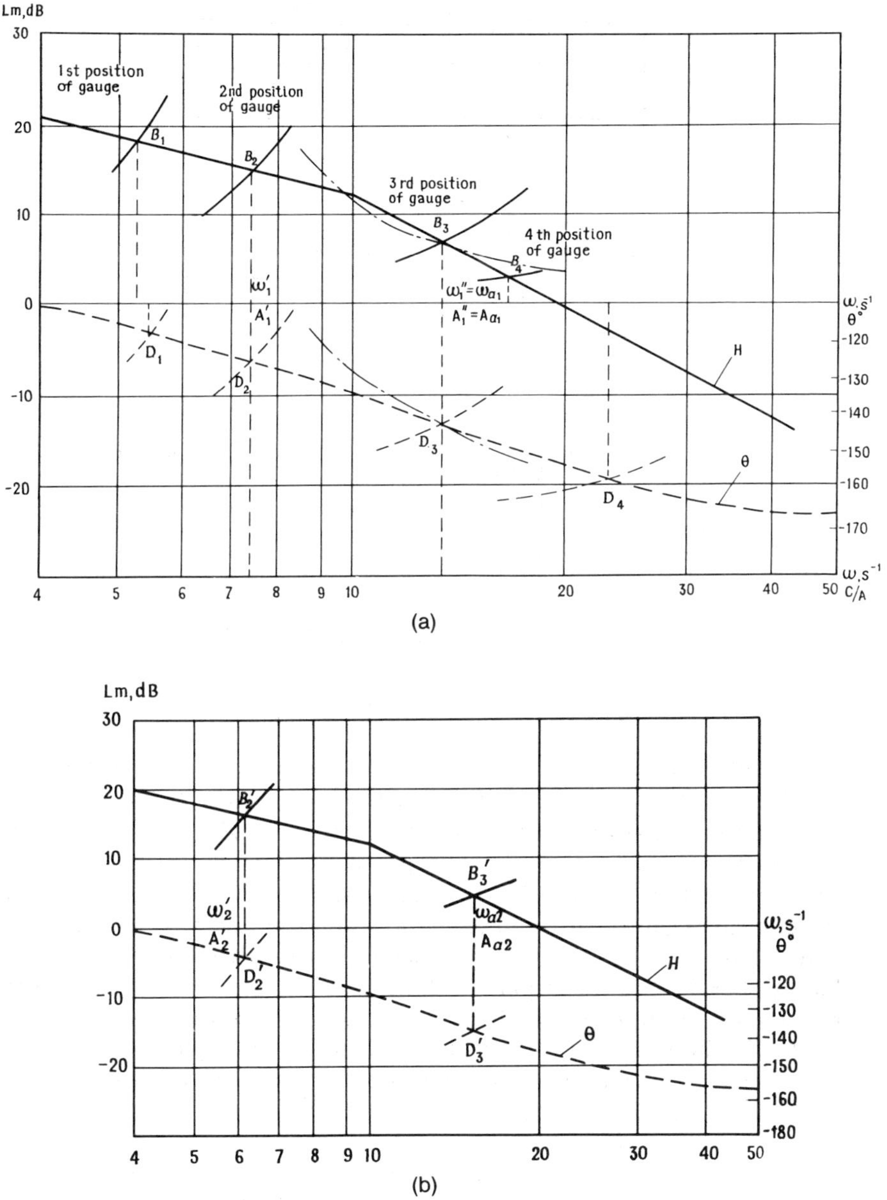

Lm,dB
1st position of gauge
2nd position of gauge
3rd position of gauge
4th position of gauge
B_1
B_2
B_3
B_4
ω_1'
A_1'
$\omega_1''=\omega_{a1}$
$A_1''=A_{a1}$
D_1
D_2
D_3
D_4
H
θ
ω,s⁻¹
θ°
C/A
(a)
Lm,dB
B_2'
B_3'
ω_2'
A_2'
ω_{a2}
A_{a2}
D_2'
D_3'
H
θ
ω,s⁻¹
θ°
(b)

Fig. 4.2

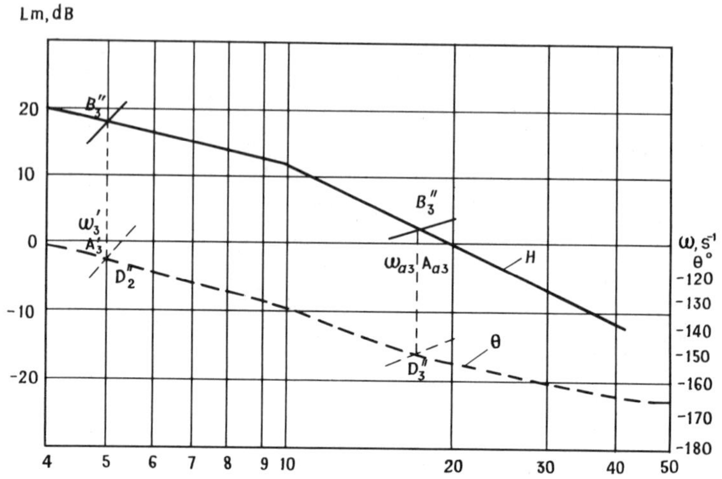

Fig. 4.2(c)

$-180° - \mu$ of the gauge intersect the respective amplitude H and phase θ plots at points B_1 and D_1 (first position of the gauge).

Because these intercepts B_1 and D_1 do not lie on the same vertical, the conditions (4.48) and (4.49) are not satisfied and the nonlinear system does not experience an oscillatory mode.

Now we shift the gauge from left to right until the intersections B_2 and D_2 (second position of the gauge) appear on the same vertical. This implies that the conditions (4.48, 49) are met and the system under study experiences an oscillatory mode. The frequency of these oscillations ω_1' can be read on the ω axis, and the amplitude of oscillations A_1' can be easily obtained from the values of C/A, read on the C/A axis, and the known C.

A further translation of the gauge again locates points B_3 and D_3 (third position of the gauge) where the conditions (4.48, 49) are met and enables evaluation of another pair ω'' and A''. Any further translation of the gauge does not produce intersections lying on the same vertical.

Having obtained two pairs of frequencies and amplitudes, we now need to determine which of them corresponds to self-excited and which to unstable oscillations. A criterion to solve this problem will be as follows. Nonlinear systems with two-valued nonlinearities experience unstable oscillations (self-excited oscillations) when an increment ΔA of the amplitude A shifts the intercept of the characteristics $\mathrm{Lm}[1/q(C/A)]$ and $\mathrm{Lm}H(\omega)$, lying on the same vertical with the intercept of the curves $-180° - \mu$ and $\theta(\omega)$, outside the region formed by the logarithm of the

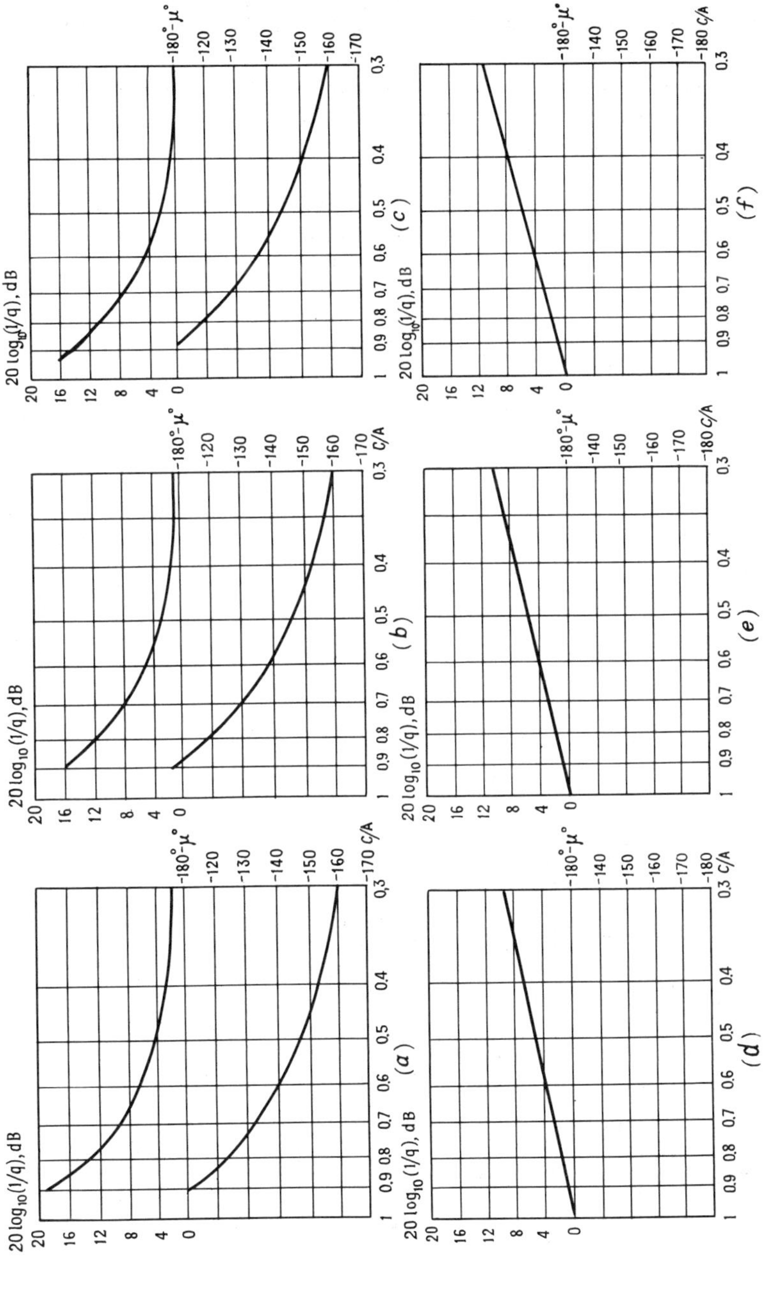

Fig. 4.3

magnitude and the frequency axis. Conversely, nonlinear system oscillations will be unstable if an increase of the amplitude makes this point enter the aforementioned region.

From Fig. 4.2*a* it follows that $\omega_1'' = \omega_{s1} = 13.5\ \mathrm{s}^{-1}$ and $A_1'' = A_{s1} = 0.0067$ rad are consistent with stable oscillations, while ω_1 and A_1' correspond to unstable oscillations.

Example 4.2. Determine the frequency ω_s and amplitude A_s of self-excited oscillations in the nonlinear system with saturation as shown in Fig. 4.1*b* and the transfer function of the linear element

$$W(s) = \frac{K_l(T_2 s + 1)^2}{s(T_1 s + 1)^2(T_3 s + 1)} \tag{4.74}$$

where K_l can vary from 100 to 1000 s^{-1}, $T_1 = 2.88$ s, $T_2 = 0.4$ s and $T_3 = 0.025$ s.

By letting $K_l = 600\ \mathrm{s}^{-1}$ and $s = j\omega$ we construct the Bode diagram with the log-magnitude $H(\omega)$ and log-phase $\theta(\omega)$ plots for this transfer function (Fig. 4.4). In points D_1 and D_2 corresponding to the second condition of harmonic balance (4.72) we erect verticals to cross the log-magnitude plot $20 \log H(\omega)$ at B_1 and B_2. At B_1 and B_2 we put the curve gauge for the nonlinear log-magnitude plot of $20 \log[1/q(C/A)]$. As can be seen, in this

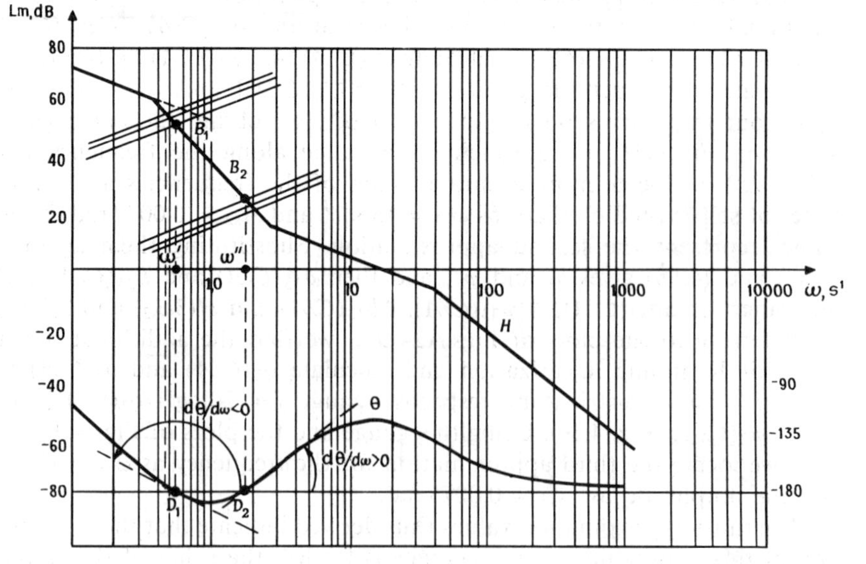

Fig. 4.4

case the first condition of harmonic balance is satisfied. Consequently, the system oscillates at frequencies ω' and ω'' with amplitudes A' and A''.

In order to establish the type of oscillation in systems with saturation, we draw tangents to the plot $\theta(\omega)$ at D_1 and D_2 and determine the signs of the derivatives $d\theta/d\omega$ about the $-180°$ axis. The negative sign of this derivative at D_1 means that the system experiences self-excited oscillations at $\omega' = \omega_{s1}$ with the amplitude $A = A_{s1}$, whereas the positiveness of $d\theta/d\omega$ implies unstable oscillations.

Example 4.3. Determine the frequencies and amplitudes of self-excited oscillations in the system specified in Example 4.1 with the aid of the second and third approximations.

We recall that in a first approximation (Fig. 4.2*a*) we found $\omega_{s1} = 13.5\ s^{-1}$ and $A_{s1} = 0.0067$ rad. To a second approximation, we resort to the dependences of the basic coefficients of harmonic linearization presented in Figs A2-3*b*, *c* and find $a'_{33}(C/A) = 0$ and $b'_{33}(C/A) = -0.2$. Then we let $s = 3j\omega_{s1}$ and calculate $H(3\omega_{s1})$ and $\theta(3\omega_{s1})$ for the linear element, using the transfer function (4.73), as

$$
\begin{aligned}
H(3\omega_{s1}) &= K_l/\{3\omega_{s1}[(3\omega_{s1}T)^2 + 1]^{1/2}\} \\
\theta(3\omega_{s1}) &= -\pi/2 - \arctan(3\omega_{s1}T)
\end{aligned}
\tag{4.75}
$$

Substituting the found values in (4.67) and (4.68) yields δ_3 and φ_3.

With reference to the curves plotted in Figs A2-3*a*-*e* we determine the additional coefficients of harmonic linearization $\Delta a_{13}(C/A)$, $\Delta' a_{13}(C/A)$, $\Delta' b_{13}(C/A)$, and $\Delta b_{13}(C/A)$, calculate $q_1(C/A)$ and $\mu_1(C/A)$ by (4.56) and (4.57), and construct the plots of $20\log[1/q(C/A)]$ and $-\pi - \mu_1(C/A)$. Lying out this curves on transparent paper, we obtain the curve gauge shown in Fig. 4.3*b*. By advancing the gauge along the frequency axis (Fig. 4.2*b*) we locate more accurate values of the frequencies and amplitudes of self-excited oscillations $\omega_{s2} = 15\ s^{-1}$ and $A_{s2} = 0.0072$ rad. These values represent the second approximation. Substituting them again in (4.67) and (4.68) gives δ'_3 and φ'_3, and then $a''_{33}(C/A)$ and $b''_{33}(C/A)$ with allowance for $\Delta a_{33}(C/A)$, $\Delta' a_{33}(C/A)$, $\Delta b_{33}(C/A)$ and $H(3\omega_{s2})$ and $\theta(3\omega_{s2})$.

Referring to the plots of Figs A2-3*a*-*c*, we find the additional coefficients of harmonic linearization and calculate $q_2(C/A)$ and $\mu_2(C/A)$ by (4.56, 57) and construct the respective gauge for $20\log[1/q_2(C/A)]$ and $-\pi - \mu_2(C/A)$ (Fig. 4.3*c*). Using this gauge and the plots of Figs 4.3*c* and 4.2*c*, we locate the third approximation for the frequency as $\omega_{s3} = 17\ s^{-1}$ and the amplitude as $A_{s3} = 0.0075$ rad.

The third-approximation values thus derived indicate that the contribution of this approximation to accuracy is far less than that of the second approximation.

Example 4.4. Determine the values of frequency and amplitude of self-excited oscillations for the system specified in Example 4.2 by means of the second and third approximation procedures.

We recall that in a first approximation we obtained $\omega_{s1} = 0.56\ s^{-1}$ and $A_{s1} = 415$ V. Resorting to the second approximation, we look up $a'_{33}(C/A) = 0.42$ and $b'_{33}(C/A) = 0.42$ in Fig. A1-3*a* to calculate

$$\begin{aligned} H(3\omega_{s1}) &= \frac{K_l[(3\omega_{s1}T_2)^2 + 1]}{3\omega_{s1}[(3\omega_{s1}T_1)^2 + 1][(3\omega_{s1}T_3)^2 + 1]^{1/2}} \\ \theta(3\omega_{s1}) &= -\pi/2 - 2\arctan(3\omega_{s1}T_1) + 2\arctan(3\omega_{s1}T_2) \\ &\quad - \arctan(3\omega_{s1}T_3) \end{aligned} \tag{4.76}$$

and determine δ_3 and φ_3 by (4.69) and (4.70).

The additional coefficients of harmonic linearization $\Delta a_{13}(C/A)$ and $\Delta b_{13}(C/A)$ are determined in Fig. A1-3*b*. Substituting the values determined by (4.76) into (4.67) and (4.68) yields δ_3 and φ_3. Then we determine $q_1(C/A)$ and $\mu_1(C/A)$ by (4.69) and (4.70) and plot on transparent paper the curves $20\log[1/q_1(C/A)]$ and $-180° - \mu_1(C/A)$ (Fig. 4.3*d*, *f*). The plot $-\pi - \mu_1(C/A)$ crosses the phase plot $\theta(\omega)$ at points D'_1 and B'_1 above the $-180°$ axis. Drawing vertical lines through these points, we find that they intersect the $H(\omega)$ plot at B'_1 and B'_2. Drawing the curves $20\log(1/q)$ and $-180° - \mu'_1(C/A)$ through B'_1 and B'_2, we obtain $\omega_{s2} = 0.5\ s^{-1}$ and $A_{s2} = 435$ V at their intersections with the axes.

The values of $q_2(C/A)$ and $\mu_2(C/A)$ are determined graphically and points B''_2 and D''_2 are located. Using these points we can find the frequency $\omega_{s3} = 0.49\ s^{-1}$ and amplitude $A_{s3} = 440$ V of the third approximation (Fig. 4.4).

Figure 4.4 suggests that the procedure of parameter evaluation may be terminated at the second approximation, since the contribution of the third approximation to the parameter accuracy is of low significance.

Example 4.5. Determine the value of amplitudes and frequencies of self-excited oscillations in the nonlinear system specified by Fig. 4.1*c* in first, second, and third approximations. The system parameters are $k_1 = 20\,000$ V/rad; $k_3 = 0.005$; $k_4 = 0.5$ V s/rad; $C = 1$; $k_n = 1$; $T_1 = 0.02$, 0.01, and 0.005 s; and $k_2 = 125$, 140, 210, and 250 rad/V s.

First consider the system with $k_2 = 210$ rad/V s and $T_1 = 0.005$ s. The characteristic equation is

$$1 + J(C/A)\left[\frac{k_1k_2}{s(T_1s+1)(T_2s+1)}\right]\left[1 + \frac{k_4s(T_1s+1)}{k_1k_3}\right] = 0 \tag{4.77}$$

or, by designating the last two bracketed cofactors by $W_1(s)$ and $W_2(s)$,

$$1 + J(C/A)W_1(s)W_2(s) = 0 \tag{4.78}$$

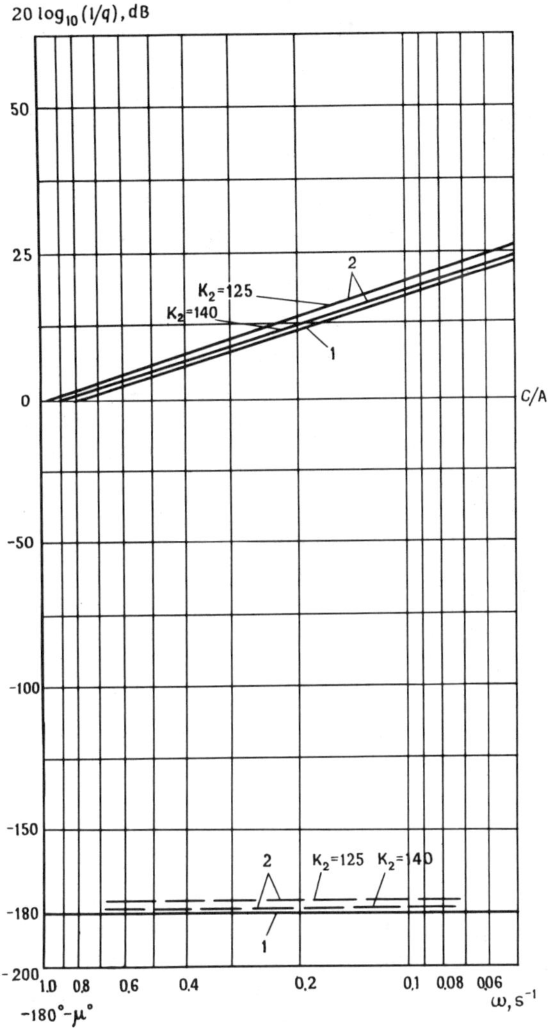

Fig. 4.5

The relevant harmonic balance equations are

$$20 \log | W_1(j\omega) W_2(j\omega) | = 20 \log[1/q(C/A)] \tag{4.79}$$

$$\theta_1(\omega) + \theta_2(\omega) = -\pi \tag{4.80}$$

Figure 4.5 shows the curve gauge 20 log(1/q) constructed by the formulas of the first approximation, and Fig. 4.6 shows the log-amplitude plot

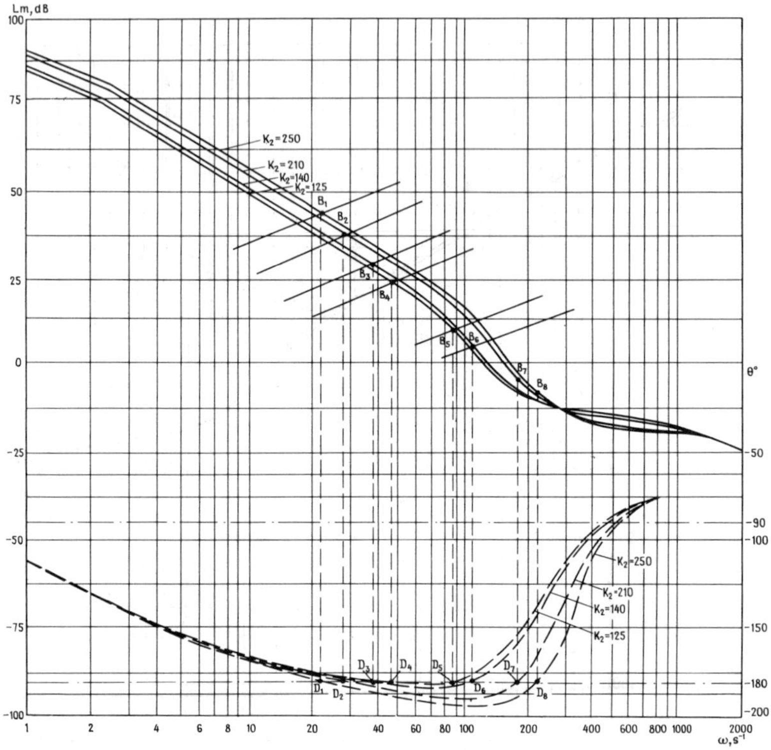

Fig. 4.6

$|W_1(j\omega)W_2(j\omega)|$ (solid lines) and phase plots $\theta_1(\omega) + \theta_2(\omega)$ (dashed lines) for $T_1 = 0.005$ s and $k_2 = 125, 140, 210, 250$ rad/V s. The plot intersects the $-180°$ axis at points D_1-D_4 thus implying the presence of oscillatory modes. At points D_1-D_4, $d\theta/d\omega < 0$ which means self-excited oscillations, whereas at D_5-D_6, $d\theta/d\omega > 0$ which implies unstable oscillations.

Vertical lines drawn through points D_1-D_8 intersect the log-magnitude plots at points B_1-B_8. Through these points we draw the straight lines corresponding to saturation. Intersections of vertical lines B_i, D_i, $i = 1, \ldots, 6$, with the frequency axis give the family of frequencies $\omega_{s1}^{(i)}$, s^{-1}, and those with the C/A axis give the values of amplitudes of self-excited oscillations $A_{s1}^{(i)}$, rad.

Now we use the procedure of Example 4.4 to refine the amplitudes and frequencies of oscillations accurate to a second approximation (Fig. 4.7).

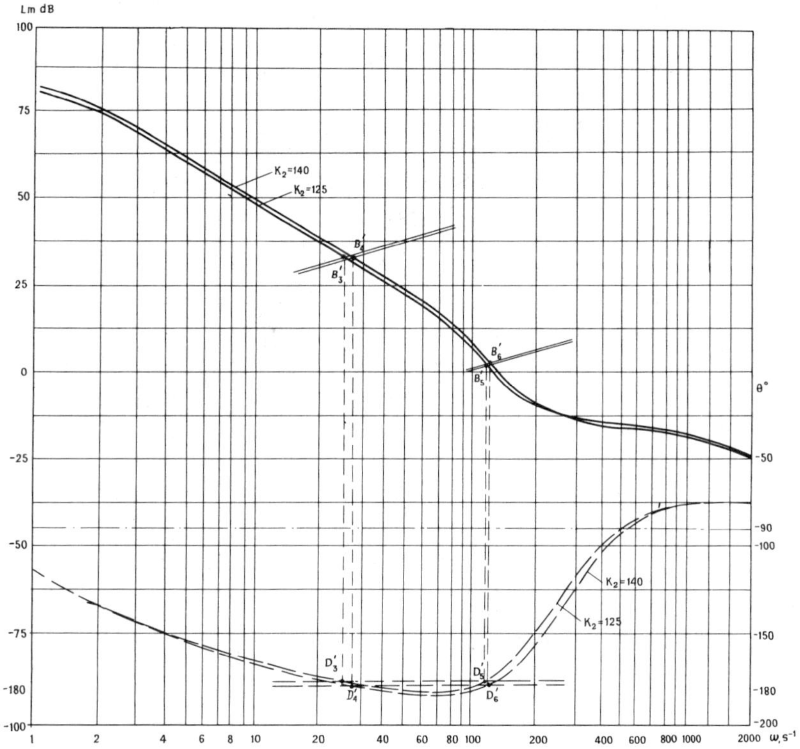

Fig. 4.7

The phase balance equation has the form

$$\theta_1(\omega_{s1}^{(i)}) + \theta_2(\omega_{s1}^{(i)}) = -\pi + \mu_1(C/A_{s1}^{(i)}) \tag{4.81}$$

This equation holds when the third harmonic produces an additional phase shift $\mu_1(C/A_{s1}^{(i)})$. From this equation it follows that the plots of the curve gauge lie above the $-180°$ axis depending on the value of k_2. In Fig. 4.5, the gauge plots are represented by curves *2* constructed to the scales of $\mathrm{Lm}(1/q)$ and $-\pi + \mu(C/A_{s1})$. By applying this gauge over Fig. 4.7 we recognize that the frequencies and amplitudes of the second approximation markedly differ from the first-approximation values, the difference increasing for greater T_1.

4.4 Regions of Stable States and Self-Excited Oscillations: Evaluation by Generalized Coefficients of Linearization

The graphical method may be used to evaluate the curves of self-excited and unstable oscillations as functions of system parameters in the first quadrant. Figure 4.8 shows Bode plots constructed for four values of K_l, namely, $K_{l1} = 20\ \mathrm{s}^{-1}$ (Fig. 4.2*a-c*), $K_{l2} = 23.7\ \mathrm{s}^{-1}$, $K_{l3} = 40\ \mathrm{s}^{-1}$, and $K_{l4} = 100\ \mathrm{s}^{-1}$. The numerals in Fig. 4.8 indicate the gauges of the first (*1*), second (*2*) and third (*3*) approximations. The points thus obtained are used to construct regions of stable states and oscillations in Fig. 4.9*a*. As can be seen from this figure, at higher gain K_l of the linear element the first and third approximation curves separating stable regions from unstable become closer. This implies that as the gain K_l increases, the effect of higher harmonics on the frequencies and amplitudes of self-excited oscillations decreases. The immediate implication of this figure is that the regions of stable states derived in first, second, and third approximations cannot exceed, respectively, $K_{l,\mathrm{b}1} = 23.7\ \mathrm{s}^{-1}$, $K_{l,\mathrm{b}2} = 20\ \mathrm{s}^{-1}$, and $K_{l,\mathrm{b}3} = 19.8\ \mathrm{s}^{-1}$.

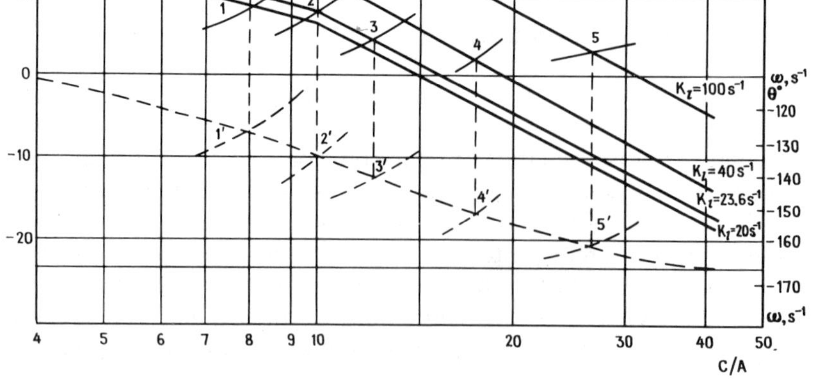

Fig. 4.8

The difference due to the harmonics effect between the bound values of K_l is seen to lie within 20%.

Figure 4.9*b* shows the bounds separating the regions of stable and unstable states in amplitude A for $T = 0.1$ s. (The curve separating the regions in the amplitude of self-excited oscillations is indicated by converging arrows, and the curve separating the regions of self-excited oscillations and unstable oscillations is indicated by diverging arrows [4.6].) As can be seen, the effect of higher harmonics on the amplitudes decreases at higher gains of the linear element.

Example 4.6. Consider the evaluation of domains of stable and unstable states depending on the time constant T of the aperiodic element.

Figure 4.9*c* shows the first approximation plots as functions of ω_s at $K_l = 40\ \mathrm{s}^{-1}$. In this case the bound values are $T_{b1} = 0.065$ s, $T_{b2} = 0.026$ s, and $T_{b3} = 0.033$ s.

Figure 4.9*d* presents the regions of stable and unstable states as functions of the amplitude A_s. From Fig. 4.9*c*, *d* it follows that the effect of higher harmonics on the regions of stable states decreases also with the growth of T. For the given parameters, however, the regions of stable states

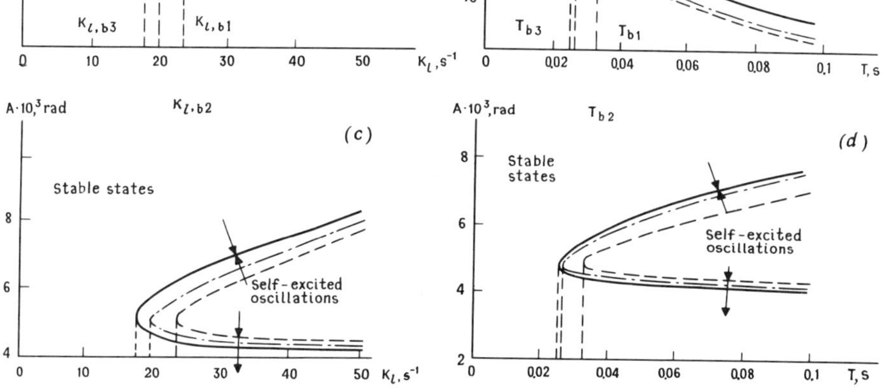

Fig. 4.9

enlarge 1.32 times. Even though in practice such phenomenon is rare, they ought to be taken into account, for the accuracy of system operation will reduce if compensating means fail to damp oscillations at certain values of C/A. The behavior of stability regions as functions of T remains essentially the same.

If control systems involve symmetric nonlinearities, the effect of higher harmonics on the regions of stable and unstable states increases and the errors in measuring frequencies and amplitudes of self-excited oscillations rise to 40% and higher since one has to allow for the effect of the first harmonic on the second, then on the third, and so on.

We resort to the gauge technique to determine the bounds of stability regions for the system specified in Example 4.4 by taking into account the effect of higher harmonics. Figure 4.4 indicates that at $K_l = 600\,s^{-1}$ this system exhibits self-excited oscillations. By varying the gain K_l we obtain the bounds separating the regions of stable and unstable states as functions of K_l.

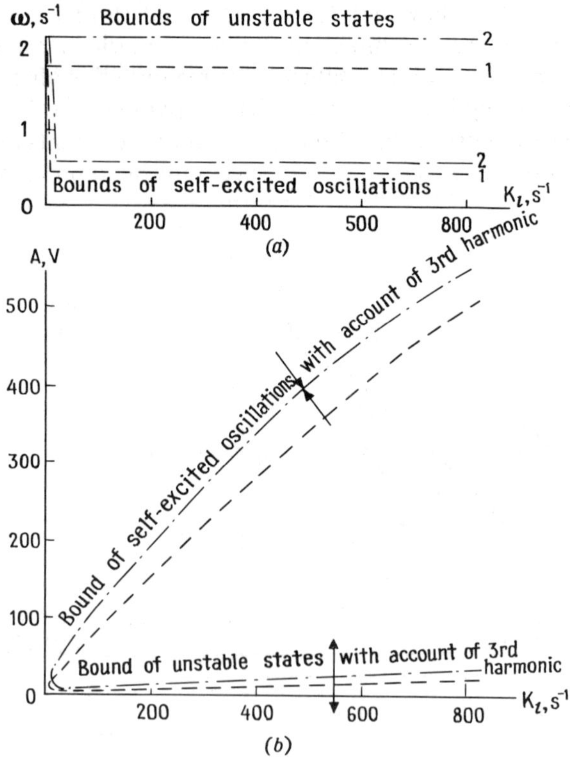

Fig. 4.10

Figure 4.10*a* shows these bounds constructed at $C = 1$ as functions of ω_s, and Fig. 4.10*b* shows the bounds as functions of the amplitude A_s of self-excited oscillations. Curve *1* corresponds to oscillations in the first harmonic, and curve *2* takes into account the effect of the third harmonic on the first.

This example indicates that at $K_l = 600\,s^{-1}$ the third harmonic extends the regions of self-excited oscillations by 25% in frequency and by 15% in amplitude. At lower gain values of the linear element K_l, the bound of self-excited-oscillation regions increases in amplitude by 50% and higher. This depends on the upward shift of the plot $\mu(C/A)$ with respect to the $-180°$ axis but in the narrow frequency range ω'-ω''.

Example 4.7. Construct the regions of stable and unstable states and bounds of self-excited oscillations in first and third approximations for variable parameters K_1 and T_1 in the system specified in Example 4.5 (Fig. 4.11). Dashed lines indicate the bounds of regions of stable and unstable states constructed with account of the first harmonic, and solid lines are constructed with allowance for the first and third harmonics.

Thus, control systems exhibit self-excited oscillations due to the effect of higher harmonics, which reduce the region of stable states. This phenomenon is of significance since at certain parameter values these oscillations may have a detrimental effect on the control elements and the plant. Suppression of self-excited oscillations is a complicated problem which may

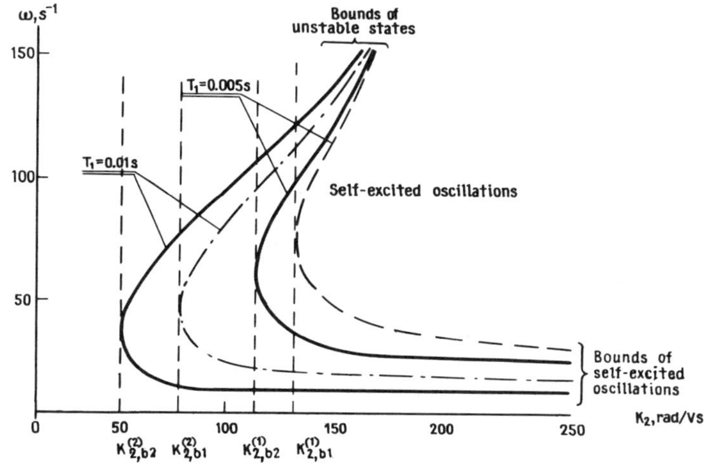

Fig. 4.11

Table 4.3 Frequencies ω_s of Self-Excited Oscillations and Bound Values of K_2 at $C = 1$

Time constant T, s	Analysis				Experiment
	in 1st harmonic		with 3rd harmonic		
	ω_{s1}, s^{-1}	$K_{2,b1}$, V	ω_{s2}, s^{-1}	$K_{2,b2}$, V	ω_s, s^{-1}
0.005	80	129	63	113	62
0.01	46	80	33	55	32

be solved by incorporating additional compensating elements, unfortunately, at the expence of system performance and accuracy.

Table 4.3 summarizes analytical and experimental data of the system discussed in Example 4.5. The experimental data in this table are very close to the analytical data derived exclusively with the account of the frequency and amplitude of the third harmonic on the first.

In all systems of automatic control unstable oscillations turn into self-excited oscillations. Figure 4.12 illustrates how self-excited oscillations set up in a nonlinear system described by the block diagram of Fig. 4.1*a*.

In order to suppress self-excited oscillations, linear and nonlinear compensating elements are sometimes incorporated in the system. These elements, however, disturb the harmonic balance conditions in a wide range of gains. Another way is to select a compensating function that prevents any periodic modes, thus ensuring stable operation for the entire system.

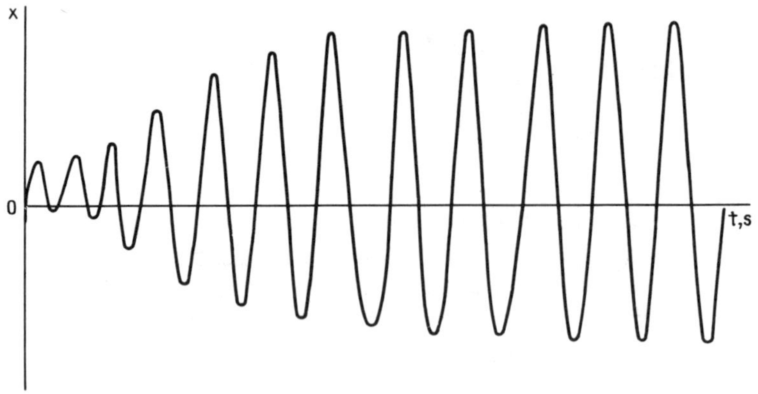

Fig. 4.12

The major setback of additional linear compensating elements is associated with a rise of the system cutoff frequency and an increase of the noise level. Nonlinear compensating elements are incorporated in feedback loops by means of tacho generators or bridges. These compensating facilities complicate selection of their parameters [4.10]. Suppression of self-excited oscillations is best when cascade (linear) and parallel (nonlinear) compensating elements are simultaneously incorporated in the system.

4.5 Forced Oscillations

Breakdown of the superposition principle in nonlinear systems complicates analysis of forced oscillations. Such an analysis can be carried out only with the method of harmonic linearization allowing for the effect of higher harmonics.

The equation of system dynamics may be written in the form [4.5]

$$x(t) + W(p)\mathscr{F}(x, px) = W_F(p)F(t) \tag{4.80}$$

where $x(t)$ is the system output, $W(p)$ and $W_F(p)$ are polynomials in $p = \mathrm{d}/\mathrm{d}t$, and $F(t)$ is the system input, or excitation.

Let the system input (excitation) be described by the function

$$F(t) = B_e \sin \omega_e t \tag{4.81}$$

with the amplitude B_e and frequency ω_e.

We make use of the condition of synchronization when the system exhibits oscillations at the frequency of excitation. This phenomenon is commonly called entrainment and described as

$$x(t) = A_e \sin(\omega_e t + \varphi) \tag{4.82}$$

To solve this problem, A_e and ω_e must be determined to first, second and subsequent approximations.

In control theory expression (4.81) is commonly written as

$$F(t) = B_e \sin[(\omega_e t + \varphi) - \varphi] = -B_e \sin \varphi \cos(\omega_e t + \varphi) + B_e \sin(\omega_e t + \varphi)\cos \varphi \tag{4.83}$$

By differentiating (4.82) we have

$$px(t) = A_e\omega_e \cos(\omega_e t + \varphi) \tag{4.84}$$

whence in view of (4.83) we find

$$F(t) = \frac{B_e}{A_e}\left(\cos \varphi - \frac{\sin \varphi}{\omega_e} p\right) x(t) \tag{4.85}$$

Substituting this equation into (4.80) yields

$$\left[1 - W_F(p)\,\frac{B_e}{A_e}\left(\cos\varphi - \frac{\sin\varphi}{\omega_e}\,p\right)\right] x(t) + W(p)F(x,\ px) = 0 \tag{4.86}$$

To simplify the problem on hand let

$$\int_0^{2\pi} \mathscr{F}(A_e \sin\psi,\ A_e\omega_e \cos\psi)\mathrm{d}\psi = 0$$

where $\psi = \omega_e t + \varphi$.

Now the harmonic linearization of the nonlinearity is straightforward

$$\mathscr{F}(x,\ px) = a(A_e,\ \omega_e)x + \frac{b(A_e,\ \omega_e)}{\omega_e}\,px \tag{4.87}$$

where

$$\begin{aligned} a(A_e,\ \omega_e) &= \frac{1}{\pi A_e}\int_0^{2\pi} \mathscr{F}(A_e \sin\psi,\ A_e\omega_e \cos\psi)\sin\psi\ \mathrm{d}\psi \\ b(A_e,\ \omega_e) &= \frac{1}{\pi A_e}\int_0^{2\pi} \mathscr{F}(A_e \sin\psi,\ A_e\omega_e \cos\psi)\cos\psi\ \mathrm{d}\psi \end{aligned} \tag{4.88}$$

Substituting (4.87) into (4.86) gives

$$1 - W_F(p)\,\frac{B_e}{A_e}\left(\cos\varphi - \frac{\sin\varphi}{\omega_e}\,p\right)$$

$$+ W(p)\left[a(A_e,\ \omega_e) + \frac{b(A_e,\ \omega_e)}{\omega_e}\,p\right] = 0 \tag{4.89}$$

We rewrite (4.89) in operator form observing that $\cos\varphi - j\sin\varphi = \exp(-j\varphi)$ and $s = j\omega_e$, *viz.*,

$$A_e\,\frac{1 + W(j\omega_e)[a(A_e,\ \omega_e) + jb(A_e,\ \omega_e)]}{W_F(j\omega_e)} = B_e \exp(-j\varphi) \tag{4.90}$$

Representing (4.90) in the form

$$\frac{A_e}{B_e} e^{j\varphi} = \frac{W_F(j\omega_e)}{1 + W(j\omega_e)[a(A_e, \omega_e) + jb(A_e, \omega_e)]} \tag{4.91}$$

and substituting

$$J(A_e, j\omega_e) = a(A_e, \omega_e) + jb(A_e, \omega_e) \tag{4.92}$$

yields

$$\frac{A_e}{B_e} e^{j\varphi} = \frac{[J(A_e, \omega_e) W(j\omega_e)]^{-1}}{1 + [J(A_e, \omega_e) W(j\omega_e)]^{-1}} W_F(j\omega_e) \tag{4.93}$$

whence

$$20 \log \frac{A_e}{B_e} - 20 \log |W_F(j\omega_e)| = 20 \log \left| \frac{[J(A_e, \omega_e) W(j\omega_e)]^{-1}}{1 + [J(A_e, \omega_e) W(j\omega_e)]^{-1}} \right| = D \tag{4.94}$$

$$\varphi_e - \arg[W_F(j\omega_e)] = \arg \left\{ \frac{[J(A_e, \omega_e) W(j\omega_e)]^{-1}}{1 + [J(A_e, \omega_e) W(j\omega_e)]^{-1}} \right\} \tag{4.95}$$

Denote

$$\left.\begin{aligned} |J(A_e, j\omega_e)| &= q(A_e, \omega_e) \\ \arg[J(A_e, j\omega_e)] &= \mu(A_e, \omega_e) \end{aligned}\right\} \tag{4.96}$$

implying

$$J(A_e, j\omega_e) = q(A_e, \omega_e) \exp \{j[\pi - \mu(A_e, \omega_e)]\} \tag{4.97}$$

Then, in view of (4.94) and (4.95),

$$20 \log \left| \frac{[J(A_e, j\omega_e) W(j\omega_e)]^{-1}}{1 + [J(A_e, j\omega_e) W(j\omega_e)]^{-1}} \right| = 20 \log |\Phi_\varepsilon(A_e, j\omega_e)| \tag{4.98}$$

$$\arg \left\{ \frac{[J(A_e, j\omega_e) W(j\omega_e)]^{-1}}{1 + [J(A_e, j\omega_e) W(j\omega_e)]^{-1}} \right\} = \arg [\Phi_\varepsilon(A_e, j\omega_e)] \tag{4.99}$$

For graphical analysis we draw the functions $1/W(j\omega_e)$ and $1/J(A_e, j\omega_e)$ on transparent paper (Fig. 4.13*a*) and apply these plots over the Nichols chart (Fig. 4.13*b*) so that the origin appears on the plot $1/J(A_e, j\omega_e)$ at a point satisfying the condition

$$20 \log \frac{A_e}{B_e} - 20 \log |W_F(j\omega_e)| = 20 \log |\Phi_\varepsilon(A_e, j\omega_e)| \tag{4.100}$$

By allowing the transparent paper with the plots of $1/J(A_e, j\omega_e)$ and $1/W(j\omega_e)$ to travel along the chart (Fig. 4.13*b*) we arrive at the plot $A_e = A_e(\omega_e)$ shown in Fig. 4.14 by dashed line.

Let us consider how the effect of higher harmonics on forced oscillations is taken into account in nonlinear systems. For this purpose in equations (4.96, 98, 100) we need to allow for the effect of the third harmonic on the first, that is, we must apply equations (4.67-70) for single-valued nonlinearities and find δ_3, φ_3, $q_1(C/A)$, and $\mu_1(C/A)$. Under the circumstances the harmonic linearization equation (4.97) should be written in the

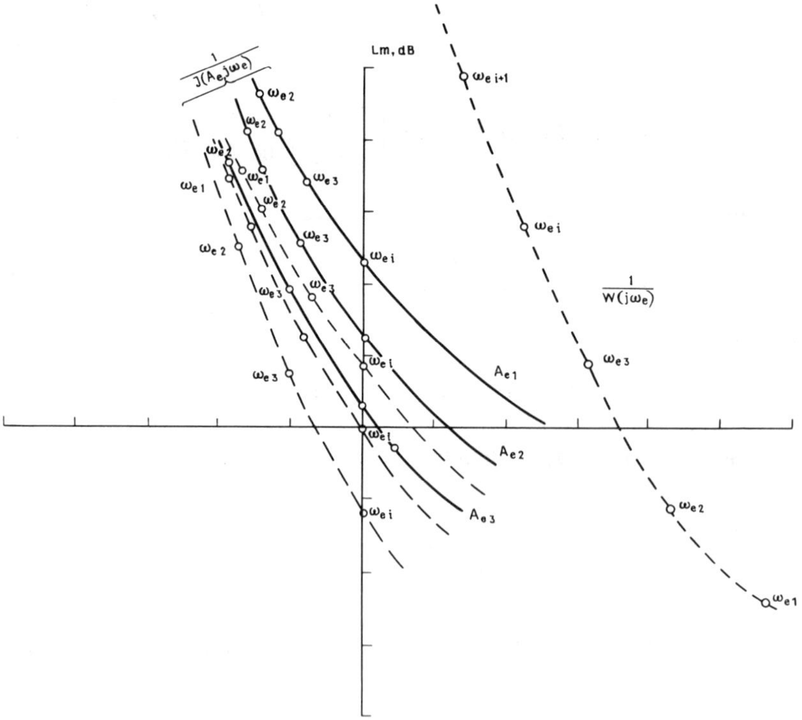

Fig. 4.13(a)

form

$$J_1(A_e, j\omega_e) = q_1(A_e, \omega_e) \exp\{j[\pi - \mu_1(A_e, \omega_e)]\} \tag{4.101}$$

and (4.98, 99) take the form

$$20 \log \left| \frac{[J_1(A_e, j\omega_e) W(j\omega_e)]^{-1}}{1 + [J_1(A_e, j\omega_e) W(j\omega_e)]^{-1}} \right| = 20 \log |\Phi_{\varepsilon 1}(A_e, j\omega_e)| \tag{4.102}$$

$$\arg \left\{ \frac{[J_1(A_e, j\omega_e) W(j\omega_e)]^{-1}}{1 + [J_1(A_e, j\omega_e) W(j\omega_e)]^{-1}} \right\} = \arg \{\Phi_{\varepsilon 1}(A_e, j\omega_e)\} \tag{4.103}$$

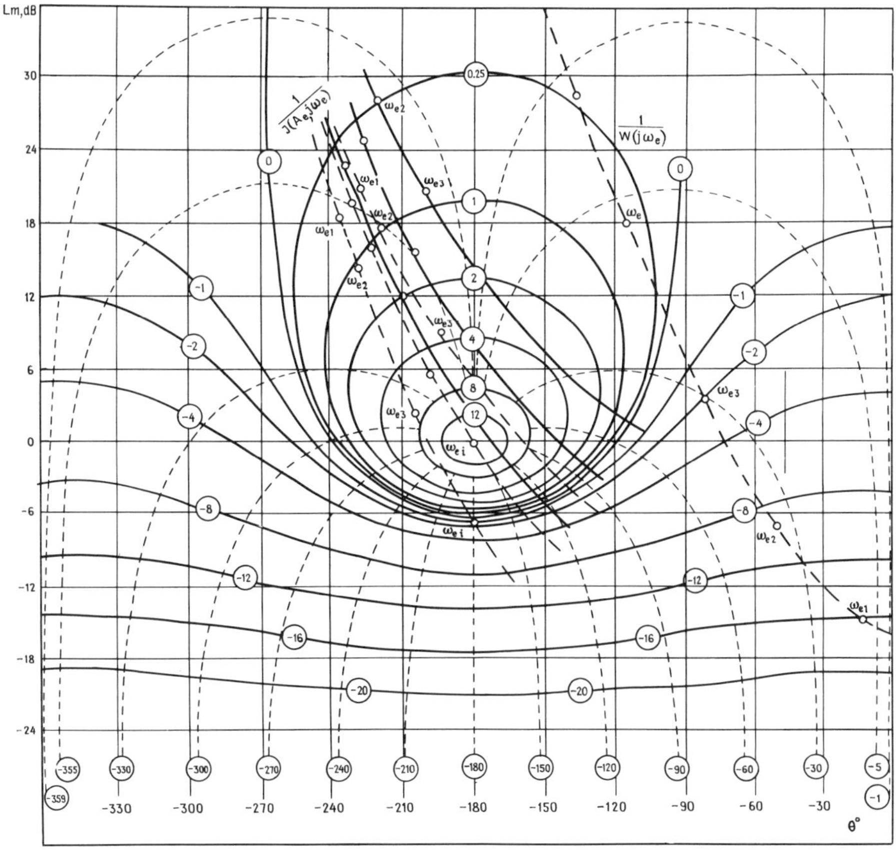

Fig. 4.13(b)

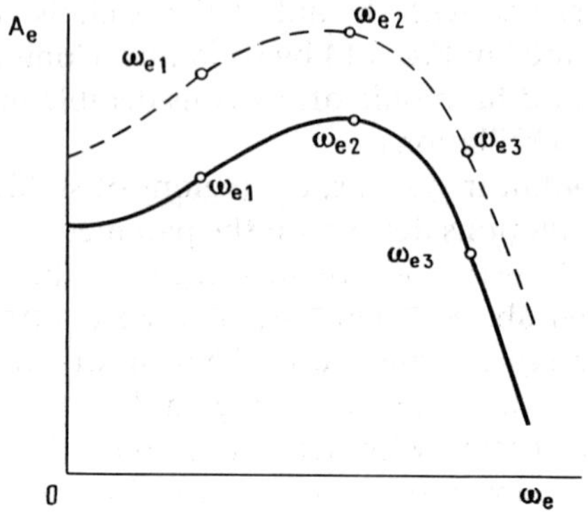

Fig. 4.14

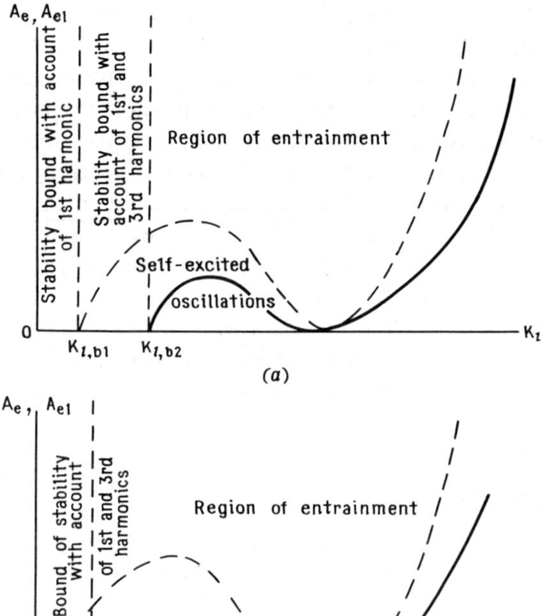

Fig. 4.15

For graphical analysis we construct the functions $1/W(j\omega_e)$ and $1/J_1(A_e, j\omega_e)$ on transparent paper and with the aid of the Nichols chart derive the plot $A_{e1} = A_{e1}(\omega_e)$ presented in Fig. 4.14 by solid line. Comparison of these plots suggests that the third harmonic offers considerable influence on the amplitude A_e of forced oscillations.

For some types of nonlinear systems, the regions of stable state entrainment and self-excited oscillations depend on the parameters and magnitude of system excitation B_e. Construction of such regions relies on the mathematical relations outlined above. Therefore, here we confine ourselves to summarizing the final results represented in the form of curves derived for Example 4.2 as functions of linear element gain K_l.

The said regions constructed with account of only the first harmonic are indicated in Fig. 4.15*a* by dashed line. The solid line corresponds to the boundary separating the regions of stable oscillations, entrainment and self-excited oscillations as constructed with account of the effect of the third harmonic on the first for the system specified in Example 4.4. Incorporation of the third harmonic effect is seen to enlarge the region of stable states and to reduce the region of self-excited oscillations. By changing the parameters of the linear element and replacing the nonlinear function by another this system may be converted to a form where the bounds obtained with account of the fundamental component (dashed line in Fig. 4.15*b*) separate only two regions, namely, the regions of entrainment and self-excited oscillations. When the third harmonic is taken into account, the region of stable states also occurs. It is worth noting that there exist nonlinear systems where the region of locking can be contracted.

Discussions of this chapter indicate that there is a need for account of the effect of higher harmonics on the behavior of autonomous and nonautonomous nonlinear systems. Higher harmonics may be not only a cause of self-excited oscillations but also a source of changes in forced oscillations.

References

4.1 N. M. Krylov and N. N. Bogoljubov: *Novye metody nelineinoi mekhaniki* (New methods of nonlinear mechanics). ONTI GTTI, Moscow-Leningrad, 1934.

4.2 L. S. Goldfarb: *Bulleten VEI.* No. 3, 44-47 (1944).

4.3 R. J. Kochenburger: *Trans. AIEE.* **69**, 270-284 (1950).

4.4 E. C. Johnson: *Applications and Trans.* AIEE **1**, July, 169-181 (1952).

4.5 E. P. Popov: *Dokl. Akad. Nauk SSSR.* **5**(5), 3-38 (1952).
E. P. Popov: *Dynamics of Automatic Control Systems.* Pergamon Press, Oxford, 1958.

4.6 E. P. Popov and I. P. Paltov: *Priblizhennye metody issledovaniya nelineinykh sistem avtomaticheskogo upravleniya.* Fizmatgiz, Moscow, 1960 [English transl: *Approximate Methods for the Investigation of Non-Linear Control Systems.* Akademsche Verlag Gesellschaft, Leipzig, 1963].

4.7 J. C. West: *Analytical Techniques for Non-Linear Control Systems.* Van Nostrand, Princeton, New Jersey, 1961.
4.8 Yu. I. Topcheyev: in *Avtomaticheskiye regulyatory i sledyashchiye sistemy* (Automatic controllers and servo systems). Gostekhizdat, Moscow, 1963, pp. 422-539.
4.9 Yu. I. Topcheyev: in *Sovremennye metody proyektirovaniya sistem avtomaticheskogo upravleniya. Analiz i sintez* (Modern methods of control system design. Analysis and synthesis), ed. by B. N. Petrov, V. S. Solodovnikov, and Yu. I. Topcheyev. Mashinostroyeniye, Moscow, 1967, pp. 388-423.
4.10 N. N. Ivashchenko: *Avtomaticheskoe regulirovanie. Teoriya i elementy sistem* (Automatic control. Theory and system elements). 4th ed. Mashinostroyeniye, Moscow, 1978.

Appendixes to Chapter 4

Appendix 1. Single-Valued Nonlinearities: Basic and Additional Coefficients of Harmonic linearization

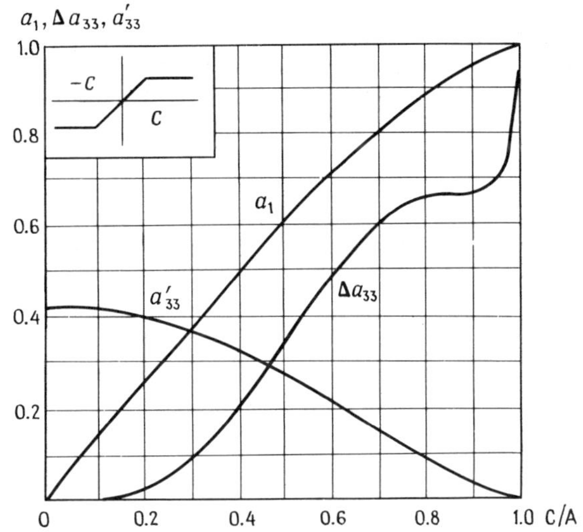

Fig. A1-1a

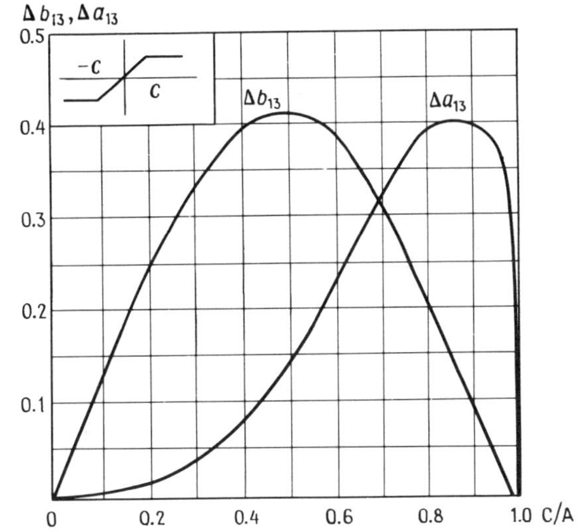

Fig. A1-1b

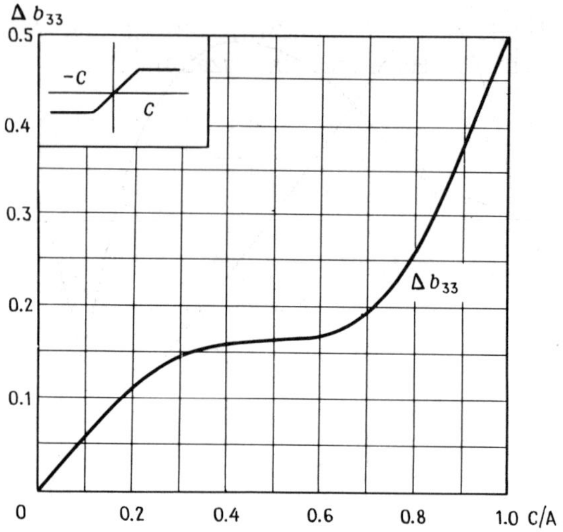

Fig. A1-1c

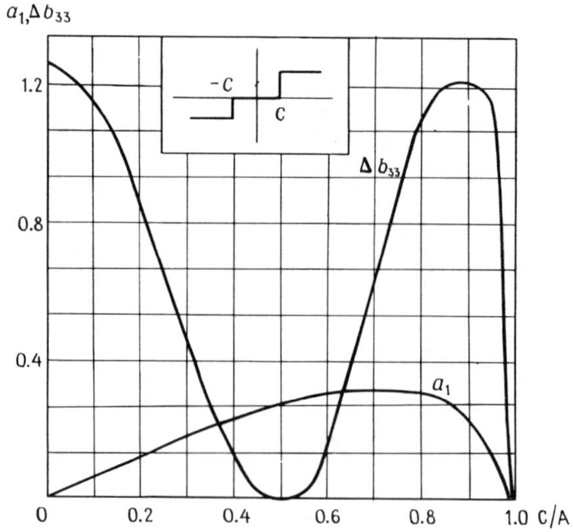

Fig. A1-2a

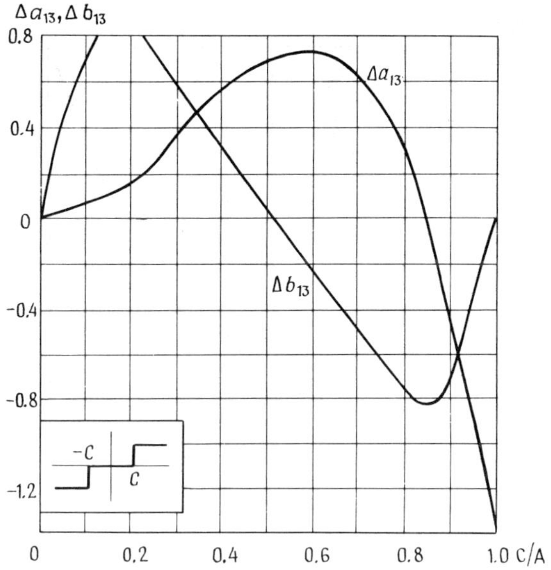

Fig. A1-2b

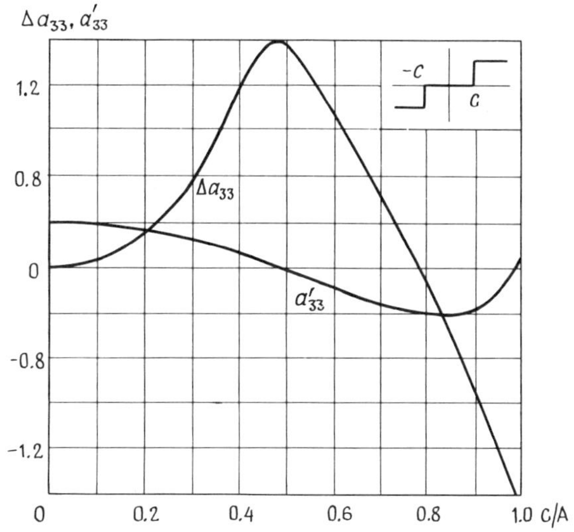

Fig. A1-2c

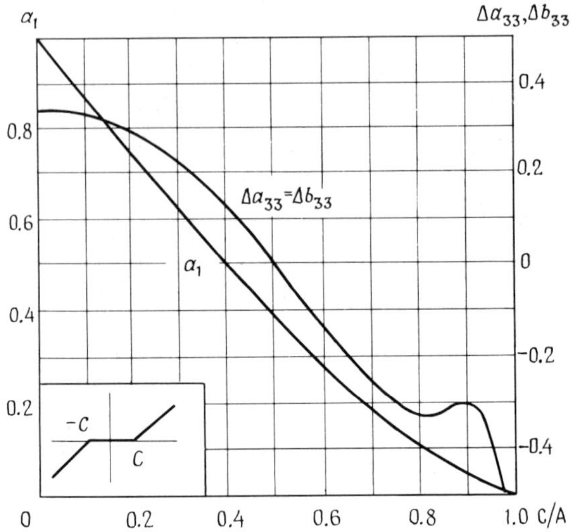

Fig. A1-3a

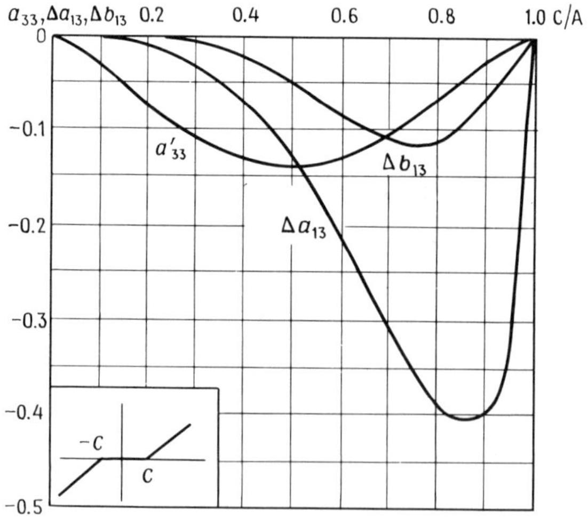

Fig. A1-3b

Appendix 2. Two-Valued Nonlinearities: Basic and Additional Coefficients of Harmonic Linearization

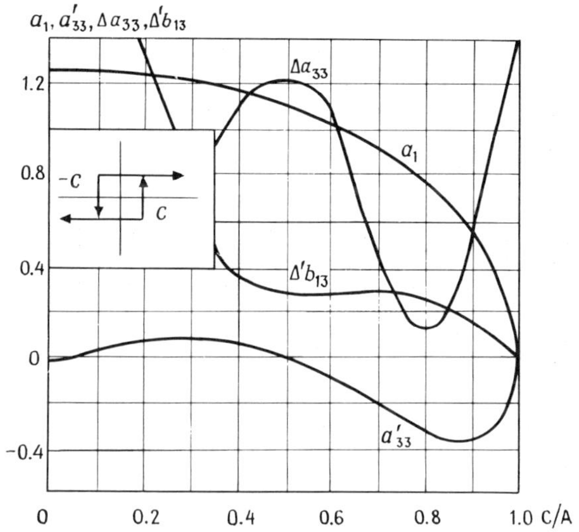

Fig. A2-1a

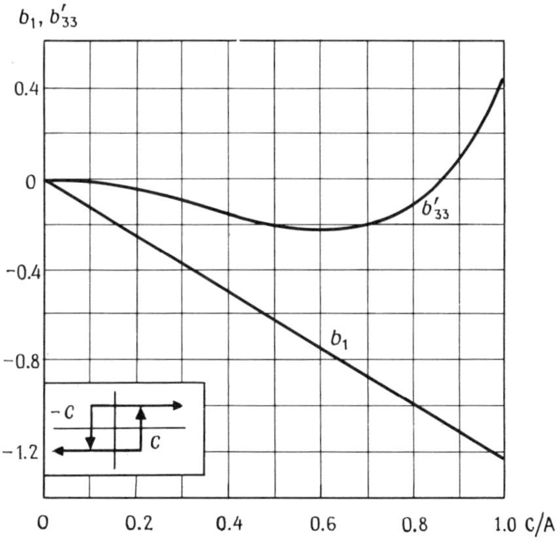

Fig. A2-1b

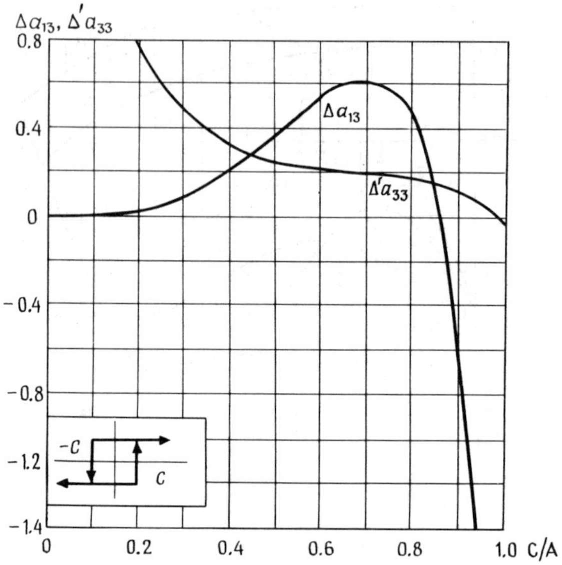

Fig. A2-1c

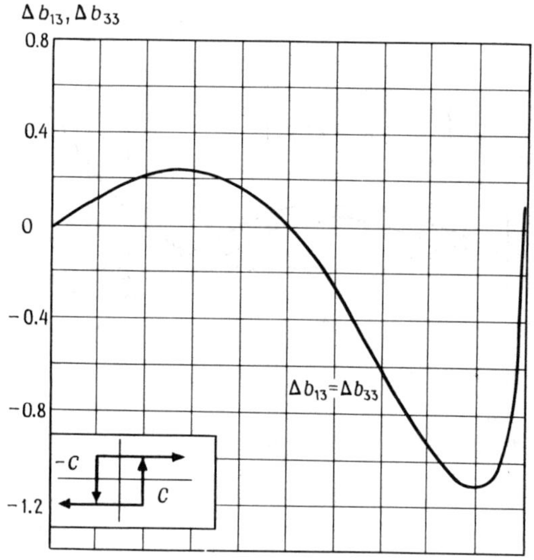

Fig. A2-1d

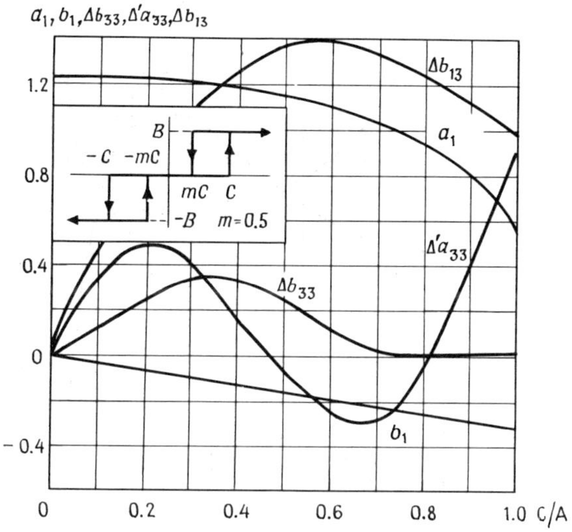

Fig. A2-2a

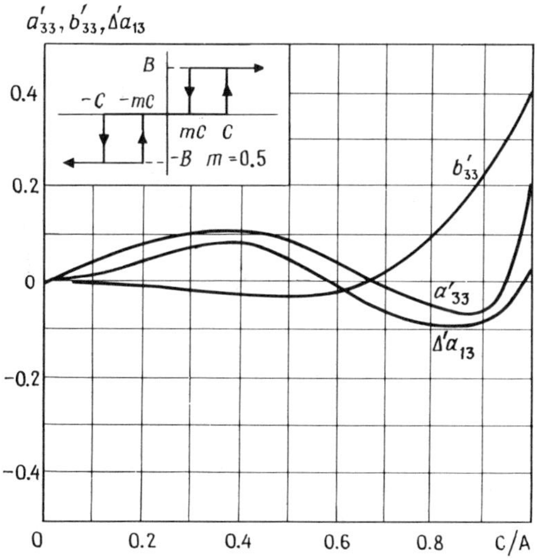

Fig. A2-2b

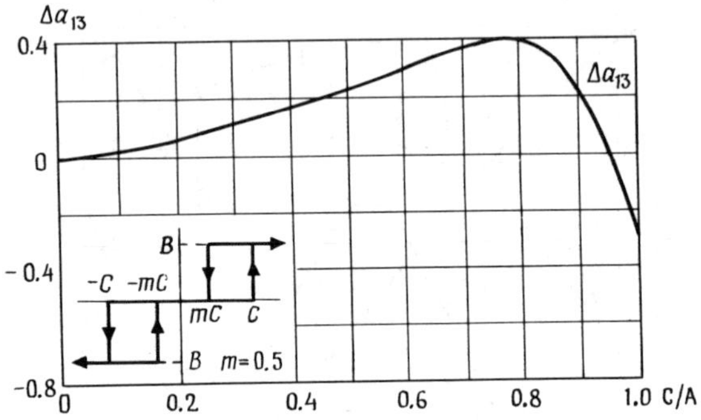

Fig. A2-2c

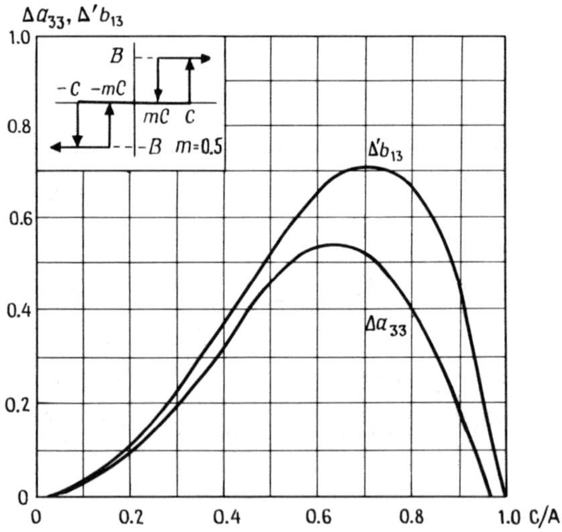

Fig. A2-2d

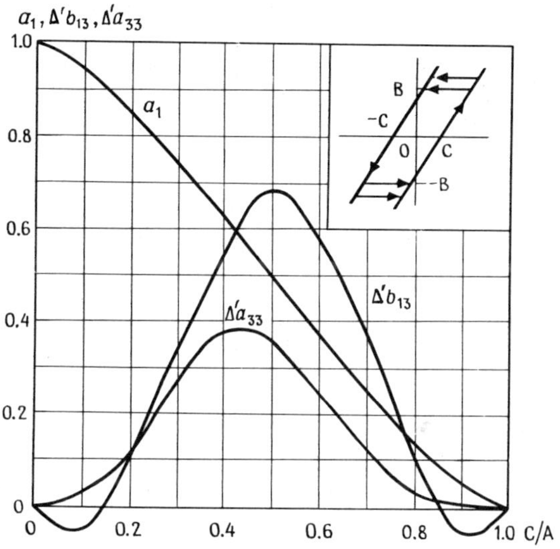

Fig. A2-3a

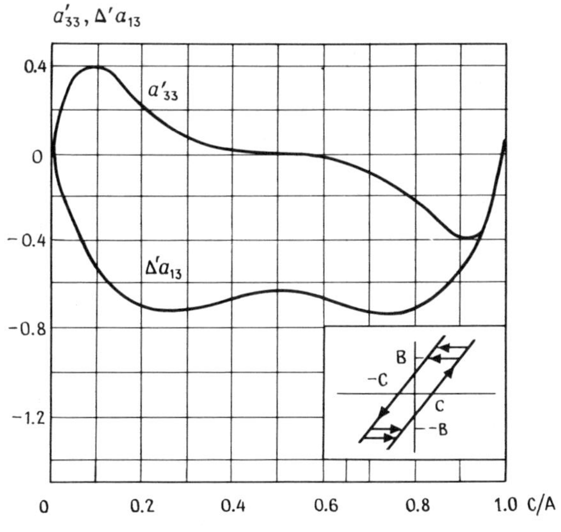

Fig. A2-3b

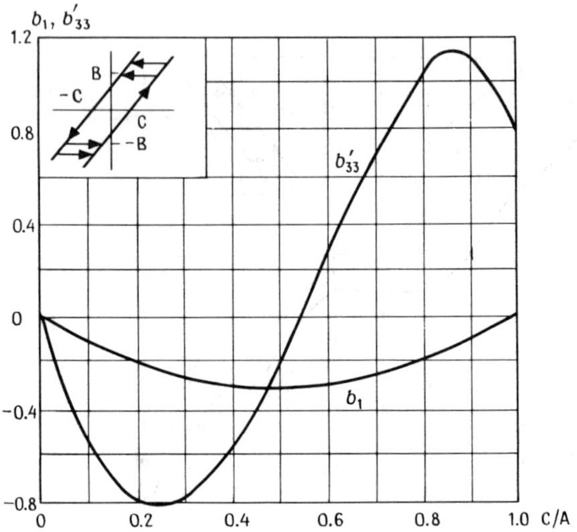

Fig. A2-3c

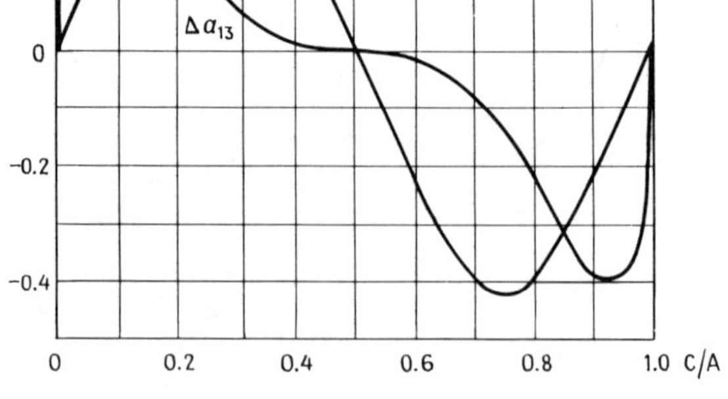

Fig. A2-3d

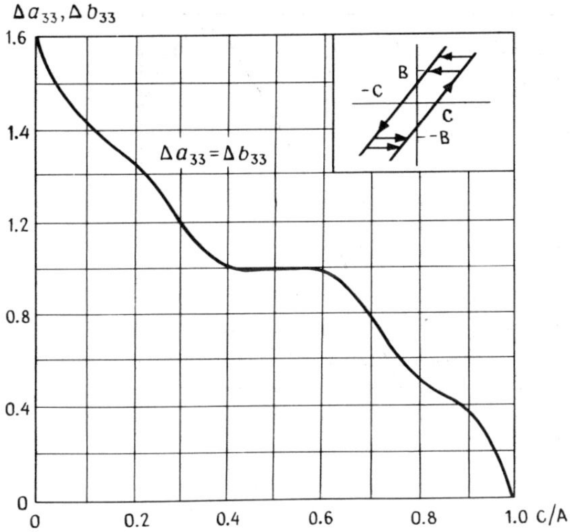

Fig. A2-3e

5 Synthesis of Piecewise-Linear Compensating Elements

B. F. Shumilov and Yu. Yu. Shumilov

For a considerable class of nonlinear time-varying control systems, design poses grave difficulties due to self-excited oscillations. These oscillations may be suppressed only with the aid of nonlinear compensating elements of piecewise-linear type. Such facilities consist of a multitude of standard linear components. Consistent operation of these components is achieved as a function of the second derivative of the input, therefore the input into the nonlinear compensating facility and its first derivative are continuous functions. This approach enables the synthesis of compensating units for which the principle of superposition approximately holds for some types of input.

Synthesis of piecewise-linear elements is of particular significance for time-varying systems that can thereby be brought closer to the time-invariant systems. These elements help eliminate unstable oscillations, thus ensuring stable behavior and dynamics of the closed-loop system. Elements designed in this way are structurally simple and need no retuning of the coefficients.

This chapter discusses methods of reduction of nonlinear elements to a form more convenient for design. It covers methods of analysis of piecewise-linear control laws under deterministic and random external stimuli. Consideration of closed-loop stability problems results in piecewise-linear elements possessing the specified pseudolinear properties.

5.1 Analytical Description and Transformation of Piecewise-Linear Functions

A piecewise-linear function of one variable [5.1]

$$F(x) = a_0 + a_1 x + \sum_{j=1}^{N} [b_j |x - x_j| + c_j \operatorname{sgn}(x - x_j)] \tag{5.1}$$

where

$$a_1 = (k_0 + k_n)/2 \tag{5.2}$$
$$b_j = (k_j - k_{j-1})/2 \tag{5.3}$$
$$c_j = [F^+(x_j) - F^-(x_j)]/2 \tag{5.4}$$

$$a_0 = F(0) - \sum_{j=1}^{n} [b_j |x_j| - c_j \operatorname{sgn}(x_j)] \tag{5.5}$$

The family of functions $y = F(x, p)$, where p is a parameter, may be represented in a similar manner.

Because each piecewise-linear function $F(x, p_i)$ with a fixed p_i has n nodes of the variable $x_j(p_i)$, $i, j = 1, \ldots, N$, it may also be represented in the form (5.1). Then the entire family of functions can be described as

$$F(x, p) = a_0(p) + a_1(p)x + \sum_{j=1}^{n} \{b_j(p)|x - x_j(p)| + c_j(p)\operatorname{sgn}[x - x_j(p)]\} \tag{5.6}$$

where

$$a_1(p) = [k_0(p) + k_n(p)]/2 \tag{5.7}$$
$$b_j(p) = [k_j(p) - k_{j-1}(p)]/2 \tag{5.8}$$
$$c_j(p) = [F^+(x_j(p), p) - F^-(x_j(p), p)]/2 \tag{5.9}$$

$$a_0(p) = F(0, p) = \sum_{j=1}^{n} \{b_j(p)|x_j(p)| - c_j(p)\operatorname{sgn}[x_j(p)]\} \tag{5.10}$$

For $p \neq p_i$ such that $p_{i-1} < p < p_i$, (5.6) yields a piecewise-linear function confined between $y = F(x, p_{i-1})$ and $y = F(x, p_i)$. These findings may be readily extended to the case of piecewise-linear representations of discontinuous n-dimensional single-valued functions.

Let an n-dimensional function be specified on a parallelepiped of $N_1 \cdot N_2 \ldots N_n$ points. Then it can be written in the form

$$F(x_1, \ldots, x_n) = \sum_{k_1=1}^{3N_1-4} \cdots \sum_{k_n=1}^{3N_n-4} \alpha(k_1, \ldots, k_n) \prod_{j=1}^{n} \varphi_{k_j}(x_j) \tag{5.11}$$

where

$$\varphi_1(x_j) = 1, \quad \varphi_2(x_j) = x_j, \quad \varphi_3(x_j) = |x_j - x_{j1}|, \ \ldots, \ \varphi_{N_j}(x_j) = |x_j - x_{jN_j-2}|,$$
$$\varphi_{N_j+1}(x_j) = \operatorname{sgn}(x_j - x_{j1}), \ \ldots, \ \varphi_{2N_j-2}(x_j) = \operatorname{sgn}(x_j - x_{jN_j-2}),$$
$$\varphi_{2N_j-1}(x_j) = \delta(x_j - x_{jN_j-2})$$

In contrast to (5.1), here each section is augmented by a δ-function with the coefficient $F(x_j) - F^-(x_j)$. The values $\alpha(k_1, \ldots, k_n)$ are the constant coefficients in the descriptions of unidimensional cross sections of the n-dimensional function; if the summation index k_i varies in inverse proportion to its ordinal number i, then $\alpha(k_1, \ldots, k_n)$ takes on in turn the values of these coefficients in the order they emerge in description. These expressions are applicable in describing and transforming any single-valued and multivalued nonlinear functions of arbitrary number of parameters.

In nonlinear system analysis we shall use the concept of nonlinear element rotation. If the original nonlinearity has the form given by (5.1), then upon rotation through an angle $\Delta\alpha$ it becomes

$$F'(x') = a_0' + a_1'x' + \sum_{j=1}^{k} b_j'|x' - x_j'| \tag{5.12}$$

where

$$\begin{aligned}
a_1' &= (k_0' + k_k')/2 \\
b_j' &= (k_j' - k_{j-1}')/2, \quad j = 1, \ldots, k \\
a_0' &= F_0'(0) - \sum_{j=1}^{k} b_j'|x_j'| \\
k_j' &= \tan(\alpha_j + \Delta\alpha), \quad j = 0, 1 \\
x_j' &= x_j \cos \Delta\alpha - \left[F(0) + k_{j-1}x_j - 2\sum_{i=1}^{j-1} b_j|x_i|\right]\sin \Delta\alpha
\end{aligned}$$

The transformations between the systems of coordinates XOY and $X'OY'$ are given by

$$\begin{aligned}
x' &= x \cos \alpha - y \sin \alpha & \quad x &= x' \cos \alpha + y' \sin \alpha \\
y' &= x \sin \alpha + y \cos \alpha & \quad y &= -x' \sin \alpha + y' \cos \alpha
\end{aligned} \tag{5.13}$$

whence

$$y' = (x' \sin \alpha + y)/\cos \alpha$$

With reference to relations (5.12) and (5.13) we represent the total rotation of a nonlinearity as the block diagram in Fig. 5.1 at (a) and the transformed diagram of rotation at (b). Similar transformations can be carried

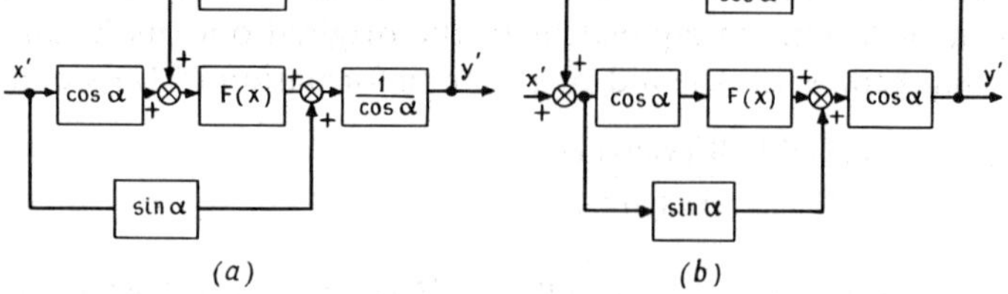

Fig. 5.1

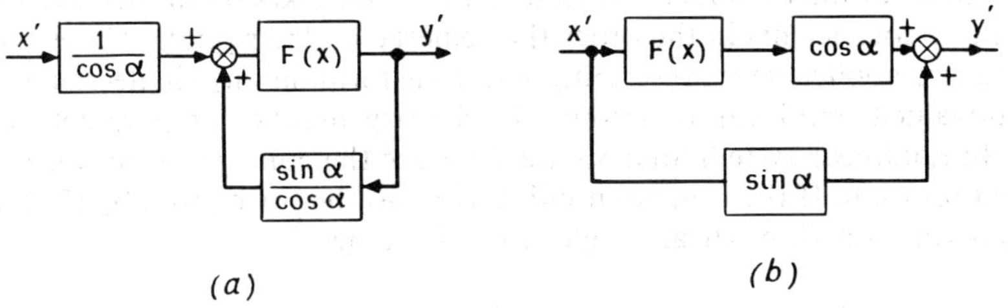

Fig. 5.2

out to reflect variation of one of the coordinates, x or y. The original nonlinearity $y = F(x)$ can be transformed to $F_1'(x')$ or $y' = F_2'(x)$. Figure 5.2 shows block diagrams corresponding to partial rotations in x at (*a*) and in y at (*b*).

A rotation of a nonlinearity may be carried out only under certain conditions. A sufficient conditions for rotation feasibility is that both $y = F(x)$ and $y' = F(x')$ be single valued. For block diagrams shown in Fig. 5.3 this condition guarantees rotation of a single-valued function with ordinary discontinuities. The total rotation of nonlinearity is shown at (*a*); a partial rotation in y and in x is shown in the diagram at (*b*) and (*c*); and the total rotation through $\alpha = 90°$ at (*d*). However, a reverse rotation based on the considered schemes is impossible.

We shall assume that the transformations displayed in Fig. 5.3 are feasible if the rotation is reversible, that is, a rotation of the transformed nonlinearity through the angle $-\alpha$ yields the original nonlinearity $y = F(x)$. This is a sufficient condition for equivalency of the block diagrams shown in Figs. 5.3 and 5.4.

Consider a system consisting of a nonlinear element $F(x)$ and a linear element desribed by the rational function $W(s) = R(s)/Q(s)$, where

$$R(s) = \sum_{i=0}^{m} r_i s^i \quad \text{and} \quad Q(s) = \sum_{j=0}^{n} q_j s^j$$

A rotation of $F(x)$ changes the initial structure of the scheme. In order for the system to remain equivalent to the original one, its linear element must be changed. With the aid of the transformation (5.12) we may write

$$W'(s) = \frac{\sin\alpha + W(s)\cos\alpha}{\cos\alpha - W(s)\sin\alpha}$$

$$= \frac{(q_n s^n + \ldots + q_0)\sin\alpha + (r_m s^m + \ldots + r_0)\cos\alpha}{(q_n s^n + \ldots + q_0)\cos\alpha - (r_m s^m + \ldots + r_0)\sin\alpha} \tag{5.14}$$

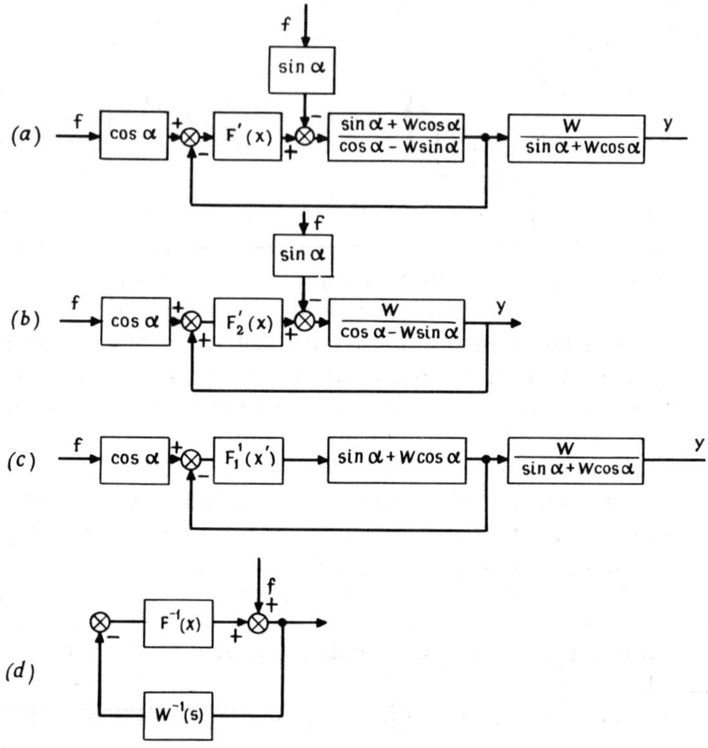

Fig. 5.3

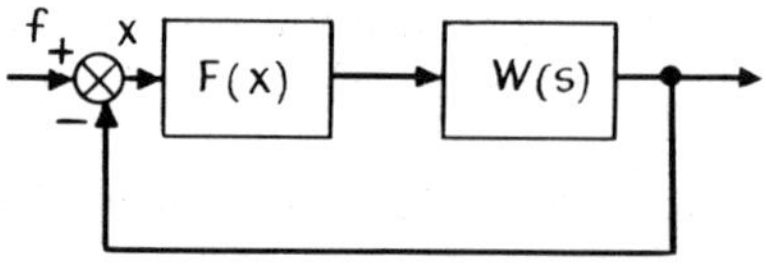

Fig. 5.4

The equivalent system is shown in Fig. 5.3*a*. From (5.14) it follows that the total rotation through an angle α leads to alteration of both zeros and poles of the original transfer function, whereas a partial rotation produces changes of either poles or zeros of $W(s)$ (see Fig. 5.3*b*, *c*).

Example 5.1. Consider a particular case of rotation through $\alpha = 90°$ illustrated in Fig. 5.3*d*.

Let a system with a limiting nonlinear element (Fig. 5.4) be of order $n = m$. Transform its transfer function

$$W(s) = \left(\sum_{i=0}^{m} r_i s^i\right) \Big/ \left(\sum_{j=0}^{n} q_j s^j\right) = r_n/q_n + \left(\sum_{i=0}^{n-1} r_i' s^i\right) \Big/ \left(\sum_{j=0}^{n-1} q_j s^j\right) \quad (5.15)$$

where $r_i' = r_i - (r_n/q_n)q_i$.

The limiting function, to which feedback is applied, is a bang-bang function if $r_n > 0$ and $\tan\alpha \geqslant 1$ ($\alpha = -\arctan(r_n/q_n)$). We reduce the original nonlinear system to a simple-relay system to simplify its analysis.

Rotational transformation is a convenient tool in analysis of nonlinear systems by the method of harmonic linearization. In some situations this method may be in error due to instability of the linear element, or inappropriate pole-zero configuration. The domain of application may be extended by carrying out rotations with the transformations represented by the block diagrams in Fig. 5.3.

Assume that the polynomials $R(s)$ and $Q(s)$ are of the same degree, $n = m$. Then from (5.14) it follows that

$$q_n \sin\alpha + r_n \cos\alpha = 0$$

from which $\alpha = -\arctan(r_n/q_n)$.

Upon a rotation through this angle the degree of the polynomial $R(s)$ decreases by unity and the method of harmonic linearization becomes applicable. In analyzing piecewise-linear compensating elements, certain rules of differentiation and integration need be stated for the piecewise-linear functions. For a function $F(x)$ given by (5.1) these rules are as follows:

$$\left[a_0 + a_1 x + \sum_{j=1}^{n} b_j |x - x_j|\right]_x' = a_1 + \sum_{j=1}^{n} b_j \operatorname{sgn}(x - x_j)$$

and

$$\int (a_0 + a_1 x + \sum_{j=1}^{n} b_j |x - x_j| \, dx = a_0 x + \frac{1}{2} a_1 x^2 + \frac{1}{2} \sum_{j=1}^{n} b_j |x - x_j| (x - x_j) + C$$

$$\int |x - x_j| x \, dx = \frac{1}{3}(x - x_j/2)|x - x_j|(x - x_j) + C$$

Integrals of the type

$$\int |x - x_j| x^i dx$$

for positive integer i are calculated by parts.

5.2 An Analytical Rule for Nonlinearity Compensation

Time-varying control systems can contain nonlinearities that impair system dynamics. One of the methods used to improve the behavior of such systems consists in compensating these nonlinearities. This problem is solved by finding the inverse nonlinearity function F^{-1} [5.1, 2]. For the class of functions under consideration, explicit piecewise-linear description of compensating nonlinearities may be derived.

If (5.1) is a continuous, strictly increasing, piecewise-linear function, then its inversion takes the form

$$x = F^{-1}(y) = \hat{a}_0 + \hat{a}_1 y + \sum_{j=1}^{n} \hat{b}_j |y - y_j| \tag{5.16}$$

where

$$\hat{a}_1 = a_1 \Big/ \left[a_1^2 - \left(\sum_{j=1}^{n} b_j\right)^2\right] \tag{5.17}$$

$$\hat{b}_k = -b_k \Big/ \left[\left(a_1 + \sum_{j=1}^{k-1} b_j - \sum_{j=k+1}^{n} b_j\right)^2 - b_k^2\right] \tag{5.18}$$

$$\hat{a}_0 = F_0^{-1} - \sum_{j=1}^{n} \hat{b}_j |F(x_j)|, \quad k = 1, \ldots, n \tag{5.19}$$

Let k_k and $\hat{k}_k$ be respectively the angle coefficients of F and F^{-1} in the kth interval, $k = 0, 1, \ldots, n$. Then

$$\hat{k}_k = 1/k_k = \left(a_1 + \sum_{j=1}^{k} b_j - \sum_{j=k+1}^{n} b_j\right)^{-1} \tag{5.20}$$

$$\hat{a}_1 = \frac{1}{2}\left(\frac{1}{k_0} + \frac{1}{k_n}\right) = a_1 \left[a_1^2 - \left(\sum_{j=1}^{n} b_j\right)^2\right]^{-1} \tag{5.21}$$

$$\begin{aligned} \hat{b}_k &= \frac{1}{2}(k_k^{-1} - k_{k-1}^{-1}) = \frac{1}{2}\left[\left(a_1 + \sum_{j=1}^{k} b_j - \sum_{j=k+1}^{n} b_j\right)^{-1} \right. \\ &\quad \left. - \left(a_1 + \sum_{j=1}^{k-1} b_j - \sum_{j=k}^{n} b_j\right)^{-1}\right] \\ &= -b_k \Big/ \left[\left(a_1 + \sum_{j=1}^{k-1} b_j - \sum_{j=k+1}^{n} b_j\right)^2 - b_k^2\right] \end{aligned} \tag{5.22}$$

from which it follows that

$$\hat{a} = F^{-1}(0) - \sum_{j=1}^{n} \hat{b}_j |F(x_j)|$$

This argument may be extended to strictly monotonous functions with jump discontinuities. In this case the points of discontinuity are joined by vertical lines so that $F(x)$ becomes a multivalued function. However, the inverse $F^{-1}(x)$ is a correctly defined single-valued function, since the inversion of a vertical segment is a horizontal segment.

Assume that $F(x)$ has a single discontinuity x_k where $y_k^- = F(x_k^-)$ and $y_k^+ = F(x_k^+)$, i.e.,

$$y = F(x) = a_0 + a_1 x + \sum_{j=1}^{n} b_j |x - x_j| + c_k \operatorname{sgn}(x - x_k) \quad (5.23)$$

while the inverse function is given by

$$x = F^{-1}(y) = \tilde{a}_0 + \tilde{a}_1 y + \sum_{j=1,\ j \neq k}^{n} \tilde{b}_j |y - y_j| + \tilde{b}_k^- |y - y_k^-| + b_k^+ |y - y_k^-| \quad (5.24)$$

The coefficients $\tilde{a}_1$ and $\tilde{b}_{j \neq k}$ are also calculated by equations (5.17, 18), but the remaining coefficients are determined by

$$\tilde{b}_k^- = -1/2 \left[a_1 + \sum_{j=1}^{k-1} b_j - \sum_{j=k}^{n} b_j \right] \quad (5.25)$$

$$\tilde{b}_k^+ = 1/2 \left[a_1 + \sum_{j=1}^{k} b_j - \sum_{j=k+1}^{n} b_j \right] \quad (5.26)$$

$$\tilde{a}_0 = F^{-1}(0) - \left[\sum_{j=1,\ j \neq k}^{n} \tilde{b}_j |y_j| + \tilde{b}_k^- |y_k| + \tilde{b}_k^+ |y_k| \right] \quad (5.27)$$

Note in passing that the original function is the inverse of the function describing the compensating nonlinearity. Therefore, a discontinuous function may be not only compensated but also compensating.

Expressions (5.1, 16, 24) may be used for a function with jump discontinuities, all other classes of nonlinearities call for analytical description of the implicit functions in solving compensation problems.

To provide a rigorous proof of this method we derive the formulas (5.25, 26). Equation (5.16), unlike (5.24), has one more node point owing to the horizontal segment with $\tilde{k}_k = 0$. Let $\tilde{k}_{k-1} = 1/k_{k-1}$ and $\tilde{k}_{k+1} = 1/k_k$, then

$$\tilde{b}_k = \tilde{b}_k^- = \frac{1}{2}(\tilde{k}_k - \tilde{k}_{k-1}) = \frac{1}{2}\left[0 - \left(a_1 + \sum_{j=1}^{k-1} b_j - \sum_{j=k}^{n} b_j \right)^{-1} \right]$$

$$= 1/2 \left(a_1 + \sum_{j=1}^{k} b_j - \sum_{j=k+1}^{n} b_j \right)$$

likewise

$$\tilde{b}_{k+1} = \tilde{b}_k^+ = \frac{1}{2}(\tilde{k}_{k+1} + \tilde{k}_k) = 1/2\left(a_1 + \sum_{j=1}^{k} b_j - \sum_{j=k+1}^{n} b_j\right)$$

and from the earlier relations it follows that

$$\begin{aligned}\tilde{a}_0 &= F^{-1}(0) - \sum_{j=1}^{n+1} \tilde{b}_j |y_j| \\ &= F^{-1}(0) - \tilde{b}_1 |y_1| - \ldots - \tilde{b}_{k-1} |y_{k-1}| - \tilde{b}_k^- |y_k^-| \\ &\quad - \tilde{b}_k^+ |y_k^+| - \tilde{b}_{k+1} |y_{k+1}| - \ldots - \tilde{b}_n |y_n| \\ &= F^{-1}(0) - \sum_{j=1,\ j\neq k}^{n} b_j |y_j| + \tilde{b}_k^- |y_k^-| + \tilde{b}_k^+ |y_k^+| \end{aligned} \tag{5.28}$$

Thus, relations (5.25-27) hold true.

Consider a control system with single-valued, nonmonotonous, piecewise-linear function specified analytically by (5.1). Then the inverse compensating nonlinearity cannot be obtained in an explicit form since this function becomes multivalued. By the formal substitution of x for y and y for x we get

$$-x + a_0 + a_1 y + \sum_{j=1}^{n} b_j |y - x_j| = 0 \tag{5.29}$$

that is, an implicit description of a compensating nonlinearity in the initial coordinates.

If no analytical description can be given to the nonlinearity (5.1), then it would be convenient to resort to graphical inversion of nonlinearities. For this purpose the original system of coordinates *XOY* should be rotated through 45° about the origin, namely,

$$y' = \frac{\sqrt{2}}{2}(-x + y) \tag{5.30}$$

$$x' = \frac{\sqrt{2}}{2}(x + y) \tag{5.31}$$

Then the reflection of the original function about the X' axis should be obtained in the $X'OY'$ system. This yields the desired plot for the compensating nonlinearity in the initial coordinate system.

Example 5.2. Consider an analytical verification of the graphical method. Assume that the nonlinearity has the form

$$y = F(x) = -\frac{1}{3}x + \frac{1}{3}|x + 3| - \frac{1}{3}|x - 3| \tag{5.32}$$

In the new system of coordinates by virtue of (5.30, 31) we have

$$\frac{\sqrt{2}}{2}(x' + y') = -\frac{\sqrt{2}}{6}(x' - y') + \frac{1}{3}\left|\frac{\sqrt{2}}{2}(x' - y') + 3\right| - \frac{1}{3}\left|\frac{\sqrt{2}}{2}(x' - y') - 3\right|$$

The reflection is carried out as

$$\frac{\sqrt{2}}{3}(-x + y) = -\frac{1}{3}\left|\frac{\sqrt{2}}{2}(-x' - y') - 3\right| + \frac{1}{3}\left|\frac{\sqrt{2}}{2}(-x' - y') + 3\right| \tag{5.33}$$

and followed by the transformation to the initial system of coordinates

$$\frac{1}{3}(-x + y) = \frac{2}{3}(x + y) + \frac{1}{3}(|y - 3| + |y + 3|)$$

After the transformations we have

$$x + \frac{1}{3}y - \frac{1}{3}|-y - 3| + \frac{1}{3}|-y + 3| = 0$$

or

$$x + \frac{1}{3}y - \frac{1}{3}|y + 3| + \frac{1}{3}|y - 3| = 0 \tag{5.34}$$

This expression is the implicit description of the compensating nonlinearity.

For multivalued closed nonlinearities, rotation of the coordinate system is unable to solve the problem of obtaining a compensating nonlinear element. In order that the aforementioned transformation may be applicable, we have to break the closed characteristic into two single-valued segments with respect to some feature. Then each of these segments may be represented as

$$F^{(1)} = a_0^{(1)} + a_1^{(1)}x + \sum_{j=1}^{n} b_j^{(1)}|x - x_j^{(1)}|$$

$$F^{(2)} = a_0^{(2)} + a_1^{(2)}x + \sum_{j=1}^{n} b_j^{(2)}|x - x_j^{(2)}| \tag{5.35}$$

If a feature for breakdown is taken to be the input variation rate, then the nonlinearity may be described in the general form as follows:

$$\mathscr{F} = \frac{1}{2}F^{(1)}(1 - \operatorname{sgn}\dot{x}) + \frac{1}{2}F^{(2)}(1 + \operatorname{sgn}\dot{x}) \tag{5.36}$$

Applying to each expression in (5.35) the rules of compensation and combining the inverse function similar to (5.36), we then arrive at a relation for the compensating nonlinearity.

Example 5.3. Assume that the nonlinearity involved has a characteristic with hysteresis. We partition it into two segments, as follows:

$$F^{(1)} = y = C - \frac{1}{2} k \left(\frac{C}{k} - B\right) x - \frac{1}{2} k \left| x - \frac{C}{k} B \right| \tag{5.37}$$

$$F^{(2)} = y = -C + \frac{1}{2} k \left(\frac{C}{k} - B\right) x + \frac{1}{2} k \left| x - \left(B - \frac{C}{k}\right) \right|$$

Taking the inverse of $F^{(1)}$ and $F^{(2)}$, we obtain curves with vertical segments which imply that the compensating nonlinearity contains characteristics of a simple relay. The final description of the compensating nonlinearity will be

$$\mathcal{F}^{-1} = \frac{1}{2} \left[B + \frac{1}{k} x - \frac{1}{2} B \operatorname{sgn}(x - C) \right] (1 + \operatorname{sgn} \dot{x})$$

$$+ \frac{1}{2} \left[-B + \frac{1}{k} x - \frac{1}{2} B \operatorname{sgn}(x + C) \right] (1 - \operatorname{sgn} \dot{x}) \tag{5.38}$$

A simplified representation of the function is also feasible, viz.,

$$\mathcal{F}^{-1} = x + B \operatorname{sgn} \dot{x}$$

It is important to note that such a compensation is effective for sinusoidal input.

5.3 Synthesis of Nonlinear Compensating Elements

Automatic control systems for time-varying plants involving nonlinear controllers exhibit self-excited oscillations and considerable domains of instability. Linear and nonlinear compensating elements were suggested to eliminate the said undesirable factors [5.3], either of these means, however, required that the parameters of the system should undergo substantial alteration in changing from one operating mode to another. Accordingly, the formulation of control laws was bound to be a complicated procedure. Clearly, compensating elements were needed that could ensure robust performance, that is, tolerated rather rough tuning and be free as far as possible from the aforementioned drawbacks.

We wish to synthesize compensating elements that would possess the properties of piecewise-linear operators. Referring to (5.1) and replacing x_j

by a dynamic linear element, we get

$$\Phi = a_0 + a_1 W_0(j\omega)u + \sum_{i=1}^{n} b_j | W(j\omega)u \mp | W_i(j\omega)u | | \quad (5.39)$$

where $c_i = 0$ and u is the input to the compensating element.

Owing to the simple calculation of the coefficients a_0, a_1, and b_i and the fact that they completely define the time-invariant piecewise-linear function, there is a possibility to specify the properties of compensating elements based on the properties of the time-invariant function. Therefore, the control law (5.39) takes care of switching the linear components on and off in the order established by the designer.

Consider a specific compensating elements for $W_i = qs^2$ and $a_0 = 0$, $a_1 = 1$, $b_1 = -1$, $b_2 = 1$. This facility is based on the following property of exponential functions of the form $u = A \exp(\omega_0 t)$, where $|u| = |\omega_0^{-2}\ddot{u}|$ and ω_0 is real or imaginary. The gist of the approach is that in practically any nonlinear time-varying system at early time instants each unstable mode may be represented as $A \exp(\lambda t) \sin(\omega t)$, where λ is a small positive number, to say nothing of self-excited oscillations. As a result we have

$$\Phi = u + \frac{1}{2}(|u - \omega_0^{-2}|\ddot{u}|| - |u + \omega_0^{-2}|\ddot{u}||) \quad (5.40)$$

By letting $u(t) = A\ \mathrm{sgn}(\omega t)$ we rewrite (5.40) as

$$\Phi = |u| \left\{ \mathrm{sgn}\, u + \frac{1}{2}[|\mathrm{sgn}\, u - (\omega/\omega_0)^2| - |\mathrm{sgn}\, u + (\omega/\omega_0)^2|] \right\}$$

Consider the quantity $\mathscr{L} = \Phi(u)/|u|$ in sequential half periods. In the first half period (Fig. 5.5*a*), with $\zeta = (\omega/\omega_0)^2$

$$\mathscr{L}_2 = -1 + \frac{1}{2}(|1 + \zeta| - |\zeta - 1|)$$

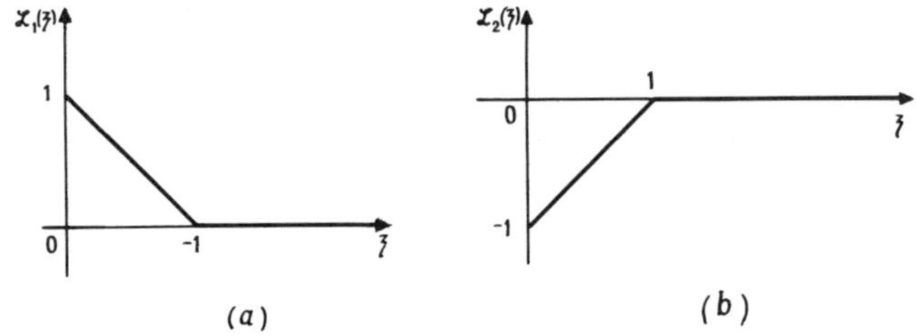

Fig. 5.5

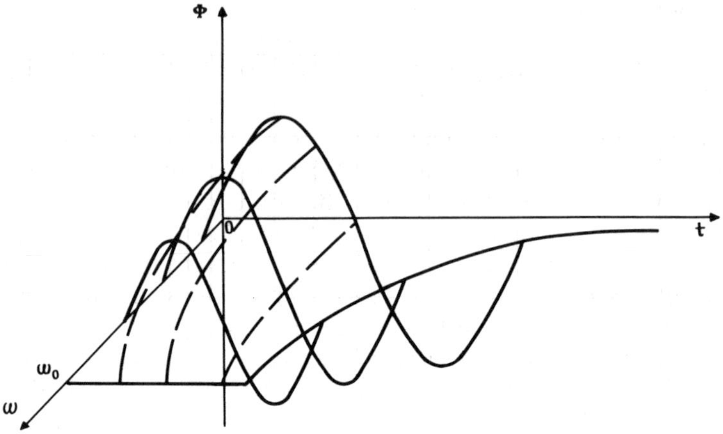

Fig. 5.6

and in the second half period (Fig. 5.5*b*)

$$\mathscr{L}_1 = 1 + \frac{1}{2}(|1 - \zeta| - |1 + \zeta|)$$

As can readily be seen, for $0 < \omega < \omega_0$ the compensating element operates as the link $1 - (\omega/\omega_0)^2$, and for $\omega \geqslant \omega_0$ a harmonic signal cutoff takes place. The three-dimensional profile of the output of the compensating element is shown in Fig. 5.6 as a function of frequency and time.

Now, we are in a position to construct the log-magnitude plot $20 \log|\mathscr{L}(\zeta(\omega))|$ of the compensating element as it is linear for harmonic signals (Fig. 5.7) and has no contribution to phase. Let us verify that there is no contribution to the phase angle in the case of a more general class of control.

Let the control be given in the form

$$\Phi = a_0 + a_1 u + \sum_{i=1}^{n} b_i |u \mp q_i |\ddot{u}|| \tag{5.41}$$

where $u = A \sin(\omega t + \varphi)$, the coefficients a_1 and b_i correspond to an odd-symmetric, piecewise-linear, time-invariant characteristic, and $a_0 \neq 0$. Then, the output is $\Phi = a_0 + k(\omega)A \sin(\omega t + \varphi)$, where $k(\omega)$ is the gain.

Rewrite (5.41) as

$$\Phi = a_0 + A|u|\left(a_1 \operatorname{sgn} u + \sum_{i=1}^{n} b_i |\operatorname{sgn} u \mp q_i \omega^2|\right)$$

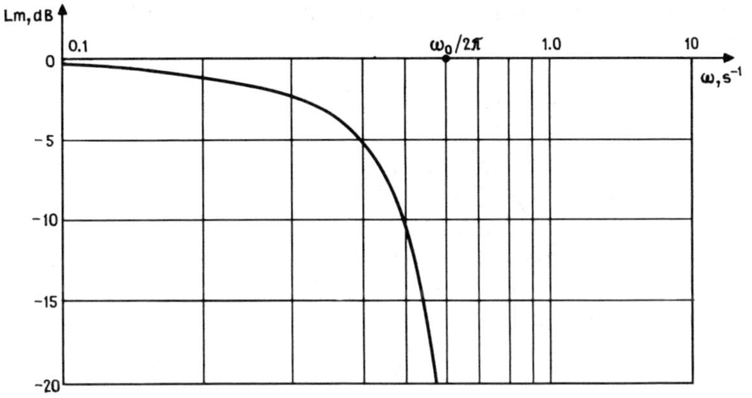

Fig. 5.7

Because the function $\Phi - a_0$ is odd-symmetric and $|u|$ is even-symmetric, then the function

$$(\Phi - a_0)/|u| = a_1 \operatorname{sgn} u + \sum_{i=1}^{n} b_i |\operatorname{sgn} u \mp q_i\omega^2|$$

will be odd-symmetric.

For $u \geqslant 0$ ($\operatorname{sgn} u = 1$) we may write

$$\frac{\Phi - a_0}{|u|} = \mathscr{L}_1 = a_1 + \sum_{i=1}^{n} b_i |1 \mp q_i\omega^2|$$

For $u < 0$ ($\operatorname{sgn} u = -1$) we have

$$\frac{\Phi - a_0}{|u|} = \mathscr{L}_2 = -a_1 + \sum_{i=1}^{n} b_i |-1 \mp q_i\omega^2|$$

Then by definition we get $\mathscr{L}_2 = -\mathscr{L}_1$

By multiplying $|u|$ by $\mathscr{L}_1$ for $u \geqslant 0$ and by $-\mathscr{L}_1$ for $u < 0$, we obtain

$$\Phi = a_0 + \left(a_1 + \sum_{i=1}^{n} b_i |1 \mp q_i\omega^2|\right) A \sin(\omega t + \varphi)$$

Since $\Phi - a_0$ is odd-symmetric,

$$k(\omega) = a_1 + \sum_{i=1}^{n} b_i |1 \mp q_i\omega^2| > 0$$

Consider the passage of deterministic signals through the compensating element.

$$(1)\ u = u_0 = \text{const};\ \Phi(u_0) = u_0 + \frac{1}{2}(|u_0| - |u_0|) = u_0.$$

This implies that the constant signal passes through the compensating element undistorted.

(2) $u = \exp(\lambda t)$. In view of $\mathrm{d}^2 \exp(\lambda t)/\mathrm{d}t^2 = \lambda^2 \exp(\lambda t)$ the compensating element handles the exponential function in much the same way as the sinusoidal signal, i.e. the dependence on λ^2/ω_0^2 is equivalent to $\mathscr{L}_1(\zeta)$. No signal passes the filter for $\lambda > \omega_0$.

(3) $u = A \exp(\lambda t) \sin \omega t$. The preferable relation between the signal parameters is as follows:

$$|1 - \exp(2\pi\lambda/\omega)| \ll A$$

In this case the signal will be treated as a sinusoidal one even at large λ.

(4) $u = u_0 + A \sin(\omega t)$ with $A \gg u_0$. If the compensating element is tuned to the error signal, then for zero static error no direct component is present in its input.

(5) $u = a + kt$; $\Phi(u(t)) = a + kt$. This implies that the property of linearity of the compensating element holds true.

$$(6)\ u = \sum_{k=1}^{l} a_k t^{k-1};$$

$$\sum_{k=1}^{l} a_k t^{k-1} - \omega_0^{-2} \left| \sum_{k=3}^{l} (k-1)(k-2) a_k t^{k-3} \right|$$

for

$$\left| \sum_{i=1}^{l} a_i t^{i-1} \right| > \omega_0^{-2} \left| \sum_{i=3}^{l} (i-1)(i-2) a_i t^{i-3} \right|$$

As follows from (5.40), $\Phi \to u$ as ω_0 tends to infinity, which corroborates the general properties of the compensating element represented by its frequency response.

The properties of the compensating element (5.40) listed above for particular types of signal are essentially a generalized property of the given control law expressed as

$$\Phi = 0 \quad \text{for} \quad |u| \leqslant \omega_0^{-2} |\ddot{u}| \tag{5.42}$$

$$\Phi = u - \omega_0^{-2} |\ddot{u}| \operatorname{sgn} u \quad \text{for} \quad |u| > 1/\omega_0^2 \tag{5.43}$$

where u is an arbitrary signal.

Particular interest is attracted to the robust performance of the compensating element, i.e., its ability to produce the response plotted in Fig. 5.7 to an arbitrary input. Assume that an odd signal is expanded in a Fourier series

$$u = \sum_{k=1}^{\infty} A_k \sin(k\omega t) \tag{5.44}$$

Let, from a certain k^*, the amplitudes A_k become small enough to neglect the terms with $k \geqslant k^*$. Then, at a sufficiently small ω_0 the condition (5.43) will be satisfied and we obtain

$$\Phi = \sum_{k=1}^{k^*} A_k \sin(k\omega t) - \omega_0^{-2} \left| \sum_{k=1}^{k^*} A_k k^2 \omega^2 \sin(k\omega t) \right| \operatorname{sgn} \left[\sum_{k=1}^{k^*} A_k \sin(k\omega t) \right] \tag{5.45}$$

For $k^2\omega^2 \approx 1$ this equation reduces to

$$\Phi \approx \sum_{k=1}^{k^*} A_k [1 - (k\omega/\omega_0)^2] \sin(k\omega t)$$

Thus, at appropriate k and ω the compensating element is close in its properties to the linear one having the log-magnitude plot shown in Fig. 5.7.

For a sufficiently large amplitude of high frequency the signal passage obeys (5.42) and no signal is passed through.

For a time-varying plant that gets unstable in some time intervals of closed-loop state, this compensating element is able to filter out the harmonic components that cause unstable motions and get the system stable. The stabilizing properties of this element are borne out by the absolute stability criterion.

Assume that the linearized closed-loop system with the nonlinear compensating element in the forward path, as shown in Fig. 5.8, has a stable linear element. We resort to the criterion of absolute stability that is valid for dynamic nonlinearities. These nonlinearities may include the compensating element that satisfies the following conditions of absolute stability:

(1) $\Phi(0) = 0$;
(2) $\Phi(\varepsilon)\varepsilon > 0$ for $\varepsilon \neq 0$, which follows from $\Phi(\varepsilon)$ being odd;

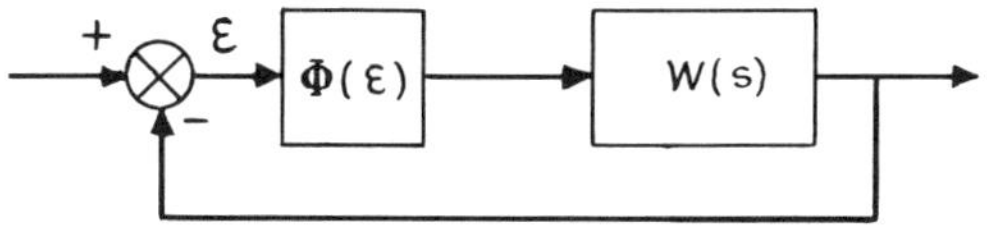

Fig. 5.8

(3) $\int_0^{\pm\infty} \Phi(\varepsilon)d\varepsilon = \pm\infty$;

(4) $0 < \Phi(\varepsilon)/\varepsilon < k$, this property is defined by the type of the piecewise-linear function; and

(5) $\mathrm{Re}\, W(j\omega) + 1/k > 0$.

Under these conditions the input is assumed bounded.

We wish to demonstrate that for any stable linear element $W(j\omega)$, and all the more for any system gain, there exists a coefficient $k(\omega_0)$ such that gets the system absolutely stable.

Because $\Phi(\varepsilon)$ is a piecewise-linear dynamic function there has to be a coefficient $k(\omega_0)$ bounding the required sector. Indeed, for $\omega_0^2 > |\ddot{\varepsilon}/\varepsilon|$, in view of the condition (5.43) we have

$$k(\omega_0) = 1 - \omega_0^{-2} |\ddot{\varepsilon}/\varepsilon|$$

This equation indicates that for one and the same domain of the quantity $|\ddot{\varepsilon}/\varepsilon|$ a decrease of ω_0 leads to an increase of the weighting coefficient ω_0^{-2} and, consequently, a decrease of $k(\omega_0)$ for an arbitrary large range of ε.

Thus, for any gain of the system a parameter ω_0 of the compensating element can be selected at which the system will be absolutely stable. The respective graphical criterion is shown in Fig. 5.9. For a linear system with a sinusoidal error the result will correspond to the frequency response of the compensating element. Conversion of the above criterion into Nyquist's for a harmonic input is shown in Fig. 5.10. This argument may be applied to demonstrate that for a bounded random input $\varepsilon = m_\varepsilon + \tilde{\varepsilon}$ a coefficient ω_0 can be selected such that the criterion of absolute stability in the mean will be satisfied subject to the condition

$$0 < \frac{\partial m_\Phi}{\partial m_\varepsilon} < k(\omega_0)$$

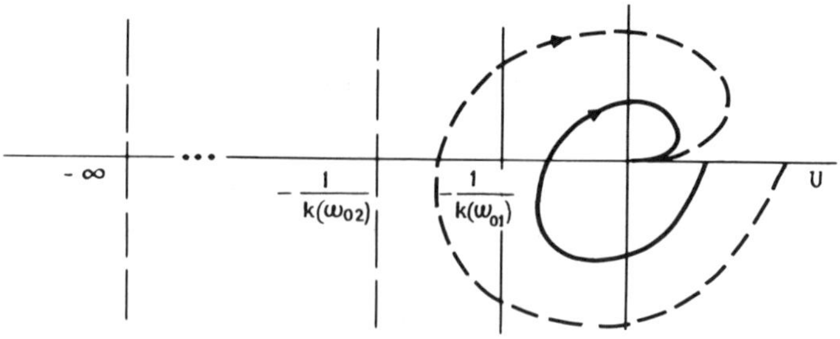

Fig. 5.9

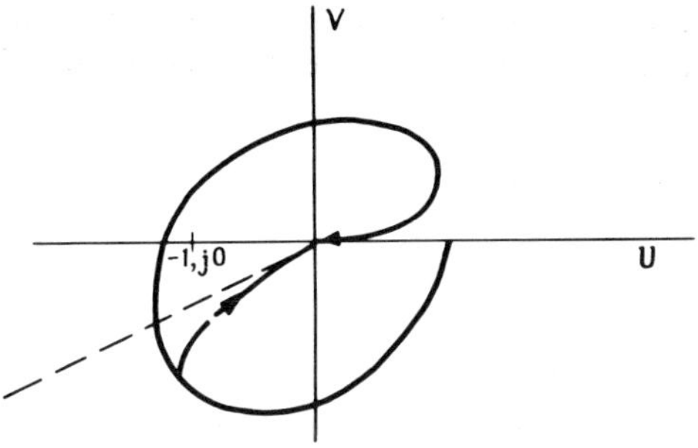

Fig. 5.10

and the criterion of absolute stability about a centered random signal will be satisfied subject to

$$0 < \frac{\partial \mathrm{E}[\Phi(\tilde{\varepsilon}(t))\tilde{\varepsilon}(t+\tau)]}{\partial \mathrm{R}_{\varepsilon}(\tau)} < k(\omega_0)$$

where R is the correlation function.

It is quite obvious that satisfying the absolute stability criterion for the linearized system ensures, for the nonlinear element, at least stability in the small. Modeling experiments for specific problems indicated that stability in the large held for the nonlinear element, too.

An implementation of the compensating facility under consideration is illustrated by the block diagram in Fig. 5.11. In this scheme $k = 1$ and s^2 have been replaced with the physically implementable components of $1/(Ts + 1)^2$ and $T^2s^2/(Ts + 1)^2$ which introduce nonzero phase distortions. An appropriate choice of the time constant T can reduce these distortions below arbitrarily low level.

Because investigation of absolute stability in the sense of mean square is rather difficult, we employ the method of statistical linearization for

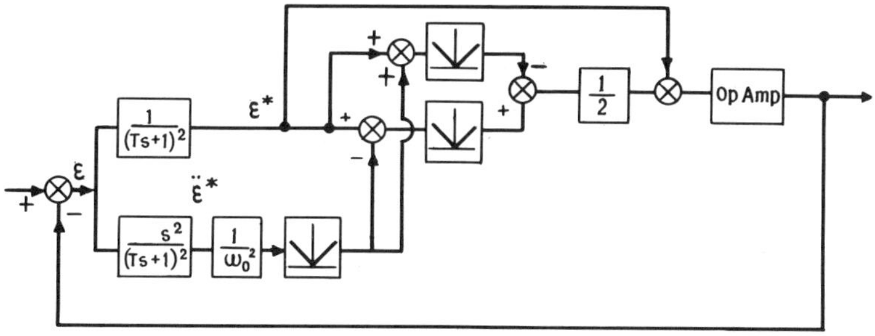

Fig. 5.11

analysis of the effect of random signals on the compensating element. Assume that the input is excited by a random normally distributed signal of the form

$$x(t) = m_x + \tilde{x}(t)$$

Then

$$\Phi = x_1 + \frac{1}{2}\{|x_1 - \omega_0^{-2}|x_2|| - |x_1 + \omega_0^{-2}|x_2||\} \tag{5.46}$$

where x_1 is the output of the element $1/(Ts + 1)^2$, and x_2 is the output of the element $s^2/(Tx + 1)^2$. The processes x_1 and x_2 are deemed noncorrelated. To ensure this property it is sufficient that the correlation function $R_{x_1x_2}(\tau)$ satisfies the condition [5.1]

$$R_{x_1x_2}(\tau)\bigg|_{\tau=0} = \frac{d^2}{d\tau^2} R_{x_1}(\tau)\bigg|_{\tau=0} = 0 \tag{5.47}$$

Clearly, there is a host of signals possessing such correlation functions.

By carrying out the satistical linearization of the nonlinear elements contained in the compensating facility one may demonstrate that the output process is as follows:

$$\Phi = x_1 - \omega_0^{-2} x_2 \tag{5.48}$$

If x_2 is a centered random signal and m_x is bounded, then as $t \to \infty$, we have $m_{x_1} \to m_x$, that is, the compensating facility passes the deterministic component undistorted. Then on the base of the cirterion of absolute stability in the mean we may conclude that for $k(\omega_0) > 1$ the closed-loop system will be absolutely stable and the required value of ω_0 always exists.

Let us look at how the signal variance varies on signal passage through the compensating unit. Let the correlation function of the signal x be approximated by the expression

$$R_x(\tau) = \mathrm{Var}_x \exp(-\mu|\tau|)$$

Then the spectral density of the signal is defined as

$$S_x(\omega) = 2\mu\, \mathrm{Var}_x/(\mu^2 + \omega^2)$$

where μ is the bound frequency of the spectrum (bandwidth), and the spectral densities of x_1 and x_2 are given by the expressions

$$S_{x_1}(\omega) = \frac{2\mu\, \mathrm{Var}_x}{(1 + T^2\omega^2)^2(\mu^2 + \omega^2)} \tag{5.49}$$

$$S_{x_2}(\omega) = \frac{2\omega^4\mu\, \mathrm{Var}_x}{(1 + T^2\omega^2)^2(\mu^2 + \omega^2)} \tag{5.50}$$

When the spectral densities are known, the variances of the input and output may be related by the expression

$$\mathrm{Var}_{\Phi} = \frac{3[1 + (\mu/\omega_0)^4]}{6 + 2\mu} \mathrm{Var}_x \tag{5.51}$$

It can be demonstrated that for $\omega_0 \geqslant \mu$, the compensating unit reduces the variance of a random signal passing through it. Thus, the system will be stable not only in the mean but also with respect to a centered random signal. For $\mu > \omega_0$ formula (5.51) is no longer useful since a more elaborate analysis is required. Modeling experiments indicated that the behavior of the compensating facility under random disturbances at $\mu > \omega_0$ well agrees with the response specification of this unit.

This compensating facility cannot be a cause of sliding modes because the elements $1/(Ts + 1)^2$ and $s^2/(Ts + 1)^2$ smooth any input signal and the nonlinear part contains continuous components only. In other words, a necessary condition for sliding modes to occur is absent, i.e., the right-hand sides of the differential equations have no discontinuities.

5.4 Harmonic Linearization of Piecewise-Linear Compensating Elements

Subject to feasibility of linear components, the piecewise-linear control law in the general case has the form

$$\Phi = a_0 + a_1 u + \sum_{i=1}^{n} b_i \,|\, W_{\mathrm{ph}}(j\omega)u - |W_{\mathrm{a}}(j\omega)u|\,|$$

where $W_{\mathrm{a}}(j\omega)$ and $W_{\mathrm{ph}}(j\omega)$ are the frequency response characteristics of the linear elements.

Consider a particular case of $n = 2$ and $a_0 = a_1 = 0$. For the input signal $x = A \sin(\omega t)$, the real (U) and imaginary (V) parts of the transfer function are [5.3]

$$U(\omega) = \frac{1}{\pi} |W_a(j\omega)| [2(\pi/2 + \theta_a - \theta_{\mathrm{ph}})\cos\theta_{\mathrm{a}} - \sin\theta_{\mathrm{a}} + \sin(2\theta_{\mathrm{ph}} - \theta_{\mathrm{a}})]$$

$$\tag{5.52}$$

$$V(\omega) = \frac{1}{\pi} |W_{\mathrm{a}}(j\omega)| [2(\pi/2 + \theta_{\mathrm{a}} - \theta_{\mathrm{ph}})\sin\theta_{\mathrm{a}} + \cos\theta_{\mathrm{a}} - \cos(2\theta_{\mathrm{ph}} - \theta_{\mathrm{a}})]$$

The total amplitude-ratio and phase response characteristics are

$$M(\omega) = \sqrt{U^2(\omega) + V^2(\omega)}$$
$$\mu(\omega) = \arctan[V(\omega)/U(\omega)]$$

Expressions (5.52) were obtained on the assumption that $W_a(s)$ is a phase-lag and $W_{ph}(s)$ is a phase-lead element. Otherwise,

$$U(\omega) = \frac{1}{\pi} \mid W_{ph}(j\omega) \mid [2(\pi/2 + \theta_a - \theta_{ph})\cos\theta_{ph} + \sin\theta_{ph} - \sin(2\theta_a - \theta_{ph})]$$

$$\tag{5.53}$$

$$V(\omega) = \frac{1}{\pi} \mid W_{ph}(j\omega) \mid [2(\pi/2 + \theta_a - \theta_{ph})\sin\theta_{ph} - \cos\theta_{ph} + \cos(2\theta_a - \theta_{ph})]$$

The coefficients of asymptotic linearization obtained for the input signal $x = A\exp(\lambda t)\sin(\omega t)$ differ from equations (5.52) solely by arguments, namely,

$$\begin{aligned} U(\lambda, \omega) = \frac{1}{\pi} \mid W_a(\lambda, \omega) \mid \{ & 2[\pi/2 + \theta_a(\lambda, \omega) - \theta_{ph}(\lambda, \omega)]\cos\theta_a(\lambda, \omega) \\ & - \sin\theta_a(\lambda, \omega) + \sin[2\theta_{ph}(\lambda, \omega) - \theta_a(\lambda, \omega)] \} \end{aligned} \tag{5.54}$$

$$\begin{aligned} V(\lambda, \omega) = \frac{1}{\pi} \mid W_a(\lambda, \omega) \mid \{ & 2[\pi/2 + \theta_a(\lambda, \omega) - \theta_{ph}(\lambda, \omega)]\sin\theta_a(\lambda, \omega) \\ & + \cos\theta_a(\lambda, \omega) - \cos[2\theta_{ph}(\lambda, \omega) - \theta_a(\lambda, \omega)] \} \end{aligned}$$

where $W_a(\lambda, \omega)$, $\theta_a(\lambda, \omega)$, and $\theta_{ph}(\lambda, \omega)$ are calculated by the respective expressions [5.3].

Basing on the expressions for the coefficients of harmonic and exponential linearization a conclusion may be drawn that in the presence of a phase shift

$$\theta_a(\omega) - \theta_{ph}(\omega) = k\pi, \quad k = 0, \pm 1, \pm 2, \ldots$$

the coefficients of the higher terms in the Fourier series of the output coordinate will be zero, and the output signal will be sinusoidal when the input contains only the fundamental component. When this condition is satisfied, there will be no effect of the signal x_0 at $x = x_0 + A\sin(\omega t)$ on the total amplitude and phase characteristics of the piecewise-linear compensating elements.

As demonstrated above, the compensating element (5.40) is linearized for the fundamental component by definition. However, if a few harmonics are present in the excitation, the additivity principle holds only approximately. Therefore, a more accurate analysis would require that higher harmonics be allowed for in calculating the coefficients of harmonic linearization. Since Φ is a piecewise-linear function of two variables u and $\ddot{u}$, it can be handled by the common procedure of calculation of harmonic linearization coefficients. Observing that $\ddot{u} = \omega^2 u$ and including ω in the amplitudes of the harmonics constituting the input signal u, we obtain a single-valued nonlinearity depending on parameter ω.

To make the subsequent computations neater we consider an improved method of harmonic linearization [5.6]. For a composite function of the form $\psi(t) = \Phi(\varphi(t))$ periodic in the time interval (t_1, t_n) with $t_n - t_1 = \tau$, we have

$$\psi(t) = \sum_{i=1}^{n-1} A_i[h(t - t_i) - h(t - t_{i+1})]\varphi_i(t), \quad i = 1, \ldots, n - 1$$

We assume that $\varphi_1(t) = \ldots = \varphi_{n-1}(t) = \varphi(t)$ are continuous over the period and possess the following property:

$$\sum_{i=1}^{n-1} [\psi(t_{i+1}) - \psi(t_i)] = \sum_{i=1}^{n} \delta_i \tag{5.55}$$

where

$$\begin{aligned} \delta_i &= \psi(t_i + 0) - \psi(t_i - 0) = A_i\varphi(t_i + 0) - A_{i-1}\varphi(t_i - 0) \\ &= (A_i - A_{i-1})\varphi(t_i), \quad t = t_i,\ i = 2, \ldots, n - 1 \end{aligned} \tag{5.56}$$

$$\delta_1 = \psi(t_1 + 0) = A_1\varphi(t_1 + 0) \tag{5.57}$$

$$\delta_n = -\psi(t_n - 0) = -A_{n-1}\varphi(t_n - 0) \tag{5.58}$$

To prove (5.55) we write

$$- \sum_{i=1}^{n} \delta_i = -\left\{\delta_1 + \delta_n + \sum_{i=2}^{n-1} \delta_i\right\} = -\left\{A_1\varphi(t_1 + 0) - A_{n-1}\varphi(t_n - 0)\right.$$

$$\left. + \sum_{i=2}^{n-1} A_i\varphi(t_i) - \sum_{i=2}^{n-1} A_{i-1}\varphi(t_i)\right\} = -\left\{A_1\varphi(t_1 + 0)\right.$$

$$\left. - A_{n-1}\varphi(t_n - 0) + \sum_{i=2}^{n-1} A_i\varphi(t_i) - \sum_{i=1}^{n-2} A_i\varphi(t_{i+1})\right\}$$

$$= -\left\{A_1\varphi(t_1 + 0) - A_{n-1}\varphi(t_n - 0) + A_{n-1}\varphi(t_{n-1}) - A_1\varphi(t_2)\right.$$

$$\left. + \sum_{i=2}^{n-2} A_i\varphi(t_i) - \sum_{i=2}^{n-2} A_i\varphi(t_{i+1})\right\}$$

whence

$$- \sum_{i=1}^{n} \delta_i = \sum_{i=1}^{n-1} [\psi(t_{i+1}) - \psi(t_i)]$$

From this expression we may draw two corollaries.

(i) If $\psi(t)$ is multiplied by an arbitrary function $\xi(t)$ continuous in the interval $(-\infty, \infty)$, then

$$\sum_{i=1}^{n-1} [\psi(t_{i+1})\xi(t_{i+1}) - \psi(t_i)\xi(t_i)] = -\sum_{i=1}^{n} \delta_i \xi(t_i)$$

The validity of this expression follows from the fact that

$$\begin{aligned} \tilde{\delta}_i &= \tilde{\psi}(t_i + 0) - \tilde{\psi}(t_i - 0) = (A_i - A_{i-1})\varphi(t_i)\xi(t_i) \\ \tilde{\delta}_1 &= A_1\varphi(t_i + 0)\xi(t_1) \\ \tilde{\delta}_n &= A_{n-1}\varphi(t_n - 0)\xi(t_n) \end{aligned}$$

that is,

$$\tilde{\delta}_i = \delta_i\xi(t_i), \ i = 1, \ldots, n$$

Returning at this point to the earlier line of the proof for (5.55), we assume that

$$\begin{aligned} \psi'(t) &= \sum_{i=1}^{n-1} \{A_i[h(t - t_i) - h(t - t_{i+1})]\}'\varphi(t) \\ &\quad + \sum_{i=1}^{n-1} A_i[h(t - t_i) - h(t - t_{i+1})]\varphi'(t) \\ &= \sum_{i=1}^{n-1} A_i[\delta(t - t_i) - \delta(t - t_{i+1})]\varphi(t) \\ &\quad + \sum_{i=1}^{n-1} A_i[h(t - t_i) - h(t - t_{i+1})]\varphi'(t) \end{aligned}$$

(ii) If $\psi(t) = \sum_{i=1}^{n-1} A_i[h(t - t_i) - h(t - t_{i+1})] \exp(-\alpha t)$, where α is a complex number, then

$$\psi'(t) = \sum_{i=1}^{n-1} A_i[\delta(t - t_i) - \delta(t - t_{i+1})]\varphi(t) - \alpha\varphi(t)$$

For

$$\xi(t) = \exp(-j2\pi t/\tau) = \exp(-j\tilde{\omega}t) \tag{5.59}$$

and

$$\psi(t_n - 0)\xi_n - \psi(t_1 + 0)\xi_1 = \sum_{i=1}^{n-1} [\psi(t_{i+1})\xi_{i+1} - \psi(t_i)\xi_i] \tag{5.60}$$

we consider the integral

$$I = \int_0^\tau \psi(t)\exp(-j\tilde{\omega}t)\mathrm{d}t = \int_0^\tau \psi(t)(-1/j\tilde{\omega})\mathrm{d}\exp(-j\tilde{\omega}t)$$

$$= -\frac{1}{j\tilde{\omega}}\exp(-j\tilde{\omega}t)\psi(t)|_0^\tau + \frac{1}{j\tilde{\omega}}\int_0^\tau \psi'(t)\exp(-j\tilde{\omega}t)\mathrm{d}t$$

$$= \frac{1}{j\tilde{\omega}}\sum_{i=1}^{n}\delta_i\exp(-j\tilde{\omega}t) + \sum_{i=1}^{n-1}A_i[\varphi(t_i)\exp(-j\tilde{\omega}t_i)$$

$$+ \varphi(t_{i+1})\exp(-j\tilde{\omega}t_{i+1})] - \frac{\alpha}{j\tilde{\omega}}\int_0^\tau \psi(t)\exp(-j\tilde{\omega}t)\mathrm{d}t$$

where $\tilde{\omega} = 2\pi n/\tau$.

Let

$$\varphi(t_i)\exp(-j\tilde{\omega}t_i) = \varphi(t_{i+k})\exp(-j\tilde{\omega}t_{i+k}), \quad i \leqslant k \leqslant n - i \tag{5.61}$$

then

$$\left(1 + \frac{\alpha}{j\tilde{\omega}}\right)\int_0^\tau \psi(t)\exp(-j\tilde{\omega}t)\mathrm{d}t = \frac{1}{j\tilde{\omega}}\sum_{i=1}^{n}\delta_i\exp(-j\tilde{\omega}t_i)$$

whence

$$\int_0^\tau \psi(t)\exp(-j\tilde{\omega}t)\mathrm{d}t = \frac{1}{\alpha + j\tilde{\omega}}\sum_{i=1}^{n}\delta_i\exp(-j\tilde{\omega}t_i) \tag{5.62}$$

The Fourier coefficient $C_n = (2/\tau)I$ may be represented as

$$C_n = \frac{2}{\tau(\alpha + j\tilde{\omega})}\sum_{i=1}^{n}\delta_i\exp(-j\tilde{\omega}t_i) \tag{5.63}$$

We shall refer to $\psi(t)$ satisfying (5.60-62) as a standard function. From (5.63) it follows that the denominator vanishes at $\alpha = -j\nu$ and $\nu = 2\pi/\tau$ ($\tilde{\omega} = 2\pi n/\tau$, $n = 1$). Therefore, the coefficient of the fundamental component is

$$C_1 = \frac{2}{\tau(j - \alpha\tau/2\pi)}\left\{\sum_{i=1}^{n}\delta_i\exp\left(-j\frac{2\pi}{\tau}t_i\right) + \frac{1}{2}\sum_{i=1}^{n-1}A_i(t_{i+1} - t_i)\right\} \tag{5.64}$$

If we let $2\pi t/\tau = \beta$, then

$$C_1 = \frac{1}{\pi(j - \alpha\tau/2\pi)}\left\{\sum_{i=1}^{n}\delta_i\exp(-j\beta_i) + \frac{1}{2}\sum_{i=1}^{n-1}A_i(\beta_{i+1} - \beta_i)\right\} \tag{5.65}$$

When the standard function is continuous and $\alpha = -j\nu$,

$$C_n = \frac{2}{j\tau(\tilde{\omega} - \nu)} \sum_{i=1}^{n} \delta_i \exp(-j\tilde{\omega}t_i) = 2\sum_{i=1}^{n} (\delta_i' + j\tilde{\omega}\delta_i)\exp(-j\tilde{\omega}t_i)/\tau(\nu^2 - \tilde{\omega}^2) \tag{5.66}$$

where δ_i' is the jump of the derivative at point t_i. In this case the fundamental component is calculated by (5.63), too.

Let us now look at the case when output of the nonlinearity is a sum of standard functions, viz.,

$$\psi(t) = \sum_{i=1}^{k} \psi_i(t)$$

then from the linearity of integral operators it follows that

$$C_{n\psi} = \sum_{i=1}^{n} C_{n\psi_i}$$

We wish to demonstrate the validity of this method for biased input $x(t) = x_0 + A\sin(\omega t)$. If for the unbiased input signal the output was partitioned in k standard functions, this number of functions remains for the input $x_0 + \exp(-\alpha t)$, namely,

$$\tilde{\psi}_j(t) = \sum_{i=1}^{n-1} A_i[h(t - \tilde{t}_i) - h(t - \tilde{t}_{i+1})]\exp\left(j\frac{2\pi}{\tau}t + x_0\right)$$

where $\tilde{t}_i$ is the root of the equation $A\exp(j2\pi t/\tau + x_0) = x_i$. However, the function $\tilde{\psi}_j$ is not standard since it does not meet condition (ii) ($\alpha_1 = -j2\pi/\tau$ and $\alpha_2 = 0$). Therefore, each of the functions $\tilde{\psi}_j$ should be broken down into two standard ones, namely,

$$\tilde{\psi}_j^{(1)}(t) = \sum_{i=1}^{n-1} A_i[h(t - \tilde{t}_i) - h(t - \tilde{t}_{i+1})]\exp(-\alpha_1 t)$$

$$\tilde{\psi}_j^{(2)}(t) = \sum_{i=1}^{n-1} x_0 A_i[h(t - \tilde{t}_i) - h(t - \tilde{t}_{i+1})]$$

If the input to the nonlinearity is a decaying or diverging signal $x(t) = A\exp(\lambda t)\sin(\omega t)$, then in agreement with (i) the function $\xi(t) = \exp(\lambda t)\exp(j\tilde{\omega}t) = \exp[(\lambda + j\tilde{\omega})t]$ does not influence its properties and

$$\hat{\psi}_j(t) = \sum_{i=1}^{n-1} A_i[h(t - \hat{t}_i) - h(t - \hat{t}_{i+1})]\exp[(\lambda - \alpha)t]$$

where $\hat{t}_i$ is the root of the equation $A\exp(\lambda t)\sin(\omega t) = x_i$.

Now, we wish to investigate how the calculation of the coefficients of higher harmonics can be simplified. Assume that the input of the odd-symmetric two-valued nonlinearity $\mathscr{F}(x, \dot{x})$ is excited by the signal

$$x = \sum_{k=1}^{n} A\delta_k \sin(k\omega t + \varphi_k)$$

with $\delta_k = A_k/A$, $A_1 = A$, and $\varphi_1 = 0$.

Denoting the first harmonic, refined by taking into account the effect of higher harmonics calculated on the basis of the action of nonrefined first harmonic, by $x^1 = A^1 \sin(\omega^1 t)$ and expanding $\mathscr{F}(x, \dot{x})$ in the Taylor series, we obtain the following coefficients [5.4]:

$$a(A) = \frac{1}{\pi A}\int_0^{2\pi}\Big\{\mathscr{F}(A \sin\psi) + \dot{\mathscr{F}}(A \sin\psi) \times \Big[\sum_{k=3}^{2n+1} A\delta_k \sin(k\psi + \varphi_k)\Big]\Big\}\sin\psi\, d\psi \tag{5.67}$$

$$b(A) = \frac{1}{\pi A}\int_0^{2\pi}\Big\{\mathscr{F}(A \sin\psi) + \dot{\mathscr{F}}(A \sin\psi) \times \Big[\sum_{k=3}^{2n+1} A\delta_k \sin(k\psi + \varphi_k)\Big]\Big\}\cos\psi\, d\psi \tag{5.68}$$

$$a_i(A) = \frac{1}{\pi A}\int_0^{2\pi}\Big\{\mathscr{F}(A \sin\psi) + \dot{\mathscr{F}}(A \sin\psi) \times \Big[\sum_{k=3}^{2n+1} A\delta_k \sin(k\psi + \varphi_k)\Big]\Big\}\sin(i\psi + \varphi_i) d\psi \tag{5.69}$$

$$b_i(A) = \frac{1}{\pi A}\int_0^{2\pi}\Big\{\mathscr{F}(A \sin\psi) + \dot{\mathscr{F}}(A \sin\psi) \times \Big[\sum_{k=3}^{2n+1} A\delta_k \sin(k\psi + \varphi_k)\Big]\Big\}\cos(i\psi + \varphi_i) d\varphi, \quad n = 1, 2, \ldots \tag{5.70}$$

Let us investigate the integrals (5.67) and (5.68). After transformation we get

$$a(A) = a_1(A) + \delta_i \cos(\varphi_i)\Delta a_{1i}(A) + \delta_i \sin(\varphi_i)\Delta a_{1i}(A) \qquad (5.71)$$

$$b(A) = b_1(A) + \delta_i \cos(\varphi_i)\Delta b_{1i}(A) - \delta_i \sin(\varphi_i)\Delta b_{1i}(A) \qquad (5.72)$$

where

$$a_1(A) = \frac{1}{\pi A}\int_0^{2\pi} \mathscr{F}(A \sin \psi) \sin \psi \, d\psi \qquad (5.73)$$

$$\Delta a_{1i}(A) = \frac{1}{\pi}\int_0^{2\pi} \dot{\mathscr{F}}(A \sin \psi)\sin(i\psi) \sin \psi \, d\psi \qquad (5.74)$$

$$\Delta' a_{1i}(A) = \frac{1}{\pi}\int_0^{2\pi} \dot{\mathscr{F}}(A \sin \psi)\cos(i\psi) \sin \psi \, d\psi \qquad (5.75)$$

$$b_1(A) = \frac{1}{\pi A}\int_0^{2\pi} \mathscr{F}(A \sin \psi) \cos \psi \, d\psi \qquad (5.76)$$

$$\Delta' b_{1i}(A) = \frac{1}{\pi}\int_0^{2\pi} \dot{\mathscr{F}}(A \sin \psi)\sin(i\psi) \cos \psi \, d\psi \qquad (5.77)$$

$$\Delta b_{1i}(A) = \frac{1}{\pi}\int_0^{2\pi} \dot{\mathscr{F}}(A \sin \psi)\cos(i\psi) \cos \psi \, d\psi \qquad (5.78)$$

The method to take the integrals (5.73-78) was indicated earlier.

We wish to demonstrate that this method is also applicable for the calculation of the integrals (5.69, 70). To simplify the mathematics involved we consider a single-valued nonlinearity $F(x)$. Then

$$\dot{F}(x) = a_1 + \sum_{j=1}^{l} b_j \operatorname{sgn}(x - x_j) + \sum_{j=1}^{l} c_j \delta(x - x_j) \qquad (5.79)$$

As a result we arrive at (5.74), say, in the form

$$\Delta a_{1i}(A) = I_1 + I_2 \qquad (5.80)$$

For

$$I_1 = \frac{1}{\pi A}\int_0^{2\pi} \left[a_1 + \sum_{k=1}^{l} b_k \operatorname{sgn}(A \sin \psi - A \sin \psi_j)\right] A \sin \psi \sin(i\psi) d\psi$$

with $x = A \sin \psi$ we have

$$\left[a_1 + \sum_{j=1}^{l} b_j \operatorname{sgn}(x - x_j)\right] x = a_1 x + \sum_{j=1}^{l} b_j |x - x_j| + \sum_{j=1}^{l} b_j x_j \operatorname{sgn}(x - x_j) \tag{5.81}$$

Now,

$$I_1 = I_{11} + I_{12} = \frac{1}{\pi A}\int_0^{2\pi}[a_1 A \sin\psi + \sum_{j=1}^{l} b_j |A \sin\psi - A \sin\psi_j|]\sin(i\psi)\mathrm{d}\psi$$

$$+ \frac{1}{\pi A}\sum_{j=1}^{l}\int_0^{2\pi}[b_j A \sin\psi_j \operatorname{sgn}(A\sin\psi - A\sin\psi_j)]\sin(i\psi)\mathrm{d}\psi$$

Here I_{11} is the readily derivable basic coefficient of harmonic linearization for the ith harmonic of the function $F(x) - \sum_{j=1}^{l} c_j \operatorname{sgn}(x - x_j)$. The integrand in I_{12} is the standard function $\sum_{j=1}^{l} b_j x_j \operatorname{sgn}(x - x_j)$ that, at $x = A \sin \psi$, is easily reducible to the form

$$f(\psi) = \sum_{j=1}^{l_A - 1} A_j[h(\psi - \psi_j) - h(\psi - \psi_{j+1})]\exp(-\alpha\psi) \tag{5.82}$$

where $\alpha = 0$; A_j is a linear combination of k_j, $j = 0, \ldots, l_A$; and A is the amplitude of the input signal. Hence, I_{12} can be computed on the lines of the method presented earlier, too.

Now we come to evaluating the second integral

$$I_2 = \frac{1}{\pi}\int_0^{2\pi}\sum_{j=1}^{l} c_j \delta(A\sin\psi - A\sin\psi_j)\sin\psi \sin(i\psi)\mathrm{d}\psi$$

$$= \frac{1}{\pi}\sum_{k=1}^{l_A} c_k \sin\psi_k \sin(i\psi_k) \tag{5.83}$$

For a single-valued function, the integral (5.75) can be calculated, too. It is taken again as a sum of two integrals. Consider the calculation of (5.69, 70). For this purpose we represent the sines and cosines of two arguments as sums of products and obtain the integrals

$$I_{21} = \frac{1}{\pi}\int_0^{2\pi} \dot{F}(A \sin \psi) \sin (k\psi) \sin (i\psi) d\psi \tag{5.84}$$

$$I_{22} = \frac{1}{\pi}\int_0^{2\pi} \dot{F}(A \sin \psi) \cos (k\psi) \sin (i\psi) d\psi \tag{5.85}$$

$$I_{23} = \frac{1}{\pi}\int_0^{2\pi} \dot{F}(A \sin \psi) \sin (k\psi) \cos (i\psi) d\psi \tag{5.86}$$

$$I_{24} = \frac{1}{\pi}\int_0^{2\pi} \dot{F}(A \sin \psi) \cos (k\psi) \cos (i\psi) d\psi \tag{5.87}$$

where $k = 3, 5, \ldots; i = 1, 3, \ldots$.

All these integrals (5.84-87) are contained in the expression

$$\tilde{I} = \frac{1}{\pi}\int_0^{2\pi} \dot{F}[A \exp(-\alpha\psi)) \exp(-j(k + i)]\psi \, d\psi \tag{5.88}$$

Representing $\dot{F}(x)$ as the sum of $\dot{F}_1(x) = a_1, \dot{F}_2(x) = \sum_{j=1}^{l} b_j \operatorname{sgn}(x - x_j)$, and $\dot{F}_3(x) = \sum_{j=1}^{l} c_j \delta(x - x_j)$ decomposes (5.88) into three integrals.

First integral

$$I_3 = \frac{a_1}{\pi}\int_0^{2\pi} \exp(-jk\psi) \exp(-ji\psi) d\psi \neq 0, \quad i = k$$

The integrals (5.85, 86) are always zero for $\dot{F}_1(x) = 0$. The integrals (5.84, 87) equal a_1.

Second integral

$$I_4 = \frac{1}{\pi}\int_0^{2\pi} \sum_{j=1}^{l_A - 1} A_j^* [h(\psi - \psi_j) - h(\psi - \psi_{j+1})] \exp(-\alpha\psi) \exp[-j(k + i)\psi] d\psi$$

By analogy with (5.82) $\alpha = 0$ and A_j^* is a new linear combination of k_j; from (5.63) we have

$$I_4 = \frac{1}{j\pi(j + k)} \sum_{r=1}^{l_A} \delta_r \exp(-jk\psi) \exp(-ji\psi) \tag{5.89}$$

When these exponential functions are disclosed by Euler's formula, the integral I_4 will consist of four addends each of which has its own formula for $\dot{F}_2(x)$ in the set (5.84-87).

Third integral

$$I_5 = \frac{1}{\pi} \sum_{r=1}^{l_A} c_r \exp(-jk\psi_r) \exp(-ji\psi_r)$$

Like in (5.83), I_5 consists of four addends corresponding to (5.84-87) this time for the function $\dot{F}_3(x)$.

Summing up I_3, I_4, and I_5 by the respective components we obtain a sum of four addends each of which has been calculated by the formulas (5.84-87).

To close this topic, we illustrate applications of this method to odd nonlinearities.

Example 5.4. Given an ideal relay with the switching function

$$C_1 = \frac{2}{j\pi A} \{B \exp(-j\psi_1) - B \exp[-j(\pi + \psi_1)]\}$$

$$= \frac{2}{j\pi A} \{B(\cos\psi_1 - j\sin\psi_1) - B \exp[\cos(\pi + \psi_1)] - j\sin(\pi + \psi_1)\}$$

$$= \frac{2B}{j\pi A} \left[\sqrt{1 - \left(\frac{C}{A}\right)^2} - j\frac{C}{A} + \sqrt{1 - \left(\frac{C}{A}\right)^2} - j\frac{C}{A}\right]$$

Here,

$$a_1 = \frac{4B}{\pi A} \sqrt{1 - \left(\frac{C}{A}\right)^2}, \quad b_1 = -\frac{4BC}{\pi A^2}$$

and

$$C_n = \frac{2}{j\pi A_n} \{B \exp(-j\psi_1 n) - B \exp[-jn(\pi + \psi_1)]\}$$

$$= \frac{2B}{j\pi A_n} \{\cos\psi_{1n} - j\sin(n\psi_1) - \cos[n(\psi_1 + \pi)] + j\sin n[(\pi + \psi_1)]\}$$

Then the n-th harmonic linearization coefficients are

$$a_n = (2B/\pi A_n)\{\cos\psi_{1n} - \cos[n(\pi + \psi_1)]\}$$
$$b_n = (2B/\pi A_n)\{-\sin(n\psi_1) + \sin[n(\pi + \psi_1)]\}$$

Example 5.5. For a three-position relay with the switching function $\psi_1 = \arcsin(C/A)$, $\psi_2 = \arcsin(mC/A)$, and

$$C_1 = \frac{2}{j\pi A}[B\exp(-j\psi_1) - B\exp(-j\psi_2)]$$
$$= \frac{2}{j\pi A}B[\cos\psi_1 - j\sin\psi_1 - \cos\psi_2 + j\sin\psi_2]$$

where

$$a_1 = \frac{2B}{\pi A}\left[\sqrt{1-\left(\frac{C}{A}\right)^2} - \sqrt{1-\left(\frac{mC}{A}\right)^2}\right]$$

$$b_1 = \frac{2}{\pi A^2}C(m-1)$$

Example 5.6. For an element with saturation we have $k = B/C$ and

$$C_{11} = \frac{1}{j\pi}\{[-k\sin\psi_1\exp(-j\psi_1) + k\sin(\pi-\psi_1)\exp[-j(\pi-\psi_1)]\}$$

$$= \frac{2k}{j\pi}(\psi_1 - \sin\psi_1\cos\psi_1)$$

$$C_{12} = \frac{2}{j\pi A}\{B\exp(-j\psi_1) + B\exp[-j(\pi-\psi_1)]\} = (4B/j\pi A)\cos\psi_1$$

$$C_\Sigma = \frac{2k}{j\pi}\left[\arcsin\frac{C}{A} + \frac{C}{A}\sqrt{1-\left(\frac{C}{A}\right)^2}\right]$$

$$C_{1n} = \frac{k}{2\pi(n+1)}\{-\exp[-j(\pi-\psi_1)(n-1)] + \exp[-j(\pi-\psi_1)(n+1)]$$
$$+ \exp[-j\psi_1(n-1)] - \exp[-j\psi_1(n+1)]\}$$

$$C_{2n} = \frac{2B}{j\pi A_n}\{\exp(-j\psi_{1n}) - \exp[-jn(\pi-\psi_1)]\}$$

$$C_{\Sigma n} = C_{1n} + C_{2n}$$

Applying this method in succession to the expressions within the internal and then external modulus bars in (5.40), we obtain the harmonic linearization of the proposed compensating element where the required number of higher harmonics is taken into account. This method is versatile, readily programmable, and yields the coefficients of harmonic linearization by the form of the output signal only.

Example 5.7. Consider the application of nonlinear control laws to time-varying control systems designed to guide landing aircraft. The system is to bring the craft to the given point on the landing trajectory with the specificed horizontal speed and zero vertical speed.

We consider one of the stages involved in landing, namely, stabilization on the equal-signal zone of the localizer beacon [5.5]. The block diagram of the longitudinal channel of the landing control system with the pitch angle ν as the control variable is given in Fig. 5.12. The compensating element is incorporated in the feedback loop. Here, ω_2 is the angular velocity, k_i are gain coefficients, T_i are time constants, and $L(t) = v_0(t - t_0)$ is the horizontal distance to the aircraft.

The system is essentially time-varying due to the coefficient $57.3/L(t)$ dependent on t in the interval 0 to t_0 equal to 100 s. We assume that the time-varying coefficient changes 100-fold and has an appreciable effect on the system transient that lasts for a few tens of seconds. Because common linear compensating elements with permanently tuned parameters have failed to ensure the stability and performance specifications, we resort to a piecewise-linear compensating element for which a particular version is shown in Fig. 5.11.

It is required to select ω_0 and T in the control law. The time constant should be chosen such that the phase response could affect the phase response of the open-loop system ($T = 0.05$ s) only insignificantly. To determine ω_0 (Fig. 5.7) we apply feedback to the system. The system coefficients are $k_1 = 1$ m/rad, $k_2 = 0.4\,\text{s}^{-1}$, $k_3 = 1$ rad/V, $k_4 = 100$ V/rad, $k_{\omega_z} = 1.3$ V s/rad, and $k_\nu = 1$. The time constants are $T_1 = T_2 = 2$ s; $T_3 = 3$ s; $T_4 = 0.5$ s; $T_5 = 1.5$ s; $\xi = 0.5$; and $v_0 = 60$ m/s.

Simulation experiments without a compensating element established that the system exhibits unstable divergent oscillations with the maximum period $T_{max} = 9.5$ s, from which $\omega_0 < 2\pi/T_{max} = 0.6\,\text{rad s}^{-1}$. Curve *1* in Fig. 5.13 illustrates the transient in glide-path angle for the system without a compensating element. In the presence of these strong oscillations, safe landing cannot be guaranteed.

After the piecewise-linear element with the selected parameters has been incorporated in the feedback loop, the time-varying system gets stable until 99th second. This case is illustrated by curve *3* in Fig. 5.13 obtained for the recommended value of $\omega_0 < 2\pi/T_{max}$. The descent along the glide path is aperiodic and ensures safe landing with high accuracy.

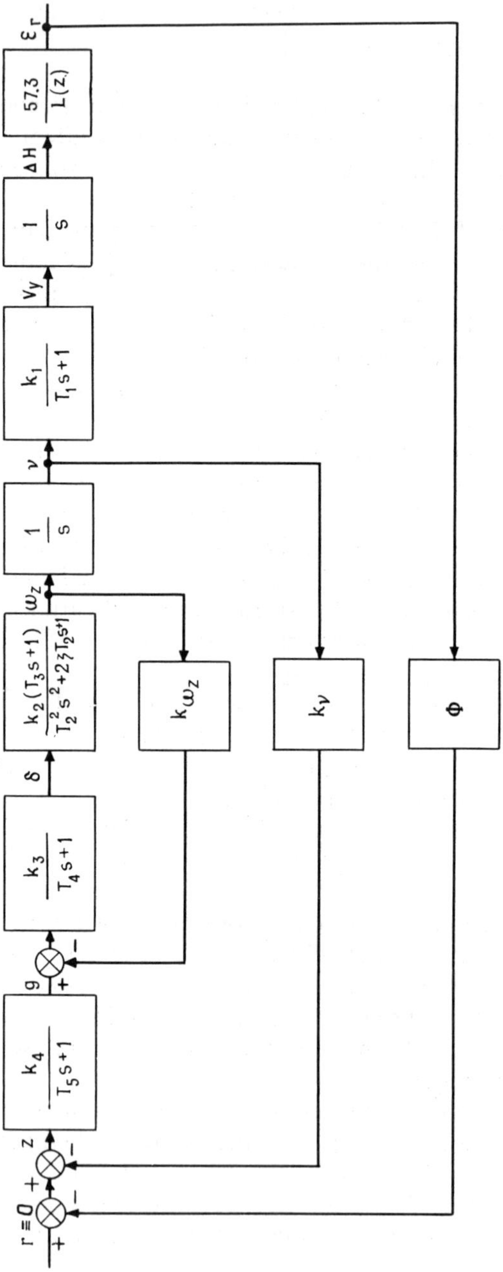
r ≡ 0
z
k4 / T5s+1
g
k3 / T4s+1
δ
k2(T3s+1) / T2²s²+2ζT2s+1
ωz
1 / s
ν
k1 / T1s+1
Vy
1 / s
ΔH
57.3 / L(z)
εr
kωz
kν
Φ

Fig. 5.12

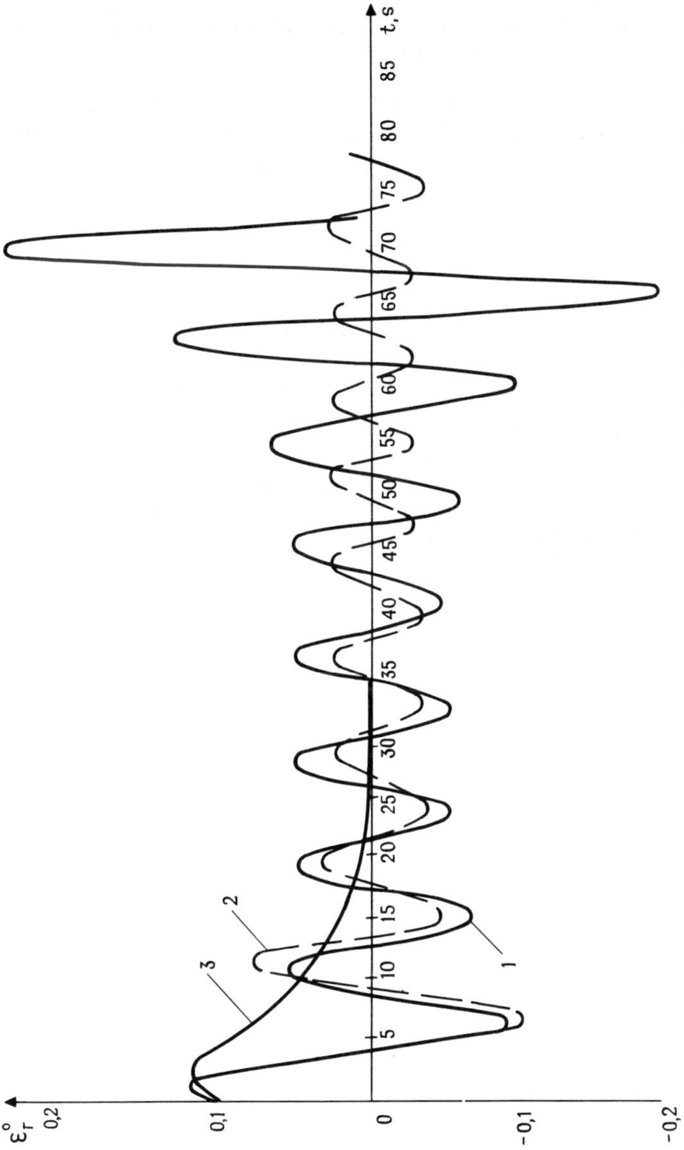

Fig. 5.13

Curve *2* in Fig. 5.13 illustrates the transient process for the system with the compensating unit when $\omega_0 > 2\pi/T_{\max}$. In this case the glide-path flight remains oscillatory, which is also inadmissible.

The above examples illustrate that the application of a piecewise-linear element with permanently tuned parameters ensures quasi-time-invariant properties for essentially time-varying control systems.

References

5.1 L. O. Chua and S. M. Kang: *Proc. IEEE.* Vol. 65, **6**, 915-922 (1977).
5.2 B. F. Shumilov and Yu. Yu. Shumilov: in *Metody issledovaniya kompleksov upravleniya* (Methods of control system analysis). Energoizdat, Moscow 1986. pp. 86-92.
5.3 E. P. Popov and Yu. I. Topcheyev (eds): *Nelineinye nestatsionarnye sistemy* (Nonlinear time-varying systems). Mashinostroyenie, Moscow 1986.
5.4 E. P. Popov and Yu. I. Topcheyev (eds): *Metod garmonicheskoi linearizatsii v proyektirovanii nelineinykh sistem avtomaticheskogo upravleniya* (Method of harmonic linearization in nonlinear control system design). Mashinostroyenie, Moscow 1970.
5.5 I. A. Mikhalev, B. I. Okoemov, and M. S. Chikulaev: *Sistemy avtomaticheskoi posadki* (Automatic landing systems). Mashinostroyenie, Moscow 1975.
5.6 M. A. Slonim: *Proc. IEEE.* Vol. 67, **6**, 952-954. (1979).

6 Multiloop Nonlinear System Design

E. P. Popov

Design of multiloop control systems by analytical methods is a formidable problem leading to complicated mathematical expressions for coefficients of harmonic linearization. For some systems the available methods produce erroneous results. For instance, actual systems exhibit self-excited oscillations, but their analysis by the method of harmonic linearization reveals no oscillatory modes. This chapter discusses a method of multiloop nonlinear system analysis involving reduction to simple block diagrams. This approach is based on a vector-matrix formalism which allows estimation of the effect of higher harmonics on system dynamics and construction of damping characteristics in the form of frequency diagrams.

The vector-matrix formalism reduces the solution of pertinent problems to algorithms readily adaptable for computer implementation. This is helpful in determining the modes of forced oscillations, regions of frequency entrainment, stable equilibrium, and self-excited oscillations. When higher harmonics and some parameters of multiloop systems are taken into account, the algorithm reveals periodic modes unavailable from the fundamental component analysis. The computer analysis evaluates the amplitude and frequency of self-excited oscillations for any number of harmonic components.

A more accurate construction of the envelopes of transient response in nonlinear systems is also obtained with the use of damping diagrams. These diagrams give the locus of constant damping factor and the frequency of oscillations as functions of the system parameters and the amplitude of the input to the nonlinear element. The diagrams reveal the regions of stability and self-excited oscillations, thus simplifying the selection of system parameters precluding the periodic modes. They may be used also as generalized characteristics locating the regions of damped, unstable, and self-excited oscillations.

Computer-aided design of control systems is a versatile method of system synthesis to performance specifications. This synthesis is carried out either on the basis of variational optimization techniques, or by choosing the system network and parameters with the damping diagrams. The second method is more attractive as it is readily apparent and does not involve unwieldy computations.

6.1 Problem Statement

A characteristic feature associated with nonlinear systems is that the principle of superposition does not hold for them. Therefore, for the system subjected to an additional excitation, the control may be substantially different. This situation is illustrated by Fig. 6.1 where the oscillation $x' = A \sin(\omega t)$ superimposed on the constant input x_0 causes the direct component y_2 of the output to be lower than the original level y_1. This implies that if x_0 is a variable signal, $x_0(t)$, the nonlinear response $y_2(x)$ to an oscillatory signal will be essentially different from the nonlinear response y_1 for nonoscillatory stimulus. This situation complicates still more if the system is excited by two superimposed aperiodic oscillations or two arbitrary stimuli. Therefore, we confine ourselves to considering a few typical situations which can be solved by generalization of the method of harmonical linearization.

In what follows we formulate four general problems reducible to solving by harmonic linearization [6.1].

1. *Evaluate the higher harmonics of periodic oscillations in the presence of self-excited oscillations and forced oscillations.* Assume that

$$x = A \sin(\omega t) + \sum_{k=2}^{\infty} \delta_k A \sin(k\omega t + \varphi_k)$$

is the motion of the nonlinear system described by the differential equation

$$Q(p)x + R(p)\,\mathscr{F}(x, px) = 0 \tag{6.1}$$

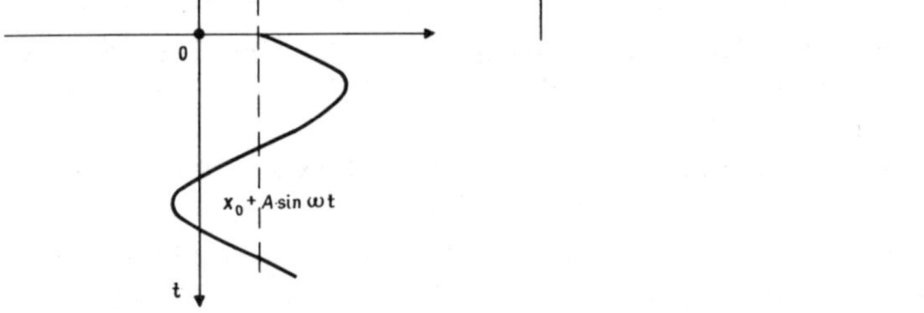

Fig. 6.1

where $Q(p)$ and $R(p)$ are polynomials in $p = \mathrm{d}/\mathrm{d}t$, and $\mathcal{F}(x, px)$ is a nonlinear function.

In the presence of self-excited oscillations we have in a certain approximation, for $k = 2, 3, \ldots, n$

$$\delta_k = \left|\frac{R(jk\omega)}{Q(jk\omega)}\right| \sqrt{a_k^2 + b_k^2}$$

$$\varphi_k = -\pi + \arg \frac{R(jk\omega)}{Q(jk\omega)} + \arctan \frac{b_k}{a_k}$$

where

$$a_k = \frac{1}{\pi A} \int_0^{2\pi} \mathcal{F}(A \sin \psi, A\omega \cos \psi) \sin (k\psi) \mathrm{d}\psi$$

$$b_k = \frac{1}{\pi A} \int_0^{2\pi} \mathcal{F}(A \sin \psi, A\omega \cos \psi) \cos (k\psi) d\psi \tag{6.2}$$

For the higher harmonics of self-excited oscillations, the quantities δ_k and φ_k are calculated in terms of the amplitude A and frequency ω of the fundamental component.

In a development of this method one can refine the quantities A and ω of the fundamental component by taking into account the quantities found for the higher harmonics [6.2].

Forced oscillations with one fundamental frequency are defined as

$$Q(p)x + R(p)\mathcal{F}(x, px) = S(p)f(t) \tag{6.3}$$

Here, $f(t)$ is a given periodic forcing function of the form

$$f(t) = B_1 \sin (\omega t) + \sum_{k=2}^{m} B_k \sin (k\omega t + \nu_k), \quad m \leqslant n, \ B_k < B_1$$

where n is the number of higher harmonics taken into account in the desired solution

$$x = A \sin (\omega t + \varphi) + \sum_{k=2}^{n} \delta_k A \sin (k\omega t + \varphi_k)$$

The frequency ω is specified and both the amplitude A and phase φ are determined by the harmonic balance method in a first approximation, that is, accurate to the effect of the fundamental component only. The quantities δ_k and φ_k are determined as [6.1]

$$\delta_k = \left| -\frac{R(jk\omega)}{Q(jk\omega)}(a_k + jb_k)\exp(jk\varphi) + \frac{S(jk\omega)}{Q(jk\omega)}\frac{B_k}{A}\exp(jk\nu_k)\right|$$

$$\varphi = -\pi + \arg\left[\frac{R(jk\omega)}{Q(jk\omega)}(a_k + jb_k)\exp(jk\varphi) + \frac{S(jk\omega)}{Q(jk\omega)}\frac{B_k}{A}\exp(jk\nu_k)\right]$$

where $k = 2, 3, \ldots, n$ for the first terms, and $k = 2, 3, \ldots, m$ for the second terms. For the last $n - m$ higher harmonics, the second addends are absent. The quantities a_k and b_k are calculated by (6.1) and (6.2).

In this case the fundamental component can be refined by allowing for the effect of higher harmonics [6.1].

2. *Give an approximate performance criterion for damped and divergent oscillatory transients with respect to the fundamental component.*

We shall seek for a solution of (6.1) in the form

$$x = A(t)\sin[\psi(t)]$$
$$\frac{dA}{dt} = A\xi(A), \quad \frac{d\psi}{dt} = \omega(A) \tag{6.4}$$

where ξ is the damping factor. In contrast to the linear systems with $x = A_0\exp(\xi t)\sin(\omega t)$ and constant ξ and ω, this time both ξ and ω are variable. It is assumed that they are varied slowly with the amplitude A and $A(t)$ of the transient. At the same time $A(t)$ may vary at a finite rate (6.4) dependent on ξ.

The functions $\xi(A)$ and $\omega(A)$ are defined by the equation [6.1]

$$Q(\xi + j\omega) + R(\xi + j\omega)[a(A, \omega, \xi) + jb(A, \omega, \xi)] = 0 \tag{6.5}$$

where

$$a = \frac{1}{\pi A}\int_0^{2\pi}\mathscr{F}(A\sin\psi, A\omega\cos\psi + A\xi\sin\psi)\sin\psi\, d\psi$$

$$b = \frac{1}{\pi A}\int_0^{2\pi}\mathscr{F}(A\sin\psi, A\omega\cos\psi + A\xi\sin\psi)\cos\psi\, d\psi$$

The functions $\xi(A)$ and $\omega(A)$ thus defined will be used to construct diagrams for transients by plotting the curves $\xi = \text{const}$ and $\omega = \text{const}$ in the k, A plane (Fig. 6.2), where k is a system parameter, say, the open-loop gain of the linear element. Such a diagram provides an insight into how fast the transient decays in the vicinity of the self-excited oscillation mode

$k > k_1$ (Fig. 6.2) and near the stable equilibrium ($A = 0$), i.e., in absence of such a mode ($k < k_1$).

By way of example, the point C corresponds to self-excited oscillations with amplitude A_C. The process of transient oscillations converges from both sides to the amplitude A_C. The quantity ξ determined by the intersections of the C vertical with the lines ξ = const may be used to judge how fast the oscillations diverge from below and decay from above. In the region of stable equilibrium, such a guess will be obtained by the intersections of the vertical M_0 with the lines ξ = const (at a given initial amplitude A_0). The variation of frequency ω with the amplitude A in various modes can be determined by the lines ω = const.

Equations (6.4) may be used, if necessary, to estimate the form of the transient process.

It is worth noting that thus far we have been focused solely on symmetric oscillations. However, the method may be extended to asymmetric transients as well [6.1].

Because the quantity k_1 (see Fig. 6.2) defines the bound of the stable equilibrium for the system, we are thus able to determine the stability boundary for nonlinear systems of the type (6.1) [6.1].

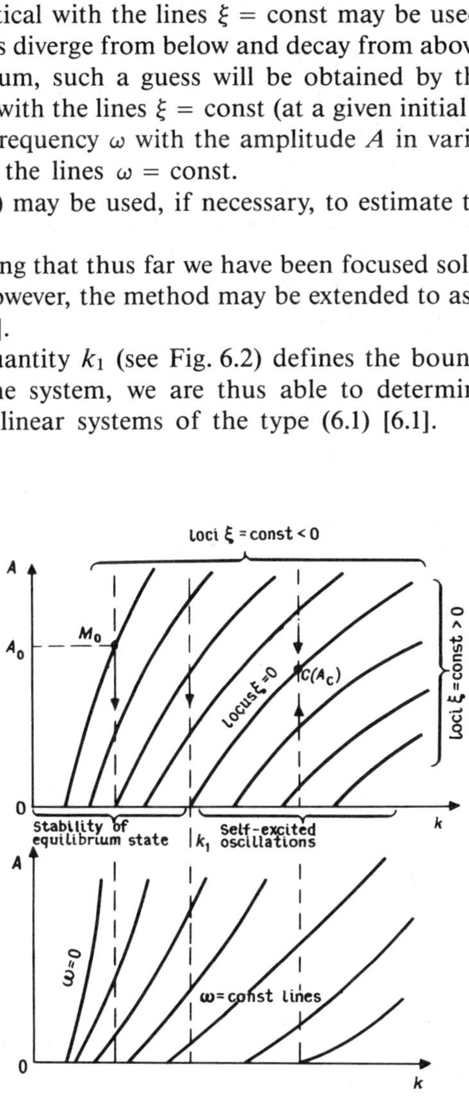

Fig. 6.2

3. *Determine the control laws in the presence of superimposed forced oscillations.*

Let the system defined by the differential equation

$$Q(p)x + R(p)\mathscr{F}(x, px) = S(p)f(t) + S_1(p)f_1(t)$$

be subjected to simultaneous inputs $f(t) = B\sin(\omega t)$ and $f_1(t)$ with the latter varying slower than the former. Then the solutions will be sought in the form

$$x = x_0(t) + x^*(t), \quad x^*(t) = A\sin(\omega t + \varphi)$$

where the component $x_0(t)$ varies slowly as compared with $x^*(t)$. This implies that we have a control $x_0(t)$ which is superimposed by the forced oscillations $x^*(t)$. Then

$$Q(p)x_0 + R(p)\mathscr{F}(x_0, A, \omega) = S_1(p)f_1(t) \tag{6.6}$$

$$\left[Q(p) - S(p)\frac{B}{A}\left(\cos\varphi - \frac{\sin\varphi}{\omega}p\right)\right]x^* + R(p)\left[a(A, \omega, x_0) + \frac{b(A, \omega, x_0)}{\omega}p\right]x^* = 0 \tag{6.7}$$

$$\mathscr{F}_0 = \frac{1}{2\pi}\int_0^{2\pi}\mathscr{F}(x_0 + A\sin\psi, A\omega\cos\psi)d\psi$$

$$a = \frac{1}{\pi A}\int_0^{2\pi}\mathscr{F}(x_0 + A\sin\psi, A\omega\cos\psi)\sin\psi\, d\psi \tag{6.8}$$

$$b = \frac{1}{\pi A}\int_0^{2\pi}\mathscr{F}(x_0 + A\sin\psi, A\omega\cos\psi)\cos\psi\, d\psi$$

Defining $A(x_0)$ and $\omega(x_0)$ from (6.7) and substituting them in the expression for $\mathscr{F}_0(x_0, A, \omega)$, we obtain

$$\mathscr{F}_0 = \Phi(x_0)$$

This function turns out to be smooth even if $\mathscr{F}(x)$ is discontinuous. This is the effect of smoothing nonlinearities by superimposed forced oscillations [6.1]. Therefore, $\Phi(x_0)$ may be subjected to usual linearization. For a nonlinear control in the presence of forced oscillations this results in the differential equation

$$[Q(p) + k_n R(p)]x_0 = S(p)f_1(t) \tag{6.9}$$

where

$$k_n = \left(\frac{\partial \mathscr{F}_0}{\partial x_0}\right)_{x_0 = 0} = k_n(A_0) \tag{6.10}$$

where A_0 is the amplitude of symmetric (at $x_0 = 0$) forced oscillations.

A noteworthy feature of this linearization is that the coefficient k_n (having a constant value for each system with specified numerical parameters) can vary with the system structure and parameters since A_0 is dependent on them. It is this circumstance that allows the behavior of the nonlinear control system to be taken into account.

The quantity k_n depends also on the amplitude B and frequency ω of the forcing function, since the amplitude of the forced oscillations also depends on these parameters. This dependence in turn leads to the one for the roots of the characteristic equation (6.9). Therefore, in nonlinear systems, the domain of stability depends on these external factors. This leads to the boundary of noise immunity of such systems [6.1] (Fig. 6.3).

4. *Investigation of control processes in systems with self-excited oscillations.*

The system is described by the equation

$$Q(p)x + R(p)\,\mathscr{F}(x, px) = S(p)f(t) \tag{6.11}$$

where $f(t)$ is a forcing function that varies slowly compared with the self-excited oscillation mode existent in the system simultaneously with the control, namely,

$$x = x_0(t) + x^*(t), \quad x^* = A \sin(\omega t)$$

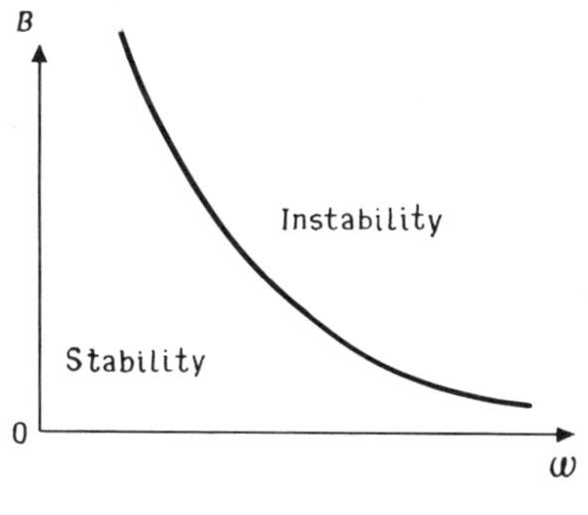

Fig. 6.3

Equation (6.11) may be divided in two equations as follows

$$Q(p)x_0 + R(p)\mathscr{F}_0(x_0, A, \omega) = S(p)f(t) \tag{6.12}$$

$$\left\{Q(p) + R(p)\left[a(A, \omega, x_0) + \frac{b(A, \omega, x_0)}{\omega}p\right]\right\}x^* = 0 \tag{6.13}$$

where $\mathscr{F}_0$, a, and b are defined by (6.8).

Eliminating $A(x_0)$ and $\omega(x_0)$ between (6.13) and the expression for $\mathscr{F}(x_0, A, \omega)$ yields

$$\mathscr{F}_0 = \Phi(x_0)$$

which implies the effect of smoothing of the nonlinearity with the aid of self-excited oscillations. Thus, we are in a position to write for the control the linearized equation

$$[Q(p) + k_n R(p)]x_0 = S(p)f(t)$$

in place of (6.12). The coefficient k_n is now defined by the formula (6.10), where A_0 is the amplitude of symmetric (for $x_0 = 0$) self-excited oscillations in the system. In this case the system designer is faced with the same feature of the system, namely, with the dependence of k_n on system structure and parameters.

The above methods are advantageous in that all the quantitative indices of control are connected directly with the structure and parameters of the nonlinear system primarily by algebraic relations. This implies the suitability of these methods for the synthesis and design of nonlinear automatic systems.

6.2 Analysis of Control Systems with Nonlinear Compensating Elements

Nonlinear compensating elements offer a much wider range of compensating possibilities than linear elements. This is especially true when nonlinear compensating elements are employed to counteract the mutilating effect of system nonlinearities, suppress self-excited oscillations and extend the range of absolute stability, enhance vibration-noise immunity, and improve the dynamic performance of control.

Consider a system whose damping is varied as a function of error. The block diagram of such a system with nonlinear feedback is shown in Fig. 6.4*a*. The system motion is described by the equations

$$x = \theta_1 - \theta_2, \quad u_1 = k_1 x, \quad u_2 = u_{fb},$$
$$(T_1 p + 1)u_3 = k_2 u_2, \quad (T_2 p + 1)p\alpha = k_3 u_3,$$

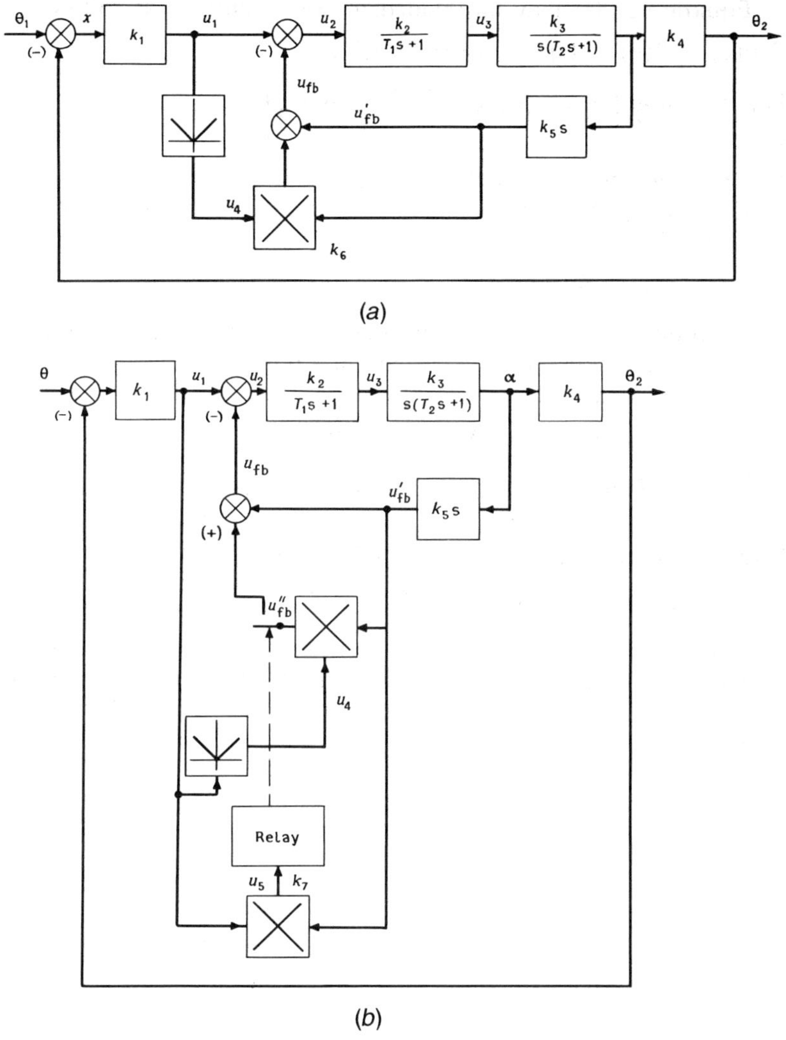

Fig. 6.4

$$\theta_2 = k_4\alpha, \quad u_{fb} = u'_{fb} - u''_{fb},$$
$$u'_{fb} = k_5 p\alpha, \quad u''_{fb} = k_0 u_4 u'_{fb}, \; u_4 = |u_1|$$

The system equation without the external forcing function $f(t)$ will have the form

$$[T_1T_2p^3 + (T_1 + T_2)p^2 + (1 + k_{fb})p + k]x - k'_{fb}|x|px = 0 \tag{6.14}$$

Hence, the nonlinearity has the form

$$\mathscr{F}(x,\ px) = |x|px$$

The harmonic linearization method yields

$$\mathscr{F} = \left(a - \frac{\xi}{\omega} b\right) x + \frac{b}{\omega} px \tag{6.15}$$

where

$$a = \frac{8}{3\pi} A\xi \quad \text{and} \quad b = \frac{4}{3\pi} A\omega \tag{6.16}$$

As a result we get

$$|x|px = \frac{4A}{3\pi} (\xi + p)x$$

Then the characteristic equation of the system (6.14) takes on the form

$$T_1 T_2 s^3 + (T_1 + T_2)s^2 + (1 + k_{fb})s + k + k'_{fb} \frac{4A}{3\pi} (s + \xi) = 0$$

The substitution $s = \xi + j\omega$ yields two equations

$$T_1 T_2 \xi^3 + (T_1 + T_2)\xi^2 + \xi \left(1 + k_{fb} - 3T_1 T_2 \omega^2 - \frac{8}{3\pi} k'_{fb} A\right) + k - (T_1 + T_2)\omega^2 = 0 \tag{6.17}$$

$$3T_1 T_2 \xi^2 \omega + 2\omega(T_1 + T_2)\xi + \left(1 + k_{fb} - \frac{4}{3\pi} k_{fb} A - T_1 T_2 \omega^2\right)\omega = 0 \tag{6.18}$$

From (6.18) it follows that

$$\omega^2 = 3\xi^2 + 2\frac{T_1 + T_2}{T_1 T_2}\xi + \frac{1 + k_{fb}}{T_1 T_2} - \frac{4k'_{fb}}{3\pi T_1 T_2} A \tag{6.19}$$

Then from (6.17) we have

$$A = \frac{3\pi}{2k'_{fb}[\xi + (T_1 + T_2)/T_1 T_2]} \left[4T_1 T_2 \xi^3 + 4(T_1 + T_2)\xi^2 + (1 + k_{fb})\xi + \frac{(T_1 + T_2)^2}{T_1 T_2}\xi - k + \frac{T_1 + T_2}{T_1 T_2}(1 + k_{fb})\right] \tag{6.20}$$

The last two formulas may be employed to construct a diagram for nonlinear transient with respect to any desired parameter, say k. For specified parameters, one may also determine the functions $\xi(A)$ and $\Omega(A)$ and obtain any performance criteria for transient processes.

Let us investigate a system with another method of damping variation (Fig. 6.4*b*). This system differs from the previous version in that the signal of nonlinear compensation u''_{fb} passes through the relay contacts.

Now, in place of (6.14) we have

$$[T_1 T_2 p^3 + (T_1 + T_2)p^2 + (1 + k_{fb})p + k]x + k'_{fb}\mathcal{F}(x, px) = 0$$

where

$$\begin{aligned} \mathcal{F}(x, px) &= -|x|px \quad \text{for} \quad xpx < 0 \\ &= 0 \quad \text{for} \quad xpx > 0 \end{aligned}$$

The coefficients of harmonic linearization vary as follows:

$$a = \frac{2}{3\pi} A|\xi| \left[\frac{(\omega/\xi)^2 + 2}{\sqrt{1 + (\omega/\xi)^2}} - 2 \operatorname{sgn} \xi\right]$$

$$b = \frac{2}{3\pi} A\omega \left[\frac{\operatorname{sgn} \xi}{\sqrt{1 + (\omega/\xi)^2}} - 1\right]$$

Substituting these expressions into (6.15) gives

$$\mathcal{F} = \frac{2A}{3\pi}\left[(\sqrt{\omega^2 + \xi^2} - \xi)x - \left(1 - \frac{\xi}{\sqrt{\omega^2 + \xi^2}}\right)px\right] \tag{6.21}$$

The characteristic equation of the harmonically linearized system will be

$$T_1 T_2 s^3 + (T_1 + T_2)s^2 + \left[1 + k_{fb} - \frac{2k'_{fb}}{3\pi} A\left(1 - \frac{\xi}{\sqrt{\omega^2 + \xi^2}}\right)\right]s + \left[k + \frac{2k'_{fb}}{3\pi} A(\sqrt{\omega^2 + \xi^2} - \xi)\right] = 0 \tag{6.22}$$

The substitution $s = \xi + j\omega$ into (6.22) results in two equations, which, when solved for A, yield the system

$$\begin{aligned} A &= \frac{[2T_1 T_2 \xi + (T_1 + T_2)](\omega^2 + \xi^2) - k}{(2k'_{fb}/3\pi)(\sqrt{\omega^2 + \xi^2} - \xi)} \\ A &= \frac{T_1 T_2 \sqrt{\omega^2 + \xi^2}}{(2k_{fb}/3\pi)(\sqrt{\omega^2 + \xi^2} - \xi)}\left(3\xi^2 + 2\frac{T_1 + T_2}{T_1 T_2}\xi + \frac{1 + k_{fb}}{T_1 T_2} - \omega^2\right) \end{aligned} \tag{6.23}$$

Solving this system, we get two functions $\omega(A)$ and $\xi(A)$ characterizing the system transient behavior at the given parameters.

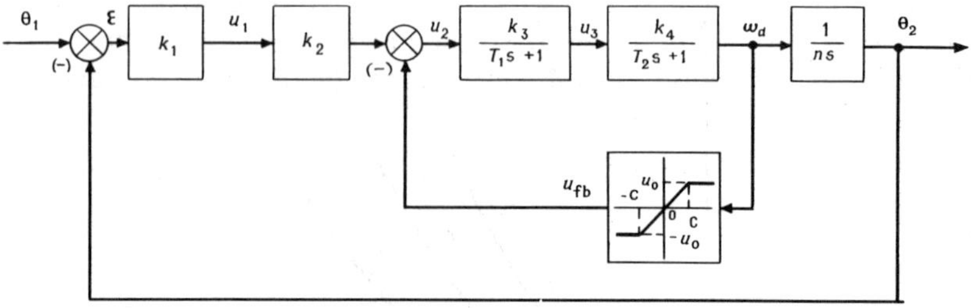

Fig. 6.5

Consider a system with limitation of feedback linearity (Fig. 6.5). This system is described by the equations

$$\begin{aligned}
&\varepsilon = \theta_1 - \theta_2, \\
&u_1 = k_1\varepsilon, \quad u_2 = k_2 u_1, \\
&(T_1 p + 1)u_3 = k_3(u_2 - u_{\text{fb}}), \quad u_{\text{fb}} = \mathscr{F}(\omega_{\text{d}}) \\
&np\theta_2 = \omega_{\text{d}}, \quad (T_2 p + 1)\omega_{\text{d}} = k_4 u_3
\end{aligned}$$

where $\mathscr{F}(\omega_{\text{d}})$ is the nonlinearity.

The method of harmonic linearization gives

$$u_{\text{fb}} = a(A)\omega_{\text{d}}$$

where

$$a(A) = \frac{2}{\pi} k_{\text{fb}} \left[\arcsin\frac{C}{A} + \frac{C}{A}\sqrt{1 - \left(\frac{C}{A}\right)^2}\right] \tag{6.24}$$

The characteristic equation of the system has the form

$$T_1 T_2 s^3 + (T_1 + T_2)s^2 + [1 + k_3 k_4 a(A)] + k = 0 \tag{6.25}$$

where $k = k_1 k_2 k_3 k_4 / n$.

From the two equations obtained upon substituting $s = \xi + j\omega$ into (6.25) we find

$$\omega^2 = \frac{k}{T_1 + T_2 + 2\xi T_1 T_2}$$

$$\xi = \frac{[(T_1 + T_2)/T_1 T_2][1 + k_3 k_4 a(A)] - k}{2[1 + k_3 k_4 a(A)] - T_1 T_2[(T_1 + T_2)/T_1 T_2 + 2\xi]^2}$$

Figure 6.6 shows $\xi(A)$ plotted for three values of half-width C of the linearity (see Fig. 6.5). At small amplitudes the damping factor is seen to be constant and independent of C. As A increases the damping factor experiences appreciable growth.

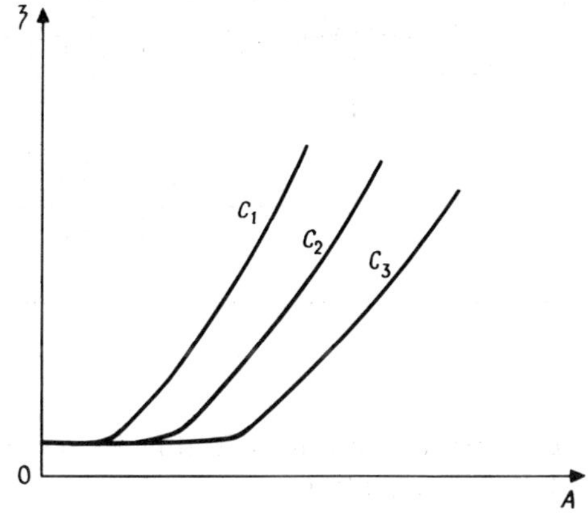

Fig. 6.6

Concluding this section, we should point out that the application of nonlinear compensating elements leads to the evaluation by various means of the diagrams for $\xi(A)$ and $\omega(A)$. Basing on these diagrams the designer can select appropriate parameters for the control system under investigation.

The following section will be devoted to the general method of construction of such diagrams. Implemented in the form of a computer algorithm, this method is an efficient tool in solving the problems of system analysis and synthesis.

6.3 Computer-Assisted Synthesis Based on Damping Diagrams

Practical implementation of performance diagrams is limited by complications involved in the solution of (6.5) for ξ and ω with arbitrary nonlinear and linear functions. Kotov and Popov [6.3] have suggested a new approach to solving harmonically linearized equations defining nonlinear system transient behavior. This method suitable for adaptation as a CAD procedure is based on a pretransformation of the initial equations to bilinear form.

We wish to demonstrate that an arbitrary polynomial $H(s)$ of degree n in real $s = \xi + \omega$ may be represented as the bilinear form

$$H(s) = H(\xi + \omega) = \sum_{i=0}^{n} h_i s^i = \mathbf{\Omega}^T \tilde{\mathbf{H}} \mathbf{\Xi} \tag{6.26}$$

where $\boldsymbol{\Omega}$ and $\boldsymbol{\Xi}$ are $(n+1)$ column vectors defined as follows

$$\boldsymbol{\Omega} = [1,\ \omega,\ \omega^2,\ \ldots,\ \omega^n]^{\mathrm{T}}$$
$$\boldsymbol{\Xi} = [1,\ \xi,\ \xi^2,\ \ldots,\ \xi^n]^{\mathrm{T}}$$

and $\tilde{\mathbf{H}}$ is a symmetric $(n+1) \times (n+1)$ matrix whose elements are uniquely defined by the coefficients of $H(s)$. For this purpose we expand $H(s)$ in a power series of ω, viz.,

$$H(s) = H(\xi,\ \omega) = \sum_{i=0}^{n} \frac{\mathrm{d}^i H}{\mathrm{d}s^i} \frac{\omega^i}{i!}\Bigg|_{s=\xi}$$

$$= \sum_{i=0}^{n} \sum_{k=0}^{n-i} h_{k+i} \frac{(k+1)(k+2)\ \ldots\ (k+i)}{i!}\ \xi^k \omega^i$$

$$= \sum_{i=0}^{n} \omega^i \sum_{k=0}^{n-i} h_{k+i} \binom{k+i}{i} \xi^k \qquad (6.27)$$

where $\binom{k+i}{i}$ is the number of combinations of $k+i$ things, i at a time. The last expression in (6.27) may be written in a compact form as (6.26), where nonzero elements of the matrix H are defined as

$$\tilde{\mathbf{H}} = \{\tilde{h}_{i+1,\ k+1}\} = \{h_{i+k}\binom{i+k}{i}\};\quad i,\ k = 0,\ 1,\ 2,\ \ldots$$

The transformation to bilinear forms of polynomials of arbitrary degree in a variable composed of two unknowns added together paves the way for operations performed on the terms in (6.5). Indeed, for the complex variable $s = \xi + j\omega$, the real and imaginary parts of $H(s)$ can readily be obtained as

$$\begin{aligned} \mathrm{Re}\{H(s)\} &= \boldsymbol{\Omega}^{\mathrm{T}} \mathbf{E}_{\mathrm{R}} \tilde{\mathbf{H}} \boldsymbol{\Xi} \\ \mathrm{Im}\{H(s)\} &= \boldsymbol{\Omega}^{\mathrm{T}} \mathbf{E}_{\mathrm{I}} \tilde{\mathbf{H}} \boldsymbol{\Xi} \end{aligned} \qquad (6.28)$$

where the diagonal matrices $\mathbf{E}_{\mathrm{R}}$ and $\mathbf{E}_{\mathrm{I}}$ are defined as follows:

$$\mathbf{E}_{\mathrm{R}} = \mathrm{diag}\{1,\ 0,\ -1,\ 0,\ 1,\ \ldots\}$$
$$\mathbf{E}_{\mathrm{I}} = \mathrm{diag}\{0,\ 1,\ 0,\ -1,\ 0,\ 1,\ \ldots\}$$

The calculation of the first and higher derivatives of the initial polynomial $H = H(s) = H(\xi + j\omega)$ with respect to one of the unknowns ξ or ω is also straightforward, namely,

$$\frac{\partial^k H}{\partial \xi^k} = \boldsymbol{\Omega}^k \tilde{\mathbf{H}} \mathbf{E}_k \boldsymbol{\Xi} \quad \text{and} \quad \frac{\partial^k H}{\partial \omega^k} = \boldsymbol{\Omega}^{\mathrm{T}} \mathbf{E}_k^{\mathrm{T}} \tilde{\mathbf{H}} \boldsymbol{\Xi}$$

Here, $\mathbf{E}_k$ is the product of k matrices of the form

$$\mathbf{E}_i = \left[\begin{array}{ccc|c} 0 & \dots & 0 & 0 \\ \hline & & & 0 \\ & \mathbf{E}_\mathrm{R} & & \vdots \\ & & & 0 \\ & & & \end{array}\right] \quad \text{or} \quad \mathbf{E}_i = \left[\begin{array}{ccc|c} 0 & \dots & 0 & 0 \\ \hline & & & 0 \\ & \mathbf{E}_\mathrm{I} & & \vdots \\ & & & 0 \\ & & & \end{array}\right]$$

that is, $\mathbf{E}_k = \prod_{i=1}^{k} \mathbf{E}_i$

Making use of (6.28), we write two equations

$$\boldsymbol{\Omega}^\mathrm{T}\tilde{\mathbf{E}}_\mathrm{R}\boldsymbol{\Xi} = 0 \tag{6.29a}$$

$$\boldsymbol{\Omega}^\mathrm{T}\tilde{\mathbf{E}}_\mathrm{I}\boldsymbol{\Xi} = 0 \tag{6.29b}$$

where $\mathbf{E}_\mathrm{R}$ and $\mathbf{E}_\mathrm{I}$ are $(n + 1) \times (n + 1)$ matrices whose elements depend on the coefficients in the polynomials $R(p)$ and $Q(p)$ in (6.1) and on the coefficients of harmonic linearization a and b.

The compact form (6.29) of these two equations is convenient in that it allows further operations to be carried out in the real rather than complex domain and simplifies the evaluation of derivatives, thus facilitating the use of numerical solution techniques.

All the variety of nonlinearities encountered in control systems can be described in terms of the following functions:

(a) single-valued odd-symmetrical;

(b) double-valued odd-symmetrical functions dependent on the input and on its derivative;

(c) single-valued and double-valued functions joined with the linear element and having the output signal parameters dependent on the amplitude and frequency of the input.

As a way to solve equation (6.29) we resort to Newton's iteration technique. If we denote the left-hand sides of (6.29*a*) and (6.29*b*) by Y_R and Y_I, respectively, then for each value of A the quantities ξ and ω are determined by the iteration procedure as follows:

$$\xi^i = \xi^{i-1} + \left(Y_\mathrm{I}\frac{\partial Y_\mathrm{R}}{\partial \omega} - Y_\mathrm{R}\frac{\partial Y_\mathrm{I}}{\partial \omega}\right)\left(\frac{\partial Y_\mathrm{R}}{\partial \xi}\frac{\partial Y_\mathrm{I}}{\partial \omega} - \frac{\partial Y_\mathrm{I}}{\partial \xi}\frac{\partial Y_\mathrm{R}}{\partial \omega}\right)^{-1} \tag{6.30}$$

$$\omega^i = \omega^{i-1} + \left(Y_\mathrm{R}\frac{\partial Y_\mathrm{R}}{\partial \xi} - Y_\mathrm{I}\frac{\partial Y_\mathrm{I}}{\partial \xi}\right)\left(\frac{\partial Y_\mathrm{R}}{\partial \xi}\frac{\partial Y_\mathrm{I}}{\partial \omega} - \frac{\partial Y_\mathrm{I}}{\partial \xi}\frac{\partial Y_\mathrm{R}}{\partial \omega}\right)^{-1}$$

As a result we obtain the dependences $\xi = \xi(A)$ and $\omega = \omega(A)$ to be used for construction of damping diagrams.

The representation (6.29) enables the use of another algorithm based on the numerical integration of differential equations. By differentiating the left-hand sides of (6.29) with respect to A we obtain

$$\begin{aligned} &\frac{d\omega}{dA}(\boldsymbol{\Omega}^T\mathbf{E}_I^T\tilde{\mathbf{E}}_R\boldsymbol{\Xi}) + \frac{d\xi}{dA}(\boldsymbol{\Omega}^T\tilde{\mathbf{E}}_R\mathbf{E}_1\boldsymbol{\Xi}) + \boldsymbol{\Omega}^T\frac{d\tilde{\mathbf{E}}_R}{dA}\boldsymbol{\Xi} = 0 \\ &\frac{d\omega}{dA}(\boldsymbol{\Omega}^T\mathbf{E}_I^T\tilde{\mathbf{E}}_I\boldsymbol{\Xi}) + \frac{d\xi}{dA}(\boldsymbol{\Omega}^T\tilde{\mathbf{E}}_I\mathbf{E}_1\boldsymbol{\Xi}) + \boldsymbol{\Omega}^T\frac{d\tilde{\mathbf{E}}_I}{dA}\boldsymbol{\Xi} = 0 \end{aligned} \tag{6.31}$$

Solving (6.31) for $d\omega/dA$ and $d\xi/dA$ yields

$$\begin{bmatrix} \dfrac{d\omega}{dA} \\ \\ \dfrac{d\xi}{dA} \end{bmatrix} = -\begin{bmatrix} \boldsymbol{\Omega}^T\mathbf{E}_I^T\tilde{\mathbf{E}}_R\boldsymbol{\Xi} & \boldsymbol{\Omega}^T\tilde{\mathbf{E}}_R\mathbf{E}_1\boldsymbol{\Xi} \\ \\ \boldsymbol{\Omega}^T\mathbf{E}_I^T\tilde{\mathbf{E}}_I\boldsymbol{\Xi} & \boldsymbol{\Omega}^T\tilde{\mathbf{E}}_I\mathbf{E}_1\boldsymbol{\Xi} \end{bmatrix}^{-1} \begin{bmatrix} \boldsymbol{\Omega}^T\dfrac{d\tilde{\mathbf{E}}_R}{dA}\boldsymbol{\Xi} \\ \\ \boldsymbol{\Omega}^T\dfrac{d\tilde{\mathbf{E}}_I}{dA}\boldsymbol{\Xi} \end{bmatrix} \tag{6.32}$$

Upon the numerical integration of the system (6.32) we also arrive at the functions $\xi(A)$ and $\omega(A)$. The methods outlined above are useful in computer-aided analysis of multiloop control systems with types (*a*) and (*b*) of nonlinearity. The procedure may be also extended to systems with type (*c*) nonlinearity by introducing the matrices $\tilde{\mathbf{Q}}$ and $\tilde{\mathbf{R}}$. In this case the amplitude and damping factor at unknown frequency can be determined by one method for all three types of nonlinearity.

By way of illustration we take $b = 0$. Then the solution can be obtained after the preliminary evaluation of A and ξ. Performing the same operations we did for the system (6.29), we can write equation (6.5) for the multiloop system in the form

$$\begin{aligned} a(A) &= -(\boldsymbol{\Omega}^T\mathbf{E}_R\tilde{\mathbf{Q}}\boldsymbol{\Xi})(\boldsymbol{\Omega}^T\mathbf{E}_R\tilde{\mathbf{R}}\boldsymbol{\Xi})^{-1} \\ \boldsymbol{\Xi}^T\tilde{\mathbf{E}}_q\boldsymbol{\Xi} &= 0 \end{aligned} \tag{6.33}$$

where $\tilde{\mathbf{E}}_q = \tilde{\mathbf{Q}}(\mathbf{E}_I\boldsymbol{\Omega}\boldsymbol{\Omega}^T\mathbf{E}_R - \mathbf{E}_R\boldsymbol{\Omega}\boldsymbol{\Omega}^T\mathbf{E}_I)\tilde{\mathbf{R}}$.

Applying the Newton method to (6.33) and recognizing that $\mathbf{E}_I$, $\mathbf{E}_R$, $\tilde{\mathbf{Q}}$, and $\tilde{\mathbf{R}}$ are all symmetrical matrices leads us to the expression for ξ, namely,

$$\xi^i = \xi^{i-1} - (\boldsymbol{\Xi}^T\tilde{\mathbf{E}}_q\boldsymbol{\Xi})\left(\frac{\partial\boldsymbol{\Xi}^T\tilde{\mathbf{E}}_q\boldsymbol{\Xi}}{\partial\xi}\right)^{-1} \tag{6.34}$$

Like for equations (6.32), the second equation in (6.33) leads to the expression

$$\frac{d\xi}{d\omega} = -\left(\boldsymbol{\Xi}\frac{d\tilde{\mathbf{E}}_q}{d\omega}\boldsymbol{\Xi}\right)[\boldsymbol{\Xi}^T(\mathbf{E}_I^T\tilde{\mathbf{E}}_q + \tilde{\mathbf{E}}_q\mathbf{E}_1)\boldsymbol{\Xi}]^{-1} \tag{6.35}$$

Solving the first equation in (6.33) and one of equations (6.34) or (6.35) simultaneously, we obtain the functions $\xi(A)$ and $\omega(A)$ which may be used to construct diagrams like those in Fig. 6.2.

When this method is adopted for the construction of diagrams of $\xi(A)$ and $\omega(A)$, one has to estimate the convergence of the iterative procedures as this is not always ensured by the recursive formulas employed for solving equations (6.5). Indeed, the outlined expressions involve inverse matrices, therefore, the right-hand side of (6.35) may be ill-defined. In such cases it is suggested to solve equations (6.5) with the use of the corollary from the theorem on the modulus maximum principle in the theory of functions of complex variable [6.4]. According to this corrollary, the modulus of an analytical function constructed over the complex domain

$$|f(s)| = \mathrm{mod}\left\{Q(s) + R(s)\left[a + \frac{b}{\omega}(s - \xi)\right]\right\}$$

cannot have either local maxima or minima unless the latter are at the zero level. This property enables us to obtain a recursive procedure for the complex variable s such that approaches the root as close as desired.

Consider the evaluation of the initial approximations in the iterations (6.30, 34) and differential equations (6.32, 35). When some parameters, say $a(A)$ and $b(A)$, vary continuously and rather slowly, then the initial approximation at each point of the recursive procedure may be taken as the final values of the unknowns determined at the previous point. To determine the initial approximation at the first point we assume that there exists an amplitude A_0 such that $a(A_0) = b(A_0) = 0$ when the system contains odd-symmetric nonlinearities. Then equation (6.5) takes the form

$$Q(s) = Q(\xi + \omega) = 0 \tag{6.36}$$

The polynomial $Q(s)$ is the denominator of the transfer function of the linear element (Fig. 6.7), and as a rule can be represented as the product

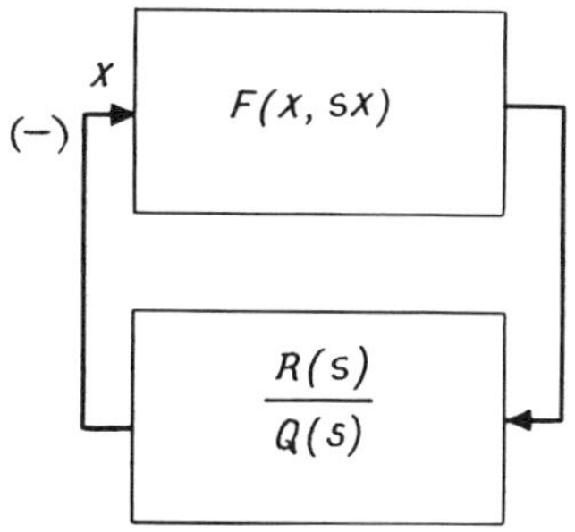

Fig. 6.7

of polynomials of lower degree. Of all complex roots of equation (6.36) the closest to the imaginary axis are determined and taken to be the initial approximations.

When studying systems with two-valued and combined nonlinearities, one should not overlook that the harmonically linearized equations involve the coefficients rather than specific nonlinearities. Therefore it is allowable to vary $a(A)$ and $b(A)$ by a linear or other polynomial law in some range of amplitudes $[A_1, A_0]$ with the end values being selected such that A_1 be at the end of the operating range and A_0 satisfy the condition $a(A_0) = b(A_0) = 0$ and ensure the existence of the derivatives $\mathrm{d}a/\mathrm{d}A$ and $\mathrm{d}b/\mathrm{d}A$ at A_1.

Thus, equation (6.36) holds and its solution obtained with the algorithm, relying in the properties of the surface of analytical function modulus, defines the initial conditions for the Newton iteration formulas (6.30, 34) or for differential equations (6.32, 35).

As the result of these procedures the linearized equation (6.5) is solved on a computer and a functional relationship is established between ξ, ω, and A. The damping factor, settling time of the transient process, overshoot, and the number of oscillations are determined with familiar expressions [6.1] as functions of system structure and parameters.

The method may be extended to the evaluation of parameters of multifrequency transients by the following formulas:

$$
\begin{aligned}
x(t) &= \sum_{i=1}^{m} x_i(t) = \sum_{i=1}^{m} A_i(t) \sin \psi_i(t) \\
\frac{\mathrm{d}A}{\mathrm{d}t} &= A_i \xi_i(A_1, A_2, \ldots, A_m) \\
\frac{\mathrm{d}\psi}{\mathrm{d}t} &= \omega_i(A_1, A_2, \ldots, A_m)
\end{aligned}
\tag{6.37}
$$

Here, m is the number of harmonics in the transient with slowly varying damping factors ξ_i and frequencies ω_i.

Performing a harmonic linearization of the nonlinearity $F(x)$, we get

$$
F(x) = \sum_{i=1}^{m} a_i x_i \tag{6.38}
$$

By way of example, at $m = 2$, we have

$$
F(x) = a_1 x_1 + a_2 x_2
$$

where

$$
a_1(A_1) = \frac{1}{2\pi^2 A_1} \int_0^{2\pi} \int_0^{2\pi} F[A_1 \sin \psi + A_2 \sin (\psi + \varphi)] \sin \psi \, \mathrm{d}\psi \, \mathrm{d}\varphi
$$

$$a_2(A_2) = \frac{1}{2\pi^2 A_2} \int_0^{2\pi} \int_0^{2\pi} F[A_1 \sin \psi + A_2 \sin (\psi + \varphi)] \sin (\psi + \varphi) \, d\psi \, d\varphi$$

Substituting (6.38) in the dynamic equation of the control system with single-valued odd-symmetric nonlinearity

$$Q(p)x + R(p)F(x) = 0$$

and replacing p with $s = \xi_i + j\omega_i$, we arrive, on the lines of the aforementioned method, at the system of $2m$ similar equations of the form

$$(\mathbf{\Omega}^{\mathrm{T}} \mathbf{E}_{\mathrm{R}}^{n} \mathbf{\Xi}_i) = 0, \quad (\mathbf{\Omega}^{\mathrm{T}} \mathbf{E}_{\mathrm{I}}^{n} \mathbf{\Xi}_i) = 0, \quad i = 1, 2, \ldots, m$$

The solutions obtained for frequencies and damping factors are then represented as multifrequency diagrams $\xi(A)$ and $\omega(A)$. For the graphical representation, the sections of the spatial multifrequency pattern are selected where the suggested solutions are valid.

6.4 Investigation of Forced Oscillations in Large Nonlinear Systems

Under forced oscillations, nonlinear control systems can exhibit a jumpwise resonance (Fig. 6.8). This phenomenon is most frequent in multiloop systems involving a few nonlinearities, as is the case with robots. The jumpwise resonance should be located and eliminated by choosing appropriate compensating elements.

Kovalev [6.5] has developed a method of suppression for this type of resonance in large systems with monoharmonic forced oscillations. This method is rather hard to implement because of the difficulties associated, firstly, with the evaluation of the amplitude of forced oscillations by the given amplitude of the periodic forcing function, and, secondly, with the

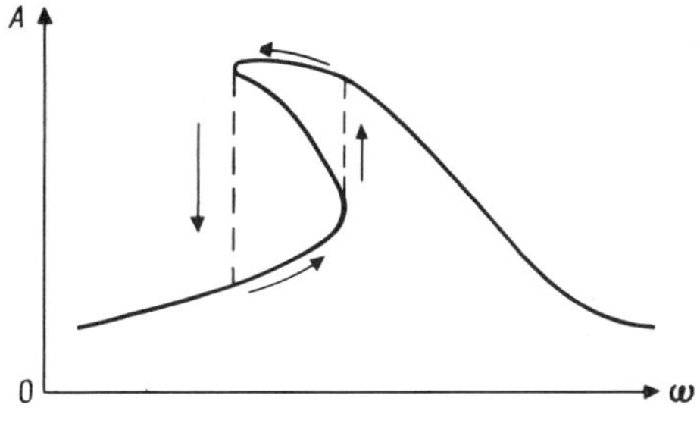

Fig. 6.8

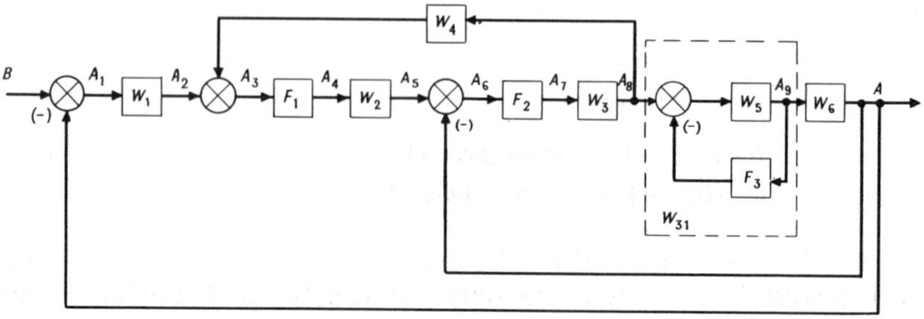

Fig. 6.9

unwieldy equations relating the amplitudes, that have to be formulated for problem solving.

The aforementioned setbacks may be overcome with a new method of constructing frequency response with respect to the fundamental component. This method relies on two procedures: (i) the successfull reduction of a multiloop system by determining an overall harmonically linearized transfer function for each closed loop with nonlinearity, and (ii) the construction of inverse frequency response characteristics.

Example 6.1. Investigate a jumpwise resonance phenomenon in a complicated multiloop system with several nonlinearities as shown in Fig. 6.9. The nomenclature of this representation is as follows: W_1 are the transfer functions of the linear elements, F_i are nonlinearities, B is the amplitude of the periodic (sinusoidal) input, and A_i are the amplitudes of the fundamental components of forced oscillations at various points of the system.

We select a point in the block diagram where for a given amplitude A_i of forced oscillations there is a possibility to eliminate crosslinks and represent the internal loops by their nonlinear overall transfer functions. Inspection of the block diagram shown in Fig. 6.9 indicates that the output signal with amplitude A should be selected as a reference signal.

To simplify the subsequent consideration we assume that all the nonlinearities F_i are single valued and may be replaced by the expressions $a_i(A_i)$ in the harmonic linearization method, although the development may hinge around a more general case of $a_i(A_i, \omega)$. Now, we specify the initial values of the amplitude A and frequency ω of forced oscillations.

Let us determine the nonlinear overall transfer function $W_{\mathrm{eq}1}$ equivalent to the internal loop of the system as

$$W_{\mathrm{eq}1}(j\omega, A) = \frac{W_1(j\omega)}{1 + W_5(j\omega)a_1(A_9)} = K_{\mathrm{eq}1}(\omega, A) \qquad (6.39)$$

where

$$A_9 = \frac{A}{|W_6(j\omega)|}$$

$$K_{eq1}(\omega, A) = |W_{eq1}(j\omega, A)|$$

$$\varphi_{eq1}(\omega, A) = \arg W_{eq1}(j\omega, A)$$

We eliminate the crosslinks by moving the takeoff point of the signal with amplitude A_8 to the system output (Fig. 6.9), and determine the amplitude A_6 by solving the transcendental equation

$$A_6 a_2(A_6) = A_7 \tag{6.40}$$

where

$$A_7 = A_9/[K_{eq1}(\omega, A)|W_3(j\omega)|]$$

and determine the nonlinear equivalent $W_{eq2}(j\omega, A)$ also in the form of modulus and argument (phase) (Fig. 6.10*a*),

$$W_{eq2}(j\omega, A) = \frac{a_2(A_6)W_3(j\omega)W_{eq1}(j\omega, A)W_6(j\omega)}{1 + a_2(A_6)W_3(j\omega)W_{eq1}(j\omega, A)W_6(j\omega)}$$

$$= K_{eq2}(\omega, A)\exp[j\varphi_{eq2}(\omega, A)] \tag{6.41}$$

where $K_{eq2}(\omega, A) = |W_{eq2}(j\omega, A)|$, $\varphi_{eq2}(\omega, A) = \arg W_{eq2}(j\omega, A)$.

As the result of these manipulations the block diagram of the original system reduces to the network shown in Fig. 6.10*b*. Next we solve the transcendental equation

$$A_3 a_1(A_3) = A_4 \tag{6.42}$$

for the amplitude A_3 of the input to the nonlinearity F_1. Here, $A_4 = A/[K_{eq2}(\omega, A)W_2(j\omega)]$.

The respective nonlinear overall transfer function is

$$W_{eq3}(j\omega, A)$$

$$= \frac{a_1(A_3)W_2(j\omega)W_{eq1}(j\omega, A)W_{eq2}(j\omega, A)W_6(j\omega)}{W_{eq1}(j\omega, A)W_6(j\omega) + a_1(A_3)W_2(j\omega)W_{eq2}(j\omega, A)W_6(j\omega)}$$

$$= K_{eq3}(\omega, A)\exp[j\varphi_{eq3}(\omega, A)] \tag{6.43}$$

where $K_{eq3}(\omega, A) = |W_{eq3}(j\omega, A)|$ and $\varphi_{eq3}(\omega, A) = \arg W_{eq3}(j\omega, A)$. We determine the modulus $K(\omega, A)$ and phase $\varphi(\omega, A)$ of the entire open-loop system and then the amplitude of the error A_1 of the overall closed-loop

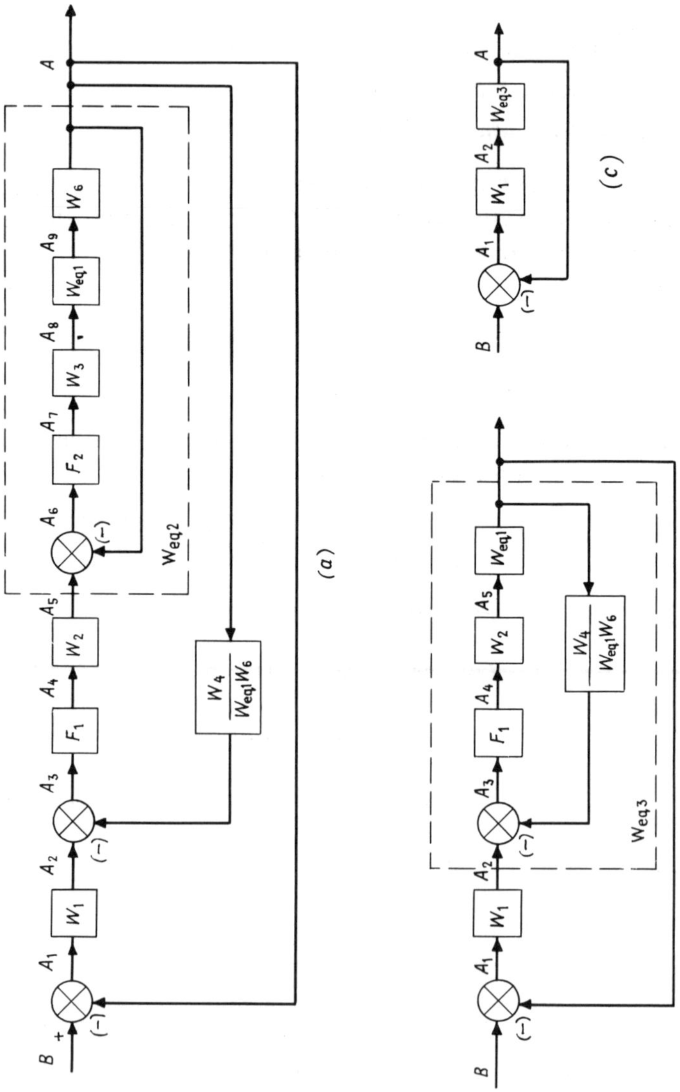
B
+
(−)
A1
W1
A2
(−)
A3
F1
A4
W2
A5
(−)
A6
F2
A7
W3
A8
Weq1
A9
W6
A
Weq2
W4
Weq1W6
(a)
(b)
Weq3
(c)

Fig. 6.10

system (Fig. 6.10*c*)

$$K(\omega, A) = K_{eq3}(\omega, A)|W_1(j\omega, A)| \tag{6.44}$$

$$\varphi(\omega, A) = \varphi_{eq3}(\omega, A) + \arg W_1(j\omega, A) \tag{6.45}$$

$$A_1 = \frac{A}{K(\omega, A)} \tag{6.46}$$

Writing the system response to the error signal as

$$\Phi_\varepsilon(j\omega, A) = \frac{1}{1 + K(\omega, A)\exp[j\varphi(\omega, A)]} \tag{6.47}$$

and

$$\Phi_\varepsilon(j\omega, A) = \frac{A_1}{B}\exp[j(\varphi_{A_1} - \varphi_B)] \tag{6.48}$$

and equating the absolute values of the right sides of these expressions, we arrive at the formula for the amplitude B

$$B = A_1|1 + K(\omega, A)\exp[j\varphi(\omega, A)]|$$

or

$$B = A_1\sqrt{1 + 2K(\omega, A)\cos\varphi(\omega, A) + K^2(\omega, A)} \tag{6.49}$$

From the transfer function of the closed-loop system

$$\Phi(j\omega, A) = \frac{K(\omega, A)\exp[j\varphi(\omega, A)]}{1 + K(\omega, A)\exp[j\varphi(\omega, A)]}$$

it follows the expression for the phase

$$\psi(\omega, A) = \varphi(\omega, A) - \arctan\frac{K(\omega, A)\sin(\omega, A)}{1 + K(\omega, A)\cos\varphi(\omega, A)} \tag{6.50}$$

We shall refer to the dependence (6.49) of the input amplitude B on the given amplitude A, at the forced-oscillation frequency, as the amplitude characteristic plot of the closed-loop system. Likewise we shall refer to the dependence (6.50) of the phase $\psi(\omega, A)$ of the closed-loop transfer function on the given amplitude A, at the forced oscillation frequency, as the phase characteristic plot. We are in a position, therefore, to construct two families of amplitude and phase characteristic plots by specifying a series of amplitude values at each frequency of forced oscillations in the frequency range under study (Fig. 6.11). The calculation of these plots can readily be relegat-

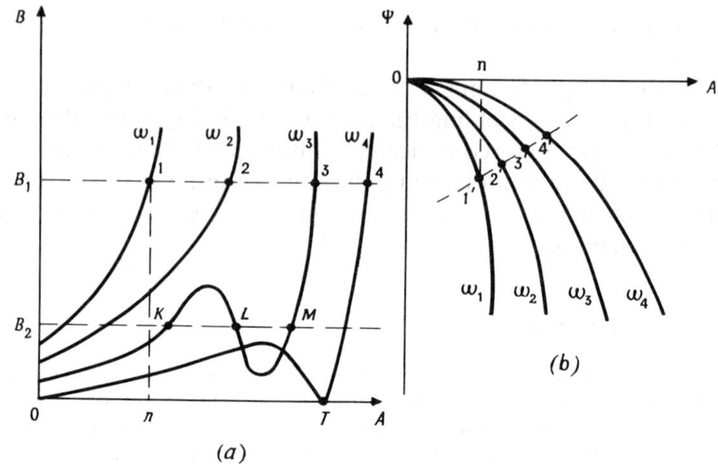

Fig. 6.11

ed to a computer. Transcendental equations (6.40) and (6.42) can then be solved by one of the known approximate techniques, say by the chord method.

In order to construct the Bode diagram for any given amplitude $B = \text{const}$, say B_1, this level is drawn in the B versus A diagram (dashed line in Fig. 6.11*a*). The intersections of this line with the family of plots define the ratios A/B corresponding to the frequencies ω_1, ω_2, etc. which form the base for construction of the Bode plot. The points of intersection *1*, *2*, etc. in the *A-B* plot are associated with the points *1'*, *2'*, etc. on the phase characteristic plot, as shown in Fig. 6.11*b*. These points define the phase response $\psi(\omega)$ of the closed loop system at $B = B_1$.

The relevant graphical procedure is clear with reference to Fig. 6.11. Point *1* located on the curve constructed for frequency ω_1 is seen to correspond to point *1'* with the same abscissa n in the $\psi(A)$ plot on the phase curve constructed for the same frequency ω_1. By evaluating the corresponding points one can easily construct the families of magnitude and phase response characteristics for a series of amplitudes B of the input signal.

The amplitude characteristic plot may also reveal a jumpwise resonance phenomenon as in Fig. 6.11*a* where to a signal with amplitude B_2 there correspond three values of amplitude A for every frequency in the band around ω_3. The points K and M belong to the stable segments of the plot satisfying the conditions $\partial B/\partial A > 0$, whereas L belongs to an unstable segment for which $\partial B/\partial A < 0$. Consequently, this method is helpful in locating stable and unstable segments of the plots with jumpwise resonance (cf. Fig. 6.8) revealed by drawing several lines of $B = \text{const}$ type. It is worth noting that the point of tangency T on the A axis corresponds to self-excited oscillations.

Combining the nonlinear block-diagram reduction procedure with the graphical technique yielding the plots shown in Fig. 6.11 substantially simplifies investigation of complicated multiloop nonlinear systems. The plots can be a handy tool in obtaining the compensating elements that eliminate undesired oscillatory modes in the nonlinear system subjected to periodic forcing functions.

References

6.1 E. P. Popov: *Prikladnaya teoriya protsessov upravleniya v nelineinykh sistemakh* (Applied theory of nonlinear system control). Nauka, Moscow 1973.

6.2 Yu. I. Topcheyev: in *Sovremennye metody proektirovaniya sistem avtomaticheskogo upravleniya* (Modern methods of control system design). Mashinostroyenie, Moscow 1967.

6.3 E. A. Kotov and E. P. Popov: *Izv. Akad. Nauk SSSR, Tekhn. Kibernetika*. No. 3. (1983).

6.4 B. V. Shabat: *Vvedenie v kompleksnyi analiz* (Introduction to complex analysis). Nauka, Moscow 1978.

6.5 B. Ye. Kovalev: *Izv. Akad Nauk SSSR. Tekhn. Kibernetika*. No. 4, (1983).

7 Logical Control of Nonlinear Systems

V. A. Gorbatov

Modern industry and research are abundant in examples of nonlinear systems of control for dynamic plants and technological processes. These are, e.g., hot rolling and zonal annealing of metal, removal of methane from the headings of coal mines, forecasting of the quality of raw material from the geochemical model of a quarry, and confinement of a plasma filament within a tokamak [7.1]. The processes in nonlinear systems exhibit such behavioral features as a possibility of a few equilibrium states, self-excited oscillations, bifurcations, etc. This behavior stems from the fact that nonlinear systems contain elements with multivalued characteristics such as backlash, hysteresis, Coulomb friction, as well as interconnected loops.

The development of nonlinear system theory has been a field of international scientific effort. Among the Soviet researchers known for their significant contributions to this field we would like to note A. A. Andronov, N. N. Bogoljubov, S. B. Emelyanov, N. M. Krylov, V. S. Kulebakin, B. N. Naumov, G. S. Pospelov, Yu. I. Topcheyev, Ya. Z. Tsypkin, A. A. Vitt, and A. A. Voronov.

The development of fitting techniques [7.2], the methods of sewing of phase plane regions, and systems with variable structure [7.3] supported by the recent advances in LSI circuit engineering have paved the way for the incorporation of the principles of logical control [7.4-10] in nonlinear control theory. In general, logical control can be implemented in a software or hardware form. A particular way of implementation is governed by the particular problem, applicational area, and the performance criteria.

Examples of hardware implementation are various analog and digital circuits that function according to the idea realized in their structure by the designer. A software implementation implies the use of computers of microprocessors. In this case the system performs the desired function by the command of a respective program which modifies the control law without having to rebuild the system network.

The major factors influencing the choice of an implementation of control are versatility and speed. The hardware implementation is less versatile but fast, whereas the software implementation is more versatile but limited in speed, though this limitation becomes less significant with the advance in LSI circuit technology. By the time of writing analog signals have been sampled at a speed of up to 5×10^6 samplings per second with an accuracy of 10-14 bits [7.1]. Available A/D and D/A converters are capable of handling bandwidths up to several tens of megahertz, thus ensuring effective logical control of nonlinear systems.

7.1 Microprocessor Implementation of Logical Control: Mathematical Description

One of the models of logical control is a control automaton (state machine) represented by a collection of six objects

$$\langle S, X, Y, \varphi, \lambda, S_0 \rangle$$

where $S = \{X_0, S_1, \ldots, S_{r-1}\}$ is the finite set of internal states with S_0 being the initial state of the automaton; $X = \{X_1, X_2, \ldots, X_n\}$ is the set of input vectors; $Y = \{Y_1, Y_2, \ldots, Y_m\}$ is the set of output vectors; and φ and λ are the functions of transitions and outputs, respectively, that associate with some pairs "internal state-input vector" an internal state $S_i = \varphi(S_j, X_k)$ of the automaton and an output vector $Y_l = \lambda(S_j, X_k)$, where S_i, $S_j \in S$, $X_k \in X$, and $Y_l \in Y$.

If the functions of transitions and outputs are given for all internal state-input vector pairs, then the automaton is said to be *completely defined*, otherwise it is *partially defined.* Depending on the current internal state and external signals, the automaton generates a sequence of control signals applied according to the realized algorithm to the controlled plant.

The automaton may be represented as a facility realizing the single-valued mapping to the right $XS^+ \rightarrow S^- Y$, where S^- is the identifier of the internal state to which the system is steered from the state S^+ by the input vector X, provided that the initial state S_0 is specified. A subsequent value of internal state is obtained by assigning to it the value S^- computed in the previous step, i.e., $S^+(t = 0) = S_0$ and $S^+(t + \tau) = S^-(t)$, where τ is the time of transition from one internal state to another.

The mapping $XS^+ \rightarrow X^- Y$ is the result of transformation (translation) of a wordy description of the designed system. Each element of the set S is a symbol of some abstract alphabet of states (nonterminal symbol). The elements of the sets X and Y are vectors whose elements are symbols of the structural alphabet of a logic used in the implementation of the designed control system. The sets X and Y (terminal symbols) define the relation of the controller with the controlled process, or plant. Each component of a vector of set X is associated on a one-to-one basis with an input channel of the automaton which conveys information from the plant to the controller. The components of vectors of set Y specify the signals defining the system response to an input signal as a function of a particular state of the automaton. Input signals of the automaton control the actuators of the nonlinear system.

The automata mapping $XS^+ \to S^- Y$ is as a rule specified as the transition graph $G_t = \langle S, (U, (X, Y))\rangle, (U \subset S^2)$ whose vertices uniquely correspond to the internal states of the automaton, and the vertices correspond to transitions between them. Each edge is weighted by a pair of vectors (X, Y). To reduce the specification of the automaton, parallel edges of the transition graph may be merged. In this case some components of input and output vectors will be denoted by a dash implying that for the transition concerned the value of signal assigned to the given component is insignificant for the system function.

Let us consider a microprocessor implementation of the logic control system. Whatever the elemental basis employed, the microprocessor system is a circuit organized on a bus basis (Fig. 7.1). For input and output of information, such systems use addressable I/O registers, or ports, whose width is decided by the width of the processor. Access to such a port is available through either I/O instructions, or through data transfer instructions with variable address modes. With this organization of data flow, information is simultaneously read, or a limited number of values of logical signals are output.

The reliability and speed of the system depends on which logic variables are conveyed to a particular port. In principle, all variables of a certain function can be sent to an individual port, then the number of ports will be equal to the number of functions. If some functions will read from the same port, then the total number of ports can be decreased in proportion. The problem of minimization of the number of ports will be taken up below.

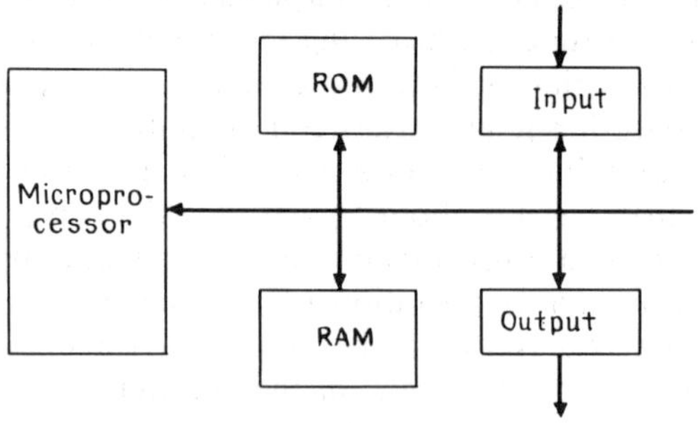

Fig. 7.1

The problem of assignment of the logical variables of Boolean functions to the input ports of the microprocessor system is formulated as follows. Given a set of variables $X = \{x_1, x_2, \ldots, x_n\}$ and a set of subsets of this set $F = \{f_1, f_2, \ldots, f_k\}$, where f_i is the set of variables on which the i-th function depends, and $f_i \subset X = \bigcup_{i=1}^{k} f_i$. A model $\psi = \langle X, F\rangle$ specifies the assignment of variables to the functions. These variables must be assigned to the ports of the microprocessor system.

The channels of the system are specified by a set Y whose decomposition into subsets $Y_1, Y_2, \ldots, Y_m$ of equal power describes the assignment of channels to the ports.

An assignment of the logical variables to the channels of a microprocessor system consists essentially in finding a function $\delta\colon X \to Y$ satisfying certain requirements, namely,

(i) all $\delta(x_i)$ with $x_i \in f_j$ and $F_j \in F$ belong to some Y_l with $l \in \{1, 2, \ldots, m\}$, i.e., all variables of some Boolean function are read from the same port;

(ii) all $\delta(x_i)$ belong to the integral number $]m/L[$ of the sets Y_l, $l \in \{1, 2, \ldots, m\}$, where $L = |Y_l|$, that is, all the variables of a Boolean function are read from the minimal number of ports that is defined by the width of the port and the number of variables in the function.

The problem may be solved under different constraints of which the ones more frequently encountered are imposed on (a) the number of channels $|Y|$ of the microprocessor system, (b) the number of ports m, and (c) the port width $|Y_l|$, $l = 1, \ldots, m$.

We shall consider the distribution of logical variables over the ports (channels) of the microprocessor system and require that all $\delta(x_i)$ be assigned to some Y_l under the constraint $|Y_l| = N$, i.e., under the assumption that any Boolean function has at most N variables and the port width N is specified [7.11].

If a Boolean function involves more than N variables, reading values of its variables may be in error when implemented in a logic circuit. To evaluate the values of a function involving more than N variables, we have to read the values of variables from more than one port, i.e., more than one operation of data input and transfer is required.

The values of some input variables may alter in the interval between input operations which last for about 5 μs in a single-chip microprocessor. This occasion can lead to erroneous evaluation of values of the Boolean function, implying unreliable operation of the system.

A model $\psi_a = \langle X, F\rangle$ is said to be *allowed* if there is a function δ: $X \to Y$ such that for all $j = 1, 2, \ldots, m$ the requirement is satisfied that all $\delta(x_i)$ with $x_i \in f_j$ belong to some Y_l with $l = 1, 2, \ldots, m$, and the constraint $Y_l = N$.

In order to determine whether or not a model is allowed we resort to the characterization analysis [7.4, 5, 7] which defines a forbidden figure in the mograph of the model $\psi = \langle X, F \rangle$ rendering this model nonallowed. The model becomes allowed when at least one element of the forbidden figure is removed or splitted.

In conformity with the methodology of the characterization analysis we determine the relation on subjugation (submodel). We shall call $\psi' = \langle X, F' \rangle$ with $F' \subset F$ a submodel of the model $\psi = \langle X, F \rangle$, i.e., a submodel covers only some part of the Boolean functions.

If a model $\psi = \langle X, F \rangle$ is allowed, then any its submodel is allowed too, for if ψ is allowed, there exists $\delta: X \to Y$. Indeed, a removal of any f_j from F such that $f_j \in F$ but $f_j \notin F'$ does not affect the constructed function δ, consequently, this function may be employed for ψ' and this submodel is allowed.

Turning to the mograph G^{M} connected with the model on a one-to-one basis, we formulate a criterion for a model to be allowed.

A model $\psi = \langle X, F \rangle$ is allowed if and only if its mograph does not contain a connected graph with the number of vertices exceeding N (port width).

By way of example, consider a system of Boolean functions of six variables $X = \{x_1, x_2, \ldots, x_6\}$. The subsets of the set X define the variables of the three functions $f_1 = \{x_1, x_2, x_4\}$, $f_2 = \{x_2, x_3, x_4, x_5\}$, and $f_3 = \{x_2, x_4, x_6\}$. The mograph corresponding to the model $\psi = \langle X, F \rangle$ is presented in Fig. 7.2*a*.

When $|Y_l| = N = 4$, the mograph G^{M} contains a connected subgraph with more than four vertices, i.e., the model is not allowed. After the splitting of the variables x_2 and x_4, we obtain an allowed model ψ_a. The mograph of this model is shown in Fig. 7.2*b*.

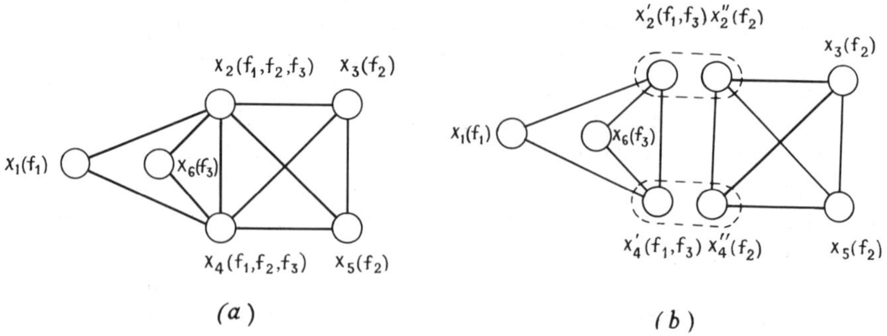

Fig. 7.2

The allowed model ψ_a uniquely defines the function δ, however for the assignment to be optimal an allowed model is required that minimizes the performance measure $\varkappa(\psi)$. When a function is constructed with a requirement that all $\delta(x_i)$ be assigned to some Y_l subject to the constraint $|Y_l| = N$, $l = 1, 2, \ldots m$, then the governing criterion is the number of ports m in the microprocessor system. The functional $\varkappa(\psi)$ is defined by the number of connected components in the model graph. Minimization of this graph is fundamental to the optimal model construction.

A model ψ_0 is *optimal* if it is allowed and the functional $\varkappa(\psi_0)$ is a minimum.

Because the optimal model is allowed, a necessary condition for its optimality is the absence of connected components of power more than N. The converse is not true since an admissible model can be nonoptimal by virtue of its performance measure being other than minimal.

The search for an optimal model is a combinatorial problem efficiently solvable on the basis of the characterization analysis.

In the model graph, a word f_α is said to ***absorb*** a word f_β if $f_\beta \subset f_\alpha$. The absorbed words do not affect the assignment of variables over the ports of the microprocessor system, for if the variables of the word f_α are distributed at port Y_l, then the variables of f_β can be read from the same port Y_l.

A vertex v_i of the graph G^M (and respectively the variable x_i) is called *nonessential*, if x_i belongs to only one port, otherwise the vertex (variable) is termed *essential*. Nonessential vertices (variables) do not influence the distribution of variables since their splitting is meaningless. On the other hand, a splitting of an essential vertex v_i (variable x_i) means that this variable will be assigned to two ports. In graphical representations, the essential vertices of mographs will be shown by full circles.

The weight α_i of a vertex v_i is the number of nonessential vertices adjacent to this vertex plus unity. For illustration, see Fig. 7.3*a*. The weight of a vertex defines the number of variables which must be carried to the port Y_l when it receives the variable x_i corresponding to vertex v_i.

The *weight* α_{ij} of a mograph edge (v_i, v_j) is the number of paths from v_i to v_j passing through exactly one nonessential vertex (Fig. 7.3*b*). The weight of an edge controls the number of variables that must be assigned to the port Y_l once it has received the variables x_i and x_j corresponding to vertices v_i and v_j.

The characterization analysis aids in arriving at a desirable assignment of variables by way of a transition from the mograph G^M specifying the distribution of variables of Boolean functions to a graph G according to the following rules:

— discard all absorbed words;

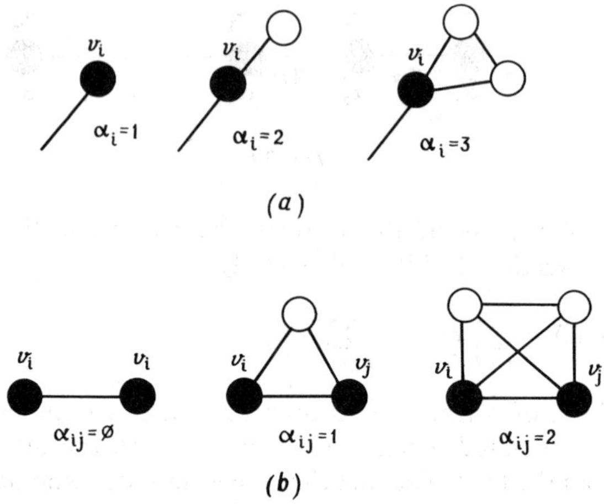

Fig. 7.3

— partition all vertices into essential and nonessential by blackening the former;

— determine the weights of all essential vertices and all edges (v_i, v_j) incident on essential vertices.

The graph G defines a model which may be used as a basis for the semantic equivalent transformation [7.5, 10] carrying this model to the optimal one.

An essential vertex v_i must be split if its degree $s(v_i) \geqslant N$, where N is the port width. Indeed, if the degree is equal or more than N, then the allocation of the respective variable to port Y_l with $|Y_l| = N$ invalidates the transfer of all required variables to this port.

In the transformation of a model defined by the graph G, *the forbidden figure* G_f^1 *is a vertex of degree* $s(v_i) \geqslant N$ (Fig. 7.4). This is a necessary condition, for even if the graph has no such vertices, it can be nonallowed and, consequently, nonoptimal (Fig. 7.5).

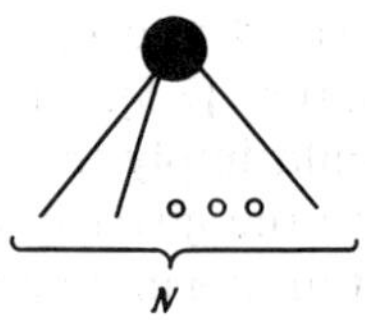

Fig. 7.4

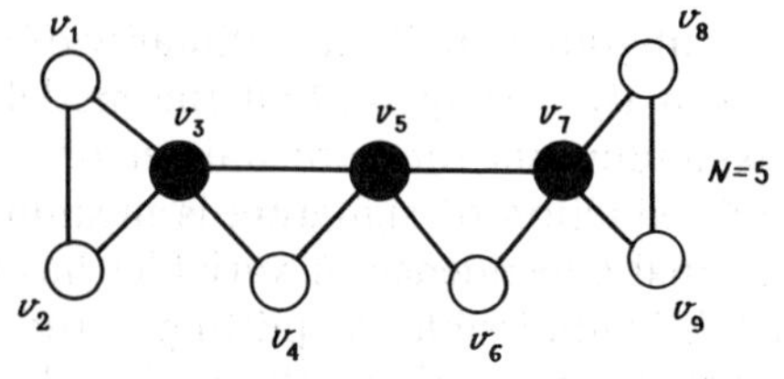

Fig. 7.5

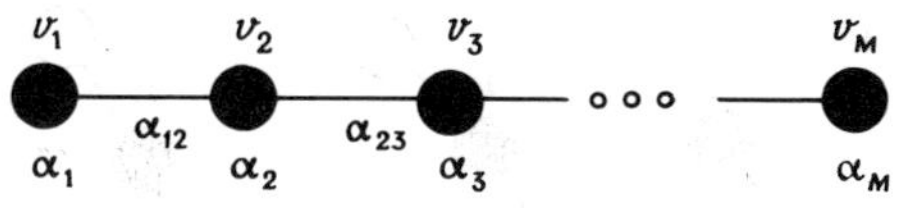

Fig. 7.6

The sum of weights of the vertices and edges constituting a trail C makes up the weight of this trail, namely,

$$\alpha(C) = \sum_{v_i \in C} \alpha_i + \sum_{(v_i,\ v_j) \in C} \alpha_{ij} \tag{7.1}$$

The weight α of a trail C is comparable with a number A, if $\alpha \geqslant A$. This relation is violated when the end vertex with the arc incident on it is excluded from the trail. The fact that the weight of some trail is comparable with A indicates that the vertices constituting the trail should be splitted when the variables are assigned to ports of width $A - 1$, since such a trail may give rise to a connected component with more than $A - 1$ vertices.

In transforming a model into optimal, the *forbidden figure* G_f^2 is a trail C constituted by essential vertices of weight $\alpha(C)$ comparable with $N + 1$ (Fig. 7.6).

The forbidden figures of the model transformation $\psi \to \psi_0$ are a handy tool in assigning the logic variables of a system of Boolean functions to the ports of the microprocessor system.

By way of example consider a system of functions with the respective graph G (Fig. 7.7*a*). The variables of these functions are to be distributed over ports of width $N = 5$. An analysis of the model for optimality reveals that it is nonoptimal since it involves a connected subgraph $\{v_1, v_2, v_3, v_4, v_5, v_6\}$ with the number of vertices exceeding five. For this example,

$$G_{f1}^1 = v_4$$
$$G_{f2}^1 = v_{12}$$
$$G_{f1}^2 = (v_4,\ v_6,\ v_8)$$
$$G_{f2}^2 = (v_8,\ v_{10},\ v_{12})$$

These subgraphs spoil the optimality of the model graph G. The characterization analysis suggests that the model can be transformed into optimal by expanding its fundamental set or reducing its signature [7.5]. In this case a reduction of signature is meaningless as this would imply that the functional dependence of variables has to be changed in the given system which is inadmissible. A splitting of the fundamental set means in practical terms that some variable should be allocated to several ports, therefore we stick to this way of optimization.

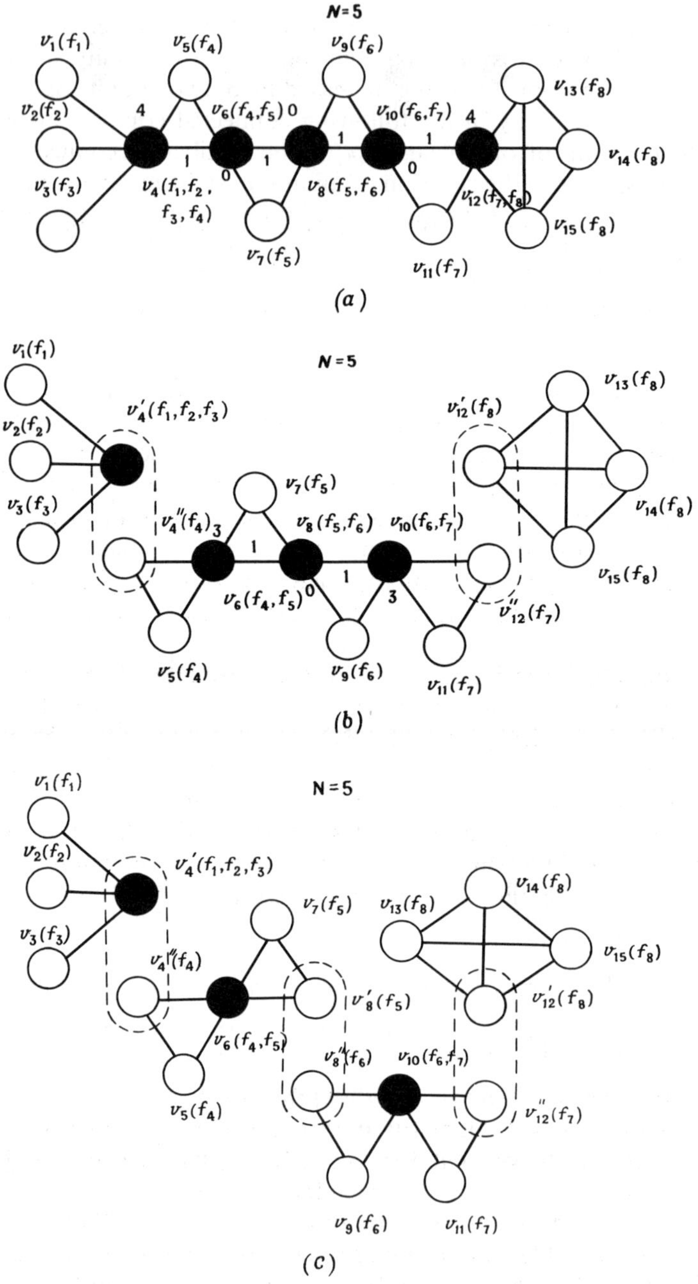
N=5
v1(f1)
v2(f2)
v3(f3)
v4(f1,f2,f3,f4)
v5(f4)
v6(f4,f5)
v7(f5)
v8(f5,f6)
v9(f6)
v10(f6,f7)
v11(f7)
v12(f7,f8)
v13(f8)
v14(f8)
v15(f8)
(a)
N=5
v'4(f1,f2,f3)
v''4(f4)
v'12(f8)
v''12(f7)
(b)
N=5
v'8(f5)
v''8(f6)
(c)

Fig. 7.7

Splitting n vertices of the graph results in that m connected components are increased by at most n units. Splitting one vertex results in the number of connected components being increased by only one unit. By performing a splitting sequentially in the following connected component, we obtain that the criterion of optimization $\varkappa$ of the number of connected components corresponds to the criterion of minimum splittings in the model transformation $\psi \rightarrow \psi_0$. Therefore, the minimum number of splittings in the graph G results in the minimum number of connected components, that is, in the optimal model.

The semantic table for this example is given below

Forbidden figure	Vertex				
	v_4	v_6	v_8	v_{10}	v_{12}
G^1_{f1}	1	0	0	0	0
G^1_{f2}	0	0	0	0	1
G^2_{f1}	1	1	1	0	0
G^2_{f2}	0	0	1	1	1

The minimum cover of the table is $\Pi = \{v_4, v_{12}\}$. This cover is associated with the splitting of the variables x_4, x_{12}; $x_4 \leftrightarrow v_4$, and $x_{12} \leftrightarrow v_{12}$ (Fig. 7.7*b*). The splitting decomposes the mograph into three connected components one of which contains the forbidden figure $G_{f2} = \{v_6, v_8, v_{10}\}$, because the weight of the trail (v_6, v_8, v_{10}), $3 + 3 + 1 + 1 = 8$, is comparable with $N + 1 = 5 + 1$. By splitting v_8, we finally arrive at the optimal model ψ_0 shown in Fig. 7.7*c*.

Consider two computer-adaptable algorithms of search for an assignment of logical variables over the input ports of the microprocessor system.

1. *Algorithm based on the semantic table cover*

For problems of small size, this algorithm can be rather easily adapted for computer implementation. The underlying idea is an exhaustive search through the connected components in the original graph. The solution is a cover of the table where rows are Boolean functions and columns are allowed figures defining sets of functions which may be allocated to one port.

The dependence of the Boolean functions on their variables is specified by the mograph G^M of the model $\psi = \langle X, F \rangle$. The width of a port $Y_l = N$.

The *allowed figure* G_ϱ is a subgraph of the graph G^M such that contains at most N vertices and includes complete subgraphs corresponding to words

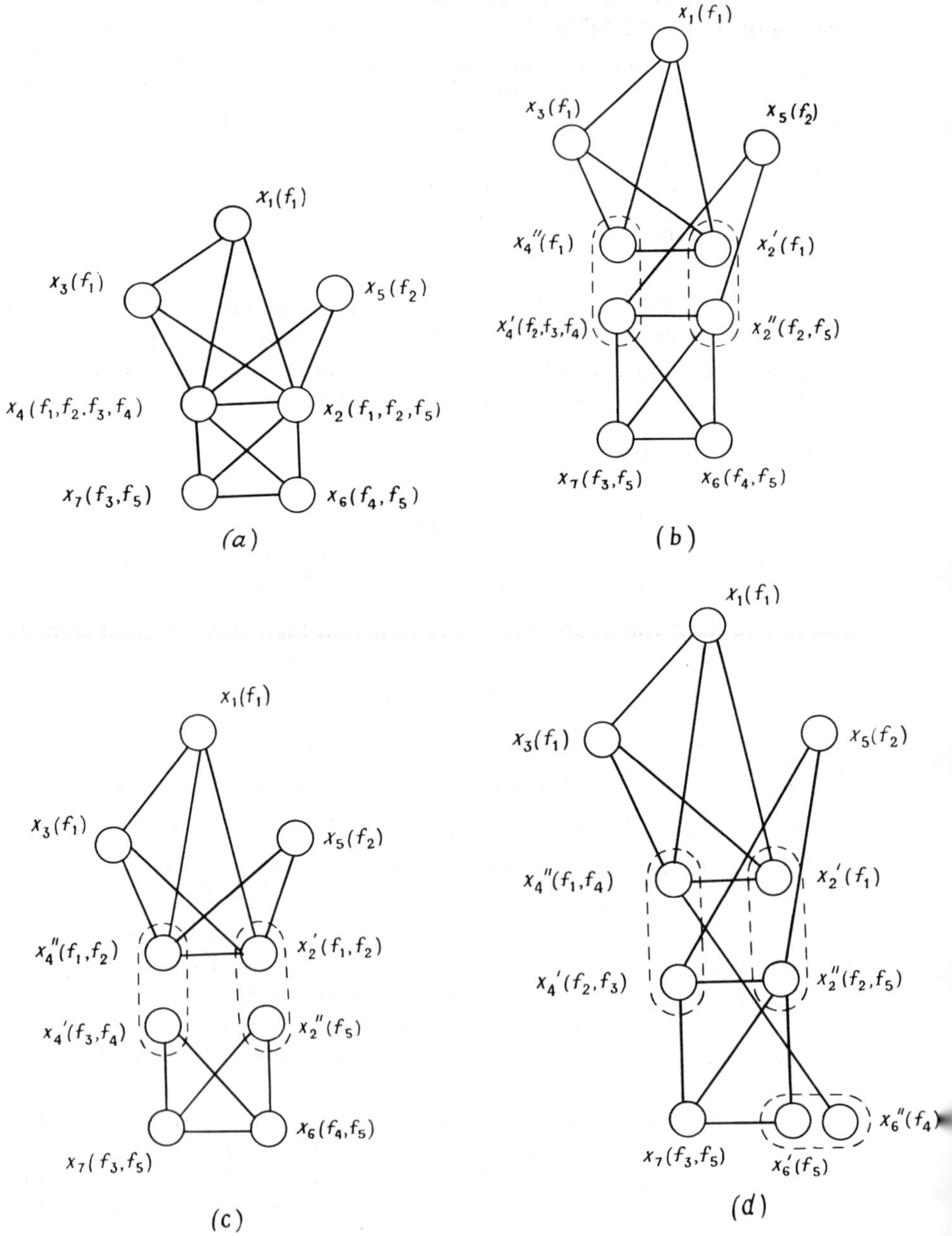

$x_1(f_1)$
$x_3(f_1)$
$x_5(f_2)$
$x_4(f_1,f_2,f_3,f_4)$
$x_2(f_1,f_2,f_5)$
$x_7(f_3,f_5)$
$x_6(f_4,f_5)$
(a)
$x_1(f_1)$
$x_3(f_1)$
$x_5(f_2)$
$x_4''(f_1)$
$x_2'(f_1)$
$x_4'(f_2,f_3,f_4)$
$x_2''(f_2,f_5)$
$x_7(f_3,f_5)$
$x_6(f_4,f_5)$
(b)
$x_1(f_1)$
$x_3(f_1)$
$x_5(f_2)$
$x_4''(f_1,f_2)$
$x_2'(f_1,f_2)$
$x_4'(f_3,f_4)$
$x_2''(f_5)$
$x_6(f_4,f_5)$
$x_7(f_3,f_5)$
(c)
$x_1(f_1)$
$x_3(f_1)$
$x_5(f_2)$
$x_4''(f_1,f_4)$
$x_2'(f_1)$
$x_4'(f_2,f_3)$
$x_2''(f_2,f_5)$
$x_6''(f_4)$
$x_7(f_3,f_5)$
$x_6'(f_5)$
(d)

Fig. 7.8

f_j if $x_i \in G$ and $x_i \in f_j$. The optimal and allowed models consist of allowed figures G_ϱ.

Partitioning G^{M} into allowed figures, we obtain components of the optimal model. In the table T

f_i	G^1 G^2 ... G^k
f_1	$\vdots$
f_2	... t_{ij} ...
$\vdots$	$\vdots$
f_s	

where $t_{ij} = 1$ if $f_i \in G^j$, a cover of rows by columns corresponds to an allowed model. Selecting the minimum cover, we arrive at the optimal model having the least number of nonconnected components, i.e., the minimum number of ports of the microprocessor system.

Example 7.1. Let

$$X = \{x_1, x_2, x_3, x_4, x_5, x_6, x_7\},$$
$$F = \{f_i, i = 1, 2, \ldots, 5\}, \qquad N = 5,$$
$$f_1 = \{x_1, x_2, x_3, x_4\}, f_2 = \{x_2, x_4, x_5\},$$
$$f_3 = \{x_4, x_7\}, f_4 = \{x_4, x_6\}, f_5 = \{x_2, x_6, x_7\}$$

The mograph of the model $\langle X, F\rangle$ is shown in Fig. 7.8*a*. We shall identify allowed figures by symbols f_j.

The allowed figures for this model are

$G_\varrho^1 = \{f_1\}$, $G_\varrho^2 = \{f_2\}$, $G_\varrho^3 = \{f_3\}$, $G_\varrho^4 = \{f_4\}$, $G_\varrho^5 = \{f_5\}$,
$G_\varrho^{12} = \{f_1, \mathrm{f}_2\}$, $G_\varrho^{13} = \{f_1, f_3\}$, $G_\varrho^{14} = \{f_1, f_4\}$, $G_\varrho^{23} = \{f_2, f_3\}$,
$G_\varrho^{24} = \{f_2, f_4\}$, $G_\varrho^{25} = \{f_2, f_5\}$, $G_\varrho^{34} = \{f_3, f_4\}$, $G_\varrho^{35} = \{f_3, f_5\}$,
$G_\varrho^{234} = \{f_2, f_3, f_4\}$, $G_\varrho^{235} = \{f_2, f_3, f_5\}$, $G_\varrho^{345} = \{f_3, f_4, f_5\}$,
$G_\varrho^{2345} = \{f_2, f_3, f_4, f_5\}$

with the table T of the form

f_i	1	2	3	4	5	12	13	14	23	24	25	34	35	234	235	345	2345
f_1	1					1	1	1									
f_2		1				1			1	1	1			1	1		1
f_3			1				1		1			1	1	1	1	1	1
f_4				1				1		1		1		1		1	1
f_5					1						1		1		1	1	1

The covers {1, 2345}, {12, 345}, and {14, 235} represented in Fig. 7.8*b*, *c*, and *d*, respectively, define the optimal models with two ports, namely,

$$\{1, 2345\} \to G_\varrho^{1} \cap G_\varrho^{2345} = \{f_1\} \cap \{f_2, f_3, f_4, f_5\} = \{x_1, x_2, x_3, x_4\} \cap \{x_2, x_4, x_5, x_6, x_7\} = \{x_2, x_4\}$$

$$\{12, 345\} \to G_\varrho^{12} \cap G_\varrho^{345} = \{f_1, f_2\} \cap \{f_3, f_4, f_5\} = \{x_1, x_2, x_3, x_4, x_5\} \cap \{x_2, x_4, x_6, x_7\} = \{x_2, x_4\}$$

$$\{14, 235\} \to G_\varrho^{14} \cap G_\varrho^{235} = \{f_1, f_4\} \cap \{f_2, f_3, f_5\} = \{x_1, x_2, x_3, x_4, x_6\} \cap \{x_2, x_4, x_5, x_6, x_7\} = \{x_2, x_4, x_6\}$$

For size reduction in an implementation of the algorithm, the construction of table T may be reduced to the evaluation of its right side only. In other words, evaluation of allowed figures G_ϱ should be started with subgraphs involving larger number of words f_j proceeding in decreasing order. When the cover is found, it will determine the optimal model with the least number of connected components.

2. *Approximate algorithm*

This algorithm joins variables entering most of the Boolean functions to assign them to one port. It is based on the frequency relation matrix **F** [7.12]. The incidence matrix **Q** [7.4] specifies a system of Boolean functions as follows: each row corresponds to a logic function and each column corresponds to the joined set X.

Initially $X = \{x_j\}$, which implies that a column is associated with a variable from the joined set X for all functions. The algorithm unites columns subject to the criterion max f_{ij} and forms a new joined set X for the generated column. This process keeps repeating until only one column of the incidence matrix remains.

The steps involved in this algorithm are as follows.

Step 1. Construct the frequency relation matrix **F** by the matrix **Q** as $\mathbf{F} = \mathbf{Q}^{\mathrm{T}}\mathbf{Q}$.

Step 2. Choose max f_{ij} $(i \neq j)$. If more than one such element exists, then choose the one ensuring more uniform powers of the joined sets.

Step 3. Build the joined set $X_i \cup X_j$ for the selected f_{ij}. If $|X_i \cup X_j| < N$, then the columns are joined and the joined set $X_i \cup X_j$ is considered in place of X_i and X_j. If $|X_i \cup X_j| = N$, then the columns are deleted and the union $X_i \cup X_j$ represents variables addressed to one port. If $|X_i \cup X_j| > N$, then delete row i and column j and go to Step 2.

Step 4. If Q contains more than one column (reduction of the number of columns results from Step 3), then construct a new matrix **Q** and go to Step 1, otherwise to Step 5.

Step 5. End of the algorithm.

Example 7.2. Given: five functions of seven variables

$$f_1 = \{x_1, x_2, x_3, x_4\},$$
$$f_2 = \{x_2, x_4, x_5\},$$
$$f_3 = \{x_4, x_7\},$$
$$f_4 = \{x_4, x_6\},$$
$$f_5 = \{x_2, x_6, x_7\}$$

that must be assigned to ports of width $N = 5$.

The matrix $\mathbf{Q}$ has the form

$$\mathbf{Q} = \begin{array}{c} \begin{matrix} x_1 & x_2 & x_3 & x_4 & x_5 & x_6 & x_7 \end{matrix} \\ \begin{bmatrix} 1 & 1 & 1 & 1 & 0 & 0 & 0 \\ 0 & 1 & 0 & 1 & 1 & 0 & 0 \\ 0 & 0 & 0 & 1 & 0 & 0 & 1 \\ 0 & 0 & 0 & 1 & 0 & 1 & 0 \\ 0 & 1 & 0 & 0 & 0 & 1 & 1 \end{bmatrix} \end{array} \quad \begin{matrix} 1 \\ 2 \\ 3 \\ 4 \\ 5 \end{matrix}$$

Step 1. Construct the frequency relation matrix $\mathbf{F}$,

$$\mathbf{F} = \begin{array}{c} \begin{matrix} x_1 & x_2 & x_3 & x_4 & x_5 & x_6 & x_7 \end{matrix} \\ \begin{bmatrix} 1 & 1 & 1 & 1 & 0 & 0 & 0 \\ 1 & 3 & 1 & \textcircled{2} & 1 & 1 & 1 \\ 1 & 1 & 1 & 1 & 0 & 0 & 0 \\ 1 & 2 & 1 & 4 & 1 & 1 & 1 \\ 0 & 1 & 0 & 1 & 1 & 0 & 0 \\ 0 & 1 & 0 & 1 & 0 & 2 & 1 \\ 0 & 1 & 0 & 1 & 0 & 1 & 2 \end{bmatrix} \end{array} \begin{matrix} x_1 \\ x_2 \\ x_3 \\ x_4 \\ x_5 \\ x_6 \\ x_7 \end{matrix}$$

Steps 2 and 3. Choose f_{24} and unite $X_2 = \{x_2\}$ and $X_4 = \{x_4\}$.
Steps 4 and 1. Construct $\mathbf{Q}^{\mathrm{i}}$ and $\mathbf{F}^{\mathrm{i}}$,

$$\mathbf{Q}^{\mathrm{i}} = \begin{bmatrix} 1 & 1 & 1 & 0 & 0 & 0 \\ 0 & 1 & 0 & 1 & 0 & 0 \\ 0 & 1 & 0 & 0 & 0 & 1 \\ 0 & 1 & 0 & 0 & 1 & 0 \\ 0 & 1 & 0 & 0 & 1 & 1 \end{bmatrix} \qquad \mathbf{F}_{\mathrm{i}} = \begin{bmatrix} 1 & 1 & 1 & 0 & 0 & 0 \\ 1 & 5 & 1 & 1 & \textcircled{2} & 2 \\ 1 & 1 & 1 & 0 & 0 & 0 \\ 0 & 1 & 0 & 1 & 0 & 0 \\ 0 & 2 & 0 & 0 & 2 & 1 \\ 0 & 2 & 0 & 0 & 1 & 2 \end{bmatrix} \begin{matrix} x_1 \\ \{x_2, x_4\} \\ x_3 \\ x_5 \\ x_6 \\ x_7 \end{matrix}$$

Steps 2 and 3. Choose f^{i}_{25} since f^{i}_{25} and f^{i}_{26} both provide unifying sets of identical power.

Steps 4 and 1. Construct matrices $\mathbf{Q}^{\text{ii}}$ and $\mathbf{F}^{\text{ii}}$,

$$\mathbf{Q}^{\text{ii}} = \begin{bmatrix} 1 & 1 & 1 & 0 & 0 \\ 0 & 1 & 0 & 1 & 0 \\ 0 & 1 & 0 & 0 & 1 \\ 0 & 1 & 0 & 0 & 0 \\ 0 & 1 & 0 & 0 & 1 \end{bmatrix} \quad \mathbf{F}^{\text{ii}} = \begin{bmatrix} 1 & 1 & 1 & 0 & 0 \\ 1 & 5 & 1 & 1 & \textcircled{2} \\ 1 & 1 & 1 & 0 & 0 \\ 0 & 1 & 0 & 1 & 0 \\ 0 & 2 & 0 & 0 & 2 \end{bmatrix} \begin{matrix} x_1 \\ \{x_2,\ x_4,\ x_6\} \\ x_3 \\ x_5 \\ x_7 \end{matrix}$$

Steps 2 and 3. Choose f_{25}^{ii} and join $\{x_2, x_4, x_6\}$ and $\{x_7\}$.

Steps 4 and 1. Compile $\mathbf{Q}^{\text{iii}}$ and $\mathbf{F}^{\text{iii}}$,

$$\mathbf{Q}^{\text{iii}} = \begin{bmatrix} 1 & 1 & 1 & 0 \\ 0 & 1 & 0 & 1 \\ 0 & 1 & 0 & 0 \\ 0 & 1 & 0 & 0 \\ 0 & 1 & 0 & 0 \end{bmatrix} \quad \mathbf{F}^{\text{iii}} = \begin{bmatrix} 1 & 1 & \textcircled{1} & 0 \\ 1 & 4 & 1 & 1 \\ 1 & 1 & 1 & 0 \\ 0 & 1 & 0 & 1 \end{bmatrix} \begin{matrix} x_1 \\ \{x_2,\ x_4,\ x_6,\ x_7\} \\ x_3 \\ x_5 \end{matrix}$$

Steps 2 and 3. Of the frequencies f_{12}^{iii}, f_{13}^{iii}, f_{23}^{iii}, and f_{24}^{iii} we select f_{13}^{iii}, for it gives more uniform unifying sets. Unite $\{x_1\}$ and $\{x_3\}$.

Steps 4 and 1. Construct $\mathbf{Q}^{\text{iv}}$ and $\mathbf{F}^{\text{iv}}$,

$$\mathbf{Q}^{\text{iv}} = \begin{bmatrix} 1 & 1 & 0 \\ 0 & 1 & 1 \\ 0 & 1 & 0 \\ 0 & 1 & 0 \\ 0 & 1 & 0 \end{bmatrix} \quad \mathbf{F}^{\text{iv}} = \begin{bmatrix} 1 & 1 & 0 \\ 1 & 4 & \textcircled{1} \\ 0 & 1 & 1 \end{bmatrix} \begin{matrix} \{x_1,\ x_3\} \\ \{x_2,\ x_4,\ x_6,\ x_7\} \\ x_5 \end{matrix}$$

Steps 2 and 3. The choice of frequency f_{12}^{iv} leads to a joined set whose power exceeds 5. Therefore, we select f_{23}^{iv}, unite $\{x_2, x_4, x_6, x_7\}$ and $\{x_5\}$, and delete the column thus obtained. Now, the variables $\{x_2, x_4, x_5, x_6, x_7\}$ will be allocated to the first port.

Steps 4 and 1. Construct $\mathbf{Q}^{\text{v}}$ and $\mathbf{F}^{\text{v}}$,

$$\mathbf{Q}^{\text{v}} = \begin{bmatrix} 1 & 0 \\ 0 & 1 \\ 0 & 0 \\ 0 & 0 \\ 0 & 0 \end{bmatrix} \quad \mathbf{F}^{\text{v}} = \begin{bmatrix} 1 & \textcircled{0} \\ 0 & 1 \end{bmatrix} \begin{matrix} \{x_1,\ x_3\} \\ x_5 \end{matrix}$$

Steps 2 and 3. Unite $\{x_1, x_3\}$ and $\{x_5\}$. These steps terminate the algorithm as the matrix $\mathbf{Q}$ consists of one column only.

The algorithm results in two sets $\{x_2, x_4, x_5, x_6, x_7\}$ and $\{x_1, x_3, x_5\}$. The functions f_2, f_3, f_4, and f_5 do not require additional expansion of this result. To implement f_1, one has to include x_2 and x_4 in the second set.

This finding differs from the optimal one $\{x_1, x_2, x_3, x_4, x_5\}$ and $\{x_2, x_4, x_6, x_7\}$ by only one extra splitting of x_5, but coincides with this optimal result in the number of ports, a major criterion.

The procedure based on table covering is applicable to assignment problems if the number of functions does not exceed eight, and the total number of variables is within 100. The approximate algorithm completes the assignment for 100 functions and 200 variables in a reasonable time of 10-15 minutes.

7.2 Microprocessor Implementation of Nonlinearities

The mapping method and the input set mapping (for interval representation of a system of Boolean functions) are most appropriate for microprocessor implementation. For the K580IK80 (Intel 8080) microprocessor and the single-chip 1816VE48 series microcomputer this implementation takes on the simplest form when the system consists of eight functions. Choice of a method depends on how many intervals are taken to represent the initial system of functions. If the number of functions is at most 50, the mapping method is more economical in terms of memory space. With this method, for each input vector there is an output vector in the memory of the microprocessor system.

Provided that the table of vectors is on the page BASE, the relevant program for the K580IK80 microprocessor is as follows.

```
        MVI H, BASE
BEGIN:  IN PORT, I
        MOV L, A
        MOV A, M        load the output set into A
        OUT PORT 0
        JMP BEGIN
```

For the single-chip microcomputer, we locate the output vector table at page three, then the program has the form

```
BEGIN:  ORL P1, #FFN      set P1 to IN
        IN A, P1
        MOV P 3A, @ A     load the output set in A
        OUT L P2, A
        JMP BEGIN
```

A comparison of these two programs is given below.

Microprocessor type	K580IK80	1816VE48
Program size, byte	11	7
Output set processing cycle, μs/clock	16.8/42	25.0/10
Memory space occupied by data table, byte	256	256

When the system of Boolean functions is representable by less than 50 intervals (conjunctions), or it is impossible to use the third page of the microcomputer, it would be reasonable to resort to the reflection method. The algorithm of this method reads the input set, reveals the variables involved in a conjunction and compares the set with the vector stored in the memory.

The relevant program for the K580IK80 microprocessor is as follows.

```
        LXI BC 0800      B = 08 is the number of functions
                         C = 00 is a comparison constant
        LXI SP DATA
        IN PORT I
        MOV D, A
LOOP:   POP HL
        XRA A            end of conjunctions
        CMP I            functions ?
        JZ FUN           yes: f = 0, Cr = 0
        MOV A, D         no
        ANA H
        ORA L
        JNZ LOOP
NO:     POP PSW
        CMP C            skip the remaining conjunctions
        JNZ NO
        DEC A            f = 1, Cr = 1
FUN:    MOV A, E
        RAR              set the bit of result
        MOV E, A
        DCR B
        JNZ LOOP
        OUT PORT 0
        JMP BEGIN
```

The implementation of this method for the 1816VE48 microcomputer is as follows.

```
BEGIN:     MOV R1, Ø8
           MOV RØ, ØØ
LOOP:      CLR C
LOOP 1:    MOV A, RØ
           MOV P3 A, @ A
           INC RØ
           MOV R3, A
           ANL A, R3            end of function?
           JZ FUN               yes, f = 0
           MOV A RØ
           MOV P3 A, @ A
           INC RØ
           MOV R4, A
           ORL P1, FF
           IN A, P1
           ANL A, R3
           ORL A, R4
           JNZ LOOP 1
           CPL C                f = 1
NO:        MOV A, RØ
           MOV P3 A, @ A
           INC RØ
           MOV R3, A
           ANL A, R3
           JZ FUN
           MOV A, RØ
           MOV P3 A, @ A
           INC RØ
           JMP NO
FUN:       MOV A, RS
           RL A
           MOV RS, A
           DINZ R1, LOOP
           OUT L P2, A
           JMP BEGIN
```

Both implementations are compared below.

Processor type	K580IK80	1816VE48
Size of program, byte	39	46
Memory space occupied by data, byte	$2N + 16$	$2N + 8$

The processing cycle in both cases depends on the input set and the number of conjunctions. Here, N is the total number of intervals of all functions.

Consider a microprocessor implementation of an asynchronous automaton. The simplest implementation results if the length of input and output vectors does not exceed the width of the microprocessor bus. When this condition is violated, the time taken to process the input vector increases, which results in a reduction of the input bandwidth. In what follows we confine our consideration to the synchronous automata having self-loops at each vertex, which duplicate all other arcs incident with this vertex. A dash in the output vector of the transition graph implies retention of the preceding value of the respective output logic variable.

Let the input and output vectors be 8-types and the system be constituted by the basic components of the 8080 microprocessor chip set.

Information about each arc occupies six bytes in the system memory: two bytes are allocated to the input vector X, two to the output vector Y, and two to indicate the address A_j in the memory which corresponds to the start of the information pertinent to vertex S_j which is the end vertex of a transition from S_i performed to meet the condition X. The data on all arcs incident with a vertex are written in the memory sequentially and in random order.

To increase the speed of automaton operation, the arcs may be sorted beforehand by the prior probability of their occurrence, if available, or by the probability of occurrence derived from simulation of the designed automaton. We denote each of six bytes specifying an arc in the memory by B^n, and a k-th bit of the n-th byte by B_k^n, where $n = 1, \ldots, 6$, and $k = 1, \ldots, 8$. The input vector X is written in two-low order bytes. The first of them serves as a mask to select the required variables so that

$$\begin{aligned} B_k^1 &= 0, \text{ if } x_k = -; \quad k = 1, \ldots, 8 \\ &= 1, \text{ if } (x_k = 0) \vee (x_k = 1) \end{aligned}$$

where x_k is the k-th variable in $X = (x_1, x_2, \ldots, x_8)$.

The second byte defines the values of essential variables and

$$\begin{aligned} B_k^2 &= 0, \quad \text{if } (x_k = 0) \vee (x_k = -) \\ &= 1, \quad \text{if } x_k = 1 \qquad\qquad k = 1, \ldots, 8 \end{aligned}$$

The next two bytes B_k^3 and B_k^4 define vector Y, the mask defining the variables whose values must be kept

$$B_k^3 = 0, \quad \text{if } (x_k = 0) \vee (x_k = 1)$$
$$= 1, \quad \text{if } x_k = -1 \qquad k = 1, \ldots, 8$$

The byte B_k^4 defines the remaining variables

$$B_k^4 = 0, \quad \text{if } (x_k = 0) \vee (x_k = -)$$
$$= 1, \quad \text{if } x_k = 1 \qquad k = 1, \ldots, 8$$

The address A of a transition corresponding to vertex S_j is written in bytes B^5 and B^6.

An arc weighted, say, by a pair (0-1010-1, 01--001-) is represented in the microsystem memory as follows:

10111101 00101001 00110001 01000010 00001001 10101111

The last two high-order bytes define the address $AF09_{16}$ controlling the transition. The location of these bytes corresponds to that of 16-bit address bytes in the stack of a 8080 microprocessor. Specifying an arc by a triad $\langle X, Y, A \rangle$ uniquely defines the arc, for the absolute location of this triad in memory gives the start vertex.

If INPUT is the input vector of the automaton, then the equality of INPUT & B^1 and B^2 indicates that the condition X is satisfied. If so, the output receives the vector OUTPUT = OUTPUT & $B^3 \vee B^4$.

The algorithm of the program supervising the automaton operation includes two cycles. The first of them is executed until among the arcs leaving vertex S_i (located beginning with the address A_i) an arc is found for which the condition X is satisfied. If no such arc has been revealed, the algorithm comes to check the next arc incident with the vertex S_{i+1} and following the last arc of S_i, i.e., the automaton takes an incorrect operation mode as if after an unconditional transition to another vertex. To avoid such a situation, it is recommended to write the last arc related to S_i as an arc weighted by the pair (--------, ---------) with an address corresponding to that of vertex S_i. Such a fictitious arc will prevent incorrect operation of the automaton when input by an unforeseen input vector. The automaton will retain its previous state without changing the output.

The computer implementation of the second cycle provides a transition to a new vertex, if the condition X is met, with the use of the two-byte address A_j written in memory.

The assembly-language program for the 8080 microprocessor is as follows.

0000	LXI SP, AI	set start address of data
0003	MVI C, ØØ	reset output
0005	VERT: In ØI	enter input vector
0007	MOV B, A	save input vector
0008	ARC: MOV A, B	recover input vector
0009	POP D E	check equality
000A	ANA E	of input vector
000B	CMP D	to vector X_i
000C	POP D E	load Y_i and A_i in GPR
000D	POP H	and move within memory space
000E	JNZ ARC	transfer to next arc
0011	MOV A, C	recover preceding output vector
0012	ANA E	generate current
0013	ORA D	output vector
0014	MOV C, A	save output vector
0015	OUT ØI	output this vector
0017	SPHL	load next address of data
0018	JMP VERT	jump to next vertex

In the run time of the program, the accumulator and general purpose register (GPR) contain the following information:

A = values of automaton input and output; INPUT & B^1;
B = values of automaton input saved for the time of searching of arcs leaving a given vertex;
C = preceding value of automaton output (current value);
D = B^2; B^4;
E = B^1, B^3;
H = high-order byte of transition address A_j;
L = low-order byte of transition address A_j.

The program occupies 27 bytes of microprocessor memory, the remaining space of the direct access memory is suitable for loading information about an automaton having 10918 arcs.

Example 7.3. Build a microprocessor control system by the transition graph shown in Fig. 7.9. The system will be built around an 8080 microprocessor set.

Depending on the complexity of the implemented automaton, the memory requirement is $27 + 6L$ where L is the number of arcs in the transition graph. We assume that the low-order bits in the input octadecimal vectors of input and output will be used as inputs and outputs. The first byte of arc data is written in memory at $001B_{16}$ address. By determining

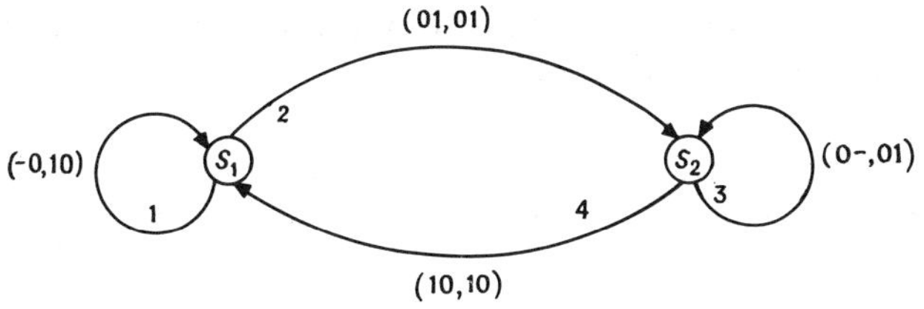

Fig. 7.9

B^1, B^2, B^3, B^4, B^5, and B^6 for each arc the following information loaded in memory is obtained.

Label	Address	Memory contents	Comment	Arc
VERT1	001B	0000 0001	X_1 = -------0	
		0000 0000		
		1111 1100	$_1$ = -------10	1
		0000 0010		
		0001 1011	A_1 = 001B (VERT1)	
		0000 0000		
	0021	0000 0011	X_2 = -------01	
		0000 0001		
		1111 1100	$_2$ = -------01	2
		0000 0001		
		0010 0111	A_2 = 0027 (VERT2)	
		0000 0000		
VERT2	0027	0000 0010	X_3 = -------0-	
		0000 0000		
		1111 1100	$_3$ = -------01	3
		0000 0001		
		0010 0111	A_3 = 0027 (VERT2)	
		0000 0000		
	002	0000 0011	X_4 = -------10	
		0000 0010		
		1111 1100	$_4$ = -------10	4
		0000 0010		
		0001 1011	A_4 = 001B (VERT1)	
		0000 0000		

The labels represent mnemonic addresses necessary to obtain absolute addresses in the hexadecimal representation. After the preparation of data, the following information represented in a hexadecimal code is written in the permanent memory (ROM)

31 1B 00 0E 00 DB 01 47 78 D1 A3 BA D1 E1 02 08
00 79 A3 B2 4F D3 01 F9 C3 05 00 01 00 FC 02 1B
00 03 01 FC 01 27 00 02 00 FC 01 27 00 03 02 FC
02 1B 00.

It is not infrequent that the nonlinear control system must execute a linear interpolation algorithm. We consider an implementation of a linear interpolation algorithm developed for the 8080 microprocessor set.

The interpolation is carried out in a relative coordinate system, that is, the position of the tool prior to the interpolation of a sequential segment is taken as the origin. The coordinates of the final point (x_k, y_k) define the segment to be interpolated. For fast travel over this segment, each step of the algorithm involves an elementary travel in one of the axes (for which the absolute value of the final point coordinate is a maximum). The condition of elementary move along the other axis defines a functional computed for the pair of the following feasible points of interpolation. The value of this functional should be determined prior to every elementary move and must be proportional to the difference of distances from the initial straight line to points of interpolation.

Let the initial line be given by the equation $y = Ax$, where $A = y_k/x_k$ (for definiteness, $x_k > y_k$), that is, each step involves translation along the x axis.

Confining our consideration to the first octant only, we find the equation of a straight line parallel to the initial line and equidistant from two points (x_i, y_i) and $(x_i, y_i + \Delta y)$, where Δy is an elementary increment in the y axis. One of this points will be used in the interpolation of the initial straight line in a current step. Since the sought line passes through $(x_i, y_i + \Delta y/2)$ and has the same slope as the initial line, its equation has the form $y = Ax + (y_i + \Delta y/2 - Ax_i)$. The sign of the expression in the parentheses defines whether the found line is above or below the initial line. A positive value implies that it is above the initial line, i.e., no move is executed in the Y axis, otherwise the line is interpolated by the point $(x_i, y_i + \Delta y)$.

For the implementation of this interpolation on an 8080 microprocessor having a limited number of instructions, we introduce the function $\gamma_i = x_k y_i - y_k x_i + \Delta y x_k/2$. The sign of this function defines whether or not an increment will be given to the travel in the Y axis. The first step computes this function for any straight line at point $x_1 = 1$ and $y_1 = 0$, i.e., points (0, 1) and (1, 1) are checked. If $\Delta y = 1$, then $\gamma_1 = x_k/2 - y_k$. For the following pairs of points, $x_{i+1} = x_i + 1$, $y_{i+1} = y_i - \operatorname{sgn} \gamma_i$, and the value of the functional for $i = 1, \ldots, n$ is

$$\begin{aligned} \gamma_{i+1} &= x_k(y_i - \operatorname{sgn} \gamma_i) - y_k(x_i + 1) + \Delta y x_k/2 \\ &= \gamma_i - x_k - \operatorname{sgn} \gamma_i - y_k \\ &= \gamma_i + x_k - y_k, \quad \text{if} \quad \gamma_i < 0 \\ &= \gamma_i - y_k, \qquad\quad \text{if} \quad \gamma_i \geqslant 0 \end{aligned}$$

This functional computed by an iterative algorithm does not require such "complicated" operations of the microcomputer as division, sign definition, and comparison of multibyte numbers. It involves only instructions for addition and subtraction and therefore is suitable for single-chip microprocessor.

The algorithm of linear interpolation built around the said functional is represented in Fig. 7.10. In general, γ varies in the range $-y_k \leqslant \gamma \leqslant x_k - y_k$, the equality is achieved at $-y_k$ if the preceding value of γ is assumed equal to $+0$, and at $x_k - y_k$, if -0. This ensures that the functional computation will not exceed the microprocessor word length.

Consider the 8080 microprocessor program realizing this algorithm. The program generates pulses at the output of 01 port, which serve as an input to the power controller of a step motor. For each octant there are two codes, code 1 and code 2. The former code is provided for travel in one axis only, and the latter for simultaneous travel in both axes. At each step one of the codes is selected and output after evaluation of sign of the functional.

Before the program starts, eight bytes of stack are allocated to each line segment. The data written in these bytes are code 1, code 2, $x_k/2 - y_k$, x_k, and $-y_k$. This organization aids in a fast preparation of data and debugging of the linear interpolation program.

The program must be loaded in RAM, since to ensure operation in real time instructions with direct operands are employed. This requires a preload at some fixed addresses of the constants: code 2, $x_k - y_k$, and $-y_k$ for each line segment.

All coordinates occupy two bytes (16 bits), the numbers being represented as 1's complements. This allows the segments with projections up to 820 mm on each axis to be interpolated for an elementary increment of 0.025 mm.

The general purpose registers contain the following information: γ in HL, β in BC, and the axial step count to check segment end in DE. The initial values are, respectively, 0, $x_k/2 - y_k$, and $x_k + 1$. Under the assumption that P has been established and there exists a job end check (all segments covered by the interpolation), the linear interpolation program has the form

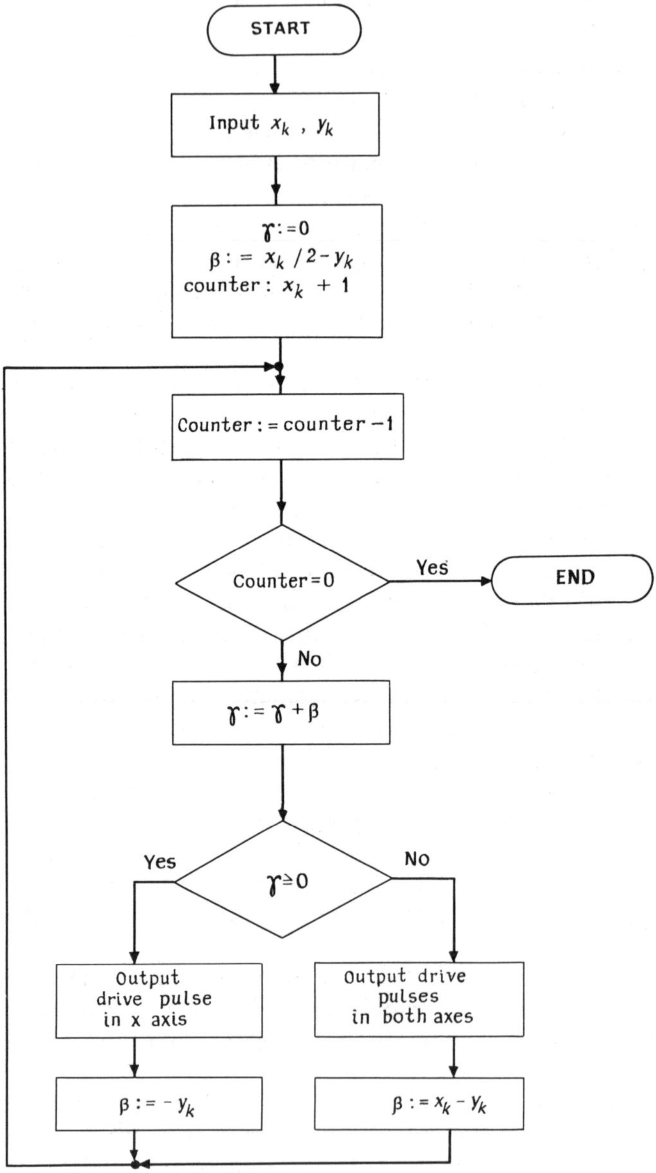
START
Input x_k , y_k
$\gamma := 0$
$\beta := x_k / 2 - y_k$
counter: $x_k + 1$
Counter := counter − 1
Counter = 0
Yes
END
No
$\gamma := \gamma + \beta$
$\gamma \geq 0$
Yes
No
Output drive pulse in x axis
Output drive pulses in both axes
$\beta := -y_k$
$\beta := x_k - y_k$

Fig. 7.10

```
BEGIN:     DCX DE          decrement x step counter
           MOV A, D        and output on completion
           ORA E
           JZ END
           XRA A           reset to input zero voltage
                           and reset ADDB attributes
           OUT Ø1          generate zero potential
           DAD BC          γ = γ + β
           ADD B           generate sign flag
           JP LAB
           DB 3E           instruction code MVI A, CODE 1
CODE 1     DB ØØ           for travel in both axes
           OUT Ø1          pulse generation
           DB Ø1           instruction code LXI BC, (x_k − y_k)
      G 1  DW ØØØØ
           JMP BEGIN       TO NEXT STEP
LAB        DB 3E           instruction code MVI A, CODE 2
CODE 2     DB ØØ           for motion in one axis
           OUT Ø1          pulse generation
           DB Ø1           instruction code LXI BC, (−y_k)
G2         DW ØØØØ
           JUMP BEGIN      TO NEXT STEP
END:       XRA A           generate zero potential
           OUT Ø1          on completion
           RET             and output
```

One step of iteration takes 99 clock cycles, i.e., with the 2.5-MHz microprocessor of 8080 series the program generates 25-kHz driving pulse sequence. For motion in other octants, the respective line segment should be referred to the first octant by the respective inversion of the signs of x_k and y_k, and evaluation and loading code 1 and code 2.

The microprocessor set may be used to advantage to simulate nonlinearities similarly to the construction of the interpolator. With a single-chip microcomputer, any nonlinearity can be software-implemented without having to modify the hardware in coming from one nonlinearity to another. Figure 7.11 presents some basic types of nonlinearities with the respective implementation algorithms.

Nonlinearity simulation can be effected to advantage by the mapping method in which the table of values is compiled by the input vector (input value) and a selected value is conveyed to output.

Consider the simulation of a relay-with-hysteresis nonlinear function on a single-chip series 1816 microcomputer. The input stimulus will be received at point P1 and form an output in port P2. The registers content is given by the following table:

RØ	R1	R2	R3
y^+	y^-	$-a$	a

```
        CLR A
        OUTL P2, A
loop    IN A, P2
        CPL A
        INC A
        ADD A, RØ
```

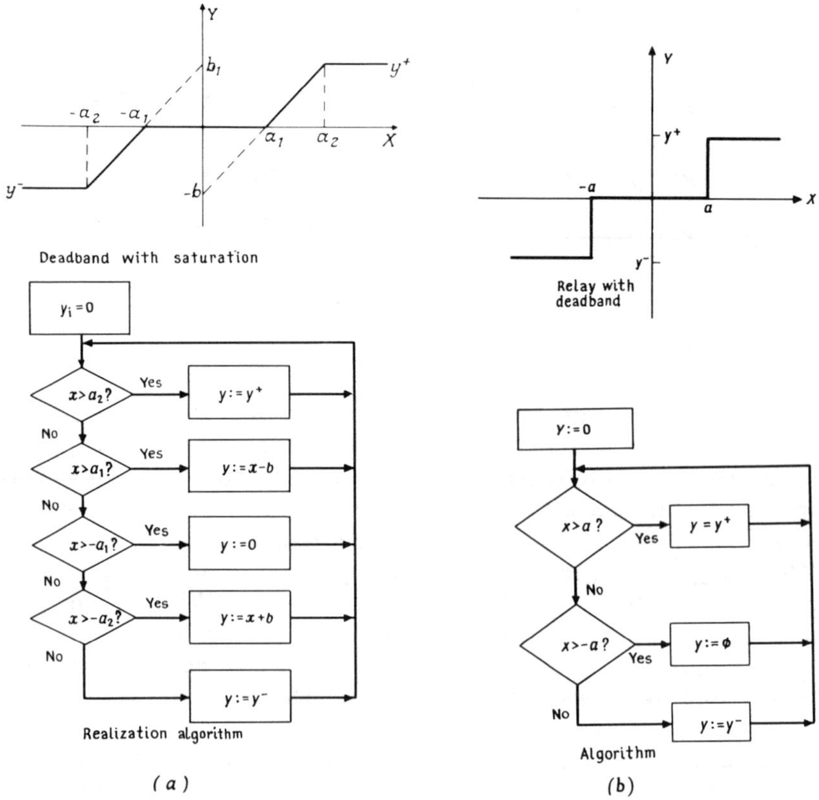

Fig. 7.11

```
          ORL P1, #FF
          IN A, P1
          CPL A
          INC A
          JZ label
          ADD R2
          INC loop
          MOV A, R1
          OUTL P2, A
          JMP loop
label     ADD R3
          JNC loop
          MOV A, RØ
          OUT P2, A
          JMP loop
```

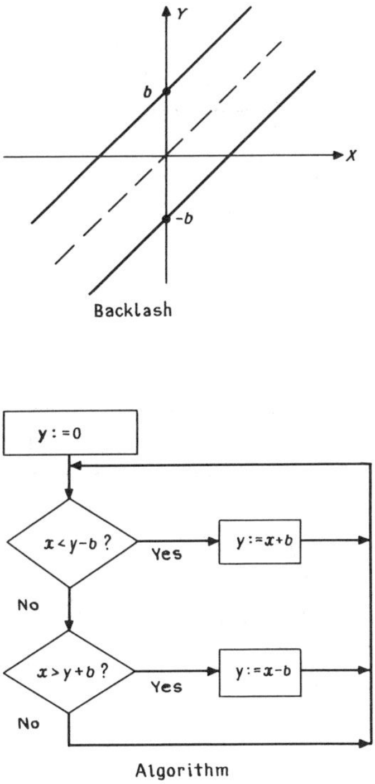

(c)

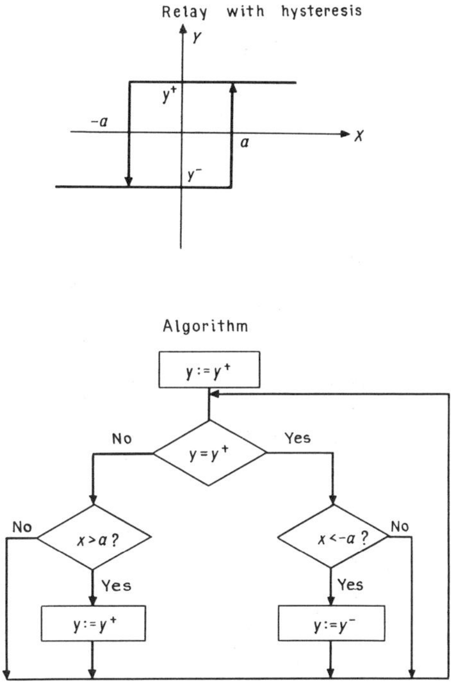

(d)

7.3 Parallelization of Microprocessor Control Systems. Estimation of Chromatic Number

A characteristic feature of nonlinear control systems is associated with a large number of input and output signals. Vectors X and Y defining the connection of the microprocessor with the plane through A/D converters have rather large dimensions. However, any output signal is, as a rule, independent of a number of inputs. The speed and reliability of such a microprocessor system can be enhanced by representing it in the form of a parallel decomposition of automata implemented by microprocessor components.

The microprocessor implementation of an automaton imposes no limitations on the number of each type of subautomaton involved but requires reduction of the number of input and output variables in each of them. This requirement stems from the fact that the microprocessors are linked with the plant through special addressable registers — I/O ports of limited width controlled by the bus width of CPU. Reliability, cost and other crucial characteristics of the controller dictate the number of subautomata required, with the dimension of input and output vectors not exceeding the bus width of CPU.

In formal terms the problem is stated as follows. Given the transition graph $G_t = \langle S(U(X, Y)) \rangle$ of the automaton and some number K, determine a decomposition into the partial Cartesian product [7.4, 7]

$$G_t = \prod_{i=1}^{n} G_i$$

of graphs $G_i = \langle S_i, (U_i(X_i, Y_i)) \rangle$, $i = 1, 2, \ldots, n$, such that for any cofactor $P(X_i) \subset P(X)$, $P(Y_i) \subset P(Y)$, $|P(X_i)| \leqslant K$, and $|P(Y_i)| \leqslant K$, where $P(Z)$ is the set of components of vector Z, and $|P(Z)|$ is the power of this set equal to the vector length. If this decomposition is feasible, carry it out so that n be a minimum.

Consider a method of parallel decomposition of automata based on the semantics of the transformations involved [7.5, 7].

When the input vector of the initial graph of transitions satisfies the condition $|P(X)| \leqslant K$, the sought decomposition always exists and the minimal number of components in this decomposition satisfying the problem conditions is defined by the dimension of Y and equals

$$n = \left]\frac{|P(Y)| - 1}{K}\right[+ 1$$

where the inverted brackets denote the integer part of the confined number.

When the dimension of input vector is increased so that $|P(X)| > K$, to search for disconnected cofactors G_i satisfying the constraints $|P(X_i)| \leqslant K$ and to describe situations where the deterministic property is violated we introduce the concept of dependence of a pair of internal states on an input variable and essential dependence of variables of the automaton. Vectors are said to be *adjacent* if they differ in only one variable. *Internal states* S_α and S_β of some automaton *depend on an input variable* x_e if one of the following conditions is met:

(1) there exists a state S_ε such that X' steers the automaton from S_ε to S_α, $S_\varepsilon X' \rightarrow S_\alpha$, and X'' steers it to S_β, $S_\varepsilon X'' \rightarrow S_\beta$, and the vectors X' and X'' are adjacent with respect to component corresponding to the variable x_e.

(2) a pair of states S_γ and S_δ depends on x_e and there exist the transitions $S_\gamma X' \rightarrow S_\alpha$, and $S_\delta X'' \rightarrow S_\beta$, where either $X' = X''$, or the vectors X' and X'' are adjacent with respect to the component corresponding to the variable x_e.

It is quite obvious that if the internal states S_α and S_β depend on some input variable, then should the component assigned to this variable be removed from the input vector, the states S_α and S_β would have to be united to save the deterministic property of the resultant transition graph.

The output variable y_r depends essentially on the input variable x_e if at least one of the following conditions is satisfied:

(1) there exists an internal state S_ε such that if the input of the automaton in state S_ε is excited by vectors adjacent in component x_e, the variable y_r assumes opposite values;

(2) a pair of internal states depends on x_e and for either $X' = X''$ or X' and X'' adjacent in the component corresponding to x_e the variable y_r assumes opposite values on the transitions $S_\alpha X'$ and $S_\beta X''$.

If we denote by $Q(y_r)$ the set of all input variables on which the output variable y_r depends essentially, then the input vector X_i corresponding to G_i realizing y_r must satisfy the relation $Q(y_r) \subset P(X_i)$. Obviously, if the automaton has an output variable that depends essentially on more than K input variables, then the sought decomposition is not feasible.

For the purpose of parallel decomposition we consider partial graphs of transitions $G_r = \langle S, (U, (X, y_r))\rangle$, ($r = 1$ to $|P(y)|$) which correspond on a one-to-one basis to output variables constituting vector Y. The sets of vertices and arcs of the graph coincide with the sets of vertices and arcs of the initial transition graph of the automaton, and each arc is weighed by input vector X and output variable y_r.

The concepts of the dependence of a pair of internal states on an input variable and the essential dependence of variables of the automaton are a handy tool in revealing the possibilities of size reduction of the input vector X in a partial transition graph. Removal of some input variable x_e leads to coalescence of pairs of G_r vertices corresponding to pairs of internal states dependent on this variable. If this does not violate the deterministic property of the resultant partial transition graph, then the graph G_r is said to be independent of the input variable x_e.

The intersection of a set of variables on which y_r depends essentially and a set of variables of which G_r is independent is an empty set. Consequently, reduction of the dimension of vector X can be carried out only at the expense of the components corresponding to variables on which y_r depends essentially.

The algorithm developed to reduce the dimension of an input vector is an iterative procedure separating one component from the vector X at each step. This corresponds to an induction of a new partial transition graph as a result of a union of pairs of internal states dependent on the variable assigned to the separated component. The procedure is repeated until none of the graphs derived in the previous step has a dependent variable.

As a result we arrive at a collection of partial transition graphs for each of which the reduction of input vector dimension cannot be continued without violating its deterministic property. If this collection contains no graph for which the input vector dimension is below K, then the initial decomposition is infeasible. Otherwise each partial transition graph satisfying these constraints may be associated with a cofactor of the sought product.

If for each output variable y_r constituting vector Y in the initial automaton, at least one partial transition graph G_r can be derived for which the input vector X dimension does not exceed K, then the first half of the problem is solved. The possibility is established to decompose the initial transition graph into $n = |P(Y)|$ disconnected cofactors satisfying the constraints of the problem. Each of the cofactors realizes one of the components of vector Y.

In order to reduce the number of cofactors we join those where the union does not rise the input and output vector length above the width of CPU. For this purpose a binary table is compiled where each column corresponds on a one-to-one basis to an output variable entering Y, and each row to an input vector of rank not exceeding the width of CPU, inducing a partial transition graph corresponding to some cofactor. An element (i, j) of the table is unity if row i is associated with a vector forming a partial transition graph corresponding to output variable j, and is zero otherwise. Each cover of the columns by the rows of the table is associated with one version of the complete decomposition of the initial automaton. Observing the constraints imposed on the number of input variables of

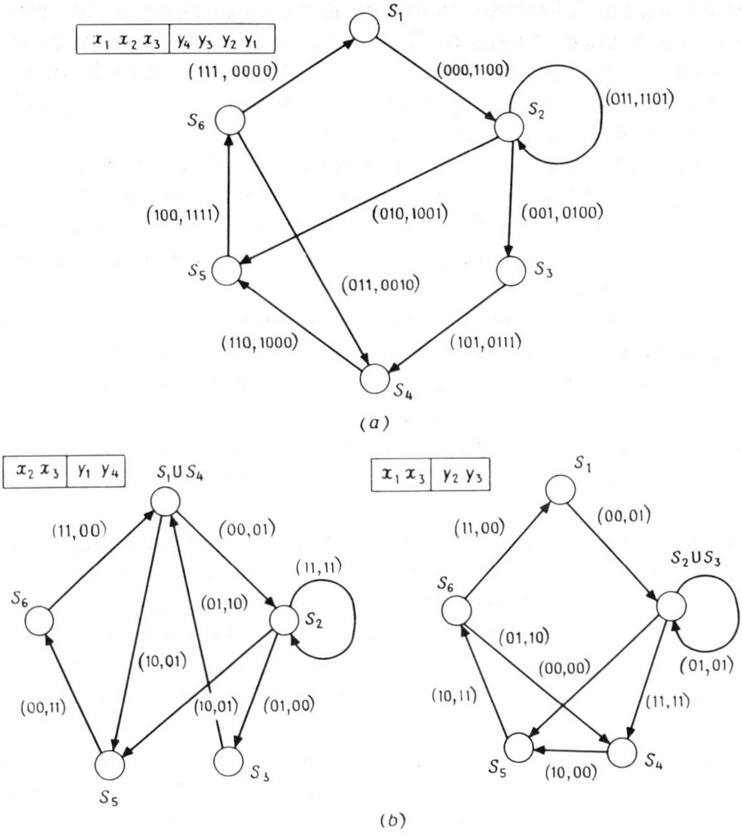

Fig. 7.12

each cofactor, only those covers are worth consideration in which the number of columns covered by one row does not exceed the width of the microprocessor. Then the cover containing the least number of elements will be consistent with the minimal number of decompositions of the automaton.

By way of example we consider a parallel decomposition of the automaton given by its transition graph in Fig. 7.12*a*. When the automaton is implemented with series K589 microprocessor 2-digit sets, the highest speed and reliability of the control system will be attained by decomposition of the transition graph into disconnected cofactors with the number of input and output variables not exceeding two.

The pair of internal states S_1 and S_4 is seen to depend on x_1, while the pairs S_2, S_3, and S_2, S_5 depend, respectively, on x_2 and x_3. The output variable y_1 is essentially dependent on x_2 and x_3, while y_2, y_3, and y_4 depend, respectively, on x_3, x_1 and x_2. The results of the maximal reduction of input vectors for each output variable are summarized in the table below.

Input sets	Output variables			
	y_1	y_2	y_3	y_4
x_2x_3	1	1	0	1
x_2	0	0	0	1
x_1	0	0	1	0
x_1x_3	0	1	1	0
x_1x_2	0	0	1	1

In view of the constraints, the table has two minimal covers, namely, $\{x_2, x_3, x_1, x_3\}$ and $\{x_2x_3, x_1x_2\}$. Figure 7.12*b* shows the transition graphs of two subautomata corresponding to the first cover. The states of these subautomata and the resultant automaton are summarized in the table [7.4] below.

First cofactor					Second cofactor					Product							
$s_{i1} \to s_{j1}$	x_2	x_3	y_1	y_4	$s_{i2} \to s_{j2}$	x_1	x_3	y_2	y_3	$s_i \to s_j$	x_1	x_2	x_3	y_4	y_3	y_2	y_1
$(s_1 \cup s_4) \to s_2$	0	0	0	1	$s_1 \to (s_2 \cup s_3)$	0	0	0	1	$s_1 \to s_2$	0	0	0	1	1	0	0
$s_2 \to s_2$	1	1	1	1	$(s_2 \cup s_3) \to (s_2 \cup s_3)$	0	1	0	1	$s_2 \to s_2$	0	1	1	1	1	0	1
$s_2 \to s_3$	0	1	0	0	$(s_2 \cup s_3) \to (s_2 \cup s_3)$	0	1	0	1	$s_2 \to s_3$	0	0	1	0	1	0	0
$s_2 \to s_5$	1	0	0	1	$(s_2 \cup s_3) \to s_5$	0	0	0	0	$s_2 \to s_5$	0	1	0	1	0	0	0
$s_3 \to (s_1 \cup s_4)$	0	1	1	0	$(s_2 \cup s_3) \to s_4$	1	1	1	1	$s_3 \to s_4$	1	0	1	0	1	1	1
$(s_1 \cup s_4) \to s_5$	1	0	0	1	$s_4 \to s_5$	1	0	0	0	$s_4 \to s_5$	1	1	0	1	0	0	0
$s_5 \to s_6$	0	0	1	1	$s_5 \to s_6$	1	0	1	1	$s_5 \to s_6$	1	0	0	1	1	1	1
$s_6 \to (s_1 \cup s_4)$	1	1	0	0	$s_6 \to s_4$	0	1	1	0	$s_6 \to s_4$	0	1	1	0	0	1	0
$s_6 \to (s_1 \cup s_4)$	1	1	0	0	$s_6 \to s_1$	1	1	0	0	$s_6 \to s_1$	1	1	1	0	0	0	0

Thus we obtained a decomposition of the initial automaton with three input variables and four output variables into two subautomata each of which contains two input and two output variables. This implementation on the basis of two series 589 microprocessors enhanced the speed of data processing.

Many optimization problems like the parallel decomposition of automata [7.7, 10] and minimization of memory space requirement in a microprocessor network [7.5] reduce to coloring of graph vertices. Gorbatov *et al.* [7.9] have suggested a bound of chromatic number $h(G)$ of a graph G nondecomposable with an additive operation in the form

$$h(G) \leqslant]s_{av}(\tilde{G})[+ 1 \tag{7.2}$$

where the subgraph $\tilde{G} \subset G$ satisfies three inequalities that define the vertex, arc, and local topological (degree of a vertex) resources, viz.,

$$|V(\tilde{G})| \geqslant 3(2^k - 1) + p(G) \tag{7.3}$$

$$2|(U(\tilde{G})| \geqslant 7 \times 3^k + 6 \times 2^k(p(G) - 3) + (p(G) - 3)^2 - p(G) + 2 \tag{7.4}$$

$$s(v_i) \geqslant p(G) + \left]\log_2 \left(\frac{V(G) - p(G)}{3} + 1\right)\right[- 1 \tag{7.5}$$

Here, $s(v_i)$ is the degree of vertex v_i, $s_{av}(\tilde{G})$ is the average degree of $\tilde{G}$, k is the order of G, and $p(G)$ is the density of G.

In refining the inequalities defining the resources of G, this estimate of the chromatic number may be used for an arbitrary graph.

If a quasi-complete graph $Q(p, k)$ [7.4] is decomposable into n addends of the form $Q(p/n, k/n)$, then the minimum power of its fundamental set and signature will be lower than that of a nondecomposable graph of the same quasidensity. The greater the number of addends in the decomposition of a quasicomplete graph, the lower the minimal values of its fundamental set and signature. Therefore, since the number of components in the decomposition lies in the range $0 \leqslant n \leqslant p/2$, the minimal powers of resources are attained if the quasicomplete graph is decomposable in $p/2$ addends for p even, and in $(p - 1)/2$ addends for p odd. In this case the components of the initial graph are quasicomplete graphs $Q(2, k_i)$ with $0 \leqslant k_i \leqslant k$. Since for any $k > 0$ the inequalities

$$2|V_{\min}(Q(2, k))| \leqslant |V_{\min}(Q(2, k - 1))| + |V_{\min}(Q(2, k + 1))|$$

$$2|U_{\min}(Q(2, k))| + (|V_{\min}(Q(2, k))|)^2 \leqslant |U_{\min}(Q(2, k - 1))| + |U_{\min}(Q(2, k + 1))| + |V_{\min}(Q(2, k - 1))||V_{\min}(Q(2, k + 1))|$$

hold and characterize the properties of the minimal resources of the fundamental set and signature of quasicomplete graphs, then to obtain minimal values of power of the fundamental set and signature the terms $Q(2, k_i)$ must be identical, or, if impossible, differ by at most unity. This statement leads to the following assertion [7.13].

A *quasicomplete graph $Q(p, k)$ of density p and order k* having minimum power and the fundamental set and signature is:

for even p

$$Q(p, k) = \sum_{i=1}^{p/2} Q(2, k_i),$$

where

$$\frac{2k}{p} - 1 < k_i < \frac{2k}{p} + 1, \quad \sum_{i=1}^{p/2} k_i = k$$

and for odd p

$$Q(p, k) = Q(p-1, k) + Q(1, 0) = \sum_{i=1}^{p-1/2} Q(2, k_i) + Q(1, 0)$$

where

$$\frac{2k}{p-1} - 1 < k_i < \frac{2k}{p-1} + 1, \quad \sum_{i=1}^{p-1/2} k_i = k$$

Basing on this assertion we can derive the minimum power of the fundamental set $|V_{\min}|$ and signature $|U_{\min}|$ of a quasi-complete graph $Q(p, k)$ of given order k and density p. By way of example we determine $|V_{\min}|$ and $|U_{\min}|$ for quasi-complete graphs of even density, namely,

$$Q(p, k) = \alpha Q(2, \gamma) + \beta Q(2, \gamma + 1)$$

where $\gamma =]2k/p[$, α and β are the roots of the system

$$\alpha + \beta = p/2$$
$$\alpha\gamma + \beta(\gamma + 1) = k$$

that is, $\alpha = (p/2)(]2k/p[+ 1) - k$ and $\beta = k - (p/2)]2k/p[$.

Since $|V_{\min}(Q(2, k))| = 3 \times 2^k - 1$, and $|U_{\min}(Q(2, k))| = (7 \times 3^k - 6 \times 2^k + 1)/2$, we have

$$|V_{\min}(Q(p, k))| = \alpha(3 \times 2^\gamma - 1) + \beta(3 \times 2^{\gamma+1} - 1)$$

$$\begin{aligned} |U_{\min}(Q(p, k))| &= \frac{\alpha}{2}(7 \times 3^\gamma - 6 \times 2^\gamma + 1) + \frac{\beta}{2}(7 \times 3^{\gamma+1} \\ &\quad - 6 \times 2^{\gamma+1} + 1) + (3 \times 2^\gamma - 1)((\alpha - 1) \\ &\quad \times (3 \times 2^\gamma - 1) + \beta(3 \times 2^{\gamma+1} - 1)) + (3 \times 2^\gamma - 1) \\ &\quad \times ((\alpha - 2)(3 \times 2^\gamma - 1) + \beta(3 \times 2^{\gamma+1} - 1)) \\ &\quad + \dots + (3 \times 2^\gamma - 1)((\alpha - \alpha)(3 \times 2^\gamma - 1) \\ &\quad + \beta(3 \times 2^{\gamma+1} - 1)) + (3 \times 2^{\gamma+1} - 1) \\ &\quad \times (\beta - 1)(3 \times 2^{\gamma+1} - 1) + \dots + (3 \times 2^{\gamma+1} - 1) \\ &\quad \times (\beta - \beta)(3 \times 2^{\gamma+1} - 1) \\ &= \frac{\alpha}{2}(7 \times 3^\gamma - 6 \times 2^\gamma + 1) + \frac{\beta}{2}(7 \times 3^{\gamma+1} \\ &\quad - 6 \times 2^{\gamma+1} + 1) + \alpha\beta(3 \times 2^\gamma - 1)(3 \times 2^{\gamma+1} - 1) \\ &\quad + \frac{\alpha(\alpha - 1)}{2}(3 \times 2^\gamma - 1)^2 + \frac{\beta(\beta - 1)}{2}(3 \times 2^{\gamma+1} - 1)^2 \end{aligned}$$

Therefore, for the graph G the system of inequalities has the form

$$\begin{aligned} V(\tilde{G}) \geqslant & \Big(\frac{p}{2}\Big(\Big]\frac{2k}{p}\Big[+ 1\Big) - k\Big)(3 \times 2^{]2k/p[} - 1) \\ & + \Big(k - \frac{p}{2}\Big]\frac{2k}{p}\Big[\Big)(3 \times 2^{]2k/p[+1} - 1) \end{aligned} \tag{7.6}$$

$$\begin{aligned} U(\tilde{G}) \geqslant & \frac{1}{2}\Big(\frac{p}{2}\Big(\Big]\frac{2k}{p}\Big[+ 1 - k\Big)(7 \times 3^{]2k/p[} - 6 \times 2^{]2k/p[} + 1) \\ & + \frac{1}{2}\Big(k - \frac{p}{2}\,]2k/p[\Big)(7 \times 3^{]2k/p[+1} - 6 \times 2^{]2k/p[+1} + 1) \\ & + \Big(\frac{p}{2}\Big(\Big]\frac{2k}{p}\Big[+ 1 - k\Big)\Big(k - \frac{p}{2}\Big]\frac{2k}{p}\Big[\Big)(3 \times 2^{]2k/p[} - 1) \\ & \times (3 \times 2^{]2k/p[+1} - 1) + \frac{1}{2}\Big(\frac{p}{2}\Big(\Big]\frac{2k}{p}\Big[+ 1\Big) - k\Big) \\ & \times \Big(\frac{p}{2}\Big(\Big]\frac{2k}{p}\Big[+ 1\Big) - k - 1\Big)(3 \times 2^{]2k/p[} - 1) + \frac{1}{2}\Big(k \\ & - \frac{p}{2}\Big]\frac{2k}{p}\Big[\Big)\Big(k - \frac{p}{2}\Big]\frac{2k}{p}\Big[- 1\Big)(3 \times 2^{]2k/p[+1} - 1) \end{aligned} \tag{7.7}$$

$$s(v_i) \geqslant p + k - 1, \quad v_i \in \tilde{G} \tag{7.8}$$

For k being multiples of $p/2$ we have

$$|V_{\min}| = \frac{p}{2}(3 \times 2^{2k/p} - 1)$$
$$|U_{\min}| = \frac{p}{4}(7 \times 3^{2k/p} - 6 \times 2^{2k/p} + 1) + \frac{p}{4}\Big(\frac{p}{2} - 1\Big)(3 \times 2^{2k/p} - 1)^2$$

For odd p, the estimates are as follows,

$$|V_{\min}(Q(p,\ k))| = |V_{\min}(Q(p - 1),\ k)| + 1$$
$$|U_{\min}(Q(p,\ k))| = |U_{\min}(Q(p - 1),\ k)| + |V_{\min}(Q(p - 1),\ k)|$$

Consequently, for an arbitrary graph G of n vertices, m arcs, density p, and order k,

$$n \geqslant \frac{p}{2}(3 \times 2^{2k/p} - 1)$$

$$h(G) = k + p, \quad k \geqslant 0, \quad n \geqslant p \geqslant 2$$

and, consequently,

$$k \leqslant \left]\frac{p}{2}\log_2\frac{2n/p+1}{3}\right[$$

Thus, the estimate of the chromatic number of G is

$$p \leqslant h(G) \leqslant p + \left]\frac{p}{2}\log_2\frac{2n/p+1}{3}\right[\tag{7.9}$$

In practical problems the graph density is often specified as boundary values, namely,

$$p_{\min} \leqslant p \leqslant p_{\max}$$

Then

$$p_{\min} \leqslant h(G) \leqslant p_{\max} + \left]\frac{p_{\max}}{2}\log_2\frac{2\tilde{n}p_{\min}+1}{3}\right[\tag{7.10}$$

where $\tilde{n}$ is the number of vertices of the graph $\tilde{G}$ obtained from the initial graph by sequential removal of vertices whose degree is less than $p_{\min} - 1$.

The values of $p_{\min}$ and $p_{\max}$ can be derived from

$$\frac{n(p-1)}{2} \leqslant m \leqslant \frac{n(n-1)}{2}$$

whence

$$p_{\min} = \left[\frac{n^2}{n^2 - 2m}\right] \quad \text{and} \quad p_{\max} = \left]\frac{2m}{n} + 1\right[\tag{7.11}$$

where the value of $p_{\min}$ is taken as it stands if integer, and is increased to the nearest integer otherwise.

Because for a quasi-complete graph the number of arcs is

$$m \geqslant \frac{p}{4}(7 \times 3^{2k/p} - 6 \times 2^{2k/p} + 1)$$

$$= \frac{p}{4}\left(\frac{p}{2} - 1\right)(3 \times 2^{2k/p} - 1)^2$$

for $p \geqslant 4$ we have

$$m \geqslant (3/8)p^2 \times 2^{4k/p} + p(p-4)/8$$

whence

$$k \leqslant \left](p/4)\log_2\frac{8m - p(p-4)}{3p^2}\right[$$

In view of (7.5) and (7.9) the estimate of chromatic number for graphs G has the form

$$p \leqslant h(G) \leqslant p + \left]\frac{p}{4}\log_2 \frac{8m - p(p-4)}{3p^2}\right[, \ p > 3$$

$$p \leqslant h(G) \leqslant p + \left]\log_2 \left(\frac{n-p}{3} + 1\right)\right[, \ p \leqslant 3$$

where m and n are the number of arcs and vertices of the graph $\tilde{G} \subset G$, the degree $s(v_i)$ of each vertex $v_i \in \tilde{G}$ being

$$s(v_i) \geqslant \left[\frac{n^2}{n^2 - 2m}\right] - 1 \tag{7.12}$$

By virtue of (7.10, 11) we obtain the estimate of chromatic number $h(G)$ of G in terms of number of vertices n and arcs m of the subgraph $\tilde{G} \subset G$ for which (7.12) holds, namely,

$$\left[\frac{n^2}{n^2 - 2m}\right] \leqslant h(G) \leqslant \left]\frac{2m+n}{n}\right[+ \left]\left]\frac{2m+n}{2n}\right[\times \log_2 \left(\frac{1}{3}\left(\frac{2n}{\frac{n^2}{n^2-2m}} + 1\right)\right)\right[\tag{7.13}$$

By way of illustration we estimate the chromatic number $h(G)$ of a graph $G = \langle V, U \rangle$, with $V = \{1, 2, \ldots, 7\}$, $n = 7$, and $U = \{\{1, 2\}, \{1, 6\}, \{1, 7\}, \{2, 3\}, \{2, 6\}, \{3, 4\}, \{3, 5\}, \{3, 7\}, \{4, 5\}, \{4, 7\}, \{5, 6\}, \{6, 7\}\}$, $m = 12$, viz.,

$$2 = \left[\frac{49}{49 - 2 \times 12}\right] \leqslant h(G) \leqslant \left]\frac{2 \times 12 + 7}{7}\right[+ \left]\left(\left]\frac{2 \times 12 + 7}{2 \times 7}\right[\right) \times \log_2 \left[\frac{1}{3}\left(\frac{2 \times 7}{49/(49 - 2 \times 12)} + 1\right)\right]\right[$$

whence $2 \leqslant h(G) \leqslant 6$. All the vertices $v_i \in G$ satisfy (7.12).

These estimates of the chromatic number of a large-dimension graph facilitate efficient decomposition of microprocessor networks.

References

7.1 I. Z. Tsypkin (ed.): *Upravlenie v slozhnykh nelineinykh sistemakh* (Control in complicated nonlinear systems). Nauka, Moscow 1984.
7.2 V. L. Kirpichev: in *Besedy o mekhanike* (Discussions on mechanics). St. Petersburg 1907.
7.3 S. V. Emelyanov and V. I. Utkin: *Dokl. Akad. Nauk SSSR* 152, No 2 (1963).
7.4 V. A. Gorbatov: *Skhemy upravleniya TsVM i grafy* (Digital computer control circuits and graphs). Energiya, Moscow 1971.
7.5 V. A. Gorbatov: *Teoriya chastichno uporyadochennykh sistem* (Theory of partially ordered systems). Sovetskoye Radio, Moscow 1976.
7.6 V. A. Gorbatov, V. V. Kafarov, and P. G. Pavlov: *Logicheskoe upravleniye technologicheskimi protsessami* (Logical process control). Energiya, Moscow 1978.
7.7 V. A. Gorbatov: *Semanticheskaya teoriya proektirovaniya avtomatov* (Semantic theory of automata design). Energiya, Moscow 1979.
7.8 V. A. Gorbatov, B. L. Ostankov, and S. A. Frolov: *Regulyarnye struktury avtomatnogo upravleniya* (Regular structures of automata control). Mashinostroyenie, Moscow 1980.
7.9 V. A. Gorbatov, P. G. Pavlov, and V. N. Chetverikov: *Logicheskoe upravleniye informatsionnymi protsessami* (Logical control of information processes). Energoatomizdat, Moscow 1984.
7.10 A. A. Voronov (ed.): *Management and Control in Large Systems.* Mir Publishers, Moscow 1986.
7.11 I. S. Khlytchiev: in *Logicheskoe upravlenie v promyshlennosti* (Logic control in industry) ed. by V. A. Gorbatov. AN SSSR Press, Tashkent 1985, p. 55-56.
7.12 V. A. Gorbatov: in *Vychislitelnaya tekhnika. Trudy MEI* (Computing. Proceedings of Moscow Power Institute). No 53, MEI Press, 1964, pp. 5-30.
7.13 V. A. Gorbatov and M. I. Smirnov: in *Logicheskoe upravleniye s pomoshchyu EVM* (Computerized logical control). AN SSSR Press, Izhevsk 1987, pp. 4-8.

8 Fundamentals of Statistical Analysis and Synthesis of Complicated Dynamic Systems

A. V. Potseluev

The term complicated dynamic systems covers a wide class of control systems, computers, and plants. Such systems are normally treated as consisting of interconnected elements. The state of these elements is governed by the input stimuli. The state of the system in turn controls its behavior at a later time. More often than not the system behavior is described in terms of output coordinates. Analytical description of these coordinates is a formidable problem. Therefore, many of recent research efforts have been devoted to numerical schemes which afford computer simulation of system behavior. This chapter discusses application of the most popular stochastic simulation procedures based on the Monte Carlo method, or the method of statistical nodes.

The method of statistical nodes is based on a targetted variation scheme for the input parameters. This type of stochastic simulation reduces the number of replications required and has proved to be computer time saving in solving problems of analysis and synthesis. However, simulation is an imprecise technique. It only compares alternatives rather than generates a system model with optimal parameters. This disadvantage can be alleviated to some degree by combining analytic and numerical methods, i.e. by making use of the method of statistical nodes. The methods to be discussed can be used for analysis and synthesis of continuous and discrete, linear and nonlinear systems.

8.1 Performance Criteria of Complex Dynamic System Behavior

A mathematical model of a complex system may be given as follows. The input processes (or stimuli) x form a subset X_a which in turn is a proper subset of a set X, i.e.

$$x \in X_a, \quad X_a \subset X$$

The outputs y form a subset Y_a, which in turn is a subset of a set Y, i.e.

$$y \in Y_a, \quad Y_a \subset Y$$

Between the elements of X_a and Y_a there is a relation γ defining how an element x of the nonempty set X is related to an element y of the nonempty set Y. The relation γ is a function mapping X onto Y, i.e.

$$\gamma\colon X \to Y \tag{8.1}$$

In the general case the relation γ of the sets $Y = (y_1, \ldots, y_n)$ and $X = (x_1, \ldots, x_m)$ may be represented in the form

$$\begin{aligned} y_1 &= \gamma_1(x_1, \ldots, x_m) \\ &\cdots\cdots\cdots\cdots \\ y_n &= \gamma_n(x_1, \ldots, x_m) \end{aligned} \tag{8.2}$$

For dynamic systems the inputs x_i, $i = 1, \ldots, m$, are functions of time t. Obviously, the outputs y_j, $j = 1, \ldots, n$, are also functions of time.

In real systems, the input stimuli $X_1(t), \ldots, X_m(t)$ are subject to some constraints. Therefore they relate to a certain class ω_i. Then the set Ω_i of feasible values of $X_i(t)$ is represented as

$$\Omega_i = \{\omega_i : T \to X_i(t)\}, \quad i = 1, \ldots, m \tag{8.3}$$

In formal terms the set Ω_i is a subset of the set $T \times X_i(t)$. For specific input processes

$$X_i(t) = \omega_i(t)$$

The mapping (1.1) relating the input $X = \{X_1, \ldots, X_m\}$ and output $Y = \{Y_1, \ldots, Y_n\}$ processes is characterized by a certain internal structure of system elements. The behavior of these elements is described by the state variables $Z = \{Z_1, \ldots, Z_r\}$ represented as

$$Z(t) = A[t, t_0, Z(t_0), \omega(t_0, t)] \tag{8.4}$$

where A is a system operator, t current time, and t_0 some preceding time instant $t_0 < t_1$ which can be viewed as the initial time.

Because a state of the system belongs to some set Ψ, the relationship (8.4) may be written in the general form as

$$\Phi = \{A : T \times T \times \Psi \times \Omega \to \Psi\} \tag{8.5}$$

where the elements of the set $T \times T \times \Psi \times \Omega$ are ordered 4-tuples t, t_0, $Z(t_0)$, ω.

Clearly, the set Γ of the output variables $Y(t)$ may be given in the form

$$\Gamma = \{B : T \times Z \to Y\} \tag{8.6}$$

or in vector form

$$Y(t) = B[t, Z(t)] \tag{8.7}$$

In real systems the input vector $X(t)$ can be deterministic or stochastic, therefore the system under consideration can be deterministic or stochastic.

Equations (8.4, 7) represent the general description of system behavior. In control systems, however, the designer commonly singles out two parts, the controller and the plant. With a continuous control function $U(t)$ generated by the controller, the behavior of the system is often characterized by a smooth transfer function φ of the form

$$\varphi(t) = A[t, t_0, z(t), \omega(t_0, t)] \tag{8.8}$$

For a discrete controller, relation (8.8) becomes

$$\varphi[k\tau] = A[k\tau, t_0, z(k\tau), \omega(k\tau)] \tag{8.9}$$

where $k = 1, 2,$ etc. is the serial number of discrete step, and τ is the duration of a step.

Solution of analysis and synthesis problems assumes a more convenient form when the system dynamics is described by differential or difference equations. The relationship (8.8) may be converted to differential equations as follows. Let an infinitesimal increment in a system state be

$$\mathrm{d}Z = \frac{\partial}{\partial t} A[t, t_0, Z(t), \omega(t_0, t)]\,\mathrm{d}t$$

or

$$\frac{\mathrm{d}Z}{\mathrm{d}t} = F[t, Z, \omega] \tag{8.10}$$

where

$$F[t, Z, \omega] = \frac{\partial}{\partial t} A[t, t_0, Z(t), \omega(t_0, t)]$$

In the general case, (8.10) is a stochastic, nonlinear, time-variable equation.

Likewise, for a discrete transition function (8.9), the equations of state of the discrete system have the form

$$Z[(k+1)\tau] = F\{k\tau, Z[k\tau], \omega\} \tag{8.11}$$

As a rule complex dynamic systems are described by high-order equations. They include a few nonlinear elements and several random disturbances serving as inputs to various interconnected elements of the system. Such system structures motivated the development of procedures for analysis and synthesis of efficient stochastic dynamic systems.

In handling problems of system analysis and synthesis, the key role is played by the performance indices $J_1, \ldots, J_l$ set up by the designer. The number l of such indices contributes to a certain degree to the complexity of system design. For the purpose of simplification of synthesis, designers prefer some generalized performance measure J_g being a function of the individual performance indices $J_1, \ldots, J_l$.

Deterministic system design makes wide use of a generalized performance criterion in the form

$$J = V[y_1(t_2), \ldots, y_n(t_2)] + \int_{t_1}^{t_2} Q(y_1, \ldots, y_n, t)\,\mathrm{d}t + \frac{1}{q}\int_{t_1}^{t_2} \sum_{j=1}^{m} \left(\frac{u_j}{k_j}\right)^q \mathrm{d}t \tag{8.12}$$

where V is a specified, positive definite function of output variables characterizing system performance at the finite time instant $t = t_2$, Q is a given positive definite function of output variables and time, $q > 1$ is a real number, k_j are given real quantities, and u_j are components of the control vector.

The second term in (8.12) describes the behavior of system transient, and the third term describes the consumption of energy.

A general stochastic performance criterion is mathematical expectation which may be written in the form

$$J = \int W(\varkappa, A) f(\varkappa)\,\mathrm{d}\varkappa \tag{8.13}$$

where W is a function of vectors $\varkappa = \varkappa(X, Y, Z)$ and the operator A of the controller, and f is a distribution function of the random process $\varkappa$.

Sometimes W is referred to as the *loss function*. By specifying a particular form of this function, one can obtain a variety of performance criteria such as, for example, the variance of an output process, the mean square error, the probability (likelihood) of an event, and the like.

8.2 Evaluation of Stochastic Performance Measures

Without loss of generality in the problem formulation for the evaluation of a stochastic performance criterion we represent (8.13) in the form

$$J = \int_a^b W(\lambda) f(\lambda)\,\mathrm{d}\lambda \tag{8.14}$$

where λ is a random variable.

Suppose that the integrand $W(\lambda)$ is hard to handle. Then the integral (8.14) can be found by replacing W by some set of functions

$$w_m(\lambda), \quad m = 1, 2, \ldots, N \tag{8.15}$$

for which the products $w_m(\lambda)f(\lambda)$ are summable in the interval $[a, b]$. Now we select the performance criterion in the form

$$J_n(W, S_n) = \int_a^b |[W(\lambda) - S_n(\lambda)]f(\lambda)|\,\mathrm{d}\lambda \tag{8.16}$$

where

$$S_n(\lambda) = \sum_{i=1}^{n} a_i w_i(\lambda)$$

is the sought linear combination of the functions given by (8.15).

We wish to find a function $S_n(\lambda)$ such that for any positive real constant ε

$$J_n(W, S_n) < \varepsilon \tag{8.17}$$

In other words, we wish to approximate the integral (8.14) with arbitrary high accuracy ε. With this problem statement, the set of functions (8.15) must be not only complete but also consistent in properties with the integrand $W(\lambda)$. To meet this condition we select the coefficients a_i in the decomposition of S_n with the aid of interpolation.

Let for given nodes μ_k (values of the random variable λ) the integrand $W(\lambda)$ and the sought function S_n have equal values, i.e., $W(\mu_k) = S_n(\mu_k)$, then by virtue of (8.16) we are able to construct the system

$$W(\mu_k) = A^{\mathrm{T}} w(\mu_k), \quad k = 1, \ldots, n \tag{8.18}$$

where $A^{\mathrm{T}} = [a_1, \ldots, a_n]$ is a row vector, $w(\mu_k)$ is a column vector with the elements $w_1(\mu_k), \ldots, w_n(\mu_k)$. Rewritten in matrix form (8.18) becomes

$$\begin{bmatrix} W(\mu_1) \\ \vdots \\ W(\mu_n) \end{bmatrix} = \begin{bmatrix} w_1(\mu_1) & \ldots & w_n(\mu_1) \\ w_1(\mu_2) & & w_n(\mu_2) \\ \ldots & \ldots & \ldots \\ w_1(\mu_n) & \ldots & w_n(\mu_n) \end{bmatrix} \mathbf{A}$$

Then the coefficient matrix $\mathbf{A}$ can be found as

$$\mathbf{A} = \begin{bmatrix} w_1(\mu_1) & \ldots & w_n(\mu_1) \\ w_1(\mu_2) & \ldots & w_n(\mu_2) \\ \ldots & \ldots & \ldots \\ w_1(\mu_n) & \ldots & w_n(\mu_n) \end{bmatrix}^{-1} \begin{bmatrix} W(\mu_1) \\ \vdots \\ W(\mu_n) \end{bmatrix} \tag{8.19}$$

From (8.19) it follows that the matrix

$$\mathbf{w} = \begin{bmatrix} w_1(\mu_1) & \ldots & w_n(\mu_1) \\ \ldots & \ldots & \ldots \\ w_n(\mu_n) & \ldots & w_n(\mu_n) \end{bmatrix}$$

must be nonsingular.

Now we represent $S_n(\lambda)$ as

$$S_n(\lambda) = A^{\mathrm{T}} w(\lambda) \tag{8.20}$$

where $w(\lambda) = [w_1(\lambda), \ldots, w_n(\lambda)]^{\mathrm{T}}$ is a column vector.

Using the equality of W and S_n in the nodes μ_k, we substitute the approximation $S_n(\lambda)$ for $W(\lambda)$ and obtain instead of (8.14)

$$J \approx \int_a^b S_n(\lambda_n) f(\lambda)\, \mathrm{d}\lambda \tag{8.21}$$

which after substitution of (8.19, 20) becomes

$$\mathbf{J} = \begin{bmatrix} W(\mu_1) \\ \vdots \\ W(\mu_n) \end{bmatrix}^T \left\{ \begin{bmatrix} w_1(\mu_1) & \dots & w_n(\mu_1) \\ \dots & \dots & \dots \\ w_1(\mu_n) & \dots & w_n(\mu_n) \end{bmatrix}^{-1} \right\}^T \begin{bmatrix} \int_a^b w_1(\lambda) f(\lambda)\, \mathrm{d}\lambda \\ \dots \\ \int_a^b w_n(\lambda) f(\lambda)\, \mathrm{d}\lambda \end{bmatrix} \tag{8.22}$$

In dynamic system analysis and synthesis it is assumed that the distribution density $f(\lambda)$ is known. Then (8.22) easily transforms to the form

$$\mathbf{J} = \mathbf{A}^{\mathrm{T}} \boldsymbol{\rho} \tag{8.23}$$

where $\boldsymbol{\rho} = [\varrho_1, \dots, \varrho_n]^{\mathrm{T}}$ with $\varrho_i = \int_a^b w_i(\lambda) f(\lambda)\, \mathrm{d}\lambda, \quad i = 1, \dots, n.$

Equation (8.22) can be rewritten as

$$J = \mathbf{W}(\mu) \gamma \tag{8.24}$$

where $\mathbf{W}(\mu) = [W(\mu_1), \dots, W(\mu_n)]$, and

$$\gamma = \left\{ \begin{bmatrix} w_1(\mu_1) & \dots & w_n(\mu_1) \\ \dots & \dots & \dots \\ w_n(\mu_n) & \dots & w_n(\mu_n) \end{bmatrix}^{-1} \right\}^T \begin{bmatrix} \int_a^b w_1(\lambda) f(\lambda)\, \mathrm{d}\lambda \\ \dots \\ \int_a^b w_n(\lambda) f(\lambda)\, \mathrm{d}\lambda \end{bmatrix}$$

By virtue of (8.24) we represent the criterion as a quadrature sum, namely,

$$J = \int_a^b W(\lambda) f(\lambda)\, \mathrm{d}\lambda \approx \sum_{k=1}^{n} W(\mu_k) \gamma_k \tag{8.25}$$

It is quite obvious that the accuracy with which the equality in (8.25) is approximated depends on the selected set of functions $w_n(\lambda)$. If the equality in (8.25) holds for n functions $w_1(\lambda), \dots, w_n(\lambda)$ rather than for $n + 1$ functions, then J is said to have a degree of accuracy n with respect to these functions.

Practical implementation of (8.25) is faced with considerable difficulties in view of the need to compute matrix $\mathbf{A}$ in (8.18). To alleviate this difficulty we suggest the following approach. Let for $k = 1, \dots, N$ nodes the equality

$$W(\mu_k) = S_n(\mu_k) = \mathbf{A}^{\mathrm{T}} w(\mu_k) \tag{8.26}$$

be satisfied.

Multiplying (8.26) by a coefficient α_k,

$$\alpha_k W(\mu_k) = \alpha_k S_n(\mu_k) \tag{8.27}$$

and summing up for all k yields

$$S = \boldsymbol{\alpha}_{\Sigma} \cdot S_{\Sigma} = \boldsymbol{\alpha}_{\Sigma} \begin{bmatrix} w_1(\mu_1) & \dots & w_N(\mu_1) \\ \dots & \dots & \dots \\ w_1(\mu_N) & \dots & w_N(\mu_N) \end{bmatrix} \mathbf{A} \tag{8.28}$$

where $\boldsymbol{\alpha}_{\Sigma} = [\alpha_1, \alpha_2, \dots, \alpha_N]$.

Let us find the conditions providing for the equality of (8.23) and (8.28), namely,

$$\mathbf{A}^{\mathrm{T}} \boldsymbol{\rho} = \boldsymbol{\alpha}_{\Sigma} \begin{bmatrix} w_1(\mu_1) & \dots & w_N(\mu_1) \\ \dots & \dots & \dots \\ w_1(\mu_N) & \dots & w_N(\mu_N) \end{bmatrix} \mathbf{A} \tag{8.29}$$

It is an easy matter to demonstrate that

$$\mathbf{A}^{\mathrm{T}} \boldsymbol{\rho} = \boldsymbol{\rho}^{\mathrm{T}} \mathbf{A} \tag{8.30}$$

In view of this equality (8.29) becomes

$$\boldsymbol{\rho}^{\mathrm{T}} \mathbf{A} = \boldsymbol{\alpha}_{\Sigma} \begin{bmatrix} w_1(\mu_1) & \dots & w_N(\mu_1) \\ \dots & \dots & \dots \\ w_1(\mu_N) & \dots & w_N(\mu_N) \end{bmatrix} \mathbf{A} \tag{8.31}$$

From (8.31) it follows that the equality of (8.23) and (8.28) will be achieved given that

$$\boldsymbol{\rho}^{\mathrm{T}} = \boldsymbol{\alpha}_{\Sigma} \begin{bmatrix} w_1(\mu_1) & \dots & w_N(\mu_1) \\ \dots & \dots & \dots \\ w_1(\mu_N) & \dots & w_N(\mu_N) \end{bmatrix} \tag{8.32}$$

whence

$$\boldsymbol{\alpha}_{\Sigma} = \boldsymbol{\rho}^{\mathrm{T}} W_N(\mu) \tag{8.33}$$

where

$$W_N(\mu) = \left\{ \begin{bmatrix} w_{11}(\mu_1) & \dots & w_N(\mu_1) \\ \dots & \dots & \dots \\ w_{N1}(\mu_1) & \dots & w_N(\mu_N) \end{bmatrix} \right\}^{-1}$$

Because for the selected set of nodes μ_k the coefficients α_{Σ} of the given functions $w_i(\lambda)$ are known, the integral (8.23) may be determined by virtue of (8.28) as

$$J \approx S = \alpha_{\Sigma} \cdot S_{\Sigma}$$

or subject to the condition (8.26)

$$J \approx \boldsymbol{\alpha}_{\Sigma} \mathbf{W}_{\Sigma}(\mu) \qquad (8.34)$$

where $\mathbf{W}_{\Sigma} = [W(\mu_1), \ldots, W(\mu_N)]^{\mathrm{T}}$.

The relation (8.34) is suitable for design. It differs from (8.24) solely in the method of evaluation of coefficients α_{Σ} (α_{Σ} in place of γ). At $N = n$, $\alpha_n = \gamma_k$, and the relations (8.34) and (8.24) yield identical results.

The computational procedure discussed above allows one to determine the integral performance cirterion (8.14) with an algebraic degree of accuracy of n ensured by the degree of the polynomials $w_n(\lambda)$.

Clearly, it would be highly desirable to find formulas for an optimal criterion of (8.14). We assume that some function R of the error

$$\delta(W) = \left| \int_a^{\delta} W(\lambda) f(\lambda)\, d\lambda - W(\mu)\gamma \right|$$

may be selected as a criterion I of a given solution to the problem, i.e.,

$$I = \min_{\mu,\gamma} R\{\delta(W)\} \qquad (8.35)$$

Then the search for a minimum of (8.35) boils down to the evaluation of an extremum of the function of parameter vectors μ and γ, viz.,

$$I = \min I\{\mu, \gamma\}$$

The solution of this problem by the method of statistical nodes will be given below. However, for some complex dynamic systems, the problem in this formulation is not always solvable. Under these circumstances, other methods are employed, including the Monte Carlo method, the method of equivalent perturbations, and such.

8.3 Evaluation of Performance Criteria for Complex Dynamic Systems

Let the set of stochastic input variables $X_1(t), \ldots, X_p(t)$ of a dynamic system be specified by a deterministic function of the vector of random variates Λ and time

$$X(t) = X(\Lambda, t) \qquad (8.36)$$

where $X(t)$ is a multidimensional stochastic process with the required stochastic properties, Λ is an m-dimensional random variable with probability distribution density $f(\lambda)$, and t is current time.

Making use of (8.36), we represent (8.13) in the form

$$J = \int_{a_1}^{b_1} \dots \int_{a_m}^{b_m} \Phi(\lambda_1, \dots, \lambda_m, t) f(\lambda_1, \dots, \lambda_m) \, d\lambda_1 \dots d\lambda_m \tag{8.37}$$

where $[a_i, b_i]$ is a finite or infinite segment on the i-th numerical axis, $i = 1, \dots, m$; $f(\lambda_1, \dots, \lambda_m)$ is the weight function (distribution density) satisfying the condition of existence of the momenta

$$\mu_{\alpha_1 \dots \alpha_m} = \int_{a_1}^{b_1} \dots \int_{a_m}^{b_m} f(\lambda_1, \dots, \lambda_m) \lambda_1^{\alpha_1}, \dots, \lambda_m^{\alpha_m} \, d\lambda_1 \dots d\lambda_m$$

and $\Phi(\lambda_1, \dots, \lambda_m, t)$ is a function for which the integral (8.37) exists.

By giving various functions Φ, one may obtain the required criteria. For example, the functions

$$\Phi = Y(\lambda_1, \dots, \lambda_m, t)$$

$$\Phi = [Y(\lambda_1, \dots, \lambda_m, t) - m_Y]^2$$

$$\begin{aligned} \Phi &= 1, \quad \text{for} \quad |Y(\lambda_1, \dots, \lambda_m, t)| < C \\ &= 0, \quad \text{for} \quad |Y(\lambda_1, \dots, \lambda_m, t)| \geqslant C \end{aligned}$$

lead, respectively, to the criteria of mathematical expectation, variance, and probability of a certain event, i.e.,

$$I = P[|Y(\lambda_1, \dots, \lambda_m, t)| < C]$$

In statistical analysis of complex dynamic systems, the function $\Phi(\lambda_1, \dots, \lambda_m, t)$ as a rule avoids analytical representation. Therefore, we have referred to it as being hard to manage.

There are two ways to solve the problem of evaluation of performance measures I for this class of systems. One of them resorts to using the method of stochastic simulation (Monte Carlo method), the other relies on representing the integrands with the aid of two- and three-dimensional Gauss-type integration formulas.

The problem of approximate evaluation of the integral

$$J = \int_E \Phi(\lambda) \, dF(\lambda) \tag{8.38}$$

where $F(\lambda)$ is a fixed function, was solved by the Gaussian quadrature

$$\int_E \Phi(\lambda) f(\lambda) \, d\lambda \approx \sum_{i=1}^{N} A_i \Phi(\lambda_i) \tag{8.39}$$

where $f(\lambda) = \partial F(\lambda)/\partial\lambda$ is the weight function, A_i are coefficients independent of $\Phi(\lambda)$, and λ_i are the nodes of (8.39).

It is quite obvious that the weight function $f(\lambda)$ may be viewed as the distribution density of the random variable Λ. Then (8.38, 39) are the mathematical expectation of the stochastic function $\Phi(\lambda)$, i.e.,

$$E[\Phi(\lambda)] = \int_E \Phi(\lambda) f(\lambda)\, d\lambda$$

The Monte Carlo method is actually a particular case of the quadrature sums (8.39). Indeed, for $A = 1/N$, where N is the number of nodes (sampling experiments), (8.39) takes the form

$$E[\Phi(\lambda)] = \frac{1}{N} \sum_{i=1}^{N} \Phi(\mu_i) \qquad (8.40)$$

It is worth emphasizing, however, that the Monte Carlo method does not require an approximate specification of $\Phi(\lambda)$ and evaluation of the coefficients of A. This method has been widely used owing to versatility and simplicity achievable through the use of sampling data (evaluation of the values of $\Phi(\mu_i)$ for the given replications). However, for certain classes of systems the algorithm (8.40) converges slowly. In orther words, some problems of statistical analysis call for a rather large number of replications. In model analysis of dynamic systems this implies that the system of equations has to be multiply integrated to determine the values of the integrand $\Phi(\mu_i)$. Precisely this circumstance has stimulated the search of other ways to the evaluation of the integral (8.38).

The approach based on the Gaussian quadrature type formulas assumes that the nodes μ_i and coefficients A_i will be specified according to some prearranged scheme and with allowance for the function $\Phi(\lambda)$. As a result the number of replications reduces substantially.

There is a variety of techniques incorporating the concept of quadrature. For a multidimensional case they may be briefly formulated as follows.

Suppose that in the domain of random disturbances described by a system of stochastic variates $\Lambda_1, \ldots, \Lambda_m$ we have a collection of nodes

$$\{\mu_{1k_1}, \ldots, \mu_{mk_m}\}, \; k_1 = 1, \ldots, q_1; \quad \ldots; \quad k_m = 1, \ldots, q_m \qquad (8.41)$$

where μ_{ik_j} is the k_j-th value of the i-th random variable, and q_i is the order of the approximating polynomial in the i-th variable.

By virtue of (8.37) we determine the system response to (8.41) with the aid of the functions $G(\mu_{1k_1}, \ldots, \mu_{mk_m}, t)$, and the criterion value by the three-dimensional Gauss-type integration formula

$$\int_{a_1}^{b_1} \ldots \int_{a_m}^{b_m} \Phi(\lambda_1, \ldots, \lambda_m, t) f(\lambda_1, \ldots, \lambda_m)\, d\lambda_1 \ldots d\lambda_m \approx \sum_{k_1, \ldots, k_m} G(\mu_{1k_1}, \ldots, \mu_{mk_m}, t) \gamma_{k_1 \ldots k_m} \qquad (8.42)$$

where $\gamma_{k_1 \ldots k_m}$ are coefficients, $G(\mu_{1k_1}, \ldots, \mu_{mk_m}, t)$ are the values of the function $G(\lambda_1, \ldots, \lambda_m, t)$ at the nodes (8.41), and the type of G depends on the specific method in which the criterion is developed.

Because the nodes (8.41) are associated with certain statistical disturbances obtained for the given system of random variates, in what follows we shall call these nodes statistical.

It has been noted that there exist a host of methods developed to determine integral criteria (8.42). They differ mainly in the way of specifying the nodes (8.41), functions G, and coefficients $\gamma_{k_1 \ldots k_m}$. For the sake of brevity we outline the general method of local statistical nodes which under certain conditions and assumptions leads to a number of software agreeable formulas [8.1] for performance measures of complicated dynamic systems.

8.4 Analysis of Complex Dynamic Systems by the Method of Statistical Nodes

The main difficulty with which the designer of complex dynamic systems is faced arises from the integrand (8.37) being not amenable to analytic evaluation. This circumstance has motivated a search for approximate representation of Φ with the aid of computer simulation, or orthogonal polynomials or expansions. Whatever the method used to derive the values $\Phi(\lambda_i)$ (8.39, 40) in analysis of complex dynamic systems, it invariably involves mathematical modeling. Below we list the major steps of solving the analysis problem.

Let a complex dynamic system be described by the vector equations

$$\begin{aligned} \frac{d\mathbf{Z}}{dt} &= F[\mathbf{Z}, \mathbf{X}_1(\mathbf{\Lambda}_1, t), t_0, t] \\ \mathbf{Y} &= \mathbf{H}[\mathbf{Z}, \mathbf{X}_2(\mathbf{\Lambda}_2, t), t] \end{aligned} \tag{8.43}$$

where F and H are the nonlinear operators containing smooth and discontinuous functions, $\mathbf{Z}$ is the vector of state, $\mathbf{X}_1$ and $\mathbf{X}_2$ are random process vectors, $\mathbf{\Lambda}_1$ and $\mathbf{\Lambda}_2$ are vectors of random variables, and $\mathbf{Y}$ is the vector of the output coordinates of the dynamic system.

We introduce a function $\Phi(Y)$ which in view of (8.43) may be transformed formally to the form $\Phi(\lambda_1, \ldots, \lambda_m, t)$, where $\lambda_1, \ldots, \lambda_m$ is a set of random variables given by the vectors Λ_1 and Λ_2. Assume that $\Phi(\lambda_1, \ldots, \lambda_m, t)$ is the integrand of the criterion (8.37), i.e., the performance measure of the dynamic system. Then the evaluation of the criterion by the method of statistical nodes reduces to evaluation of the values of functions $G(\mu_{1k_1}, \ldots, \mu_{mk_m}, t)$ and coefficients $\gamma_{k_1 \ldots k_m}$ (8.42). We dwell on this procedure in a little more detail.

The functions $G(\mu_{1k_1}, \ldots, \mu_{mk_m}, t)$ are evaluated usually by simulation, i.e., by integrating the system of equations (8.43) at the given nodes (8.41) and carrying the function $\Phi(\lambda_2, \ldots, \lambda_m, t)$ to $G(\lambda_1, \ldots, \lambda_m, t)$. Since the functions $G(\mu_{1k_1}, \ldots, \mu_{mk_m}, t)$ characterize the dynamic processes in the system and are constructed on the basis of state coordinates Z and output coordinates $Y(t)$ (8.43), they are called *replications* (or *realizations*). In evaluation of the criterion (8.42), these replications (simulation runs) consume the major proportion of the computer time. The number of replications depends on the particular method of assigning the nodes (8.41) and the method of representation of the integrand $\Phi(\lambda_1, \ldots, \lambda_m, t)$ [8.1]. However, for all methods relaying upon the approximation of the integrand an increase in the number of random variables $\Lambda_1, \ldots, \Lambda_m$ leads to considerable growth of simulation runs required.

Choosing the statistical nodes (8.41) and coefficients $\gamma_{k_1}, \ldots, \gamma_{k_m}$, the designer usually tends to achieve the highest possible accuracy in evaluating the criterion (8.37). The approximate expression

$$\int_{a_1}^{b_1} \cdots \int_{a_m}^{b_m} \Phi(\lambda_1, \ldots, \lambda_n, t) f(\lambda_1, \ldots, \lambda_m)\, d\lambda_1 \ldots d\lambda_m \approx \sum_{k_1, \ldots, k_m} \Phi(\mu_{1k_1}, \ldots, \mu_{mk_m}, t) \gamma_{k_1 \ldots k_m} \tag{8.44}$$

can be made exact by an appropriate choice of the degree of the interpolating polynomials $w_{q_i}(\lambda)$ in random variable λ_i, $(i = 1, \ldots, m)$.

Chernetsky [8.2] has demonstrated that the formula (8.44) should be an interpolation and the degree of the polynomials must not exceed $q_i - 1$. Then the equality in (8.44) is ensured for all functions $\Phi(\lambda_1, \ldots, \lambda_m, t)$ of degree at most $2q_i - 1$ in each variable λ_i, $i = 1, \ldots, m$.

Indeed, by defining the function $\Phi(\lambda_1, \ldots, \lambda_m, t)$ with the Lagrange formula,

$$\Phi(\lambda_1, \ldots, \lambda_m, t) = \sum_{k_1, \ldots, k_m} \Phi(\mu_{1k_1}, \ldots, \mu_{mk_m}, t) \prod_{i=1}^{m} \frac{w_{q_i}(\lambda_i)}{w_q'(\mu_{ik_i})(\lambda_i - \mu_{ik_i})}$$

$$k_i = 1, \ldots, q_i; \quad i = 1, \ldots, m \tag{8.45}$$

and, substituting (8.45) into (8.44), we get the exact equality in (8.44) subject to

$$\gamma_{k_1 \ldots k_m} = \int_{a_1}^{b_1} \cdots \int_{a_m}^{b_m} \prod_{j=1}^{m} \frac{w_{q_i}(\lambda_i)}{w_q'(\mu_{ik_i})(\lambda_i - \mu_{ik_i})} f(\lambda_1, \ldots, \lambda_m)\, d\lambda_1 \ldots d\lambda_m$$

where $w_{q_i}(\lambda_i)$ form an orthogonal system of polynomials weighted by $f(\lambda_1, \ldots, \lambda_m)$ on the interval $[a_i, b_i]$, $i = 1, \ldots, m$.

The statistical modes (8.41) are the roots of the equations $w_{q_i} = 0$.

Thus, in this interpolation method [8.2] the nodes are selected by the given density of distribution on the interval $[a_i, b_i]$, $i = 1, \ldots, m$.

In the method of equivalent perturbations [8.3] the nodes (8.41) and coefficients $\gamma_{k_1 \ldots k_m}$ are assigned so as to satisfy the given statistical moments. This approach reduces the number of required replications of $\Phi(\mu_{1m_1}, \ldots, \mu_{mk_m}, t)$.

Potseluev [8.1] suggests another approach allowing for the dynamic properties of the system under study on the basis of a preliminary analysis of its dynamic behavior. This in turn allows a distribution of nodes over the numerical axes λ_i, $i = 1, \ldots, m$, with due account of the real properties of the system and is advantageous in a substantial reduction of the required number of replications of $\Phi(\mu_{1k_1}, \ldots, \mu_{mk_m}, t)$. In what follows we discuss this approach in more detail.

Suppose that an analysis of the filtering properties of a complex dynamic system carried out over the domain of random stimuli $[a_{i0}, a_{in_i}]$, $i = 1, \ldots, m$, has revealed individual subregions $[a_{ij}, a_{i,j+1}]$, $i = 1, \ldots, m$; $j = 0, 1, \ldots, n - 1$, where the random processes λ_i, $i = 1, \ldots, m$, have distinct effects. This allows various number of nodes to be assigned in each of the subregions depending on the system response, thus improving the accuracy of system estimation for a given number of statistical nodes, or reduce the number of required nodes and the number of respective replications of $\Phi(\mu_{1k_1}, \ldots, \mu_{mk_m}, t)$ for a specified accuracy of estimation of the criterion.

This approach is used in the development of the method of local statistical nodes outlined below.

Using the given subregions $[a_{ij}, a_{i,j+1}]$, we may represent (8.37) as a sum of multidimensional integrals, viz.,

$$J = \sum_{j_1=0}^{n_1-1} \sum_{j_2=0}^{n_2-1} \cdots \sum_{j_m=0}^{n_m-1} \int_{a_{1j_1}}^{a_{1j_1+1}} \cdots \int_{a_{mj_m}}^{a_{mj_m+1}} \Phi(\lambda_1, \ldots, \lambda_m, t) \times f(\lambda_1, \ldots, \lambda_m)\, d\lambda_1 \ldots d\lambda_m \tag{8.46}$$

The distribution density function $f(\lambda_1, \ldots, \lambda_m)$ must be normalized on the intervals $[a_{ij}, a_{i,j+1}]$, i.e.

$$\int_{a_{1j_1}}^{a_{1j_2}} \cdots \int_{a_{mj_m}}^{a_{mj_m+1}} f\alpha_{j_1 \ldots j_m}(\lambda_1, \ldots, \lambda_m)\, d\lambda_1 \ldots d\lambda_m = 1 \tag{8.47}$$

Therefore, equation (8.46) must incorporate the normalization coefficients

$\alpha_{j_1 \ldots j_m}$ to insure the condition (8.47). Then (8.46) takes the form

$$J = \sum_{j_1=0}^{n_1-1} \sum_{j_2=0}^{n_2-1} \ldots \sum_{j_m=0}^{n_m-1} \alpha_{j_1 \ldots j_m} \int_{a_{1j_1}}^{a_{1j_2+1}} \ldots \int_{a_{mj_m}}^{a_{mj_m+1}} \Phi(\lambda_1, \ldots, \lambda_m, t)$$
$$\times f\alpha_{j_1 \ldots j_m}(\lambda_1, \ldots, \lambda_m)\, d\lambda_1 \ldots d\lambda_m \tag{8.48}$$

where

$$\alpha_{j_1 \ldots j_m} = \int_{a_{1j_1}}^{a_{1j_2}} \ldots \int_{a_{mj_m}}^{a_{mj_m+1}} f(\lambda_1, \ldots, \lambda_m)\, d\lambda_1 \ldots d\lambda_m$$

$$\sum_{j_1=0}^{n_{j_1-1}} \ldots \sum_{j_m=0}^{n_{j_m-1}} \alpha_{j_1 \ldots j_m} = 1 \quad \text{and} \quad \alpha_{j_1 \ldots j_m} \geqslant 0$$

Using the representation (8.42), equation (8.48) can be carried to the form

$$J \approx \sum_{j_1=0}^{n_{j_1-1}} \sum_{j_2=0}^{n_{j_1-1}} \ldots \sum_{j_m=0}^{n_{j_m-1}} \alpha_{j_1 \ldots j_m} \sum_{k_1 \ldots k_m} G(\mu_{1k_1}, \ldots, \mu_{mk_m}, t)\gamma_{k_1 \ldots k_m},$$
$$k_{ji} = 1, \ldots, q_{ji} \tag{8.49}$$

where q_{ji} is the number of nodes in the interval $[a_{ij_i}, a_{ij_i+1}]$.

Efficiency of this approach has been treated in detail in [8.1].

8.5 Mathematical Models of Stochastic Processes

Random process generation in analysis and synthesis of dynamic systems is especially difficult where stochastic processes with given correlation properties are required, or the generated time-variable or time-invariant processes are to possess some preassigned statistical characteristics.

Time-Invariant Random Processes with a Given Correlation Matrix

Let the dynamic system under study be subjected to m stochastic disturbances $X_1(t), \ldots, X_m(t)$ with expectations $m_{x_1}, \ldots, m_{x_m}$ and correlation matrix

$$\mathbf{R}_{ij}(\tau) = \begin{bmatrix} R_{11}(\tau) & R_{12}(\tau) & \ldots & R_{1m}(\tau) \\ R_{21}(\tau) & R_{22}(\tau) & \ldots & R_{2m}(\tau) \\ \ldots & \ldots & \ldots & \ldots \\ R_{m1}(\tau) & R_{m2}(\tau) & \ldots & R_{mm}(\tau) \end{bmatrix} \tag{8.50}$$

We intend to consider the evaluation of a Φ_{mn} transformation which forms a system of random processes $X_1(t), \ldots, X_m(t)$ possessing the above statistics with the aid of an arbitrary set of processes $X^*_{1i}(t), \ldots, X^*_{mj}(t)$, $i = 1, 2, \ldots; j = 1, 2, \ldots$. We begin by solving this problem for $m = 2$.

The correlation matrix of the desired system takes the form

$$\mathbf{R}_i(\tau) = \begin{bmatrix} R_{x_1x_1}(\tau) & R_{x_1x_2}(\tau) \\ R_{x_2x_1}(\tau) & R_{x_2x_2}(\tau) \end{bmatrix} \tag{8.51}$$

The matrix (8.51) has four elements specified for the variates X_1 and X_2, therefore we introduce four processes X_{11}, X_{12}, X_{21}, and X_{22} related by

$$\begin{aligned} X_1 &= X_{11} + X_{12} \\ X_2 &= X_{21} + X_{22} \end{aligned} \tag{8.52}$$

from which we get

$$\begin{aligned} R_{x_1x_1}(\tau) &= R_{x_{11}x_{11}}(\tau) + R_{x_{12}x_{12}}(\tau) + 2R_{x_{11}x_{12}}(\tau) \\ R_{x_2x_2}(\tau) &= R_{x_{21}x_{21}}(\tau) + R_{x_{22}x_{22}}(\tau) + 2R_{x_{21}x_{22}}(\tau) \\ R_{x_1x_2}(\tau) &= R_{x_{11}x_{21}}(\tau) + R_{x_{11}x_{22}}(\tau) + R_{x_{12}x_{21}}(\tau) + R_{x_{12}x_{22}}(\tau) \end{aligned} \tag{8.53}$$

We require that the following relations hold:

$$\begin{aligned} R_{x_1x_1}(\tau) &= R_{x_{11}x_{11}}(\tau) + R_{x_{12}x_{12}}(\tau) \\ R_{x_2x_2}(\tau) &= R_{x_{21}x_{21}}(\tau) + R_{x_{22}x_{22}}(\tau) \\ R_{x_1x_2}(\tau) &= R_{x_{12}x_{21}}(\tau) \end{aligned} \tag{8.54}$$

Clearly this will be true if the processes X_{11} and X_{12}, X_{11} and X_{22}, X_{11} and X_{21}, X_{22} and X_{12}, and X_{22} and X_{21} are not correlated. We require also that they be independent, then

$$\begin{aligned} &R_{x_{11}x_{12}}(\tau) = 0, \quad R_{x_{11}x_{21}}(\tau) = 0, \quad R_{x_{12}x_{22}}(\tau) = 0, \\ &R_{x_{21}x_{21}}(\tau) = 0, \quad R_{x_{11}x_{22}}(\tau) = 0 \end{aligned}$$

The first two equations in the set (8.54) are rather easy to satisfy — just by knowing the respective correlation functions of X_{11}, X_{12}, X_{21}, and X_{22}. Therefore, we dwell on the third line.

The cross-spectral density of the pocesses x_{12} and x_{21} may be given as

$$S_{x_{12}x_{21}}(j\omega) = \frac{1}{2\pi} \int_{-\infty}^{+\infty} R_{x_{12}x_{21}}(\tau) \mathbf{e}^{-j\omega\tau} \, d\tau \tag{8.55}$$

Let us introduce an element with transfer function $H(s)$ transforming a process x_{12} into x_{21} (Fig. 8.1). Then

$$S_{x_{12}x_{21}}(j\omega) = H(j\omega)S_{x_{12}x_{12}}(\omega) \tag{8.56}$$

Since $S_{x_{12}x_{12}}(\omega)$ is known by satisfying the first equation in (8.54), and $S_{x_{12}x_{21}}(j\omega)$ is known in agreement with (8.55), then we may determine the transfer function $H(s)$ by the relation

$$H(j\omega) = S_{x_{12}x_{21}}(j\omega)/S_{x_{12}x_{12}}(\omega) \tag{8.57}$$

In agreement with Fig. 8.1 we can write

$$X_{21} = HX_{12} \tag{8.58}$$

Eliminating X_{21} between (8.58) and (8.52), we get

$$\begin{aligned} X_1 &= X_{11} + X_{12} \\ X_2 &= X_{22} + HX_{12} \end{aligned} \tag{8.59}$$

Since the processes X_{11} and X_{22} are independent of X_{12} and HX_{12}, then X_1 and X_2 satisfy the properties of cross-correlation $R_{x_1x_2}(\tau)$.

Assuming that (8.54) hold, we give a process $X_{12}(t)$ with correlation function $R_{x_{12}x_{12}}$ subject to the condition

$$R_{x_{11}x_{11}}(\tau) = R_{x_1x_1}(\tau) - R_{x_{12}x_{12}}(\tau)$$

In view of the second equation in (8.54) we have

$$R_{x_{22}x_{22}}(\tau) = R_{x_2x_2}(\tau) - R_{x_{21}x_{21}}(\tau)$$

The cross-correlation properties are ensured by the *H*-converter.

Figure 8.2 shows the block diagram of the Φ_{23}-converter built in agreement with equations (8.59). In order that the processes (8.52) have the desired expectations m_{x_1} and m_{x_2} we require that $m_{x_1} = m_{x_{11}}$, and $m_{x_2} = m_{x_{22}}$. The complete block diagram of a converter providing the re-

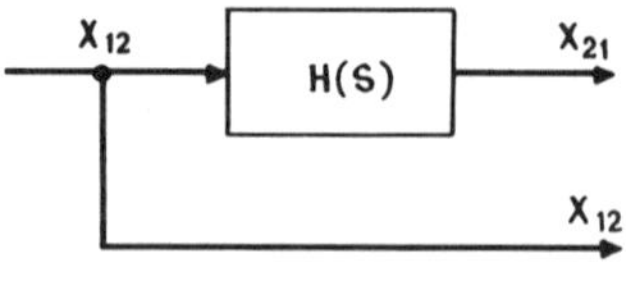

Fig. 8.1

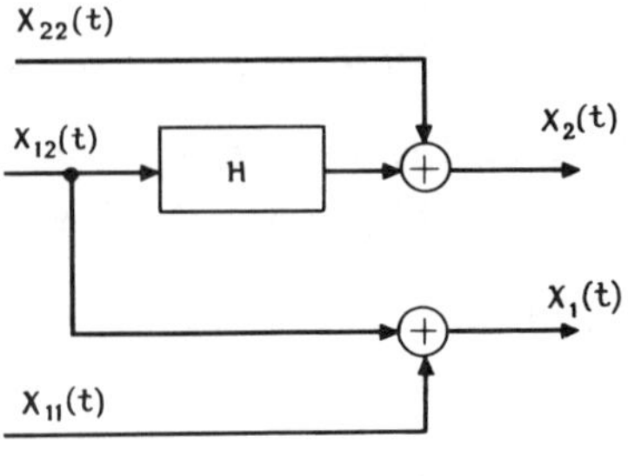

Fig. 8.2

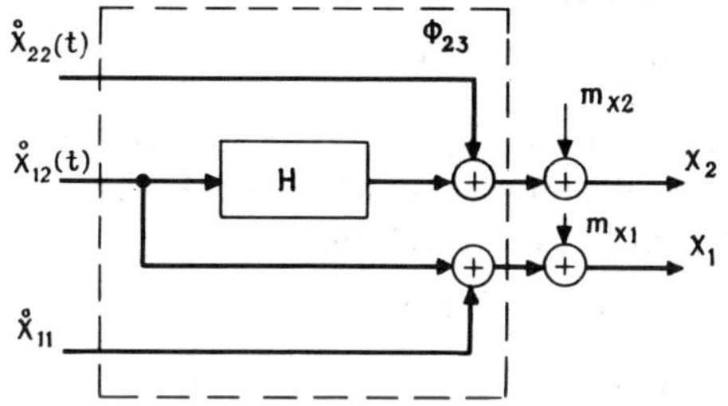

Fig. 8.3

quired correlation properties is presented in Fig. 8.3; $\mathring{X}_{11}$, $\mathring{X}_{12}$, and $\mathring{X}_{22}$ denote central random processes.

It is worth noting at this point that the choice of the system (8.52) was effected so as to satisfy the four elements of the matrix (8.51). However, in view of $R_{x_2x_1}(\tau) = R_{x_1x_2}(-\tau)$ equations (8.52) may be simplified, retaining three processes. Let

$$\begin{aligned} X_1 &= X_{11} \\ X_2 &= X_{21} + X_{22} \end{aligned} \tag{8.60}$$

then

$$R_{x_1x_2}(\tau) = R_{x_{11}x_{11}}(\tau)$$

$$R_{x_2x_2}(\tau) = R_{x_{21}x_{21}}(\tau) + R_{x_{22}x_{22}}(\tau) + 2R_{x_{21}x_{22}}(\tau)$$

$$R_{x_1x_2}(\tau) = R_{x_{11}x_{21}}(\tau) + R_{x_{11}x_{22}}(\tau)$$

Likewise require that

$$X_{22} = H_{x_{11}}$$

where H denotes a converter with complex transfer function

$$H(j\omega) = S_{x_{11}x_{22}}(j\omega)/S_{x_{11}x_{11}}(j\omega) \tag{8.61}$$

Consequently, specifying a process X_{11} with

$$R_{x_{11}x_{11}}(\tau) = R_{x_1x_1}(\tau)$$

X_{22} with

$$R_{x_{22}x_{22}}(\tau) = E[H_{11}(t)H_{11}(t + \tau)]$$

and X_{21} with

$$R_{x_{21}x_{21}}(\tau) = R_{x_2x_2}(\tau) - E[HX_{11}(t)HX_{11}(t + \tau)]$$

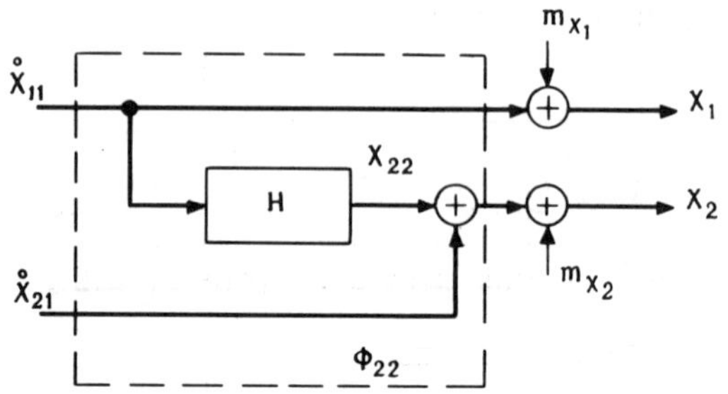

Fig. 8.4

and letting X_{21} and X_{22}, X_{11} and X_{22}, and X_{11} and X_{21} be pairwise independent, we arrive at the system of equations

$$X_1 = X_{11}$$

$$X_2 = X_{21} + HX_{11}$$

satisfying (8.51).

The block diagram of a Φ_{22}-converter producing two processes X_1 and X_2 with the given correlation functions (8.51) and expectations is depicted in Fig. 8.4.

Example 8.1. Suppose that the correlation matrix of two processes X_1 and X_2 is given in the form

$$\mathbf{R}_{ij}(\tau) = \begin{bmatrix} 2\pi g_0\delta(\tau) & R_{x_1x_2}(\tau) \\ R_{x_2x_1}(\tau) & De^{-\alpha|\tau|}\cos\beta\tau \end{bmatrix}$$

where the cross-correlation functions $R_{x_1x_2}(\tau)$ and $R_{x_2x_1}(\tau)$ may be expressed through the given cross-spectral density

$$S_{x_1x_2}(j\omega) = a/(b + j\omega)$$

The spectral density $S_{x_{11}x_{11}}(\omega) = S_{x_1x_1}(\omega)$ is found as

$$S_{x_1x_1}(\omega) = \frac{1}{2\pi}\int_{-\infty}^{\infty} 2\pi g_0\delta(\tau)e^{-j\omega\tau}\,d\tau = g_0$$

Then in agreement with (8.61) the complex transfer function is determined as

$$H(j\omega) = \frac{a}{g_0(b + j\omega)} \quad \text{or} \quad H(j\omega) = \frac{k}{Tj\omega + 1}$$

where $k = a/bg_0$ is the gain coefficient and $T = 1/b$ is the time constant.

Consequently, the transfer function of the element $H(s)$ characterizes an aperiodic element

$$H(s) = \frac{k}{Ts + 1}$$

If X_{11} is the input to this element, its output will be a process with spectral density

$$S_{x_{22}x_{22}}(\omega) = \frac{k^2}{|T\omega + 1|^2} S_{x_{11}x_{11}}(\omega) = \frac{k^2 g_0}{T^2\omega^2 + 1}$$

or

$$S_{x_{22}x_{22}}(\omega) = \frac{c}{\omega^2 + \alpha_1^2}$$

where $c = k^2 g_0/T^2$ and $\alpha_1 = 1/T$.

The correlation function will have the form

$$R_{x_{22}x_{22}}(\tau) = \int_{-\infty}^{\infty} \frac{c}{\omega^2 + \alpha_1^2} \exp(-j\omega\tau)\, d\omega = D_1 e^{-\alpha_1|\tau|}$$

where $D_1 = \pi c/\alpha_1$.

The correlation functions of two independent processes X_{11} and X_{22} are

$$R_{x_{11}x_{11}}(\tau) = R_{x_1x_1}(\tau) = 2\pi g_0 \delta(\tau)$$

$$R_{x_{21}x_{21}}(\tau) = De^{-\alpha_1|\tau|} \cos\beta\tau - D_1 e^{-\alpha|\tau|}$$

The block diagram of the Φ_{22}-converter is shown in Fig. 8.5.

In the general case, the structure of a Φ_{mn}-converter may be derived by analogy with the construction of the Φ_{22}- and Φ_{23}-converters with the

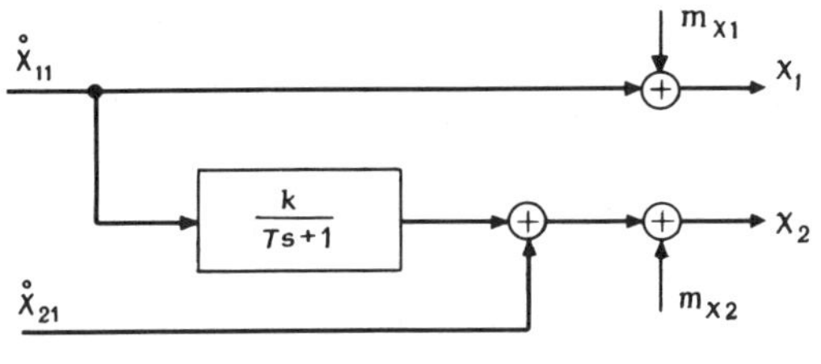

Fig. 8.5

help of the following system of constraint equations for random processes:

$$X_1(t) = \sum_{i=1}^{m} X_{1i}(t)$$

$$X_h(t) = \sum_{i=h}^{m} X_{hi}(t) + \sum_{i=1}^{h-1} H_{ih} X_{ih}(t) \qquad (8.62)$$

$$h = 2, 3, \ldots, m$$

The number of independent processes X_{hi} and X_{ii} ($i = 1, \ldots, n$; $h = 1, \ldots, n$; $h < i$) required to obtain the processes $X_1, \ldots, X_m$ with the given correlation properties depends on n correlation functions of (8.50) and is determined as

$$n = \frac{m^2 + m}{2}$$

Constructing the system of equations

$$R_{x_1 x_1}(\tau) = \sum_{i=1}^{m} R_{x_{1i} x_{1i}}(\tau)$$

$$R_{x_h x_i}(\tau) = \sum_{i=h}^{m} R_{x_{hi} x_{hi}}(\tau) + \sum_{i=1}^{h-1} \varphi_{ij}[R_{x_{ih} x_{ih}}(\tau)]$$

$$R_{x_1 x_h}(\tau) = \varphi_{1h1h}[R_{x_{1h} x_{1h}}(\tau)]$$

$$R_{x_h x_r}(\tau) = \varphi_{hr\,hr}[R_{x_h x_h}(\tau)]$$

$$h = 2, \ldots, m; \quad r = 2, \ldots, m; \quad h < r$$

and solving it for the initial matrix (8.50) one may construct, with the formalism outlined above, a Φ_{mn}-converter generating m random processes with the specified correlation properties.

Nonlinear Models of Random Processes

Let a random process $\mathscr{L}(t)$ be described by a real function of bounded variation in a finite interval having a finite mean square value $\langle |\mathscr{L}(0)|^2 \rangle$. Then such a process may be represented in the form

$$\mathscr{L}(t) = m_\alpha(t) + \sum_{k=-\infty}^{+\infty} A_k(\Lambda, t) \exp[iB_k(\Lambda, t)] \qquad (8.63)$$

where m_α is an aperiodic deterministic component, $A_k(\Lambda, t)$ is the magnitude of the k-th component, $B_k(\Lambda, t)$ is the phase of the k-th component, and Λ is the vector of random variables.

A constraint imposed on the k-th components is that the total sum yields a central random function.

After obvious manipulations (8.63) may be rewritten as

$$\mathscr{L}(t) = m_\alpha(t) + \sum_{k=1}^{\infty} [C_k(\Lambda, t) \cos B_k(\Lambda, t) + jD_k(\Lambda, t) \sin B_k(\Lambda, t)] \tag{8.64}$$

$$\mathscr{L}(t) = m_\alpha(t) + \sum_{k=1}^{\infty} A_k(\Lambda, t) \cos [B_k(\Lambda, t) + \varphi_k] \tag{8.65}$$

where φ_k is the phase shift, or

$$\mathscr{L}(t) = m_\alpha(t) + \sum_{k=1}^{\infty} A_k(\Lambda, t) \sin [B_k(\Lambda, t) + \varphi_k'] \tag{8.66}$$

The mathematical models (8.63-66) are sufficient to describe practically all random disturbances to which dynamic systems are subjected. An obvious setback of these models is the infinite number of members which limits their practical usability. In practice the number of components is reduced, for example, by assuming that

$$\mathscr{L}(t) = m_\alpha(t) + \sum_{k=1}^{\infty} A_k \cos (\omega_k t + \varphi_k) \tag{8.67}$$

Accordingly, the problem of truncation error emerges. If ω_k and φ_k are deterministic quantities and A_k are stochastic, then this representation of a process is referred to as *linear.* With random ω_k and φ_k the model is treated as nonlinear.

The practical analysis and synthesis of dynamic systems exploits a wide spectrum of nonlinear models of random processes. Some of these models are summarized in Table 8.1. A distinctive feature of nonlinear representations is a far lesser number of members involved in the mathematical model. Models 1.5, 1.6, and 1.7 contain only three random variables, φ, A and ω. The correlation function $R_{\mathscr{L}\mathscr{L}}(\tau)$ is exactly of the required form and model 1.7 generates a random process distributed to a normal law.

8.6 Generation of Dependent Random Variables

Analysis and synthesis of dynamic systems subjected to independent random disturbances do not pose large problems for the designer. This is not true when the dynamic system is excited by a collection of random variables with given statistical crosslinks as this requires generation of random disturbances. There is no way of obtaining such a collection of random variates $\Lambda_1, \ldots, \Lambda_m$ with required distribution density $f_\lambda(\lambda_1, \ldots, \lambda_m)$ with the aid of a system of measuring elements without a respec-

Table 8.1. Nonlinear Models (A, φ, and ω are independent; φ is uniform in $[0, 2\pi]$)

Model	Statistical characteristics of components	General statistical characteristics of random process
1.1 $A \sin(\omega t + \varphi)$	A and ω are deterministic	Ergodic process $f_{\mathcal{L}}(\mathcal{L}, t_1) = [\pi\sqrt{A^2 - \mathcal{L}^2}]^{-1}$ $(\lvert\mathcal{L}\rvert \leqslant A)$ $R_{\mathcal{L}\mathcal{L}}(\tau) = (A^2/2)\cos(\omega\tau)$
1.2 $A \sin(\omega t + \varphi)$	A has density $f_A(A)$ ω is deterministic	$f_{\mathcal{L}}(\mathcal{L}, t_1) = \frac{1}{\pi}\int_{-\infty}^{\infty} \frac{f_A(A)\,dA}{\sqrt{A^2 - \mathcal{L}^2}}$ $R_{\mathcal{L}\mathcal{L}}(\tau) = 0.5E[A^2]\cos(\omega\tau)$
1.3 $A \sin(\omega t + \varphi)$	$f_A(A) = A\exp(-A^2/2)$, for $A \geqslant 0$ $= 0$, for $A < 0$ ω is deterministic	Gaussian process $R_{\mathcal{L}\mathcal{L}}(\tau) = 0.5E[A^2]\cos(\omega\tau)$
1.4 $A \sin(\omega t + \varphi)$	$A = \sigma_{\mathcal{L}}\sqrt{2}$ is deterministic ω has density $f_\omega(\omega)$	$f(\mathcal{L}, t) = \frac{1}{\pi}\frac{1}{\sqrt{A^2 + \mathcal{L}^2}}$, $(\lvert\mathcal{L}\rvert A)$ $R_{\mathcal{L}\mathcal{L}}(\tau) =$ $= \frac{A^2}{2}\int_{\omega} \cos(\omega\tau) f_\omega(\omega)\,d\omega$
1.5 $A \sin(\omega t + \varphi)$	$f_\omega(\omega) = \frac{1}{\pi}\int_0^{\infty} r_{\mathcal{L}\mathcal{L}}(\tau)\cos(\omega\tau)d\tau$ $E[A^2] = R_{\mathcal{L}\mathcal{L}}(0)$	$R_{\mathcal{L}\mathcal{L}}(\tau)$ is required
1.6 $A \sin(\omega t + \varphi)$	$f_\omega(\omega) = \frac{1}{\pi}\int_0^{\infty} r_{\mathcal{L}\mathcal{L}}(\tau)\cos(\omega\tau)d\tau$ A has density function $f_A(A)$	$R_{\mathcal{L}\mathcal{L}}(\tau)$ is required $f_{\mathcal{L}}(\mathcal{L}, t_1) = \frac{1}{\pi}\int_{\mathcal{L}}^{\infty} \frac{f_A(A)\,dA}{\sqrt{A^2 - \mathcal{L}^2}}$, $\lvert\mathcal{L}\rvert \leqslant A$
1.7 $A \sin(\omega t = \varphi)$	$f_\omega(\omega) = \frac{1}{\pi}\int_0^{\infty} r_{\mathcal{L}\mathcal{L}}(\tau)\cos(\omega\tau)d\tau$ $f_A(A) = A\sigma_{\mathcal{L}}^{-2}\exp(-\mathcal{L}^2/2\sigma_{\mathcal{L}})$	$R_{\mathcal{L}\mathcal{L}}(\tau)$ is required $f_{\mathcal{L}}(\mathcal{L}, t_1)$ $= (\sqrt{2\pi}\sigma_{\mathcal{L}})^{-1}\exp(-\mathcal{L}^2/2\sigma_{\mathcal{L}})$

Table 8.1 (*concl.*)

	Model	Statistical characteristics of components	General statistical characteristics of random process
2	$A \sin(\omega t + \pi/4)$	A distributed to arbitrary law $f_\omega(\omega) = \frac{1}{\pi}\int_0^\infty r_{\mathcal{L}\mathcal{L}}(\tau)\cos(\omega\tau)d\tau$	$R_{\mathcal{L}\mathcal{L}}(\tau)$ is required
3.1	$A \cos(\omega t + \varphi)$	$A = \sigma_{\mathcal{L}}\sqrt{2}$ is deterministic $f_\omega(\omega) = \frac{1}{\pi}\int_0^\infty r_{\mathcal{L}\mathcal{L}}(\tau)\cos(\omega\tau)d\tau$	$R_{\mathcal{L}\mathcal{L}}(\tau)$ is required
3.2	$A \cos(\omega t + \varphi)$	$f_\omega(\omega) = \frac{1}{\pi}\int_0^\infty r_{\mathcal{L}\mathcal{L}}(\tau)\ldots d\tau$ A has density function $f_A(A)$	$R_{\mathcal{L}\mathcal{L}}(\tau)$ is required $f_{\mathcal{L}}(\mathcal{L}, t_1) = \frac{1}{\pi}\int_{-\infty}^{\infty} f(A)\ldots dA$
4	$A \cos(\omega t + \pi/4)$	$f_\omega(\omega) = \frac{1}{\pi}\int_0^\infty r_{\mathcal{L}\mathcal{L}}\cos(\omega\tau)d\tau$	$R_{\mathcal{L}\mathcal{L}}(\tau)$ is required
5	$C \sin(\omega t)$ $+ D\cos(\omega t)$	$E[C] = E[D] = E[\omega] = 0$ $E[C^2] = E[D^2] = R_{\mathcal{L}\mathcal{L}}(0)$ $f_\omega(\omega) = \frac{1}{\pi}\int_0^\infty r_{\mathcal{L}\mathcal{L}}(\tau)\cos(\omega\tau)d\tau$	$R_{\mathcal{L}\mathcal{L}}(\tau)$ is required
6	$\sigma_{\mathcal{L}}(\sin\omega t$ $+ A\cos\omega t)$	$E[A] = E[\omega] = 0,\ E[A^2] = 1$ $f_\omega(\omega) = \frac{1}{\pi}\int_0^\infty r_{\mathcal{L}\mathcal{L}}(\tau)\cos(\omega\tau)d\tau$	$R_{\mathcal{L}\mathcal{L}}(\tau)$ is required
7	$\sigma_{\mathcal{L}}[\sin\omega(t + T)$ $+ \lambda\cos\omega(t + T)]$	$E[\lambda] = E[\omega] = 0,\ E[\lambda^2] = 1$ $f_\omega(\omega) = \frac{1}{\pi}\int_0^\infty r_{\mathcal{L}\mathcal{L}}(\tau)\cos(\omega\tau)d\tau$ $E[\lambda^n] = C_n,\ n = 3, 4, \ldots;$ $T > 0$	$R_{\mathcal{L}\mathcal{L}}(\tau)$ is required $E[\mathring{L}^{2m+1}(t)] = 0,\ m = 1, 2, \ldots$ $E[\mathring{L}^{2n}(t)]$ are required, $n = 1, 2, \ldots$

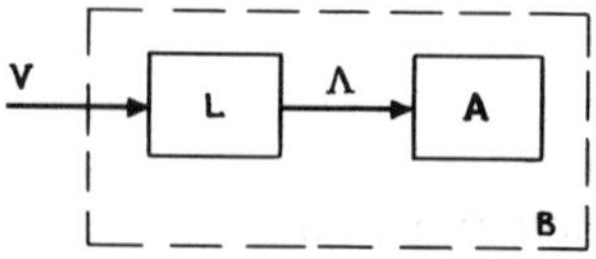

Fig. 8.6

tive matching system. Therefore a problem has emerged to develop a functional converter which would transform a set of random variables $V_1, \ldots, V_m$ with a given distribution density $f_g(v_1, \ldots, v_m)$ into a set of random variables $\Lambda_1, \ldots, \Lambda_m$ with the required distribution density. By placing such an L-converter at the input of the system under investigation (labeled A in Fig. 8.6) one would sound the system (B) with input stimuli $V_1, \ldots, V_m$ which is statistically equivalent to the initial system (A) subjected to the stimuli $\Lambda_1, \ldots, \Lambda_m$.

Mathematically this problem may be formulated as a search for a mapping L that transforms the set V into the set Λ, i.e.,

$$L : V \to \Lambda \tag{8.68}$$

In the general case the relation of the sets $\Lambda(\lambda_1, \ldots, \lambda_n)$ and $V(v_1, \ldots, v_m)$ may be represented as a set of functions

$$\begin{aligned} \lambda_1 &= \mathscr{L}_1(v_1, \ldots, v_m) \\ &\cdots\cdots\cdots\cdots\cdots \\ \lambda_n &= \mathscr{L}_n(v_1, \ldots, v_m) \end{aligned} \tag{8.69}$$

or in a compact form

$$L(\lambda_1, \ldots, \lambda_n, v_1, \ldots, v_m) = 0 \tag{8.70}$$

Generation of a System of Dependent Random Variables. General Problem Statement

Given a system of random variables $V_1, \ldots, V_m$ with known distribution density $f_v(v_1, \ldots, v_m)$. It is required to find a mapping function (converter)

$$L(\lambda_1, \ldots, \lambda_m, v_1, \ldots, v_m) = 0 \tag{8.71}$$

which generates a system of random variables $\Lambda_1, \ldots, \Lambda_m$ with the required distribution density $f_\lambda(\lambda_1, \ldots, \lambda_m)$.

Suppose that the solution (8.69) is known and given in vector-matrix form

$$\Lambda = \mathscr{L}(V) \tag{8.72}$$

Then the joint distribution function is

$$F(\lambda_1, \ldots, \lambda_m) = P\begin{bmatrix} \Lambda_i < \lambda_i \\ i = 1, \ldots, m \end{bmatrix} = P\begin{bmatrix} \mathscr{L}_i(v_1, \ldots, v_m) \\ i = 1, \ldots, m \end{bmatrix}$$

$$= \int \cdots \int_{\mathscr{L}_i(V),\, i=1, \ldots, m} f(v_1, \ldots, v_m)\, \mathrm{d}v_1 \ldots \mathrm{d}v_m \tag{8.73}$$

In the new variables

$$\eta_i = \mathscr{L}_i(v_1, \ldots, v_m), \quad i = 1, \ldots, m \tag{8.74}$$

this distribution function becomes

$$F(\lambda_1, \ldots, \lambda_m) = \int_{-\infty}^{\lambda_1} \cdots \int_{-\infty}^{\lambda_m} f_v(\psi_1, \ldots, \psi_m) \left| \frac{\partial \psi_1, \ldots, \psi_m}{\partial(\eta_1, \ldots, \eta_m)} \right| \mathrm{d}\eta_1 \ldots \mathrm{d}\eta_m \tag{8.75}$$

where $\dfrac{\partial(\psi_1, \ldots, \psi_m)}{\partial(\eta_1, \ldots, \eta_m)}$ is the Jacobian of functions of (8.74), and $\psi_i(\eta_1, \ldots, \eta_m) = v_i$ are the solutions of (8.74).

Differentiating (8.75) with respect to $\lambda_1, \ldots, \lambda_m$, we get

$$\frac{\partial F(\lambda_1, \ldots, \lambda_m)}{\partial(\lambda_1, \ldots, \lambda_m)} = f_\lambda(\lambda_1, \ldots, \lambda_m)$$

$$= f_v(\psi_1, \ldots, \psi_m) \left| \frac{\partial(\psi_1, \ldots, \psi_m)}{\partial(\lambda_1, \ldots, \lambda_m)} \right| \tag{8.76}$$

where $\dfrac{\partial(\psi_1, \ldots, \psi_m)}{\partial(\lambda_1, \ldots, \lambda_m)}$ is the Jacobian of $v_i = \psi_i(\lambda_1, \ldots, \lambda_m)$, $i = 1, \ldots, m$.

Thus, the given set of equations

$$v_i = \psi_i(\lambda_1, \ldots, \lambda_m), \quad i = 1, \ldots, m \tag{8.77}$$

defining the map of V and Λ is associated with some set of equations

$$\lambda_i = \mathscr{L}_i(v_1, \ldots, v_m), \quad i = 1, \ldots, m \tag{8.78}$$

characterizing the map of V onto Λ and the expression (8.76) relating the distribution density functions $f_v(v_1, \ldots, v_m)$, $f_\lambda(\lambda_1, \ldots, \lambda_m)$ and functions (8.77).

Equation (8.76) may be used to develop requirements to the form of functions in (8.77). Obviously, the sought functions (8.77) shall depend on the Jacobian of these functions

$$\frac{\partial(\psi_1, \ldots, \psi_m)}{\partial(\lambda_1, \ldots, \lambda_m)} = \begin{vmatrix} \dfrac{\partial\psi_1}{\partial\lambda_1} & \dfrac{\partial\psi_1}{\partial\lambda_2} & \cdots & \dfrac{\partial\psi_1}{\partial\lambda_m} \\ \dfrac{\partial\psi_2}{\partial\lambda_1} & \dfrac{\partial\psi_2}{\partial\lambda_2} & \cdots & \dfrac{\partial\psi_2}{\partial\lambda_m} \\ \cdots & \cdots & \cdots & \cdots \\ \dfrac{\partial\psi_m}{\partial\lambda_1} & \dfrac{\partial\psi_m}{\partial\lambda_2} & \cdots & \dfrac{\partial\psi_m}{\partial\lambda_m} \end{vmatrix} \tag{8.79}$$

The construction of functions in (8.77) can be simplified by representing (8.79) in the form

$$\frac{\partial(\psi_1, \ldots, \psi_m)}{\partial(\lambda_1, \ldots, \lambda_m)} = \begin{vmatrix} \dfrac{\partial\psi_1}{\partial\lambda_1} & \dfrac{\partial\psi_1}{\partial\lambda_1} & \cdots & \dfrac{\partial\psi_1}{\partial\lambda_m} \\ 0 & \dfrac{\partial\psi_2}{\partial\lambda_2} & \cdots & \dfrac{\partial\psi_2}{\partial\lambda_m} \\ \cdots & \cdots & \cdots & \cdots \\ 0 & 0 & & \dfrac{\partial\psi_m}{\partial\lambda_m} \end{vmatrix} \tag{8.80}$$

The Jacobian (8.80) corresponds to equations (8.77) in the form

$$\begin{aligned} v_1 &= \psi_1(\lambda_1, \ldots, \lambda_m) \\ v_2 &= \psi_2(\lambda_2, \ldots, \lambda_m) \\ &\cdots\cdots \\ v_m &= \psi_m(\lambda_m) \end{aligned} \tag{8.81}$$

Then the Jacobian will be

$$\frac{\partial(\psi_1, \ldots, \psi_m)}{\partial(\lambda_1, \ldots, \lambda_m)} = \prod_{i=1}^{m} \frac{\partial\psi_i}{\partial\lambda_i} \tag{8.82}$$

and (8.76) becomes

$$f_\lambda(\lambda_1, \ldots, \lambda_m) = f_v(\psi_1, \ldots, \psi_m) \left| \Pi \frac{\partial\psi_i}{\partial\lambda_i} \right| \tag{8.83}$$

Below we demonstrate how relations (8.81, 83) can be used to generate a system of dependent random variables.

Generation of a System of Dependent Random Variables by Independent Random Variates

Let equation (8.81) have the form

$$
\begin{aligned}
v_1 &= F_{v_1}^{-1} F_\lambda(\lambda_1 \mid \lambda_2, \ldots, \lambda_m) \\
v_2 &= F_{v_2}^{-1} F_\lambda(\lambda_2 \mid \lambda_3, \ldots, \lambda_m) \\
&\cdots\cdots\cdots\cdots\cdots \\
v_m &= F_{v_m}^{-1} F_\lambda(\lambda_m)
\end{aligned}
\tag{8.84}
$$

where $F_{v_i}^{-1}$ is the inverse of the distribution function $F_{v_i}(v_i)$ of the system of independent random variates $V_1, \ldots, V_m$; and $F_\lambda(\lambda_j \mid \lambda_{j+1}, \ldots, \lambda_m)$, $j = 1, \ldots, m - 1$, $F_\lambda(\lambda_m)$ are conditional distribution functions of a system of dependent random variables $\Lambda_1, \ldots, \Lambda_m$.

In this case the Jacobian (8.82) will be constructed by obtaining the partial derivatives

$$
\begin{aligned}
\frac{\partial \psi_1}{\partial \lambda_1} &= \frac{\partial F_{v_1}^{-1} F_\lambda(\lambda_1 \mid \lambda_2, \ldots, \lambda_m)}{\partial \lambda_1} \\
&\cdots\cdots\cdots\cdots\cdots \\
\frac{\partial \psi_m}{\partial \lambda_m} &= \frac{\partial F_{v_m}^{-1} F_\lambda(\lambda_m)}{\partial \lambda_m}
\end{aligned}
\tag{8.85}
$$

These derivatives will be treated as composite functions $\psi(\eta)$ viz.,

$$
v = \Psi(\eta) = F_v^{-1}(u) \tag{8.86}
$$

where $u = F_\lambda(\eta)$.

Under the circumstances,

$$
\psi'(\eta) = [F_v^{-1}(u)]'u' = [F_v^{-1}(u)]'F_\lambda'(\lambda) = [F_v(u)]'f_\lambda(\eta)
$$

Observing that the first cofactor is

$$
[F_v^{-1}(u)]' = 1/F_v'(v) = 1/F_v'(\psi(\eta)) = 1/f_v(\psi(\eta))
$$

we have

$$
\psi'(\eta) = f_\lambda(\eta)/f_v(\psi(\eta)) \tag{8.87}
$$

In view of (8.87) we transform (8.85) to the form

$$
\begin{aligned}
\frac{\partial \psi_1}{\partial \lambda_1} &= \frac{1}{f_v(\psi_1)} f_\lambda(\lambda_1 \mid \lambda_2, \ldots, \lambda_m) \\
\frac{\partial \psi_2}{\partial \lambda_2} &= \frac{1}{f_v(\psi_2)} f_\lambda(\lambda_2 \mid \lambda_3, \ldots, \lambda_m) \\
&\cdots\cdots\cdots\cdots\cdots \\
\frac{\partial \psi_m}{\partial \lambda_m} &= \frac{1}{f_v(\psi_m)} f_\lambda(\lambda_m)
\end{aligned}
$$

The Jacobian (8.80) is now representable as

$$\frac{\partial(\psi_1, \ldots, \psi_m)}{\partial(\lambda_1, \ldots, \lambda_m)} = \prod_{i=1}^{m} \frac{\partial \psi_i}{\partial \lambda_i}$$

$$= \frac{f_\lambda(\lambda_1 \mid \lambda_2, \ldots, \lambda_m) f_\lambda(\lambda_2 \mid \lambda_3, \ldots, \lambda_m) \ldots f_\lambda(\lambda_m)}{f_{v_1}(\psi_1) f_{v_2}(\psi_2) \ldots f_{v_m}(\psi_m)}$$

or

$$\frac{\partial(\psi_1, \ldots, \psi_m)}{\partial(\lambda_1, \ldots, \lambda_m)} = \frac{f_\lambda(\lambda_1, \ldots, \lambda_m)}{f_v(\psi_1, \ldots, \psi_m)} \tag{8.88}$$

By virtue of their properties the functions $f_\lambda(\lambda_1, \ldots, \lambda_m)$, $f_v(\psi_1, \ldots, \psi_m)$ have positive values, therefore the following relation holds

$$\left| \frac{\partial(\psi_1, \ldots, \psi_m)}{\partial(\lambda_1, \ldots, \lambda_m)} \right| = \frac{f_\lambda(\lambda_1, \ldots, \lambda_m)}{f_v(\psi_1, \ldots, \psi_m)}$$

and equation (8.76) assumes the form of identity, viz.,

$$f_\lambda(\lambda_1, \ldots, \lambda_m) = f_v(\psi_1, \ldots, \psi_m) \frac{f_\lambda(\lambda_1, \ldots, \lambda_m)}{f_v(\psi_1, \ldots, \psi_m)}$$

This circumstance proves that equation (8.84) satisfies the basic relation (8.76). Consequently, a system of dependent random variables $\Lambda_1, \ldots, \Lambda_m$ with distribution density $f_\lambda(\lambda_1, \ldots, \lambda_m)$ is obtainable by the system of equations (8.84) and a given set of independent random variates $V_1, \ldots, V_m$ distributed with density $f_v(v_i)$, $i = 1, \ldots, m$.

A practical solution to the problem of generation of a system of dependent random variables can be achieved as follows.

(1) Given the distribution densities $f_\lambda(\lambda_1, \ldots, \lambda_m)$ and $f_v(v_i)$, $i = 1, \ldots, m$, in agreement with (8.84) compile equations of the *L*-converter in the form

$$\begin{aligned}
\int_{-\infty}^{v_1} f_v(v_1)\, dv_1 &= \int_{-\infty}^{\lambda_1} \cdots \int_{-\infty}^{\lambda_m} f_\lambda(\lambda_1 \mid \lambda_2, \ldots, \lambda_m)\, d\lambda_1 \ldots d\lambda_m \\
\int_{-\infty}^{v_2} f_v(v_2)\, dv_2 &= \int_{-\infty}^{\lambda_2} \cdots \int_{-\infty}^{\lambda_m} f_\lambda(\lambda_2 \mid \lambda_3, \ldots, \lambda_m)\, d\lambda_2 \ldots d\lambda_m \\
&\cdots\cdots\cdots\cdots\cdots\cdots \\
\int_{-\infty}^{v_m} f_v(v_m)\, dv_m &= \int_{-\infty}^{\lambda_m} f_\lambda(\lambda_m)\, d\lambda_m
\end{aligned} \tag{8.89}$$

or

$$
\begin{aligned}
F_v(v_1) &= F_\lambda(\lambda_1 \mid \lambda_2, \ldots, \lambda_m) \\
F_v(v_2) &= F_\lambda(\lambda_2 \mid \lambda_3, \ldots, \lambda_m) \\
&\cdots\cdots\cdots\cdots\cdots \\
F_v(v_m) &= F_\lambda(\lambda_m)
\end{aligned}
\tag{8.90}
$$

Implementation of equations (8.89) or (8.90) in a computer software is straightforward.

(2) With a given ν-th sample of values

$$\{v_{1\nu}, \ldots, v_{m\nu}\}, \ \nu = 1, 2, \ldots$$

of independent random variables seek for a set of values of dependent stochastic variates

$$\{\lambda_{1\nu}, \ldots, \lambda_{m\nu}\}$$

from the system (8.89) or (8.90) to apply them to the dynamic system under investigation.

Generation of a Set of m Noncorrelated Random Variables by a Single Random Variate

Some problems of the preceding subsection where m dependent random variables with given distribution density are generated by a set of m independent random variates, suffice with the first and second moments. Then the L-converter may give way to a π-converter generating a system of m random variables by one random variate. Mathematically this problem statement implies that a rule must be evaluated such that maps a unidimensional random variable X onto a set $\Lambda(\Lambda_1, \ldots, \Lambda_m)$ of random variates, i.e.,

$$\pi\colon X \to \Lambda \tag{8.91}$$

Such a mapping of one-dimensional set X onto m-dimensional set Λ is in principle achievable for higher moments too [8.1]. In what follows we intend to dwell on the generation of m uncorrelated random variables in more detail.

The block diagram of analysis of dynamic systems A with the aid of a converter is shown in Fig. 8.7. The output of the π-converter is a set

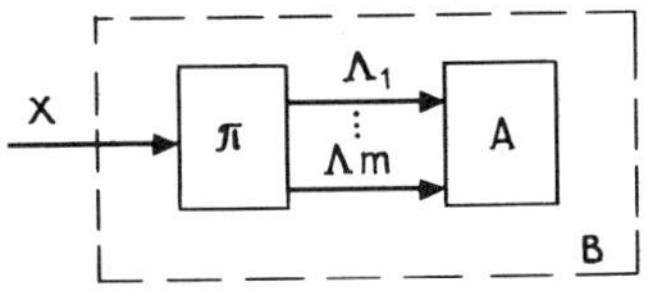

Fig. 8.7

of random variates with given statistical properties, therefore analysis is carried out for a statistically equivalent system excited by one random stimulus X.

In agreement with the procedure outlined in the preceding subsection we are to find a system of equations

$$\begin{aligned} \lambda_1 &= \pi_1(x) \\ &\cdots\cdots\cdots \\ \lambda_m &= \pi_m(x) \end{aligned} \tag{8.92}$$

which would effect the given mapping (8.91).

Assume that the system of random variables $\Lambda_1, \ldots, \Lambda_m$ is specified by the correlation matrix

$$\mathbf{R}_{ij} = \begin{bmatrix} R_{11} & R_{12} & R_{13} & \cdots & R_{1m} \\ & R_{22} & R_{23} & \cdots & R_{2m} \\ \cdots & \cdots & \cdots & \cdots & \cdots \\ & & & & R_{mm} \end{bmatrix}$$

and the system (8.92) is represented by a set of equations of the form

$$\begin{aligned} \lambda_1 &= v_1 \\ \lambda_i &= \sum_{j=1}^{i-1} a_{ij} v_j + v_i \\ \lambda_i &= b_i \lambda^{2i-1}, \quad i = 2, \ldots, m \end{aligned} \tag{8.93}$$

where $\Lambda_1 = X$ is a random variate with a normal distribution law.

Since formally the system of $2m$ equations (8.93) contains $2m$ variables $\lambda_1, \ldots, \lambda_m, v_1, \ldots, v_m$, mutual substitution can carry (8.93) into a system of m required equations (8.92).

Let us determine the coefficients a_i, b_i that ensure the conditions (8.93). For this purpose we compile the set of equations

$$\begin{aligned} & R_{\lambda_i\lambda_i} = \sum_{j=1}^{i-1} a_{ij}^2 \operatorname{Var}(v_j) + \operatorname{Var}(v_j), \quad i = 1, \ldots, m \\ & R_{\lambda_i\lambda_i} = b^2 \mu_{[(2m-1)2]}, \quad i = 2, \ldots, m \\ & a_{i1} \operatorname{Var}(v_1) = b_i \mu_{[2i]}, \quad i = 2, \ldots, m \\ & \sum_{j=1}^{\nu-1} a_{\nu j} a_{ij} \operatorname{Var}(v_j) + a_{i\nu} \operatorname{Var}(v_\nu) = b_\nu b_i \mu_{[2(\nu+i-1)]} \\ & \nu = 2, 3, \ldots, m-1; \gamma = 1, 2, \ldots, m-2; i = 3, 4, \ldots, m \end{aligned} \tag{8.94}$$

where $\gamma = i - 2$ is the number of last equations, and μ_s is the s-th order moment for random variables Λ_1. Solving the system (8.94) yields the coefficients a_i and b_i.

It has been demonstrated [8.4] that there is no need to solve the system (8.94) each time. Since the random variables $\Lambda_1, \ldots, \Lambda_m$ are not correlated, it will suffice to construct for the given variance values Var $(\lambda_1), \ldots,$ Var (λ_m) one table of samples $\lambda_1, \ldots, \lambda_m$ [8.4], and, if necessary, to compute the values of disturbances for the random variables $\Lambda_1', \ldots, \Lambda_m'$ by the relations

$$\Lambda_i' = k_i \Lambda_i, \quad i = 1, \ldots, m$$

where $k_i = \sqrt{\mathrm{Var}'(\Lambda_i)/\mathrm{Var}(\Lambda_i)}$, $i = 1, \ldots, m$.

This calculation substantially simplifies investigation of dynamic systems.

8.7 Synthesis of Statistically Optimal Complex Dynamic Systems

General Outline of the Optimal System Synthesis

The main problem facing the designer in optimizing a dynamic control system stems primarily from the lacking of a closed analytical solution to the problem of construction of optimal control algorithms. Therefore, wide acceptance in practice has been gained by methods of algorithmic form that combine analytical and numerical procedures and rely on computer-aided design techniques. These powerful methods provide the necessary prerequisites for design of efficient control systems for complex dynamic processes and plants. In the nearest future statistically optimal dynamic systems might be synthesized by including some algorithms in the overall closed loop. Three general relevant procedures deserve emphasizing.

(i) The synthesis of an optimal control program.

(ii) The synthesis of an optimal control law, that is, an algorithm generating the manipulated variable. This algorithm is obtained subject to the extremum of some functional and provided that the constraints are satisfied both for the phase coordinate of system motion and for the constructional parameters of the control system.

(iii) The optimization of the control system as a whole involving final definition of its structural network and selection of constructional parameters of the controller and the plant subject to a system of generalized criteria and constraints.

In handling specific problems, some of these procedures may be combined, say (i and ii) or (ii and iii). As a rule at the first stage of synthesis of control algorithms some system elements are handled in simplified representation, therefore procedure (iii) is an inseparable part of the synthesis on a computer.

Mathematical Formalism of the Synthesis Problem

Let the plant under control with a given proportion (completely or partially) of elements of the control system under study be described by the vector-matrix equations

$$\begin{aligned} \dot{\mathbf{Z}}_1 &= \mathbf{F}_1(\mathbf{Z}_1, \mathbf{X}_1, \mathbf{U}_1, t_0, t) \\ \mathbf{Y} &= \mathbf{H}_1(\mathbf{Z}_1, t) \\ \dot{\mathbf{Z}}_2 &= \mathbf{F}_2(\mathbf{Z}_2, \mathbf{Y}, \mathbf{X}_2, t_0, t) \end{aligned} \tag{8.95}$$

where $\dot{\mathbf{Z}}_1$ is the state vector of plant, $\mathbf{Y}$ is the vector of plant output, $\mathbf{Z}_2$ is the state vector of the known part of the control system, and $\mathbf{X}_1$, $\mathbf{X}_2$ are the vectors of perturbations and noise of the measuring elements.

At the first state of synthesis, the initial control system (8.95) is transformed to a simpler system, namely,

$$\begin{aligned} \dot{\mathbf{Z}}_1 &= \mathbf{F}_1'(\mathbf{Z}_1, \mathbf{X}_1', \mathbf{U}_1, t_0, t) \\ \mathbf{Y} &= \mathbf{H}_1'(\mathbf{Z}_1, t) \\ \dot{\mathbf{Z}}_2 &= \mathbf{F}_2'(\mathbf{Z}_2, \mathbf{Y}, \mathbf{X}_2', t_0, t) \end{aligned} \tag{8.96}$$

where $\mathbf{F}_1'$, $\mathbf{H}_1'$, $\mathbf{F}_2'$ are the simplified operators of the control (8.95), and $\mathbf{X}_1'$, $\mathbf{X}_2'$ are the vectors of perturbations and noise of lowered dimension.

Equations (8.96) are reduced to a form allowing for the use of analytical design for the synthesis of programs and laws of control.

The second state is concerned with the synthesis of control in the open-loop form

$$U = U'(t) \tag{8.97}$$

or closed-loop form

$$U = U'(\mathbf{Z}_2, t) \tag{8.98}$$

For discrete (digital) systems, these respectively become

$$U = U(kT)$$
$$U = U'[Z_2(kT), kT]$$

where k is a serial step in system operation, and T is quantization (sampling) period.

For some dynamic systems the control is given in the form of a sum

$$U = \tilde{U}(t) + \Delta U(Z_2, t)$$

where $\tilde{U}(t)$ is the program control synthesized under some simplifications of (8.96), and $\Delta U(Z_2, t)$ is the stabilizing control obtained with account of phase coordinate variation of the controls (8.96).

It would be extremely difficult to develop a structural network of the control system for the simplified equations (8.96) because of the simplifications made in coming from (8.95) to (8.96).

At the third stage, the initial structure of equations is recovered and a complete set of disturbances and noise X_1, X_2 is introduced. The algorithms of control are developed for the adopted system network and described by refined equations such as

$$U = U(kT)$$

or

$$U = U[Z_2(kT),\ kT]$$

With due account of actual system components, described analytically or incorporated into the computer-assisted simulation setup, the synthesized dynamic system becomes

$$\begin{aligned} \dot{\mathbf{Z}}_1 &= \mathbf{F}_1(\mathbf{Z}_1,\ \mathbf{X}_1,\ U,\ t_0,\ t) \\ \mathbf{Y} &= \mathbf{H}_1(\mathbf{Z}_1,\ t) \\ \dot{\mathbf{Z}}_2 &= \mathbf{F}(\mathbf{K}_1,\ \mathbf{Z}_2,\ \mathbf{Y},\ \mathbf{X}_2,\ t_0,\ t) \\ U(t) &= U(\mathbf{K}_2,\ \mathbf{Z}_2,\ t) \end{aligned} \tag{8.99}$$

where $\mathbf{K}_1$ and $\mathbf{K}_2$ are the vectors of constructional parameters to be selected.

The alternative representation for the discrete version of the system is

$$\begin{aligned} \dot{\mathbf{Z}}_1 &= \mathbf{F}_1(\mathbf{Z}_1,\ \mathbf{X}_1,\ U[kT],\ t_0,\ t) \\ \mathbf{Y} &= \mathbf{H}_1(\mathbf{Z}_1,\ t) \\ \mathbf{Z}_2[kT] &= \mathbf{F}_2(\mathbf{K}_1,\ \mathbf{Z}_2,\ \mathbf{X}_1,\ \mathbf{Y},\ t_0,\ (k-1)T) \\ U[kT] &= U[\mathbf{K}_2,\ \mathbf{Z}_2[kT],\ kT] \end{aligned} \tag{8.100}$$

This stage is concluded by the overall system optimization carried out with a generalized statistical criterion

$$I = I(K_1,\ K_2) = I(K)$$

where $K(K_1, K_2)$ is the generalized vector of constructional parameters subject to the given constraints

$$G(K,\ Z_1,\ Z_2,\ U,\ t) \leqslant 0 \tag{8.101}$$

In orther words, an optimal vector is sought

$$I(K^*) = \operatorname*{extr}_{K \in \Omega} I(K) \tag{8.102}$$

where Ω is the parameter domain. This search is effected as a rule by methods of programming.

It is worth noting that some components of vector K may influence not only the control process but also the structure of the system. This should not be overlooked in synthesis of quasioptimal control systems. Strictly speaking, the procedure of synthesis outlined above also produces a quasioptimal system, as an exact analytical solution of the synthesis problem is unobtainable in principle.

We wish to demonstrate the effect of vector K_1 on the structure of control system with the aid of a continuous system as an example.

Differentiating the equation

$$\dot{\mathbf{Z}}_2 = \mathbf{F}_2(\mathbf{K}_1, \mathbf{Z}_2, \mathbf{Y}_1, \mathbf{X}_2, t_0, t)$$

with respect to K_1, we get

$$\frac{d\psi}{dt} = \frac{\partial \mathbf{F}_2}{\partial \mathbf{Z}_2}\psi + \frac{\partial \mathbf{F}_2}{\partial \mathbf{K}_1} \tag{8.103}$$

where $\psi = \partial \mathbf{Z}_2 / \partial \mathbf{K}_1$.

From this equation it follows that by varying vector K_1 we can influence F_2 and, consequently, the structure of the control system.

Note in conclusion that at the last stage of synthesis an actual control is developed subject to the generalized criterion (8.102) and constraints (8.101). The resultant structure of the control system will be optimal in the said sense.

References

8.1 A. V. Potseluev: *Statisticheskii analiz i sintez slozhnykh dinamicheskikh sistem* (Statistical analysis and synthesis of complex dynamic systems). Mashinostroyeniye, Moscow 1984.

8.2 V. I. Chernetsky: *Analiz tochnosti nelineinykh sistem upravleniya* (Analysis of nonlinear system performance). Mashinostroyeniye, Moscow 1969.

8.3 B. G. Dostupov (ed.): *Statisticheskie metody v proyektirovanii nelineinykh sistem avtomaticheskogo upravleniya* (Statistical methods in design of nonlinear automatic control systems). Mashinostroyeniye, Moscow 1970.

8.4 A. V. Potseluev (ed.): Statisticheskoe issledovaniye i optimizatsiya sledyashchikh sistem (Statistical investigation and optimization of servo-systems). Mashinostroyeniye, Moscow 1977.

8.5 V. S. Bogdanov: *Avtomat. Telemekhan.* No. 6, 5-15, and No. 7, 25-33 (1973).

9 Stability of Nonlinear Stochastic Systems: A Path Integral Approach

N. V. Nikitin

This chapter is concerned with application of the path integral formalism to analysis of stability of nonlinear control systems subjected to random disturbances. Two ways will be explored: one is based on the calculation of performance criteria (functionals) for the nonlinear control systems, the other on the evaluation of the eigenvalues of the Kolmogorov equation. Both techniques suit the purpose well when the intensity of the disturbances is rather high. Under weak disturbances, for the unperturbed motion the time of system residence in the domain D_{cr} of the admissible values of state vector is very large. Under the circumstances, stability assessment and investigation of loss of control call for asymptotic methods of analysis. A measure of stability may then be taken in the form of a minimum of the action functional for the trajectories starting from the equilibrium point at the origin and ending at the boundary F of the domain D_{cr}. This minimum value of the functional defines the mathematical expectation of the time the system takes to leave D_{cr}. The origin is assumed to be the point of attraction of D_{cr}. For more complicated situations with many equilibrium points, specifically in the presence of limit cycles, asymptotic methods may come handy for the estimation of the probability of occurrence or change of oscillatory modes and some parameters.

In some systems bounds are imposed not only on state vector values, but also on the generalized rate of system transient behavior. Specifically, once the dynamic equations of a mechanical system have been reduced to a set of first order equations, the vector of state includes both the coordinates and the speed, while the bounds are imposed on the overshoot, i.e. on the variation rate of the state vector. For this case an uncertainty relation may be derived saying that for dynamic systems disturbed by white Gaussian noise it is impossible to monitor simultaneously and accurately the state vector and its variation rate.

9.1 Stability of Nonlinear Systems under Random Disturbances

Large efforts of control theory researchers are being devoted to tackling problems of stability of linear and nonlinear dynamic systems arising when the initial conditions or right-hand sides of the dynamic equations are perturbed. In analysis of time-invariant (autonomous) systems, classic stability theory investigates the behavior of the differential equations

$$\frac{\mathrm{d}\mathbf{x}}{\mathrm{d}t} = \mathbf{b}(\mathbf{x}) \tag{9.1}$$

where without loss of generality it is assumed that the system (9.1) has an equilibrium point $\mathbf{b}(0) = 0$ at the origin $\mathbf{x} = 0$.

This equilibrium point is *stable in the sense of Lyapunov* if for any neighborhood U_1 of $\mathbf{x} = 0$ there exists a domain $U_2 \subset U_1$ such that a solution of (9.1) starting at any point $\mathbf{x}_0 \in U_2$ remains in U_1 for all $t \geqslant 0$.

The system is *asymptotically stable* if for the trajectories starting at points x_{in} sufficiently close to the equilibrium position

$$\lim_{t \to \infty} \mathbf{x}(t) = 0 \tag{9.2}$$

The concept of stability in the sense of Lyapunov (stability with respect to perturbations of the initial conditions) is closely related to the problem of stability under continuous disturbances. In this case, the equation

$$\dot{\mathbf{x}}_b = \mathbf{b}(\mathbf{x}_b) + \mathbf{w}(t) \tag{9.3}$$

is considered as well as (9.1) with $\mathbf{w}(t)$ being some bound continuous function of time defined on R^N.

The investigation of stability of the solutions of (9.3) with respect to small perturbations $\mathbf{w}(t)$ reduces to the evaluation of conditions which the field $\mathbf{b}(\mathbf{x})$ must obey to ensure that a solution $\mathbf{x}_b(t)$ tend to zero uniformly in t as

$$\|\mathbf{x}_{\text{in}}\| \to 0 \quad \text{and} \quad \sup_{0 \leqslant t < \infty} \|\mathbf{w}(t)\| \to 0 \tag{9.4}$$

The uniform asymptotic stability in the sense of Lyapunov (when the condition (9.2) is satisfied uniformly in $\mathbf{x}_{\text{in}}$ for some neighborhood of $\mathbf{x}_{\text{in}}$) guarantees stability to the system (9.3) with respect to small consistent perturbations [9.1].

This situation is no longer true when the system is subject to random disturbances. The concept of smallness of these perturbations requires then additional refinement.

If we assume that the perturbations $\mathbf{w}(t)$ are uniformly small in probability, i.e. for finite times $\sup_{0 \leqslant t < \infty} \|\mathbf{w}(t)\|$ tends to zero in probability, then the situation becomes similar to the deterministic case.

More frequent practical situations, however, are those where the system is subject to perturbations in the form of white Gaussian noise. In this case the measure of smallness may be imposed as the assumption that the elements of the noise intensity matrix or some other characteristics of the stochastic process tend to zero. This assumption makes, on the one hand, unlikely large random perturbations over a long time interval and, on the other hand, allows for short-time perturbations of any magnitude. Under these assumptions the trajectory sooner or later escapes any neighborhood of the equilibrium position. Moreover, if

$$\mathbf{w}(t) = \sigma(\mathbf{x})\mathbf{v}(t)$$

where $\mathbf{v}(t)$ is the unit vector of white Gaussian noise, then however small are the values of

$$E\|\mathbf{w}(t)\mathbf{w}^{\mathrm{T}}(t)\| = \|\sigma(\mathbf{x})\sigma^{\mathrm{T}}(\mathbf{x})\| = \|\mathbf{G}(\mathbf{x})\| = \varepsilon > 0$$

in the time interval $0 \leqslant t < \infty$, the trajectories $\mathbf{x}_b(t)$ deviate from the equilibrium position at $\mathbf{x} = 0$ arbitrarily far with unit probability even for bound $\|\mathbf{b}(\mathbf{x})\|$ [9.2]. This implies that the idea underlying the concept of stability of Lyapunov becomes meaningless for the systems subjected to random perturbations in the form of white noise.

We shall assume that the behavior of the controlled system subjected to the said perturbations is described by a random process $\mathbf{X}(t)$ whose characteristics are to be defined. One method of specifying $\mathbf{X}(t)$ is by means of the stochastic differential equation

$$\frac{\mathrm{d}\mathbf{X}(t)}{\mathrm{d}t} = \mathbf{b}(\mathbf{X}, \mathbf{u}(\mathbf{X})) + \sigma(\mathbf{X})\mathbf{v}(t) \tag{9.5}$$

where $\mathbf{b}(\mathbf{X}, \mathbf{u}(\mathbf{X})) = \{b_k(\mathbf{X}, \mathbf{u}(\mathbf{X}))\}$ is an N-dimensional deterministic vector function, $\mathbf{u}(\mathbf{X})$ is the vector of control variables, $\sigma(\mathbf{X}) = \{\sigma_{ik}(\mathbf{X})\}$ is an $N \times N$ nonlinear deterministic matrix, $\mathbf{v}(t) = \{v_k(t)\}$ is the N-dimensional vector of independent central processes of unit Gaussian white noise [9.3], and $\sigma(\mathbf{X})\mathbf{v}(t)$ is the perturbation vector.

Assume that there exists a region $D_{\mathrm{cr}} \subset R^N$ including the equilibrium point $\mathbf{x} = 0$ such that the conditions of normal performance of the plant under control are satisfied so long as the trajectory of process $\mathbf{X}(t)$ lies within D_{cr}, and the control breaks down when $\mathbf{X}(t)$ escapes D_{cr}. We shall call D_{cr} the *critical region*. The breakdown (loss) of control will be understood as a situation when the plant and the control system terminate their operation, or a sharp change of the control laws occurs.

We introduce the time of first escape of the random process $\mathbf{X}(t)$ from the critical region D_{cr}, provided that at the initial time instant the system was at $\mathbf{x}_{\mathrm{in}}$, as the random variable

$$\tau(\mathbf{x}_{\mathrm{in}}) = \min\{t\colon \mathbf{X}(t) \notin D_{\mathrm{cr}} | \mathbf{X}(t_{\mathrm{in}}) = x_{\mathrm{in}}\}$$

For each realization of $\mathbf{X}(t)$ the random variable $\tau(\mathbf{x}_{\mathrm{in}})$ assumes its own value which varies from one realization to another in a random manner.

Consequently $\tau(\mathbf{x}_{in})$ is a functional specified on the trajectories of the random process $\mathbf{X}(t)$.

As a measure of stability of a system subjected to random perturbations in the critical region D_{cr}, it would be natural to use probability characteristics of $\tau(\mathbf{x}_{in})$. The choice of a characteristic should be defined by the statement of the specifical problem.

If the system is to operate during a given time interval $(0, T)$, then the measure of stability may be taken as the probability $P_{ins}(\mathbf{x}_{in}, T)$ that no loss of control occurs in this time interval, viz.,

$$P_{ins}(\mathbf{x}_{in}, T) = P\{\tau(\mathbf{x}_{in}) \geqslant T\} \tag{9.6}$$

i.e. the probability that $\tau(\mathbf{x}_{in}) \geqslant T$.

When the time of system operation is not specified in advance, the measure of stability may be taken as $\tau_1(\mathbf{x}_{in})$, i.e. the mathematical expectation of the first escape of the system from D_{cr},

$$\tau_1(\mathbf{x}_{in}) = E[\tau(\mathbf{x}_{in})] \tag{9.7}$$

The probabilistic characteristics (9.6) and (9.7) may be expressed through path integrals (discussed in the following section) or computed with the aid of statistical tests.

Another approach relies on evaluation of $P_{ins}(\mathbf{x}_{in}, T)$ and $\tau_1(\mathbf{x}_{in})$ on the basis of the probability density of the transition from the initial to final state $f(\mathbf{x}_f, t_f; \mathbf{x}_{in}, t_{in})$, which is a solution of the Kolmogorov equation for the region D_{cr} with the absorbing boundary F. At $t_f = T$, the integral of $f(\mathbf{x}_f, T; \mathbf{x}_{in}, 0)$ with respect to $\mathbf{x}_f$ over D_{cr} is the probability that until the time instant T the system has had no escape beyond the boundary F of D_{cr}, viz.,

$$P_{ins}(\mathbf{x}_{in}, T) = \int_{D_{cr}} f(\mathbf{x}_f, T; \mathbf{x}_{in}, 0)\,\mathrm{d}\mathbf{x}_f \tag{9.8}$$

The probability of control breakdown by the instant T is determined as $P_{cb}(\mathbf{x}_{in}, T) = 1 - P_{ins}(\mathbf{x}_{in}, T)$. The distribution of control breakdown probability density with respect to time has the form

$$\varphi_{cb}(\mathbf{x}_{in}, t) = \frac{\mathrm{d}P_{cb}(\mathbf{x}_{in}, t)}{\mathrm{d}t} = -\frac{\mathrm{d}P_{ins}(\mathbf{x}_{in}, t)}{\mathrm{d}t}$$

Accordingly, the mathematical expectation of the first system escape from D_{cr} is given as

$$\tau_1(\mathbf{x}_{in}) = -\int_0^\infty t\,\frac{\mathrm{d}P_{ins}(\mathbf{x}_{in}, t)}{\mathrm{d}t}\,\mathrm{d}t = tP_{ins}(t)\Big|_0^\infty + \int_0^\infty P_{ins}(\mathbf{x}_{in}, t)\,\mathrm{d}t \tag{9.9}$$

The first term on the right side of (9.9) vanishes since, as will be shown below, $P_{ins}(t)$ decays exponentially in t as $t \to \infty$. Thus,

$$\tau_1(\mathbf{x}_{in}) = \int_0^\infty \left[\int_{D_{cr}} f(\mathbf{x}_f,\ t;\ \mathbf{x}_{in},\ 0)\,d\mathbf{x}_f \right] dt$$

The probability characteristics (9.6, 7) depend on the state $\mathbf{x}_{in}$ which the system occupies in the initial time instant. For some problems this state is itself random. It would be natural, therefore, to select as a measure of stability a parameter independent of $\mathbf{x}_{in}$. This condition is satisfied by $T_X = \lambda_1^{-1}$, where λ_1 is the least eigenvalue of the Kolmogorov equation solved as the boundary-value problem with an absorbing boundary F of the region D_{cr}. In this case the eigenvalues of the Kolmogorov equation form a discrete spectrum of eigenvalues λ_i, namely, $0 < \lambda_1 \leqslant \lambda_2 \leqslant \lambda_3 \ldots$, associated with the eigenfunctions $\psi_1(\mathbf{x})$, $\psi_2(\mathbf{x})$, etc. Then, for the said boundary value problem, the probability density of a transition from initial to final state is given in the form of a series [9.4]

$$f(\mathbf{x}_f,\ T;\ \mathbf{x}_{in},\ 0) = \sum_{j=1} \psi_j(\mathbf{x}_f)\psi_j(\mathbf{x}_{in}) \exp(-\lambda_j T) \tag{9.10}$$

At large T, all members of the sum, except the leading term associated with the least eigenvalue λ_1, decay. Substituting the first term of (9.10) in (9.8, 9), we see that at large T

$$\begin{aligned} &P_{ins}(\mathbf{x}_{in},\ T) \sim \psi_1(\mathbf{x}_{in}) \exp(-\lambda_1 T) \\ &\tau_1(\mathbf{x}_{in}) \sim \psi_1(\mathbf{x}_{in})/\lambda_1 \end{aligned} \tag{9.11}$$

From (9.11) it follows that an increase of the parameter $T_X = \lambda_1^{-1}$ leads to a growth of the probability that no loss of control occurs in the time interval $(0,\ T)$. The mathematical expectation of the first escape from D_{cr} also increases. (The calculation of eigenvalues of the Kolmogorov equation with the path integral formalism will be outlined in Sect. 9.3.)

For same large systems, loss of control is a rare phenomenon. The emergency protection system of nuclear power stations, for example, operates on average once in a few months. Nevertheless the investigation of loss of control problems at low frequency of escape of $\mathbf{X}(t)$ from D_{cr} is of considerable interest because breakdown of control can lead to undesirable effects and result in plant failure. At such a low failure rate, usual methods of loss of control analysis are no longer applicable. One cannot even resort to the practically universal method of statistical modeling because of tremendous difficulties. Indeed, a simulation of a few-months random process behavior would require an exceedingly long computer time span to say nothing of the accumulation of error during the run and the need for statistical data acquisition by multiple sampling. A way out is seen in resorting to asymptotic methods of analysis of dynamic systems subject to small random perturbations (discussed in Sect. 9.4).

We have an intention therefore to consider problems involved in the representation $\tau(\mathbf{x}_{in})$ through path integrals and in the evaluation of functions $P_{ins}(\mathbf{x}_{in}, T)$ and $\tau_1(\mathbf{x}_{in})$ characterizing the degree of stability of dynamic systems subjected to random disturbances.

9.2 Path Integral Representation of Performance Indices

In many applications [9.3] the change of state of the controlled system can be described by a random process $X(t)$ satisfying the stochastic differential equation (9.5). This process relates to the class of diffusion Markov processes and is described in terms of the transition probability density $f(\mathbf{x}, t; \mathbf{x}_{in}, t_{in})$ satisfying the direct Kolmogorov equation [9.3]

$$\frac{\partial f(\mathbf{x}, t; \mathbf{x}_{in}, t_{in})}{\partial t} = -\sum_{k=1}^{N} \frac{\partial}{\partial x_k} a_k(\mathbf{x}, \mathbf{u}(x)) f(\mathbf{x}, t; \mathbf{x}_{in}, t_{in}) + \frac{1}{2} \sum_{k,l=1}^{N} \frac{\partial^2}{\partial x_k \partial x_l} G_{kl}(\mathbf{x}) f(\mathbf{x}, t; \mathbf{x}_{in}, t_{in}) \tag{9.12}$$

and the condition

$$\lim_{t \to t_{in}} f(\mathbf{x}, t; \mathbf{x}_{in}, t_{in}) = \delta(\mathbf{x} - \mathbf{x}_{in})$$

where the elements of the drag vector, $a_k(\mathbf{x})$, and diffusion coefficient matrix, $G_{kl}(\mathbf{x})$, coinciding with the perturbation intensity matrix, are [9.3]

$$\begin{aligned} a_k(\mathbf{x}, \mathbf{u}(\mathbf{x})) &= b_k(\mathbf{x}, \mathbf{u}(\mathbf{x})) + \frac{1}{2} \sum_{j,i=1}^{N} \sigma_{ji}(\mathbf{x}) \frac{\partial}{\partial x_j} \sigma_{ki}(\mathbf{x}) \\ G_{kl}(\mathbf{x}) &= \sum_{j=1}^{N} \sigma_{kj}(\mathbf{x}) \sigma_{lj}(\mathbf{x}) \end{aligned} \tag{9.13}$$

The transition probability density $f(\mathbf{x}, t; \mathbf{x}_{in}, t_{in})$ completely describes the dynamics of processes in the system. If the probability distribution density $\varphi(\mathbf{x}_{in}, t_{in})$ of the system is known at the initial time instant t_{in}, then, having acquired $f(\mathbf{x}, t; \mathbf{x}_{in}, t_{in})$, we are able to find $\varphi(\mathbf{x}, t)$ for any subsequent time instant $t \geqslant t_{in}$, viz.,

$$\varphi(\mathbf{x}, t) = \int_{-\infty}^{+\infty} (\mathbf{x}_{in}, t_{in}) f(\mathbf{x}, t; \mathbf{x}_{in}, t_{in}) \, d\mathbf{x}_{in} \tag{9.14}$$

With $\varphi(\mathbf{x}, t)$ at hand we may determine the mathematical expectation for an arbitrary function $F(\mathbf{x}(T))$ of the state vector $\mathbf{x}(T)$ at a given time T, the matrix of correlation moments, the probability of finding the system in a given region, and such.

However, equation (9.12) is amenable to analytical treatment for only linear and piecewise linear laws of $\mathbf{a}(\mathbf{x}, t, \mathbf{u}(\mathbf{x}, t))$ under the assumption that the intensity of noise perturbing the system is a function of time only [9.5]. Analytical solutions to (9.12) are known also for a number of one-dimensional systems ($N = 1$), namely, for absorbing and reflecting screens [9.5], and a relay law of $\mathbf{a}(\mathbf{x})$ [9.6].

This situation has motivated the development of numerical solvers for Kolmogorov's equation. One approach relies on the mathematical formalism of path integrals. The basic idea underlying this approach consists in representing the transition probability density and performance criteria (functionals) of the system in the form of path integrals followed by application of numerical schemes based on multiple integral approximations of these integrals or other approximate computational schemes [9.7].

In order to obtain a path integral representation for $f(\mathbf{x}_{\mathrm{f}}, t_{\mathrm{f}}; \mathbf{x}_{\mathrm{in}}, t_{\mathrm{in}})$ we break the time interval $[t_{\mathrm{in}}, t_{\mathrm{f}}]$ into m segments by points $t_1, t_2, \ldots, t_{m-1}$ spaced from one another by the time stretch $\varepsilon_T = (t_{\mathrm{in}} - t_{\mathrm{f}})/m$. The transition probability density $f(\mathbf{x}_b, t_b; \mathbf{x}_a, t_a)$ must satisfy the Markov equation

$$f(\mathbf{x}_b, t_b; \mathbf{x}_a, t_a) = \int_{R^N} f(\mathbf{x}_b, t_b; \mathbf{x}_c, t_c) f(\mathbf{x}_c, t_c; \mathbf{x}_a, t_a)\, \mathrm{d}\mathbf{x}_c \tag{9.15}$$

where t_c is an intermediate time instant, $t_a < t_c < t_b$.

By multiple application of (9.15) to time instants $t_1, t_2, \ldots, t_{m-1}$ we obtain

$$f(\mathbf{x}_{\mathrm{f}}, t_{\mathrm{f}}; \mathbf{x}_{\mathrm{in}}, t_{\mathrm{in}}) = \int_{R^{N(m-1)}} \prod_{j=0}^{m-1} f(\mathbf{x}_{j+1}, t_{j+1}; \mathbf{x}_j, t_j)\, \mathrm{d}\mathbf{x}_1 \mathrm{d}\mathbf{x}_2 \ldots \mathrm{d}\mathbf{x}_{m-1} \tag{9.16}$$

where t_j are time points spaced at ε_T, so that $t_j = t_{\mathrm{in}} + j\varepsilon_T$.

Substituting the approximate expression $f_\varepsilon(\mathbf{x}_{j+1}, t_{j+1}; \mathbf{x}_j, t_j)$ valid for small ε_T for $f(\mathbf{x}_{j+1}, t_{j+1}; \mathbf{x}_j, t_j)$ in (9.16) yields the m-th order approximation of $f(\mathbf{x}_{\mathrm{f}}, t_{\mathrm{f}}; \mathbf{x}_{\mathrm{in}}, t_{\mathrm{in}})$, i.e.

$$f_m(\mathbf{x}_{\mathrm{f}}, t_{\mathrm{f}}; \mathbf{x}_{\mathrm{in}}, t_{\mathrm{in}}) = \int_{R^{N(m-1)}} \prod_{j=0}^{m-1} f_\varepsilon(\mathbf{x}_{j+1}, t_{j+1}; \mathbf{x}_j, t_j)\, \mathrm{d}\mathbf{x}_1\, \mathrm{d}\mathbf{x}_2 \ldots \mathrm{d}\mathbf{x}_{m-1} \tag{9.17}$$

If an increase in the order of the approximation m does not result in accumulation of errors associated with the substitution of $f_\varepsilon(\mathbf{x}_{j+1}, t_{j+1}; \mathbf{x}_j, t_j)$ for $f(\mathbf{x}_{j+1}, t_{j+1}; \mathbf{x}_j, t_j)$, then

$$f(\mathbf{x}_{\mathrm{f}}, t_{\mathrm{f}}; \mathbf{x}_{\mathrm{in}}, t_{\mathrm{in}}) = \lim_{m \to \infty} f_m(\mathbf{x}_{\mathrm{f}}, t_{\mathrm{f}}; \mathbf{x}_{\mathrm{in}}, t_{\mathrm{in}}) \tag{9.18}$$

Now we proceed to develop an expression for $f_\varepsilon(\mathbf{x}_{j+1}, t_{j+1}; \mathbf{x}_j, t_j)$ valid for small ε_T. Consider a case of constant elements of the drag coefficient vector $\mathbf{a}_0$ and diffusion coefficient matrix $\mathbf{G}_0$. If so, the solution of (9.12) will be written [9.3] as

$$f_0(\mathbf{x}_{j+1}, t_j + \varepsilon_T; \mathbf{x}_j, t_j) = \frac{1}{\sqrt{(2\pi\varepsilon_T)^N \det \mathbf{G}_0}}$$
$$\times \exp\left\{-\frac{\varepsilon_T}{2}\left(\frac{\mathbf{x}_{j+1}-\mathbf{x}_j}{\varepsilon_T} - \mathbf{a}_0\right)^{\mathrm{T}} \mathbf{G}_0^{-1}\left(\frac{\mathbf{x}_{j+1}-\mathbf{x}_j}{\varepsilon_T} - \mathbf{a}_0\right)\right\}$$
$$= \frac{1}{\sqrt{(2\pi\varepsilon_T)^N \det \mathbf{G}_0}} \exp\left\{-\varepsilon_T L^{(0)}\left(\frac{\mathbf{x}_{j+1}-\mathbf{x}_j}{\varepsilon_T}\right)\right\} \tag{9.19}$$

The function $L^{(0)}[(\mathbf{x}_{j+1} - \mathbf{x}_j)/\varepsilon_T]$ may be given physical sense of the Lagrange function for a material point with some effective mass moving at a constant velocity $\mathbf{x} = (\mathbf{x}_{j+1} - \mathbf{x}_j)/\varepsilon_T$ along the straight line connecting in the $\mathbf{x}$, t space points $(\mathbf{x}_j, t_j)$ and $(\mathbf{x}_{j+1}, t_j + \varepsilon_T)$. In one-dimensional case with $a_0 = 0$, the Lagrange function has the form $L^{(0)} = g_0^{-1}x^2/2$ which implies that for systems subjected to random perturbations, the mass analog of the material point m in the Lagrange function is the quantity inverse with the intensity of perturbations g_0 acting on the system.

From (9.19) it is clear that at $\mathbf{a}_0 = 0$ the $L^{(0)}$ is a minimum and, consequently, the transition probability is a maximum at $\dot{\mathbf{x}} = \dot{\mathbf{x}}_* = 0$. For nonzero values of the drag coefficient vector, the trajectory $\dot{\mathbf{x}}_*$ satisfies the equation $\dot{\mathbf{x}}_* = \mathbf{a}_0$.

As $\varepsilon_T \to 0$, $\varepsilon_T L^{(0)} \sim (\mathbf{x}_{j+1} - \mathbf{x}_j)^2/\varepsilon_T$ which is large for all points $\mathbf{x}_{j+1}$ spaced from $\mathbf{x}_j$ by a value exceeding about $\sqrt{\varepsilon_T}$. Thus, as ε_T tends to zero, the region of $\mathbf{x}_{j+1}$ values for which $f(\mathbf{x}_{j+1}, t_j + \varepsilon_T, \mathbf{x}_j, t_j)$ substantially deviates from zero, contracts to point $\mathbf{x}_j$. Therefore if $\mathbf{a}(\mathbf{x})$ is a smooth function in $\mathbf{x}$ and t, the expression for $f_\varepsilon(\mathbf{x}_{j+1}, t_j + \varepsilon_T; \mathbf{x}_j, t_j)$ for small ε_T may be approximately written in the form (9.19) substituting there $\mathbf{a}(\mathbf{x}_j, t_j)$ for $\mathbf{a}_0$. Substituting $f_\varepsilon(\mathbf{x}_{j+1}, t_j + \varepsilon_T; \mathbf{x}_j, t_j)$ in (9.17), we get

$$f_m(\mathbf{x}_\mathrm{f}, t_\mathrm{f}; \mathbf{x}_\mathrm{in}, t_\mathrm{in}) = C_m \int_{x_\mathrm{in}}^{x_\mathrm{f}} \exp\{-S^{(0)}[\mathbf{x}_\mathrm{in}, \mathbf{x}_1, \ldots, \mathbf{x}_{m-1}, \mathbf{x}_\mathrm{f}]\, D\mathbf{x}_W(t)$$
$$= C_m \int_{-\infty}^{+\infty} \cdots \int \exp\left\{-\frac{\varepsilon_T}{2}\sum_{j=0}^{m-1}\left(\left[\frac{\mathbf{x}_{j+1}-\mathbf{x}_j}{\varepsilon_T} - \mathbf{a}(\mathbf{x}_j, \mathbf{u}(\mathbf{x}_j))\right]^{\mathrm{T}}\right.\right.$$
$$\left.\left.\times\, \mathbf{G}^{-1}(t_j)\left[\frac{\mathbf{x}_{j+1}-\mathbf{x}_j}{\varepsilon_T} - \mathbf{a}(\mathbf{x}_j, \mathbf{u}(\mathbf{x}_j))\right]\right)\right\} \prod_{j=1}^{m-1} d\mathbf{x}_j \tag{9.20}$$

where

$$C_m = \prod_{j=1}^{m-1} \{(2\pi\varepsilon_T)^N \det \mathbf{G}(t_j)\}^{-1/2}$$

$$\varepsilon_T = T/m, \quad = (t_f - t_{in})/m$$

It is quite natural that the transition of the case of constant drag coefficient vector $\mathbf{a}_0$ to the case allowing for the dependence of $\mathbf{a}(\mathbf{x})$ of spatial coordinates needs a more rigorous justification. A detailed discussion of this topic may be found in Ref. [9.8].

The limit of multiple integrals on the right hand side of (9.18) taken for $m - 1$ tending to infinity is called a *path integral*. In agreement with the Fomin theorem [9.9] this limit coincides with the functional integral with respect to the conditional Wiener measure defined on the domain of continuous functions.

For small ε_T, the summation over j in (9.20) may be replaced by integration with respect to time t. Then

$$f_m(\mathbf{x}_f, t_f; \mathbf{x}_{in}, t_{in})$$

$$= C_m \int_{-\infty}^{+\infty} \cdots \int \exp\left\{-\frac{1}{2}\int_{t_{in}}^{t_f} \left[\frac{d\mathbf{x}_W^m(t)}{dt} - \mathbf{a}(\mathbf{x}_W^m(t), \mathbf{u}(\mathbf{x}_W^m(t)))\right]^T \mathbf{G}^{-1}(t)\right.$$

$$\left.\times \left[\frac{d\mathbf{x}_W^m}{dt} - \mathbf{a}(\mathbf{x}_W^m(t), \mathbf{u}(\mathbf{x}_W^m(t)))\right] dt\right\} \prod_{j=1}^{m-1} d\mathbf{x}_j$$

$$= C_m \int_{-\infty}^{+\infty} \exp\{-S^{(0)}[\mathbf{x}_W^m(t)]\} \prod_{j=1}^{m-1} d\mathbf{x}_j \tag{9.21}$$

Here, $\mathbf{x}_W^m(t)$ is a trajectory in the $\mathbf{x}$, t space which has the shape of a broken line with knees at points $(\mathbf{x}_j, t_j)$. The curve $\mathbf{x}_W^m(t)$ emanates from point $(\mathbf{x}_0, t_0) \equiv (\mathbf{x}_{in}, t_{in})$ and terminates at point $(\mathbf{x}_m, t_m) \equiv (\mathbf{x}_f, t_f)$. The equation of this broken line is

$$\mathbf{x}_W^m(\theta) = \mathbf{x}_j + (\mathbf{x}_{j+1} - \mathbf{x}_j)(\theta - j), \quad j \leqslant \theta < j + 1 \tag{9.22}$$

where $\theta = m(t - t_{in})/(t_f - t_{in})$ and j is an integer-valued parameter taking on the values of 0, 1, . . . , $m - 1$.

In the expression (9.21)

$$S^{(0)}[\mathbf{x}_W^m(t)] = \frac{1}{2}\int_{t_{in}}^{t_f} \left[\frac{d\mathbf{x}_W^m(t)}{dt} - \mathbf{a}(\mathbf{x}_W^m(t), \mathbf{u}(\mathbf{x}_W^m(t)))\right]^T \mathbf{G}^{-1}(t)$$

$$\times \left[\frac{d\mathbf{x}_W^m}{dt} - \mathbf{a}(\mathbf{x}_W^m(t), \mathbf{u}(\mathbf{x}_W^m(t)))\right] dt \tag{9.23}$$

The functional $S^{(0)}[\mathbf{x}_W^m(t)]$ is an integral with respect to time of the function $L^{(0)}(\dot{\mathbf{x}}(t); \mathbf{x}(t), t)$ taken over the trajectory $\mathbf{x}(t) = \mathbf{x}_W^m(t)$. If we interpret $L^{(0)}(\dot{\mathbf{x}}(t), \mathbf{x}(t), t)$ as the system Lagrangian, then the functional $S^{(0)}[\mathbf{x}_W^m(t)]$ has the physical sense or action [9.10].

The action functional $S[\mathbf{x}(t)]$ defines the relative contribution of trajectories close to $\mathbf{x}(t)$ in the expression for the transition probability density, namely, the smaller the value of $S[\mathbf{x}(t)]$, the larger the probability of system moving along a trajectory close to $\mathbf{x}(t)$.

When $m \to \infty$, the trajectory $\mathbf{x}_W^m(t)$ tends to a certain trajectory $\mathbf{x}_W(t)$ of a continuous Markov process. Accordingly, when m tends to infinity, the expression (9.23) contains stochastic integrals of the form

$$\int_{t_{in}}^{t_f} \left(\frac{d\mathbf{x}_W}{dt}\right)^T \mathbf{G}^{-1}(t)\mathbf{a}(\mathbf{x}_W(t), \mathbf{u}(\mathbf{x}_W(t)))\, dt$$

and

$$\int_{t_{in}}^{t_f} \mathbf{a}^T(\mathbf{x}_W(t), \mathbf{u}(\mathbf{x}_W(t)))\mathbf{G}^{-1}(t) \frac{d\mathbf{x}_W}{dt}\, dt$$

The principal difference of a stochastic integral from a Riemann integral is that the result of replacement of a stochastic integral with the Darboux integral depends on the position of the integration nodes τ_j within subintervals (t_j, t_{j+1}).

If we specify $\tau_j^{(\xi)}$ and $\varkappa_j^{(\xi)}$ in the form

$$\begin{aligned} \tau_j^{(\xi)} &= t_j + \xi\varepsilon_T \\ \varkappa_j^{(\xi)} &= \mathbf{x}_j + \xi(\mathbf{x}_{j+1} - \mathbf{x}_j) \end{aligned} \tag{9.24}$$

with $0 \leqslant \xi \leqslant 1$, then to the value $\xi = 0$ there corresponds the stochastic Ito integral $I^{(0)}$ and to $\xi = 1/2$ the Stratanovich integral $I^{(1/2)}$. For the stochastic integral

$$I = \int_{t_{in}}^{t_f} \mathbf{\Phi}(\mathbf{x}_W(t), t) \frac{d\mathbf{x}_W}{dt}\, dt \tag{9.25}$$

by definition

$$I^{(\xi)} - I^{(0)} = \underset{m\to\infty}{\text{l.i.m.}} \left\{ \varepsilon_T \sum_{j=0}^{m-1} [\Phi(\varkappa_j^{(\xi)}, \tau_j^{(\xi)}) - \Phi(\varkappa_j^{(0)}, \tau_j^{(0)})] \right\}$$

The value of the limit in the mean on the right-hand side of (9.25) is known [9.11]. This limit may be used to derive from (9.21) an expression for the functional $S^{(\xi)}[\mathbf{x}_W(t)]$ for arbitrary value of ξ, viz.,

$$S^{(\xi)}[\mathbf{x}_W(t)] = \frac{1}{2}\int_{t_{in}}^{t_f}\left\{\left[\frac{d\mathbf{x}_W}{dt} - \mathbf{a}(\mathbf{x}_W(t), t, \mathbf{u}(\mathbf{x}_W(t), t))\right]^T \times \mathbf{G}^{-1}(t)\left[\frac{d\mathbf{x}_W}{dt} - \mathbf{a}(\mathbf{x}_W(t), t, \mathbf{u}(\mathbf{x}_W(t), t))\right] + 2\xi \operatorname{div} \mathbf{a}(\mathbf{x}, t, \mathbf{u}(\mathbf{x}, t))\big|_{\mathbf{x}=\mathbf{x}_W(t)}\right\} dt \tag{9.26}$$

By virtue of (9.26) the expression (9.20) may be written as

$$f_m^{(\xi)}(\mathbf{x}_f, t_f; \mathbf{x}_{in}, t_{in}) = \prod_{j=0}^{m-1} [(2\pi\varepsilon_T)^N \det \mathbf{G}(\tau_j^{(\xi)})]^{-1/2}$$

$$\times \int_{-\infty}^{+\infty} \cdots \int \exp\left\{-\frac{1}{2}\varepsilon_T \sum_{j=0}^{m-1}\left(\left[\frac{\mathbf{x}_{j+1} - \mathbf{x}_j}{\varepsilon_T} - \mathbf{a}(\varkappa_j^{(\xi)}, \mathbf{u}(\varkappa_j^{(\xi)}))\right]^T\right.\right.$$

$$\times \mathbf{G}^{-1}(\tau_j^{(\xi)})\left[\frac{\mathbf{x}_{j+1} - \mathbf{x}_j}{T} - \mathbf{a}(\varkappa_j^{(\xi)}, \mathbf{u}(\varkappa_j^{(\xi)}))\right]$$

$$\left.\left. + 2\xi \operatorname{div} \mathbf{a}(\mathbf{x}, \mathbf{u}(\mathbf{x}))\right|_{\mathbf{x}=\varkappa_j^{(\xi)}}\right)\right\} \prod_{j=1}^{m-1} d\mathbf{x}_j \equiv \prod_{j=0}^{m-1} [(2\pi\varepsilon_T)^N \det \mathbf{G}(\tau_j^{(\xi)})]^{-1/2}$$

$$\div \int_{-\infty}^{+\infty} \cdots \int \exp\{-S^{(\xi)}[\mathbf{x}_0, \mathbf{x}_1, \ldots, \mathbf{x}_{m-1}, \mathbf{x}_m]\} \prod_{j=1}^{m} d\mathbf{x}_j \tag{9.27}$$

Giving $\xi = 1/2$ this formula yields the (symmetrized) integral of Stratonovich. This integral possesses the property of stability with respect to the limit transition from the processes with piecewise differentiable trajectories (sublimit models of Brownian motion) to diffusion Markov processes whose trajectories are nondifferentiable with unit probability. (Henceforth we shall drop the subscript $\xi = 1/2$ when using the Stratonovich integral.)

Following Topcheyev and Nikitin [9.11], we may demonstrate that, observing the dependence of the perturbation intensity matrix on the state vector $\mathbf{x}$, the expression (9.27) takes the form

$$f_m^{(\xi)}(\mathbf{x}_{\mathrm{f}},\ t_{\mathrm{f}};\ \mathbf{x}_{\mathrm{in}},\ t_{\mathrm{in}})$$

$$= \int_{-\infty}^{+\infty} \cdots \int C_m^{(\xi)} \exp\left\{ -\frac{\varepsilon_T}{2} \sum_{j=0}^{m-1} \left(\left[\frac{\mathbf{x}_{j+1} - \mathbf{x}_j}{\varepsilon_T} - \mathbf{a}(\varkappa_j^{(\xi)},\ \tau_j^{(\xi)}) \right.\right.\right.$$

$$\left. + \xi G_{\mathrm{d}}(\varkappa_j^{(\xi)},\ \tau_j^{(\xi)})\, \mathrm{Div}\,(\mathbf{G}(\varkappa_j^{(\xi)},\ \tau_j^{(\xi)})/\mathbf{G}_{\mathrm{d}}(\varkappa_j^{(\xi)},\ \tau_j^{(\xi)})) \right]^{\mathrm{T}} \mathbf{G}^{-1}(\varkappa_j^{(\xi)},\ \tau_j^{(\xi)})$$

$$\times \left[\frac{\mathbf{x}_{j+1} - \mathbf{x}_j}{\varepsilon_T} - \mathbf{a}(\varkappa_j^{(\xi)},\ \tau_j^{(\xi)}) + \xi G_{\mathrm{d}}(\varkappa_j^{(\xi)},\ \tau_j^{(\xi)}) \right.$$

$$\left. \times\, \mathrm{Div}\,(\mathbf{G}(\varkappa_j^{(\xi)},\ \tau_j^{(\xi)})/G_{\mathrm{d}}(\varkappa_j^{(\xi)},\ \tau_j^{(\xi)})) \right] + 2\xi G_{\mathrm{d}}(\varkappa_j^{(\xi)},\ \tau_j^{(\xi)})\, \mathrm{div}\,[\mathbf{a}(\varkappa_j^{(\xi)},\ \tau_j^{(\xi)})$$

$$\left.\left. \times\, G_{\mathrm{d}}^{-1}(\varkappa_j^{\xi},\ \tau_j^{(\xi)}) - \xi\, \mathrm{Div}\,(\mathbf{G}(\varkappa_j^{(\xi)},\ \tau_j^{(\xi)}) G_{\mathrm{d}}^{-1}(\varkappa_j^{(\xi)},\ \tau_j^{(\xi)}))] \right) \right\} \prod_{j=1}^{m-1} \mathrm{d}\mathbf{x}_j \qquad (9.28)$$

Here,

$$C_m^{(\xi)} = \prod_{j=0}^{m-1} [(2\pi\varepsilon_T)^{N/2} G_{\mathrm{d}}(\varkappa_j^{(\xi)},\ \tau_j^{(\xi)})]^{-1}$$

and

$$G_{\mathrm{d}}(\varkappa_j^{(\xi)},\ \tau_j^{(\xi)}) = [\det \mathbf{G}(\varkappa_j^{(\xi)},\ \tau_j^{(\xi)})]^{1/2}$$

In (9.28) the result of the operator Div action on some matrix $\mathbf{H}(\mathbf{x}) = \{h_{jk}(\mathbf{x})\}$ is a vector whose α-th element is

$$(\mathrm{Div}\, \mathbf{H}(\mathbf{x}))_\alpha = \sum_{k=1}^{N} \frac{\partial h_{\alpha k}(\mathbf{x})}{\partial x_k}$$

By analogy with (9.27) the exponent in (9.28) may be written as $-S^{(\xi)}[\mathbf{x}_0,\ \mathbf{x}_1,\ \ldots,\ \mathbf{x}_m]$.

Using (9.27) and (9.28), one may obtain the representation of performance indices for nonlinear systems through path integrals.

As such indices characterizing the performance of a control system under investigation, one may in the general case take functionals $Y_k[\mathbf{X}(t)]$ dependent on the trajectory of the process $\mathbf{X}(t)$ in the state space on the time interval $(0,\ T)$. Since the trajectory $\mathbf{X}(t)$ cannot be predicted in advance, the functional $Y_k[\mathbf{X}(t)]$ is a random quantity characterized by its expectation

$$J_k = E\{Y_k[\mathbf{X}(t)]\} = \lim_{m\to\infty} \int_{-\infty}^{+\infty} \cdots \int C_m Y_k[\mathbf{x}_{\mathrm{W}}^m(t)] \varphi(\mathbf{x}_{\mathrm{in}})$$

$$\times \exp\{-S[\mathbf{x}_{\mathrm{W}}^m(t)]\} \prod_{j=0}^{m} \mathrm{d}\mathbf{x}_j \qquad (9.29)$$

The probability $P_*(t_f)$ that at time $t = t_f$ the system is within some region D_{cr} is

$$P_*(t_f) \to Y_*[\mathbf{x}_W^m(t)] = \pi_D(\mathbf{x}_f) = 1 \quad \text{for} \quad \mathbf{x}_f \in D_{cr}$$
$$= 0 \quad \text{for} \quad \mathbf{x}_f \notin D_{cr} \tag{9.30}$$

where the arrow means correspondence.

The probability that during the time interval $(0, T)$ the system has never escaped the region D_{cr} provided that $x_{in} = x_0 \in D_{cr}$ is

$$P_{ins}(\mathbf{x}_{in}, t) \to Y_{ins}[\mathbf{x}_W^m(t)] = \pi_D[\mathbf{x}_W^m(t)]$$
$$\varphi(\mathbf{x}_{in}) = \delta(\mathbf{x}_{in} - \mathbf{x}_0) \tag{9.31}$$
$$\pi_D[\mathbf{x}_W^m(t)] = 1, \quad \text{if} \quad \mathbf{x}_W^m(t) \in D_{cr}$$
$$= 0, \quad \text{if} \quad \mathbf{x}_W^m(t) \quad \text{escapes} \quad D_{cr}$$

The expectation of the first system escape time from D_{cr} equal to $\tau_1(\mathbf{x}_{in})$ may be found by formulas (9.9) and (9.31).

It is important to emphasize that the expression (9.29) is applicable when the determinant of the perturbation intensity matrix is nonzero. However, practical situations are often encountered with perturbations acting only in some system coordinates $i = 1, 2, \ldots, N$. If we set the intensities of perturbations in other coordinates $i = N + 1, \ldots, L$ equal to zero, then the expressions (9.27, 28) acquire singularities of delta-function type. These singularities owe their existence to the fact that the dependences $X_j(t)$ for $j = N + 1, \ldots, L$ are determined by a known realization $X_i(t)$ in a deterministic manner. Therefore such situation call for a separate consideration.

We assume that the equations of motion have the form

$$\frac{dX_i}{dt} = a_i(\mathbf{X}, t, \mathbf{u}(\mathbf{X}, t)) + v_i(\mathbf{X}, t), \quad i = 1, \ldots, N \tag{9.32}$$

$$\frac{dX_j}{dt} = a_j(\mathbf{X}, t, \mathbf{u}(\mathbf{X}, t)), \quad j = N + 1, \ldots, L \tag{9.33}$$

If the trajectory $\mathbf{x}_W^m$ is given by (9.22) in the form of a broken line in the N-dimensional space $\{x_1, \ldots, x_N\}$, then substituting this line into (9.33) carries this equation to a closed system of $L - N$ ordinary differential equations. For many cases the solution $\mathbf{y}_W^m(t)$ of this system can be obtained in analytical form [9.11]. If so, then the expression (9.29) may be employed for the determination of characteristics of the system (9.32, 33), the only difference being that each trajectory $\mathbf{x}_W^m$ is associated with a dependence $\mathbf{y}_W^m(t)$.

The outlined representation of nonlinear system performance criteria through path integrals may serve as a basis for the development of a number of analytical and numerical techniques for evaluation of the characteristics of dynamic systems subjected to random perturbations. A large number of illustrative examples and case-studies of real systems can be found in Refs. [9.7, 11].

9.3 Eigenvalues of the Kolmogorov Equation and System Residence Time in a Given Region

The evaluation of eigenvalues of the Kolmogorov equation is of considerable theoretical and practical interest [9.4, 12], for it gives a clue for finding the characteristic time of system residence in, or escape from, a given region, the asymptotic probability distribution density as $T \to \infty$, for estimating the dependence of the loss of control probability on time. This section deals with a procedure of evaluation of the spectrum λ_1, $\lambda_2, \ldots,$ on the basis of the approximate representation of $f(\mathbf{x}_f, T; \mathbf{x}_{in}, 0)$ in the form (9.27, 28).

If we represent formally T as $i\theta$ ($i = \sqrt{-1}$) in (9.10), then with the real parameter λ shifted for an infinitesimal value to the upper half of the complex plane and $x_{in} = x_f = x_0$ we have

$$\operatorname{Re}[D(\lambda, \mathbf{x}_0)] \equiv \operatorname{Re}\left[\int_0^\infty f(\mathbf{x}_0, i\theta; \mathbf{x}_0, 0) \exp(i\lambda\theta)\, d\theta\right]$$

$$= \sum_j |\psi_j(\mathbf{x}_0)|^2 \pi\delta(\lambda - \lambda_j) \tag{9.34}$$

where $D(\lambda, \mathbf{x}_0)$ is the Green function of the kernel of the evolution operator in the Kolmogorov equation at imaginary time, and $\delta(\lambda - \lambda_j)$ is a delta function.

If we substitute $f_m(\mathbf{x}_0, i\theta; \mathbf{x}_0, 0)$ for $f(\mathbf{x}_0, i\theta; \mathbf{x}_0, 0)$ in (9.34), we obtain the approximation of order $m - 1$ of the exact function $D(\lambda, \mathbf{x}_0)$, namely,

$$D_{m-1}(\lambda, \mathbf{x}_0) = \int_0^\infty f_m(\mathbf{x}_0, i\theta; \mathbf{x}_0, 0) \exp(i\lambda\theta)\, d\theta \tag{9.35}$$

For $m \to \infty$ the real part of $D_{m-1}(\lambda, \mathbf{x}_0)$ tends, in agreement with (9.18, 34), to a sum of δ-functions on the right side of (9.34). At finite m and fixed vector $\mathbf{x}_0$, the real part $\operatorname{Re}[D_{m-1}(\lambda, \mathbf{x}_0)]$ should have peaks near the exact values of λ_j. As m increases, the width of the peaks should decrease and the height increase.

Consider for the beginning the case of $N = 1$, where we assume that the perturbations acting on the system are of constant intensity, i.e.,

$G(\mathbf{x}, t) = g_0$. Letting $t_{in} = 0$, $t_f = T$, and $T = i\theta$ in (9.21) and substituting this expression in (9.35) yields

$$D_{m-1}(\lambda, g_0) = d_0^{m-1} C_m \int_0^{\infty} d\theta (i\theta)^{-m/2} \left(\int_{-\infty}^{+\infty} \prod_{j=1}^{m-1} dq_j \right.$$

$$\left. \times \exp\left\{ i\left[\beta\sigma_m/\theta - \frac{A_0}{d_0}\theta f(q_0, q_1, \ldots, q_{m-1}) + \lambda\theta \right]\right\} \right) \tag{9.36}$$

where d_0 and A_0 are characteristic scales; $q_j = x_j/d_0$; $a(x) = A_0 a(x/d_0) = A_0 a(q)$; $C_m = (m/2\pi g_0)^{m/2}$, $\beta = d_0^2 m/2g_0$;

$$f(q_0, q_1, \ldots, q_{m-1}) = \int_0^1 \varphi(q_W^m(\tau))\, d\tau; \quad \tau = t/T;$$

$$\varphi(q_W^m(\tau)) = \frac{1}{2}\left(\frac{\partial a(q)}{\partial q} + \frac{A_0 d_0}{g_0} a^2(q) \right)\Bigg|_{q = q_W^m(\tau)};$$

$$q_{in} = q_0; \quad q_m = q_f = q_0; \quad \sigma_m = \sum_{j=0}^{m-1} (q_{j+1} - q_j)^2;$$

and $q_W^m(\tau) = q_j + (q_{j+1} - q_j)(m\tau - j)$ with $j \leqslant m\tau \leqslant j + 1$ and $j = 0, 1, \ldots, m-1$.

Changing the order of integration with respect to q_i and θ in (9.36) and evaluating the integral with respect to θ analytically on the basis of the tabulated integral

$$\int_0^{\infty} x^{\nu-1} \exp\left[\frac{i\mu}{2}\left(x + \frac{\beta_1^2}{x} \right)\right] dx = i\pi\beta^{\nu} \exp(-i\nu\pi/2) H_{-\nu}^{(1)}(\beta_1\mu) \tag{9.37}$$

where $\mathrm{Im}(\beta\mu) > 0$ and $\mathrm{Im}(\beta^2\mu) > 0$, we have

$$D_{m-1}(\lambda', q_0) = B_m \int_{-\infty}^{+\infty} dq_j (z/\sigma_m)^{\frac{m}{2}-1} H_{\frac{m}{2}-1}^{(1)}(z)$$

Here, $B_m = \pi d_0^{m-1} C_m (2\beta)^{1-\frac{m}{2}}$; $H_{\frac{m}{2}-1}^{(1)}(z)$ is the Hankel function of the first kind;

$$z = \gamma\{\sigma_m[\lambda' - f(q_0, q_1, \ldots, q_{m-1})]\}^{1/2}$$

$$\gamma = 2(\beta A_0/d_0)^{1/2}; \quad \text{and} \quad \lambda' = \lambda d_0/A_0$$

Accurate to the factor B_m the real part of (9.37) may be written as

$$R_{m-1}(\lambda', q_0) = \mathrm{Re}\,[D_{m-1}(\lambda', q_0)/B_m] = \int_V \prod_{j=1}^{m-1} \mathrm{d}q_j (z/\sigma_m)^{\frac{m}{2}-1} J_{\frac{m}{2}-1}(z) \tag{9.38}$$

where J is the Bessel function, and the domain of integration V is defined by the condition

$$f(q_0, q_1, \ldots, q_{m-1}) \leqslant \lambda' \tag{9.39}$$

If this inequality is violated, the integrand in (9.37) is an imaginary function with zero contribution in $R_{m-1}(\lambda', q_0)$.

For odd $m = 2k + 3$ ($k = 0, 1$, etc.) the Bessel function $J_{m/2-1}(z)$ with a half-integer index $m/2 - 1 = k + 1/2$ reduces to elementary functions. For $m = 2k + 3$ the relation (9.38) takes the form

$$R_{m-1}(\lambda', q_0) = \left(\frac{2}{\pi}\right)^{1/2} \int_V \prod_{j=1}^{m-1} \mathrm{d}q_j \sigma_m^{1-\frac{m}{2}} z^{m-2} \left(-\frac{1}{z}\frac{\mathrm{d}}{\mathrm{d}z}\right)^{\frac{m}{2}-\frac{3}{2}} \frac{\sin z}{z} \tag{9.40}$$

By way of example, for $k = 0$ and $m = 3$ we have

$$R_2(\lambda', q_0) = \left(\frac{2}{\pi}\right)^{1/2} \int_{-V} \mathrm{d}q_1\, \mathrm{d}q_2\, \sigma_3^{-1/2} \sin z$$

and at $k = 1$ and $m = 5$,

$$R_4(\lambda', q_0) = \left(\frac{2}{\pi}\right)^{1/2} \int_V \prod_{j=1}^{4} \mathrm{d}q_j \sigma_5^{-3/2}(\sin z - z\cos z)$$

The relation (9.21) suggests that other complete sets of functions may be used for the approximation of the Wiener process instead of broken lines $x_W^m(t)$, for instance,

$$x_s^m(t) = x_*(t) + \sum_{j=1}^{m-1} b_j \sin(\pi jt/T) \tag{9.41}$$

where $x_*(t) = (x_f - x_{in})t/T + x_{in}$.

Such a set of functions would retain the expression (9.21) accurate to a factor C_m which gives way to C_{ms} for the decomposion (9.41). This stems from the fact that transition from $x_W^m(t)$ to $x_s^m(t)$ in (9.21) is similar to the change of integration variables in a multiple integral. This replacement entails the appearance of a Jacobian, i.e., the change of C_m for C_{ms}. The factor C_{ms} should not depend on the form of $\mathbf{a}(\mathbf{x})$. It can be derived on the lines of the following argument. By direct substitution one may readily verify that for $\mathbf{a}(\mathbf{x}) = 0$ (9.21) yields for all values of m an exact expression for transition probability density of unidimensional Wiener process. We require that this condition should be retained in replacing $x_{\mathrm{W}}^m(t)$ with $x_{\mathrm{s}}^m(t)$. Then

$$f_m(x_{\mathrm{f}}, T; x_{\mathrm{in}}, 0) = C_{ms} T^{-m/2} \int_{-\infty}^{+\infty} \prod_{j=1}^{m-1} \mathrm{d}b_j$$

$$\times \exp\left\{ -\frac{1}{2} \int_0^T \left[\frac{1}{g_0} \left(\frac{\mathrm{d}x_{\mathrm{s}}^m}{\mathrm{d}t} - a(x_{\mathrm{s}}^m(t)) \right)^2 + \frac{\partial}{\partial x} a(x) \bigg|_{x = x_{\mathrm{s}}^m(t)} \right] \mathrm{d}t \right\} \tag{9.42}$$

where

$$C_{ms} = \frac{(m-1)!}{\sqrt{2\pi}} \frac{\pi^{(m-1)/2}}{2^{m-1} g_0^{m/2}}$$

Based on this argument we may derive an expression for $f_m(x_{\mathrm{f}}, T; x_{\mathrm{in}}, 0)$ with an arbitrary complete set of functions by constructing $x_l^m(t)$ similar to (9.41). In this case the form of (9.42) will be retained accurate to the change of C_{ms} for C_{ml} and $x_{\mathrm{s}}^m(t)$ for $x_l^m(t)$; and (9.36) takes the form

$$D_{m-1,l}(\lambda, q_0) = d_0^{m-1} C_{ml} \int_0^\infty \mathrm{d}\theta (i\theta)^{-m/2} \left(\int_{-\infty}^{\infty} \prod_{j=1}^{m-1} \mathrm{d}q_j \right.$$

$$\left. \times \exp\left\{ i[\beta_l \sigma_{ml}/\theta - \frac{A_0}{d_0} \theta f_l(q_0, q_1, \ldots, q_{m-1}) + \lambda\theta] \right\} \right) \tag{9.43}$$

where

$$f_l(q_0, q_1, \ldots, q_{m-1}) = \int_0^1 \varphi(q_l^m(\tau))\, \mathrm{d}\tau$$

$$\varphi(q_l^m(\tau)) = \frac{1}{2} \left(\frac{\partial a(q)}{\partial q} + \frac{A_0 d_0}{g_0} a^2(q) \right) \bigg|_{q = q_l^m(\tau)}$$

At $l = s$, we get $\beta_s = \pi^2 d_0^2 / 4g_0$,

$$\sigma_{ms} = \sum_{j=1}^{m} j^2 q_j^2, \quad q_j = b_j/d_0, \quad \text{and} \quad q_{\mathrm{s}}^m(\tau) = q_0 + \sum_{j=1}^{m-1} q_j \sin(\pi j \tau)$$

Integrating (9.43) with respect to θ, we obtain a formula for the real part of $D_{m-1,l}(\lambda, q_0)$, viz.,

$$R_{m-1,l}(\lambda', q_0) = \int_V \prod_{j=1}^{m-1} \mathrm{d}q_j (z/\sigma_{ml})^{\frac{m}{2}-1} J_{m/2-1}(z) \tag{9.44}$$

where $J_{m/2-1}$ is the Bessel function,

$$z = \gamma_l\{\sigma_{ml}[\lambda - f_l(q_0, q_1, \ldots, q_{m-1})]\}^{1/2}$$

$$\gamma_l = 2(\beta_l A_0/d_0)^{1/2} \quad \text{and} \quad \lambda' = \lambda d_0/A_0$$

Consider a few examples of application of (9.44). With the drag coefficient law

$$a(q) = -q^k, \quad k = 1, 3, 5, \text{ etc.} \tag{9.45}$$

for $m = 2$, $q_0 = 0$, $l = s$, and $l = W$ the expression (9.44) becomes

$$R_{1W}(\lambda_a, 0) = \left(\frac{g_0}{A_0 d_0}\right)^{1/(k+1)} \int_V J_0\left(2\sqrt{2}q\left\{\lambda_a - \frac{1}{2}\left[\frac{q^{2k}}{2k+1} - q^{k-1}\right]\right\}^{1/2}\right) dq \tag{9.46}$$

$$R_{1s}(\lambda_a, 0) = \left(\frac{g_0}{A_0 d_0}\right)^{1/(k+1)} \int_V J_0\left(\pi q\left\{\lambda_a - \frac{1}{2}[B_{2k}q^{2k} - kB_{k-1}q^{k-1}]\right\}^{1/2}\right) dq \tag{9.47}$$

where J_0 is the Bessel function with zero subscript and

$$\lambda_a = \lambda'\left(\frac{A_0 d_0}{g_0}\right)^{\frac{k-1}{k+1}}$$

The domain of integration V in (9.46, 47) is defined on the condition that $\{\cdot\} > 0$, and the coefficients B_M on the right-hand side of (9.47) are

$$B_M = \frac{(M-1)!!}{2^{M/2}(M/2)!} \quad \text{for even } M$$

We note that the minus sign and odd values of exponent k in (9.45) are chosen on the condition of stability of the equilibrium point $q_0 = 0$.

For the linear law of $a(q)$, i.e. for $k = 1$, the integrals in (9.46, 47) allow analytical treatment, and on integration we have

$$R_{1W}(\lambda', 0) = \left(\frac{g_0}{A_0 d_0}\right)^{1/2} \sqrt{\frac{3}{2}\left(\lambda' - \frac{1}{2}\right)}\, \pi \cos(\pi/4) \times J_{-1/4}\left(\sqrt{3}\left(\lambda' + \frac{1}{2}\right)\right) J_{1/4}\left(\sqrt{3}\left(\lambda' + \frac{1}{2}\right)\right) \tag{9.48}$$

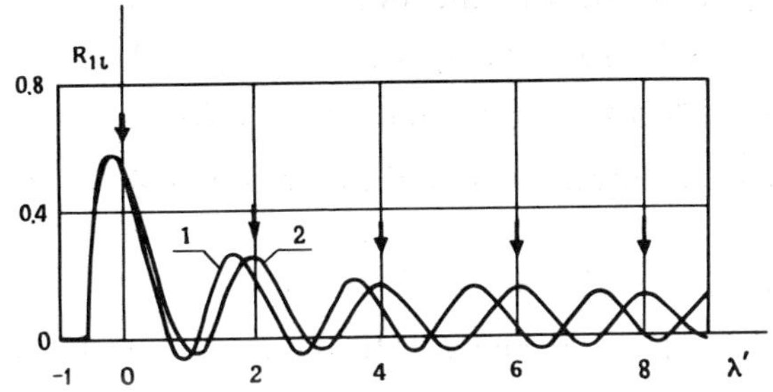

Fig. 9.1

$$R_{1s}(\lambda', 0) = \left(\frac{g_0}{A_0 d_0}\right)^{1/2} \sqrt{\lambda' + \frac{1}{2}}\ \pi \cos(\pi/4)$$
$$\times J_{-1/4}\left(\frac{\pi}{2}\left(\lambda' + \frac{1}{2}\right)\right) J_{1/4}\left(\frac{\pi}{4}\left(\lambda' + \frac{1}{2}\right)\right) \quad (9.49)$$

where J are the Bessel functions with the subscripts $1/4$ and $-1/4$.

Figure 9.1 shows the plots of $R_{1\mathrm{w}}(\lambda', 0)$, labeled *1*, and $R_{1s}(\lambda', 0)$ labeled *2*, constructed by the formulas (9.48) and (9.49), respectively. The arrows point to the exact even eigenvalues ($j = 0$, 2, 4, 6, and 8) of the Kolmogorov equation for $a(q) = -q$. The plots show no maxima corresponding to odd j. This may be attributed to the fact that the amplitude of δ-functions entering the right-hand side of (9.34) is proportional to $|\psi_j(x_0)|^2$, and for odd j the eigenfunctions $\psi_j(x_0)$ vanish at $x_0 = 0$.

For the second-order approximation, $m - 1 = 2$, with $l = W$ and $l = s$, we obtain from (9.44) accurate to a dimensionless factor independent of λ'

$$R_{2\mathrm{w}}(\lambda', 0) = \int_V dq_1\, dq_2 (q_1^2 + q_2^2 - q_1 q_2)^{-1/2} \sin\left\{\sqrt{12}(q_1^2 + q_2^2 - q_1 q_2)^{1/2}\right.$$
$$\times \left[\lambda' + \frac{1}{6}\left(q_1^{k-1} + q_2^{k-1} + \frac{q_2^k - q_1^k}{q_2 - q_1}\right) - \frac{1}{6(2k+1)}\right.$$
$$\left.\left.\times \left(q_1^{2k} + q_2^{2k} + \frac{q_2^{2k+1} - q_1^{2k+1}}{q_2 - q_1}\right)\right]^{1/2}\right\} \quad (9.50)$$

$$R_{2s}(\lambda', 0) = \int_V dq_1\, dq_2 (q_1^2 + 4q_2^2)^{-1/2} \sin\left\{\pi(q_1^2 + 4q_2^2)^{1/2}\right.$$
$$\left.\times \left[\lambda' + \frac{k}{2} P_{k-1}(q_1, q_2) - \frac{1}{2} P_{2k}(q_1, q_2)\right]^{1/2}\right\} \quad (9.51)$$

where $P_{2M}(q_1, q_2)$ in (9.51) is a polynomial of degree $2M$ homogeneous in q_1 and q_2, defined as follows:

$$P_{2M}(y_1, y_2) = \sum_{j=0}^{2M} B_{2M,j} q_1^{2M-j} q_2^j$$

with the coefficients $B_{2M,j}$ being zero for odd j and given as

$$B_{2M,j} = \frac{[(2M)!]^2}{2^{2M}(2M-j)!M!(j/2)!(M+j/2)!}$$

for even j.

Figure 9.2 shows the plot of $R_{2s}(\lambda', 0)$ as a function of λ' for $k = 1$. The position of maxima of $R_{2s}(\lambda', 0)$ corresponds, accurate to within 5%, to the exact values of λ_j indicated by arrows.

Problems of loss of control often require the study of system behavior within a region confined by absorbing screens. We wish to illustrate application of the formula (9.44) to such situations.

Let $a(q)$ be specified as

$$\begin{aligned} a(q) &= +\infty, \quad \text{for } q > 1 \\ &= 0, \quad \text{for } |q| \leqslant 1 \\ &= -\infty, \quad \text{for } q < -1 \end{aligned} \tag{9.52}$$

i.e., with two absorbing screens at $q = +1$ and $q = -1$. The domain of integration V with respect to q_j at $l = W$ in (9.44) is confined by -1 and

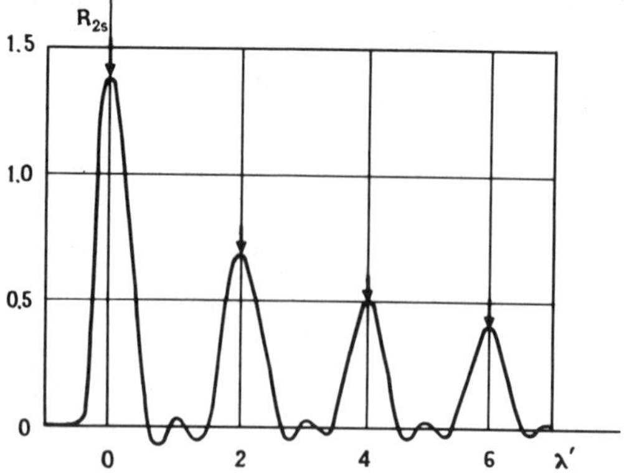

Fig. 9.2

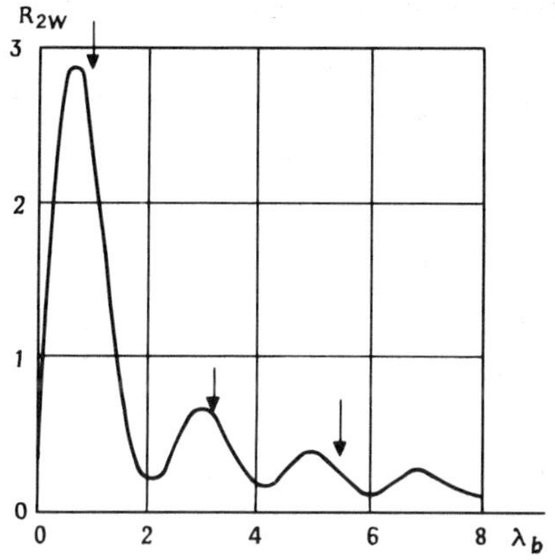

Fig. 9.3

+1, therefore,

$$R_{2W}(\lambda_b,\ 0) = \pi^{-1/2} \int_{-1}^{1} \int_{-1}^{1} \mathrm{d}q_1\, \mathrm{d}q_2 (q_1^2 + q_2^2 - q_1 q_2)^{-1/2} \times \sin\{\sqrt{2}\lambda_b (q_1^2 + q_2^2 - q_1 q_2)^{1/2}\} \tag{9.53}$$

where $\lambda_b = \sqrt{A_0 d_0 \lambda' / g_0} = \sqrt{d_0^2 \lambda / g_0}$.

The plot of $R_{2W}(\lambda_b, 0)$ computed with (9.53) is shown in Fig. 9.3. A qualitative agreement between the position of maxima of $R_{2W}(\lambda_b, 0)$ and exact values of λ_j indicated by arrows is evident.

Now we wish to extend the above formula to the general multidimensional case. Let the vector of drag coefficients $\mathbf{a}(\mathbf{x})$ and the matrix of perturbation intensities $\mathbf{G}$, assumed independent of $\mathbf{x}$ and t, have the form

$$\begin{aligned} \mathbf{a}(\mathbf{x}) &= A_0 \mathbf{a}(\mathbf{x}/d_0) = A_0 \mathbf{a}(\mathbf{q}) \\ \mathbf{G} &= g_0 \mathbf{G}_1 \end{aligned} \tag{9.54}$$

For the multidimensional case,

$$f_m(\mathbf{x}_\mathrm{f},\ T;\ \mathbf{x}_\mathrm{in},\ 0) = C_{mlN} T^{-Nm/2} \int_{-\infty}^{+\infty} \prod_{j=1}^{m-1} \mathrm{d}\mathbf{x}_j \times \exp\left\{ -\frac{1}{2} \int_0^T \left(\left[\frac{\mathrm{d}\mathbf{x}_l(t)}{\mathrm{d}t} - \mathbf{a}(\mathbf{x}_l(t)) \right]^\mathrm{T} \times \mathbf{G}^{-1} \left[\frac{\mathrm{d}\mathbf{x}_l(t)}{\mathrm{d}t} - \mathbf{a}(\mathbf{x}_l(t)) \right] + \operatorname{div} \mathbf{a}(\mathbf{x}) \Big|_{\mathbf{x}=\mathbf{x}_l(t)} \right) \mathrm{d}t \right\} \tag{9.55}$$

Substituting this expression into (9.35) yields

$$D_{m-1,l}(\lambda, \mathbf{q}_0) = d_0^{N(m-1)} C_{mlN} \int_0^{\infty} d\theta (i\theta)^{-Nm/2} \int_{-\infty}^{+\infty} \prod_{j=1}^{m-1} d\mathbf{q}_j$$

$$\times \exp\left\{ i\left[\frac{\beta_{lN}\sigma_{mlN}}{\theta} - \frac{A_0}{d_0}\theta f_l(\mathbf{q}_0, \mathbf{q}_1, \ldots, \mathbf{q}_{m-1}) + \lambda\theta \right] + \frac{A_0 d_0}{g_0} \zeta_l(\mathbf{q}_0, \mathbf{q}_1, \ldots, \mathbf{q}_{m-1}) \right\} \tag{9.56}$$

$$f_l(\mathbf{q}_0, \mathbf{q}_1, \ldots, \mathbf{q}_{m-1}) = \int_0^1 \varphi(\mathbf{q}_l(\tau))\, d\tau$$

$$\varphi(\mathbf{q}_l(\tau)) = \frac{1}{2}\left[\operatorname{div} \mathbf{a}(\mathbf{q}) \Big|_{\mathbf{q}=\mathbf{q}_l(\tau)} + \frac{A_0 d_0}{g_0} \mathbf{a}^{\mathrm{T}}(\mathbf{q}_l(\tau)\mathbf{G}_1^{-1}\mathbf{a}(\mathrm{q}_l(\tau)) \right]$$

$$\zeta_l(\mathbf{q}_0, \mathbf{q}_1, \ldots, \mathbf{q}_{m-1}) = \frac{1}{2}\int_0^1 \left[\frac{d\mathbf{q}_l^{\mathrm{T}}(\tau)}{d\tau} \mathbf{G}_1^{-1}\mathbf{a}(\mathbf{q}_l(\tau)) + \mathbf{a}^{\mathrm{T}}(\mathbf{q}_l(\tau))\mathbf{G}_1^{-1} \frac{d\mathbf{q}_l(\tau)}{d\tau} \right] d\tau$$

For $l = W$ we have

$\beta_{WN} = d_0^2 m/2g_0$;

$\sigma_{mWN} = \sum_{j=0}^{m-1} (\mathbf{q}_{j+1} - \mathbf{q}_j)\mathbf{G}_1^{-1}(\mathbf{q}_{j+1} - \mathbf{q}_j)$;

$\mathbf{q}_W(\tau) = \mathbf{q}_j + (\mathbf{q}_{j+1} - \mathbf{q}_j)[m\tau - j]; \quad m\tau \in [j, j+1]$;

$j = 0, 1, \ldots, m-1$; and

$C_{mWN} = \{m^N/(2\pi)^N \det \mathbf{G}\}^{m/2}$.

For $l = s$ we have

$\beta_{sN} = \pi^2 d_0/4g_0$;

$\mathbf{q}_s(\tau) = \mathbf{q}_0 + \sum_{j=1}^{m-1} \mathbf{q}_j \sin(\pi j\tau)$;

$\sigma_{msN} = \sum_{j=1}^{m-1} j^2 \mathbf{q}_j^{\mathrm{T}}\mathbf{G}_1^{-1}\mathbf{q}_j$; and

$$C_{msN} = \left[\frac{(m-1)!\,\pi^{(m-1)/2}}{\sqrt{2\pi}\, 2^{m-1}} \right]^N (\det \mathbf{G})^{-m/2}$$

Here we note that in the one-dimensional case of $N = 1$, the expression (9.43) lacks the term $\zeta_l(\mathbf{q}_0, \mathbf{q}_1, \ldots, \mathbf{q}_{m-1})$. We demonstrate that this term is zero at $N = 1$. Setting $N = 1$, we get

$$\zeta_l(q_0, \ldots, q_{m-1}) = \int_0^1 a(q_l(\tau)) \frac{\mathrm{d}q_0(\tau)}{\mathrm{d}\tau} \mathrm{d}\tau = \int_{q_{\mathrm{in}}}^{q_{\mathrm{f}}} a(q)\,\mathrm{d}q \tag{9.57}$$

If $q_{\mathrm{in}} = q_{\mathrm{f}} = q_0$, then the integral on the right side is zero.
On integration with respect to θ the relation (9.56) becomes

$$D_{m-1,l}(\lambda, \mathbf{q}_0) = B_{mlN} \int_{-\infty}^{+\infty} \prod_{j=1}^{m-1} \mathrm{d}\mathbf{q}_j (z_N/\sigma_{mlN})^{\frac{Nm}{2}-1} H^{(1)}_{Nm/2-1}(z_N)$$
$$\times \exp\left[\frac{A_0 d_0}{g_0} \zeta_{Nl}(\mathbf{q}_0, \ldots, \mathbf{q}_{m-1})\right] \tag{9.58}$$

where

$$B_{mlN} = \pi d_0^{Nm} C_{mlN}(2\beta_{lN})^{1-\frac{Nm}{2}}$$
$$z_N = \gamma_{lN}\{\sigma_{mlN}[\lambda' - f_l(\mathbf{q}_0, \ldots, \mathbf{q}_{m-1})]\}^{1/2}$$
$$\gamma_{lN} = 2(\beta_{lN}A_0/d_0)^{1/2}, \quad \text{and} \quad \lambda' = d_0\lambda/A_0$$

Accurate to the factor B_{mlN} the real part of $D_{m-1,l}(\lambda', \mathbf{q}_0)$ is

$$R_{m-1,l}(\lambda', \mathbf{q}_0) = \mathrm{Re}\,[D_{m-1,l}(\lambda', \mathbf{q}_0)/B_{mlN}]$$
$$= \int_V \prod_{j=1}^{m-1} \mathrm{d}\mathbf{q}_j (z_N/\sigma_{mlN})^{\frac{Nm}{2}-1} J_{\frac{Nm}{2}-1}(z_N)$$
$$\times \exp\left[\frac{A_0 d_0}{g_0} \zeta_{Nl}(\mathbf{q}_0, \ldots, \mathbf{q}_{m-1})\right] \tag{9.59}$$

where the boundary of the domain of integration is defined by the condition

$$\lambda' - f_l(\mathbf{q}_0, \mathbf{q}_1, \ldots, \mathbf{q}_m) \geqslant 0 \tag{9.60}$$

As an example of application of the formula (9.59) we take the case of $N = 2$ with the drag coefficient vector defined by the law

$$\begin{aligned} \mathbf{a}(\mathbf{q}) &= q\mathbf{e}_\varphi, \quad \text{for } |\mathbf{q}| \leqslant 1 \\ &= 0, \quad \text{for } |\mathbf{q}| > 1 \end{aligned} \tag{9.61}$$

where $\mathbf{e}_\varphi$ is the unit vector at φ in cylindrical coordinates with the axis perpendicular to and piercing the plane (q_x, q_y) at point (0, 0).

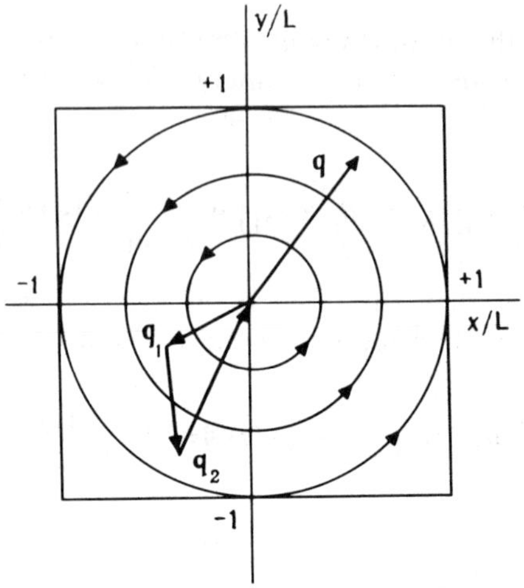

Fig. 9.4

Suppose also that $d_0 = L$, $A_0 = \omega_0 L$, where ω_0 is some constant with the dimension of pulsatance, s^{-1}. The vector $\mathbf{q}$ in (9.61) has two projections $\mathbf{q} = \{q_x, q_y\}$ with $q_x = x/L$ and $q_y = y/L$, and the modulus $|\mathbf{q}| = (q_x^2 + q_y^2)^{1/2} = q$.

We shall seek the solution of the boundary value problem with the drag coefficient $\mathbf{a}(\mathbf{q})$ given by (9.61) in a region D_{cr} confined by absorbing screens and shaped as a square of side $2L$ (Fig. 9.4), namely,

$$D_{cr} = \{|q_x| \leqslant 1, \ |q_y| \leqslant 1\}$$

Substituting (9.61) into (9.59) at $m = 2$, we get

$$R_1(\lambda_c, 0) = \int\limits_V d\mathbf{q}_1 \left(\frac{z_2}{\sigma_{2W2}}\right) J_1(z_2) \exp\{\pi_1 \zeta_{2W}(\mathbf{q}_0, \mathbf{q}_1)\} \qquad (9.62)$$

where $\lambda_c = 2\lambda d_0^2/g_0$ and $\pi_1 = A_0 d_0/g_0$.

From (9.56, 58) at $m = 2$ it follows that

$$z_2 = 2\{\mathbf{q}_1^2[\lambda_c - 2\pi_1^2 f(\mathbf{q}_0, \mathbf{q}_1)]\}^{1/2}$$

$$\sigma_{2W2} = 2\mathbf{q}_1^2, \ \operatorname{div}\mathbf{a}(\mathbf{q}) = 0$$

$$\zeta_{2W}(\mathbf{q}_0, \mathbf{q}_1) = \oint\limits_\Gamma \mathbf{a}(\mathbf{q}_W)\, d\mathbf{q}_W = 0 \quad \text{at } m = 2$$

and

$$f(\mathbf{q}_0, \mathbf{q}_1) = \int\limits_0^1 \varphi(\mathbf{q}_W(\tau))\, d\tau = \frac{1}{2}\int\limits_0^1 \mathbf{a}^2(\mathbf{q}_W(\tau))\, d\tau$$

Depending on the modulus of $\mathbf{q}_1$, the function $f(\mathbf{q}_0, \mathbf{q}_1)$ and the expression (9.62) have different form, namely,

$$f(\mathbf{q}_0, \mathbf{q}_1) = \frac{1}{6}\mathbf{q}_1^2$$

$$R_1(\lambda_c, 0) = \int_V d\mathbf{q}_1 \frac{\{\mathbf{q}_1^2[\lambda_c - \pi_1^2\mathbf{q}_1^2/3]\}^{1/2}}{\mathbf{q}_1^2} J_1(2\mathbf{q}_1[\lambda_c - \pi_1^2\mathbf{q}_1^2/3]^{1/2}) \qquad (9.63)$$

for $|\mathbf{q}_1| \leqslant 1$, $\mathbf{q}_0 = 0$; and

$$f(\mathbf{q}_0, \mathbf{q}_1) = \frac{1}{2}\left(\int_0^{\tau_1} \mathbf{q}_W^2(\tau)\, d\tau + \int_{\tau_2}^{1} \mathbf{q}_W^2(\tau)\, d\tau\right) = \frac{1}{6|\mathbf{q}_1|}$$

$$|\mathbf{q}_W(\tau_1)| = |\mathbf{q}_W(\tau_2)| = 1, \quad \tau_1 = 1/(2|\mathbf{q}_1|), \quad \tau_2 = 1 - \tau_1$$

$$R_1(\lambda_c, 0) = \int_V d\mathbf{q}_1 \frac{\{\mathbf{q}_1^2[\lambda_c - \pi_1^2/3|\mathbf{q}_1|]\}^{1/2}}{\mathbf{q}_1^2} J_1(2\mathbf{q}_1[\lambda_c - \pi_1^2/3|\mathbf{q}_1|]^{1/2}) \quad (9.64)$$

for $|\mathbf{q}_1| > 1$ and $\mathbf{q}_0 = 0$.

The plots of $R_1(\lambda_c, 0)$ calculated for $\pi_1 = 0$, 1, 2, 3, and 4 (labeled *1-5* respectively) are presented in Fig. 9.5. The arrow points to the first exact eigenvalue of the respective boundary value problem found with $\pi_1 = 0$.

For $m = 3$, from (9.59) it follows that

$$R_2(\lambda_c, 0) = \int_V d\mathbf{q}_1\, d\mathbf{q}_2 \left(\frac{z_2}{\sigma_{3W2}}\right)^2 J_2(z_2) \exp[\pi_1 \zeta_{2W}(\mathbf{q}_0, \mathbf{q}_1, \mathbf{q}_2)] \qquad (9.65)$$

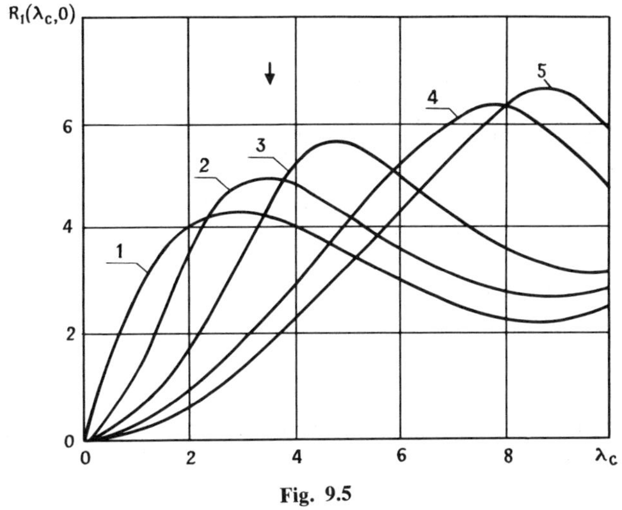

Fig. 9.5

By virtue of (9.56, 58)

$$\sigma_{3W2} = 2(\mathbf{q}_1^2 - \mathbf{q}_1\mathbf{q}_2 + \mathbf{q}_2^2)$$

$$z_2 = \{3\sigma_{3W2}[\lambda_c - 2\pi_1 f(\mathbf{q}_0, \mathbf{q}_1, \mathbf{q}_2)]\}^{1/2}$$

$$\zeta_{2W}(\mathbf{q}_0, \mathbf{q}_1, \mathbf{q}_2) = \oint \mathbf{a}(\mathbf{q}_W)\, d\mathbf{q}_W = \oint_\Gamma \mathbf{a}(\mathbf{q})\, d\mathbf{q}$$

where for $m = 3$ the contour Γ is a triangle with the vertices (0, 0), (q_{1x}, q_{1y}), (q_{2x}, q_{2y}) in the q_x, q_y plane (Fig. 9.4). By the Stokes theorem the integral of a vector $\mathbf{a}(\mathbf{q})$ over the contour Γ may be transformed to the integral of curl $\mathbf{a}(\mathbf{q})$ over the surface S within the boundary Γ, namely,

$$\oint_\Gamma \mathbf{a}(\mathbf{q})\, d\mathbf{q} = \int_S \operatorname{curl} \mathbf{a}(\mathbf{q})\, d\mathbf{s} \tag{9.66}$$

In view of (9.62)

$$\begin{aligned} \operatorname{curl} \mathbf{a}(\mathbf{q}) &= 2\mathbf{e}_z, \quad \text{for } |\mathbf{q}| \leqslant 1 \\ &= 0, \quad \text{for } |\mathbf{q}| > 1 \end{aligned} \tag{9.67}$$

The relations (9.66, 67) substantially simplify the computation of $\zeta_{2W}(\mathbf{q}_0, \mathbf{q}_1, \mathbf{q}_2)$ for the problem at hand.

Five situations should be observed in calculating $f(0, \mathbf{q}_1, \mathbf{q}_2)$ and $\zeta_{2W}(0, \mathbf{q}_1, \mathbf{q}_2)$ as follows (Fig. 9.6):

(1) $|\mathbf{q}_1| \leqslant 1, \quad |\mathbf{q}_2| \leqslant 1$
(2) $|\mathbf{q}_1| > 1, \quad |\mathbf{q}_2| \leqslant 1$
(3) $|\mathbf{q}_1| < 1, \quad |\mathbf{q}_2| > 1$
(4) $|\mathbf{q}_1| > 1, \quad |\mathbf{q}_2| > 1, \quad |\mathbf{q}_W(\tau)| > 1, \quad 1/3 \leqslant \tau \leqslant 2/3$
(5) $|\mathbf{q}_1| > 1, \quad |\mathbf{q}_2| > 1, \quad |\mathbf{q}_W(\tau)| < 1, \quad 1/3 \leqslant \tau \leqslant 2/3$

For case (1) we have

$$\zeta_{2W}(0, \mathbf{q}_1, \mathbf{q}_2) = q_{1x}q_{2y} - q_{2x}q_{1y}$$

$$f(0, \mathbf{q}_1, \mathbf{q}_2) = \frac{1}{18}(2\mathbf{q}_1^2 + \mathbf{q}_1\mathbf{q}_2 + 2\mathbf{q}_2^2)$$

$$R_2(\lambda_c, 0) = \frac{3}{2}\int_V d\mathbf{q}_1\, d\mathbf{q}_2\, \frac{[\lambda_c - \pi_1^2(2\mathbf{q}_1^2 + \mathbf{q}_1\mathbf{q}_2 + 2\mathbf{q}_2^2)/9]}{\mathbf{q}_1^2 - \mathbf{q}_1\mathbf{q}_2 + \mathbf{q}_2^2}$$

$$\times J_2(\{6(\mathbf{q}_1^2 - \mathbf{q}_1\mathbf{q}_2 + \mathbf{q}_2^2)[\lambda_c - \pi_1^2(2\mathbf{q}_1^2 + \mathbf{q}_1\mathbf{q}_2 + 2\mathbf{q}_2^2)/9]\}^{1/2}) \exp[\pi_1(\mathbf{q}_{1x}\mathbf{q}_{2y} - \mathbf{q}_{2x}\mathbf{q}_{1y})]$$

The equations for the remaining four cases are quite unwieldy and will not be given here.

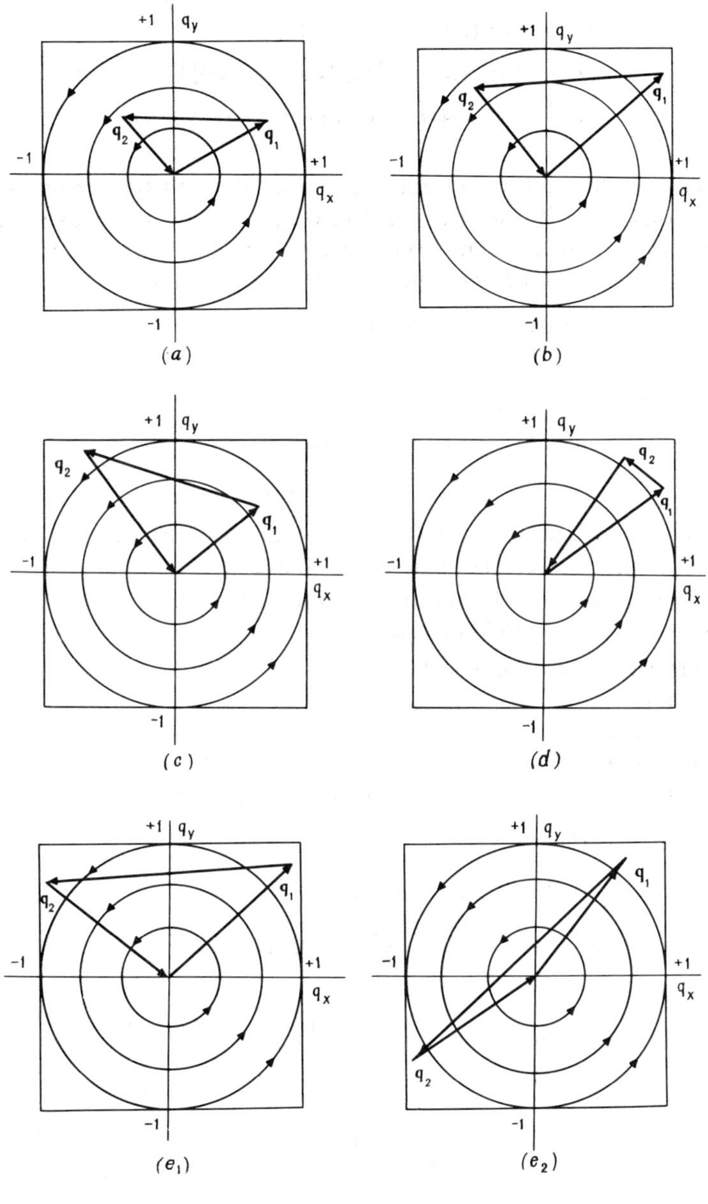

Fig. 9.6

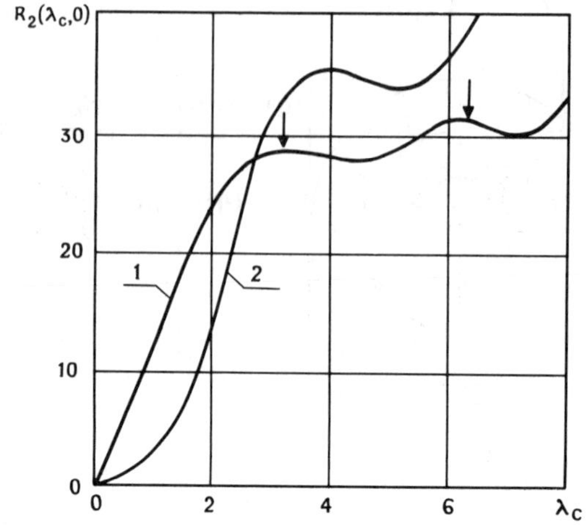

Fig. 9.7

Figure 9.7 shows the plots of $R_2(\lambda_c, 0)$ obtained for $\pi_1 = 0$ and $\pi_1 = 3$ (curves *1* and *2* respectively). At $\pi_1 = 0$ the position of maxima of $R_2(\lambda_c, 0)$ corresponds well to the exact eigenvalues, indicated by the arrows.

Now we wish to generalize the formulas (9.56, 59) for the case allowing for the dependence of the perturbation intensities on the system state vector. Substituting the expression for $f_m(\mathbf{x}_f, t_f; \mathbf{x}_{in}, t_{in})$ in the form (9.28) into (9.35) yields

$$D_{m-1,W}(\lambda, \mathbf{q}_0) = d_0^{N(m-1)}[m/(2\pi g_0)]^{Nm/2} \int_0^\infty d\theta (i\theta)^{-Nm/2}$$

$$\times \left(\int_{-\infty}^{+\infty} \prod_{j=1}^{m-1} d\mathbf{q}_j \prod_{j=1}^{m-1} [\det \mathbf{G}_1(\varkappa_k)]^{-1/2} \exp \left\{ i \left[\Omega_{mWN}/\theta \right.\right.\right.$$

$$\left.\left.\left. - \frac{A_0}{d_0} \theta f_W(\mathbf{q}_0, \ldots, \mathbf{q}_{m-1}) + \lambda\theta \right] + \frac{A_0 d_0}{g_0} \zeta_{NW}(\mathbf{q}_0, \ldots, \mathbf{q}_{m-1}) \right\} \right) \quad (9.68)$$

where

$$\Omega_{mWN} = \frac{d_0^2}{2g_0} \int_0^1 \frac{d\mathbf{q}_W^{\mathrm{T}}(\tau)}{d\tau} \mathbf{G}_1^{-1}(\mathbf{q}_W(\tau)) \frac{d\mathbf{q}_W(\tau)}{d\tau} d\tau$$

$$f_W(\mathbf{q}_0, \ldots, \mathbf{q}_{m-1}) = \int_0^1 \varphi(\mathbf{q}_W(\tau)) \, d\tau \quad (9.69)$$

$$\varphi(\mathbf{q}_W(\tau)) = \frac{1}{2}\left\{G_{d_1}(\mathbf{q})\,\mathrm{div}\left[\mathbf{a}(\mathbf{q})G_{d_1}^{-1}(\mathbf{q}) - \frac{1}{2}\frac{g_0}{A_0 d_0}\,\mathrm{Div}\,(\mathbf{G}(\mathbf{q})G_{d_1}^{-1}(\mathbf{q}))\right]\right.$$
$$+ \frac{A_0 d_0}{g_0}\left[\mathbf{a}(\mathbf{q}) - \frac{1}{2}\frac{g_0}{A_0 d_0}G_{d_1}(\mathbf{q})\,\mathrm{Div}\,(\mathbf{G}_1(\mathbf{q})/G_{d_1}(\mathbf{q}))\right]^{\mathrm{T}}\mathbf{G}_1^{-1}(\mathbf{q})$$
$$\left.\times\left[\mathbf{a}(\mathbf{q}) - \frac{1}{2}\frac{g_0}{A_0 d_0}G_{d_1}(\mathbf{q})\,\mathrm{Div}\,(\mathbf{G}_1(\mathbf{q})/G_{d_1}(\mathbf{q}))\right]\right\}\Bigg|_{\mathbf{q}=\mathbf{q}_W(\tau)} \tag{9.70}$$

$$G_{d_1}(\mathbf{q}) = [\det \mathbf{G}_1(\mathbf{q})]^{1/2}, \quad \varkappa_k = (\mathbf{q}_k + \mathbf{q}_{k+1})/2,$$

$$\zeta_{NW}(\mathbf{q}_0, \ldots, \mathbf{q}_{m-1}) = \frac{1}{2}\int_0^1 \left\{\frac{\mathrm{d}\mathbf{q}_W^{\mathrm{T}}(\tau)}{\mathrm{d}\tau}\mathbf{G}_1^{-1}(\mathbf{q})\left[\mathbf{a}(\mathbf{q}) - \frac{1}{2}\frac{g_0}{A_0 d_0}G_{d_1}(\mathbf{q})\right.\right.$$
$$\left.\times\,\mathrm{Div}\,(\mathbf{G}_1(\mathbf{q})/G_{d_1}(\mathbf{q}))\right] + \left[\mathbf{a}(\mathbf{q}) - \frac{1}{2}\frac{g_0}{A_0 d_0}G_{d_1}(\mathbf{q})\,\mathrm{Div}\,(\mathbf{G}_1(\mathbf{q})/G_{d_1}(\mathbf{q}))\right]^{\mathrm{T}}$$
$$\left.\times\,\mathbf{G}_1^{-1}(\mathbf{q})\frac{\mathrm{d}\mathbf{q}_W(\tau)}{\mathrm{d}\tau}\right\}\Bigg|_{\mathbf{q}=\mathbf{q}_W(\tau)}\mathrm{d}\tau \tag{9.71}$$

Integrating (9.68) with respect to θ and taking the real part, we get

$$R_{m-1,W}(\lambda', \mathbf{q}_0) = \int_V \prod_{j=1}^{m-1} \mathrm{d}\mathbf{q}_j \prod_{k=0}^{m-1} [\det G_1(\varkappa_k)]^{-1/2}$$
$$\times\,(d_0 z_N/\Omega_m W_N A_0)^{\frac{Nm}{2}-1} J_{\frac{Nm}{2}-1}(z_N)$$
$$\times\,\exp\left[\frac{A_0 d_0}{g_0}\zeta_{NW}(\mathbf{q}_0, \ldots, \mathbf{q}_{m-1})\right] \tag{9.72}$$

where

$$z_N = 2\left[\frac{\Omega_m W_N A_0}{d_0}(\lambda' - f_W(\mathbf{q}_0, \ldots, \mathbf{q}_{m-1}))\right]^{1/2}$$
$$\lambda' = d_0/A_0.$$

Consider a particular case of $N = 1$ and $m = 2$. Let

$$G_1(q) = 1 + \beta q^2, \quad a(q) = \frac{1}{4\pi_1}\frac{\partial}{\partial q}G_1(q),$$
$$\mathbf{q}_0 = 0, \quad \text{and} \quad \pi_1 = A_0 d_0/g_0 \tag{9.73}$$

With this definition the expressions (9.69-71) take the form

$$\Omega_{2W1} = \frac{d_0^2}{g_0}\int_0^{1/2} \frac{4(q_1 - q_0)^2}{1 + \beta(q_0 + (q_1 - q_0)2\tau)^2}\,\mathrm{d}\tau = \frac{d_0^2}{g_0}\int_0^{1/2} \frac{4q_1^2}{1 + 4\beta\tau^2 q_1^2}\,\mathrm{d}\tau$$

$$= \frac{2q_1 d_0^2}{g_0\sqrt{\beta}} \arctan(q_1\sqrt{\beta})$$

$\varphi(q_W(\tau)) = 0,\quad f_W(q_0, q_1) = \zeta_W(q_0, q_1) = 0,$ and

$$R_W(\lambda', 0) = \int_{-\infty}^{\infty} \mathrm{d}q_1 \left[1 + \frac{\beta}{4} q_1^2\right]^{-1} J_0\left(2\sqrt{2}\left[\pi_1 \frac{q_1}{\sqrt{\beta}} \lambda' \arctan(q_1\sqrt{\beta})\right]^{1/2}\right)$$

The expressions (9.69, 72) may be used to determine the characteristic time of system residence (escape) in a given region D_{cr}.

By way of example we consider the equation of dynamics of a nuclear reactor in the point approximation with one effective group of delayed neutrons, viz.,

$$\frac{\mathrm{d}n}{\mathrm{d}t} = \frac{(\delta k - \beta)n}{l} + \lambda c$$
$$\frac{\mathrm{d}c}{\mathrm{d}\tau} = \frac{\beta n}{l} - \lambda c \tag{9.74}$$

where n is the number of neutrons in the reactor, l is the neutron lifetime (generation time), c and λ are the number and the decay constant of the parent nuclei of the delayed neutrons, β is the fraction of the delayed neutrons, and δk is the reactivity.

The quantity δk is the sum of the reactivity due to the means of control denoted by δk_u, and the perturbation in reactivity proper, δk_p. The dependence of δk_u upon n is defined by the law of control being exercised. Assuming that δk_p has the form of white noise and that

$$\frac{\delta k_u}{\beta} = u((n - n_0)/n_0)$$

where n_0 is the number of neutrons available in the nominal operating mode, we reduce the system (9.74) to the canonical form

$$\tau_n \frac{\mathrm{d}X_1}{\mathrm{d}t} = -X_1 + u(X_1)(1 + X_1) + X_2 + \sigma_0(1 + X_1)v_1(t) \tag{9.75}$$

$$\tau_c \frac{\mathrm{d}X_2}{\mathrm{d}t} = X_1 - X_2 \tag{9.76}$$

where $u(X_1) = -\gamma X_1$, X_1 and X_2 are the dimensionless deviations in the number of neutrons and parent nuclei from the respective levels in the nominal operating mode, γ is the gain factor, $\tau_n = l/\beta$ and $\tau_c = \lambda^{-1}$ are the time constants, σ_0 is a coefficient characterizing the intensity of perturbations in reactivity, and $v_1(t)$ is a white noise of unit intensity.

Substituting $X_1(t)$ in the form of a broken line $x_W^m(\theta)$ given by (9.22) into (9.76) and integrating gives the analytical solution

$$\begin{aligned} y_W^m(t) &= y_j \exp\left(-\frac{\theta - j}{T_0}\right) + x_j\left[1 - \exp\left(-\frac{\theta - j}{T_0}\right)\right] \\ &\quad + (x_{j+1} - x_j)T_0\left[\frac{\theta - j}{T_0} + \exp\left(-\frac{\theta - j}{T_0}\right) + 1\right], \\ j \leqslant \theta &\leqslant j + 1, \quad j = 0, 1, \ldots, m - 1 \end{aligned} \tag{9.77}$$

where $T_0 = \tau_c m/(t_f - t_{in})$, $\theta = m(t - t_{in})/(t_f - t_{in})$, x_j are the vertices of the broken line $x_W^m(t)$, and y_j are points of the solution $y_W^m(t)$ at $\theta = j$. In agreement with this expression, the recurrence relation for y_{j+1} is as follows:

$$\begin{aligned} y_{j+1} &= y_j \exp(-T_0^{-1}) + x_j(1 - \exp(-T_0^{-1})) \\ &\quad + (x_{j+1} - x_j)[1 - T_0 + T_0 \exp(-T_0^{-1})], \\ j &= 0, 1, \ldots, m - 1 \end{aligned} \tag{9.78}$$

The value of y_0 in this relation at $j = 0$ is defined by the initial conditions, namely, if at the initial time instant $X_2(t)|_{t=t_{in}} = x_2(0)$, then $y_0 = x_2(0)$.

Loss of control (operation of the emergency protection system) occurs when $|X_1| \geqslant \Delta$. We suppose that the control law is linear, viz., $u(X_1) = -\gamma_1 X_1 - \gamma_2$.

If we neglect the effect of delayed neutrons in (9.75), then in agreement with (9.13), i.e., allowing for the dependence of perturbation intensity on X_1, we have

$$\begin{aligned} a(x) &= \frac{1}{\tau_n}[u(x)(1 + x) - x] + \frac{1}{2}\frac{\sigma_0^2}{\tau_n^2}(1 + x) \\ G_1(x) &= \frac{\sigma_0^2}{\tau_n^2}(1 + x)^2 \end{aligned} \tag{9.79}$$

For the system operating in the normal mode, the range of x is bounded by Δ which as a rule is far below unity. Therefore we may safely let $1 + x \cong 1$ in (9.79) and for the linear law $u(x) = -\gamma_1 x - \gamma_2$ we obtain

$$\begin{aligned} &a(q) = -C_1 q + C_2, \ C_1 = 1 + \gamma_1, \ C_2 = \frac{\sigma_0^2}{2\tau_n \Delta} - \gamma_2 \\ &A_0 = \Delta\tau_n^{-1}, \ d_0 = \Delta, \ x/\Delta = q, \ g_0 = \sigma_0^2/\tau_n^2 \end{aligned} \tag{9.80}$$

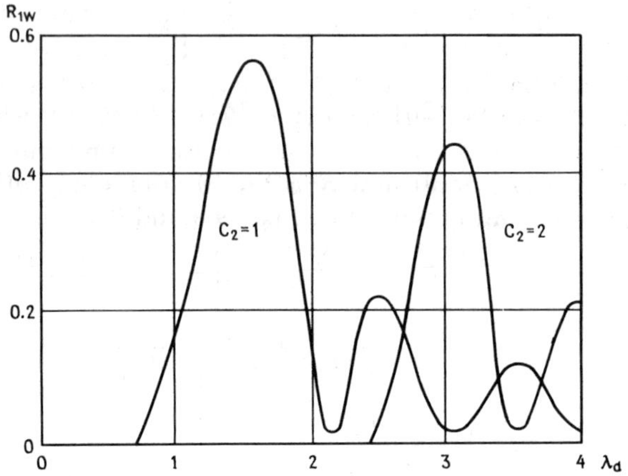

Fig. 9.8

$R_{m-1}(\lambda_d', q_0)$ is defined by (9.38), where

$$z = \sqrt{2}bm^{1/2}\left\{\sum_{i=0}^{m-1}(q_{i+1}-q_i)^2\left[\lambda_d' - \frac{b^2C_2^2 - C_1}{2}\right.\right.$$

$$\left.\left. - \frac{b^2C_1}{6m}\sum_{j=0}^{m-1}(C_1(q_j^2 + q_jq_{j+1} + q_{j+1}^2) - 3C_2(q_{j+1}+q_j))\right]\right\}^{1/2} \tag{9.81}$$

and

$$\sigma_m = \sum_{j=0}^{m-1}(q_{j+1}-q_j)^2,\ b^2 = \Delta^2\tau_n/\sigma_0^2,\ \lambda_d' = \lambda\tau_n$$

The domain of integration V is defined by two simultaneous conditions $[\cdot] > 0$ and $|q_j| < 1$.

For $m = 2$ and $m = 3$, from (9.38) and (9.81) it follows respectively

$$R_{1W}(\lambda_d', 0) = \int_V J_0(2^{3/2}bq_1\left[\lambda_d' - \frac{b^2C_2^2 - C_1}{2}\right.$$

$$\left. - \frac{b^2C_1}{6}(C_1q_1^2 - 3C_2q_1)\right]^{1/2}dq_1$$

$$R_{2W}(\lambda_d', 0) = \pi^{-1/2}\int_V dq_1\,dq_2(q_1^2 - q_1q_2 + q_2^2)^{-1/2}$$

$$\times \sin\left\{\sqrt{12}b(q_1^2 - q_1q_2 + q_2^2)^{1/2}\left[\lambda_d' - \frac{b^2C_2^2 - C_1}{2} - \frac{b^2C_2}{18}\,(C_1(2q_1^2 + q_1q_2 + 2q_2^2) - 6C_2(q_1 + q_2))\right]^{1/2}\right\}$$

The plots of $R_{1W}(\lambda_d, 0)$ and $R_{2W}(\lambda_d, 0)$ computed with $\lambda_d = \sqrt{\lambda_d'}$, $b = 1$, and $C_1 = 1.1$ are presented in Figs. 9.8 and 9.9. Figure 9.10 shows the plots of the first eigenvalue λ_{d1}' versus C_2 computed for three different

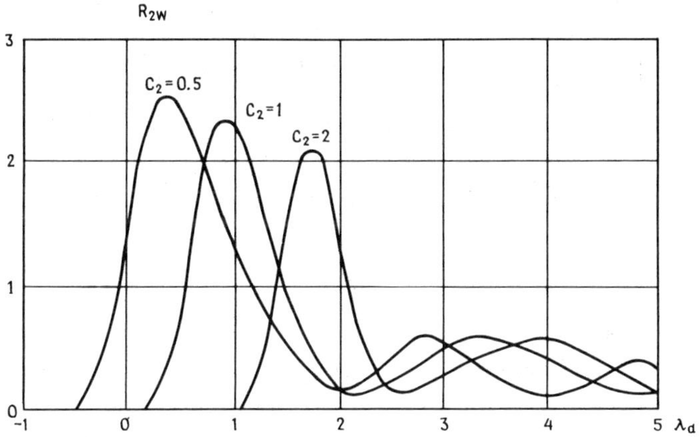

Fig. 9.9

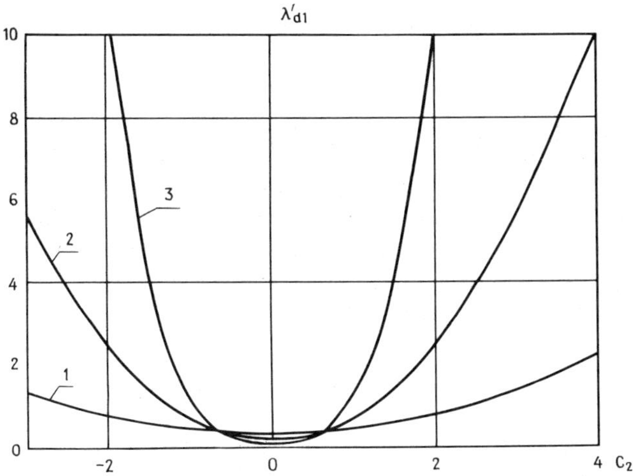

Fig. 9.10

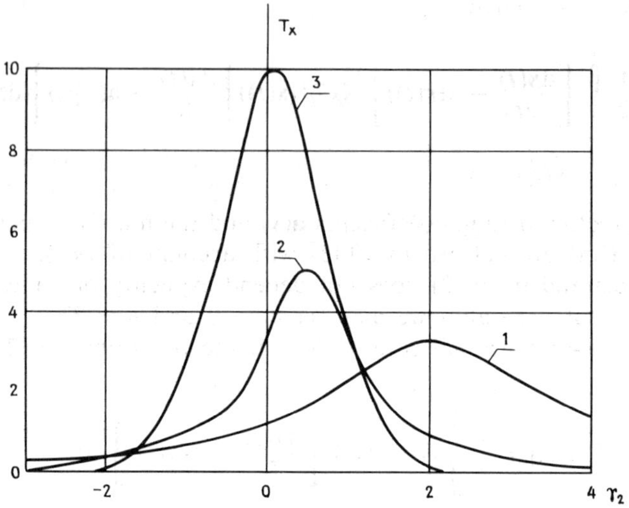

Fig. 9.11

intensities of perturbation g_0 corresponding to $b = 0.5$, 1, and 2 (curves *1-3*). As g_0 decreases (with growth of b), λ'_{d1} approaches zero at $C_2 = 0$. The plot of the characteristic time $T_c = (\lambda'_{d1})^{-1}$ the system resides in the region D_{cr}: $|X_1| < \Delta$, or takes to escape from this region, is shown in Fig. 9.11. The parameter values are the same as in Fig. 9.10. The maximum value of $T_c(\gamma_2)$ is achieved at $C_2 = 0$, i.e. for $\gamma_2 = \sigma_0^2/2\Delta\tau_n$.

Taking into account the delayed neutrons in the dynamic equations (9.75, 76) results in a decrease of λ'_{d1}, however the profile of $T_c(\gamma_2)$ and the peak position of γ_{2*} do not change.

This evidence suggests that path integrals provide a footing for the development of efficient techniques for the evaluation of eigenvalues of the Kolmogorov equation. These eigenvalues are useful in determining the characteristic time of system escape from D_{cr}. However, when the intensity of perturbation tends to zero, so does the first eigenvalue λ_1. In view of finite width of the peaks of $R_{ml}(\lambda, 0)$, the accuracy of location of λ_1 with this dependence drops accordingly.

9.4 Weak Perturbation Stability Analysis

As the intensity of perturbation tends to zero, the elements of the matrix $\mathbf{G}^{-1}(\mathbf{x})$ tend to infinity, therefore in the expression (9.28) for the action functional $S[\mathbf{x}(t)]$ we should confine ourselves to only the terms linear in $\mathbf{G}^{-1}(\mathbf{x})$. Then

$$S[\mathbf{x}(t)] = \frac{1}{2}\int_{t_{\text{in}}}^{t_{\text{f}}} \left[\frac{\mathrm{d}\mathbf{x}(t)}{\mathrm{d}t} - \mathbf{a}(\mathbf{x}(t))\right]^{\mathrm{T}} \mathbf{G}^{-1}(\mathbf{x}(t)) \left[\frac{\mathrm{d}\mathbf{x}(t)}{\mathrm{d}t} - \mathbf{a}(\mathbf{x}(t))\right] \mathrm{d}t \tag{9.82}$$

$$\mathbf{x}(t_{\text{in}}) = \mathbf{x}_{\text{in}}, \quad \mathbf{x}(t_{\text{f}}) = \mathbf{x}_{\text{f}}$$

where the vector of drag coefficients $\mathbf{a}(\mathbf{x})$ and the matrix of perturbation intensities $\mathbf{G}(\mathbf{x})$ are defined by (9.13) with account of (9.5).

The integrand in (9.82) does not depend explicitly on time, therefore without loss of generality we may set $t_{\text{in}} = 0$ and $t_{\text{f}} = T$.

In the one-dimensional case of $N = 1$, the functional (9.82) takes on the form

$$S[x(t)] = \frac{1}{2}\int_{0}^{T} G^{-1}(x(t)) \left[\frac{\mathrm{d}x(t)}{\mathrm{d}t} - a(x(t))\right]^{2} \mathrm{d}t \tag{9.83}$$

When $\mathbf{G}^{-1}(\mathbf{x})$ tends to infinity, the functional $S[\mathbf{x}(t)]$ goes to infinity for all trajectories $\mathbf{x}(t)$ which make the expression in brackets on the right-hand side of (9.82) other than zero. In agreement with the physical meaning of the action functional, the greater $S[\mathbf{x}(t)]$, the lower the probability of the system moving in a trajectory close to $\mathbf{x}(t)$. Therefore, for $\|\mathbf{G}(\mathbf{x})\| \to 0$ the transition probability density of the dynamic system will be essentially nonzero only for points $\mathbf{x}_{\text{f}}$ close to the trajectory $\mathbf{x}_{\text{d}}(t)$ satisfying the equation

$$\frac{\mathrm{d}\mathbf{x}_{\text{d}}(t)}{\mathrm{d}t} = \mathbf{a}(\mathbf{x}_{\text{d}}(t)); \quad \mathbf{x}_{\text{d}}(t)_{\text{in}} = \mathbf{x}_{\text{in}} \tag{9.84}$$

Thus, the approach to investigation of a measure of system stability under weak intensity of perturbations changes qualitatively depending on whether or not a trajectory $\mathbf{x}_{\text{d}}(t)$ starting at $\mathbf{x}_{\text{in}}$ reaches the boundary F of the region D_{cr} in a finite time.

If in a finite time a trajectory $\mathbf{x}_{\text{d}}(t)$ defined by equation (9.84) reaches the boundary F of D_{cr} (Fig. 9.12*a*), then, as $\|\mathbf{G}(\mathbf{x})\| \to 0$, the mathematical expectation of the time the system remains in D_{cr} for a given initial state $\mathbf{x}_{\text{in}}$ is defined by the instant t_F when the trajectory $\mathbf{x}_{\text{d}}(t)$ reaches the boundary F, i.e.,

$$\lim_{\|\mathbf{G}(\mathbf{x}(t))\| \to 0} \tau(\mathbf{x}_{\text{in}}) = t_{\text{F}} \tag{9.85}$$

In this case with probability tending to unity the point of system's escape to the boundary F will lie in a small neighborhood of point $\mathbf{x}_{\text{d}}(t_F)$.

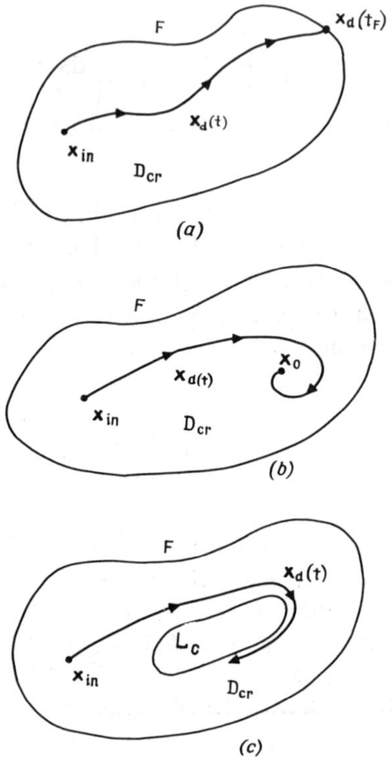

Fig. 9.12

The situation changes when $\mathbf{x}_d(t)$ fails to reach F in a finite time. For example, $\mathbf{x}_d(t)$ can tend to some attraction center $\mathbf{x}_0$ within D_{cr} (Fig. 9.12*b*) or approach a limit cycle $\mathbf{x}_{lc}(t)$ which passes through a set of points L_c entirely belonging to D_{cr} (Fig. 9.12*c*).

Let us assume that for all points $\mathbf{x}$ belonging to the boundary F the scalar (dot) product of $\mathbf{a}(\mathbf{x})$ and the outer normal $\mathbf{n}(\mathbf{x})$ to the boundary F is negative, i.e. $\mathbf{a}(\mathbf{x})\cdot\mathbf{n}(\mathbf{x}) < 0$, in other words assume that for $\mathbf{x}_{in} \in D_{cr}$ the trajectories $\mathbf{x}_d(t)$ cannot escape D_{cr}. This condition implies that no loss of control occurs in the absence of perturbations. However in this case too when perturbations are brought into the picture the trajectory of a random process $\mathbf{X}(t)$ starting at $\mathbf{x}_{in} \in D_{cr}$ at $\|\mathbf{G}(\mathbf{x})\| \neq 0$ escapes D_{cr} in a finite time with probability unity. The point of escape of $\mathbf{X}(t)$ from D_{cr} and the time taken to reach the boundary F depend on the form of the field $\mathbf{a}(\mathbf{x})$ in the entire region D_{cr} rather than on the trajectory $\mathbf{x}_d(t)$.

Consider the case when D_{cr} is a domain of attraction for $\mathbf{x}_0$ and the unperturbed system is asymptotically stable, i.e., any trajectory $\mathbf{x}_d(t)$ starting at $\mathbf{x}_{in} \in D_{cr}$ tends to the equilibrium position $\mathbf{x}_0$ as $t \to \infty$, remaining in D_{cr}.

Following the terminology of Wentzel and Freidlin [9.10], we call the function

$$V(\mathbf{x}_0, \mathbf{x}_f) = \inf\{S[\mathbf{x}(t)] : \mathbf{x}(t) \in C_{0T}(R^N); \quad \mathbf{x}(0) = \mathbf{x}_0, \quad \mathbf{x}(T) = \mathbf{x}_f\} \qquad (9.86)$$

the quasipotential of x_f with respect to x_0. Here $C_{0T}(R^N)$ is the space of functions continuous on $[0, T]$ with the range in R^N. The time interval T is not fixed for a trajectory $\mathbf{x}(t)$. Since $S[\mathbf{x}(t)] \geqslant 0$, $V(\mathbf{x}_0, \mathbf{x}_f) \geqslant 0$. Also, $V(\mathbf{x}_0, \mathbf{x}_0) = 0$ and the function $V(\mathbf{x}_0, \mathbf{x}_f)$ is continuous in $\mathbf{x}_f$. The same properties may be found in the potentials of electric and other fields. Moreover, $V(\mathbf{x}_0, \mathbf{x}_f)$ is a measure of difficulty involved in the system transition from $\mathbf{x}_0$ to $\mathbf{x}_f$ [9.10]. This explains the choice of quasipotential for the name.

Suppose there exists a sole point $\mathbf{x}_* \in F$ for which the quasipotential $V(\mathbf{x}_0, \mathbf{x}_f)$ is a minimum over all points $\mathbf{x}_f$ belonging to F, i.e.,

$$V(\mathbf{x}_0, \mathbf{x}_*) = \min_{\mathbf{x}_f \in F} V(\mathbf{x}_0, \mathbf{x}_f) \qquad (9.87)$$

For small values of $\|\mathbf{G}(x)\|$, the mathematical expectation of the time of the first escape of $X(t)$ to the boundary F, i.e., the expectation of the time to failure for the system is [9.10]

$$\tau_1(\mathbf{x}_{in}) = t_0 \exp\{V(\mathbf{x}_0, \mathbf{x}_*)\} \qquad (9.88)$$

In addition, as $\|\mathbf{G}(\mathbf{x})\| \to 0$, the point of the first escape of $\mathbf{X}(t)$ to the boundary F lies with unit probability in a small vicinity of $\mathbf{x}_*$, i.e., for any scalar $\delta > 0$

$$\lim_{\|\mathbf{G}(\mathbf{x})\| \to 0} P_{\mathbf{X}(t)}\{\varrho(\mathbf{X}(t_F), \mathbf{x}_*) < \delta\} = 1 \qquad (9.89)$$

Here, $\mathbf{X}(t_F)$ represents the coordinates of the first escape of the random process $\mathbf{X}(t)$ to the boundary F at time t_F, $\varrho(\mathbf{x}, \mathbf{y}) = \left(\sum_{i=1}^{N} (x_i - y_i)^2\right)^{1/2}$ is the distance between $\mathbf{x}$ and $\mathbf{y}$ in the metric space R^N.

The proof of (9.88, 89) (see [9.10]) is based on the fact that if $\mathbf{x}_0$ is a stable equilibrium position of the system (9.84), D_{cr} is the domain of attraction for $\mathbf{x}_0$, and $\mathbf{a}(\mathbf{x})\cdot\mathbf{n}(\mathbf{x}) < 0$ for $\mathbf{x} \in F$, then, as $\|\mathbf{G}(\mathbf{x})\| \to 0$, before reaching the boundary F the trajectory of the Markov random process $\mathbf{X}(t)$ with probability tending to unity appears in a small neighborhood of $\mathbf{x}_0$. Therefore, it is sufficient to study the escape from D_{cr} of trajectories starting in a small neighborhood of $\mathbf{x}_0$.

Thus in agreement with (9.88, 89), see [9.10], the value of $V(\mathbf{x}_0, \mathbf{x}_*)$ defines the time of system operation to the first failure and $\mathbf{x}_*$ the most probable point of system escape to the boundary of D_{cr} under small intensity of perturbations acting on the system.

In the general case search of a form of quasipotential $V(\mathbf{x}_0, \mathbf{x}_*)$ is a formidable problem. However, under some rather substantial assumptions (special form of the right side), this problem is amenable to analytical treatment. We assume that the vector of drag coefficients has the form

$$\mathbf{a}(\mathbf{x}) = -\nabla U(\mathbf{x}) + \mathbf{l}(\mathbf{x}) \tag{9.90}$$

where $U(\mathbf{x}_0) = 0$ and for $\mathbf{x}$ in D_{cr} $U(\mathbf{x}) \geqslant 0$; $\mathbf{l}(\mathbf{x})$ is orthogonal with the gradient of $U(\mathbf{x})$ at every point, i.e. $\nabla U(\mathbf{x}) \cdot \mathbf{l}(\mathbf{x}) = 0$ for all $\mathbf{x} \in D_{\mathrm{cr}}$. Also it is assumed that the intensity matrix of perturbations $\mathbf{G}(\mathbf{x})$ in (9.82) is independent of $\mathbf{x}$ and equals

$$\mathbf{G}(\mathbf{x}) = g_0 \mathbf{I}$$

Under these assumptions (9.82) simplifies to

$$S[\mathbf{x}(t)] = \frac{1}{2g_0} \int_0^T \left(\frac{\mathrm{d}\mathbf{x}(t)}{\mathrm{d}t} + \nabla U(\mathbf{x}(t)) - \mathbf{l}(\mathbf{x}(t)) \right)^2 \mathrm{d}t \tag{9.91}$$

and the quasipotential reduces to the form

$$\mathbf{V}(\mathbf{x}_0, \mathbf{x}_{\mathrm{f}}) = \inf \{ S[\mathbf{x}(t)] \} = \frac{2}{g_0} U(\mathbf{x}_{\mathrm{f}}) \tag{9.92}$$

By way of example, consider a nonlinear system at $N = 1$ assuming that

$$\frac{\mathrm{d}x}{\mathrm{d}t} = a(x) + \sqrt{g_0}\, v(t), \quad a(x_0) = 0 \tag{9.93}$$

$$a(x) > 0, \quad \text{for} \quad x < x_0$$

$$a(x) < 0, \quad \text{for} \quad x > x_0$$

If we denote

$$U(x) = -\int_{x_0}^{x} a(\eta)\, \mathrm{d}\eta$$

then $a(x) = -\nabla U(x) = -\mathrm{d}U(x)/\mathrm{d}x$. It follows that in the one-dimensional case, for any nonlinear system having one equilibrium point, the drag coefficient may be represented in the form of (9.90). The quasipotential of $\mathbf{x}_{\mathrm{f}}$ in this case is

$$V(\mathbf{x}_0, \mathbf{x}_f) = -\frac{2}{g_0}\int_{x_0}^{x_f} a(\eta)\,\mathrm{d}\eta \geqslant 0 \tag{9.94}$$

Assuming that the region D_{cr} is a line segment $[x_-, x_+]$ containing point $\mathbf{x}_0$ and letting for definiteness that $U(x_+) < U(x_-)$, we obtain for the expectation of the time to the first system failure

$$\tau_1(x_0) = t_0 \exp\left\{-\frac{2}{g_0}\int_{x_0}^{x_+} a(\eta)\,\mathrm{d}\eta\right\}$$

A second example may be a linear stochastic system with constant coefficients,

$$\mathbf{x} = \mathbf{B}\mathbf{x} + \sqrt{g_0}\,\mathbf{I}\mathbf{v}(t)$$

Suppose that the matrix $\mathbf{B}$ of system coefficients is normal, i.e.,

$$\mathbf{B}\mathbf{B}^{\mathrm{T}} = \mathbf{B}^{\mathrm{T}}\mathbf{B}$$

and the symmetric matrix $\mathbf{B} + \mathbf{B}^{\mathrm{T}}$ is negative definite. In this case the origin $\mathbf{x}_0 = 0$ is an asymptotically stable equilibrium position. Under the above assumptions the vector field $\mathbf{a}(\mathbf{x}) = \mathbf{B}\mathbf{x}$ may be given as

$$\mathbf{a}(\mathbf{x}) = \mathbf{B}\mathbf{x} = -\nabla\left(-\frac{\mathbf{B} + \mathbf{B}^{\mathrm{T}}}{4}\,\mathbf{x}\cdot\mathbf{x}\right) + \frac{\mathbf{B} - \mathbf{B}^{\mathrm{T}}}{2}\,\mathbf{x}$$

Thus the quasipotential is

$$V(0, \mathbf{x}_f) = -\frac{((\mathbf{B} + \mathbf{B}^{\mathrm{T}})\mathbf{x}_f, \mathbf{x}_f)}{2g_0} \tag{9.95}$$

As a third example, we consider the case where the vector field $\mathbf{a}(\mathbf{x})$ possesses cylindrical or spherical symmetry (generalization for a case of $N > 3$ is straightforward). If the projection of $\mathbf{a}(\mathbf{x})$ onto $\mathbf{e}_r$ depends on r only and is expressed through the derivative of a sign-invariant function, then $\mathbf{a}(\mathbf{x})$ may be represented in the form (9.90). For instance, letting $N = 2$ and

$$\left.\begin{aligned} \mathbf{a}(\mathbf{x}) &= a_r(r)\mathbf{e}_r + a_\varphi(r, \varphi)\mathbf{e}_\varphi \\ a_r(r) &= -\nabla\psi(r) = -\mathrm{d}\psi(r)/\mathrm{d}r \\ \psi(0) &= 0, \quad \psi(r) \geqslant 0 \quad \text{for} \quad r > 0 \end{aligned}\right\} \tag{9.96}$$

we obtain the quasipotential in the form

$$V(0, \mathbf{x}_f) = \frac{2}{g_0}\,\psi(|\mathbf{x}_f|)$$

The expressions (9.96) correspond to the following system of differential equations in x_1, x_2:

$$\frac{dx_1}{dt} = a_r(r)\cos\varphi - a_\varphi(r, \varphi)\sin\varphi + \sqrt{g_0}\, v_1(t)$$

$$\frac{dx_2}{dt} = a_r(r)\sin\varphi + a_\varphi(r, \varphi)\cos\varphi + \sqrt{g_0}\, v_2(t)$$

where

$$r = \sqrt{x_1^2 + x_2^2}, \quad \cos\varphi = x_1/r, \quad \sin\varphi = x_2/r$$

Specifically, letting

$$\psi(r) = r^2/2, \quad a_r(r) = -r, \quad \text{and} \quad a_\varphi(r, \varphi) = r$$

we get

$$\left.\begin{aligned} dx_1/dt &= -x_1 - x_2 + \sqrt{g_0}\, v_1(t) \\ dx_2/dt &= x_1 - x_2 + \sqrt{g_0}\, v_2(t) \end{aligned}\right\} \tag{9.97}$$

The quasipotential for this system is

$$V(0, \mathbf{x}_f) = \mathbf{x}_f^2/g_0 = (x_{1f}^2 + x_{2f}^2)/g_0$$

It is worth noting that this expression may be obtained from (9.95) recognizing that for the system (9.97)

$$\mathbf{B} = \begin{bmatrix} -1 & -1 \\ 1 & -1 \end{bmatrix}, \quad \mathbf{B}^T = \begin{bmatrix} -1 & 1 \\ -1 & -1 \end{bmatrix}, \quad \mathbf{B} + \mathbf{B}^T = \begin{bmatrix} 2 & 0 \\ 0 & 2 \end{bmatrix}$$

When the dependence of the perturbation intensity matrix $\mathbf{G}(\mathbf{x})$ upon the state vector $\mathbf{x}$ is essential, or when the intensity of perturbations active in different coordinates is distinct, the quasipotential $V(\mathbf{x}_0, \mathbf{x}_f)$ denies analytical evaluation. Approximate evaluation may be carried out with a computational procedure based on a search of the lower bound in (9.86) over the class of trajectories $\mathbf{x}_W^m(t)$ being broken lines in the $\mathbf{x}$, t space. The m-th order approximation of the quasipotential $V(\mathbf{x}_0, \mathbf{x}_f)$ has the form

$$V_m(\mathbf{x}_0, \mathbf{x}_f) = \inf\{S[\mathbf{x}_W^m(t)] : \mathbf{x}(t_{in}) = \mathbf{x}_0, \ \mathbf{x}(t_f) = \mathbf{x}_f \,|\, (t_f - t_{in}) = T\} \tag{9.98}$$

If we reduce the time step $\varepsilon_T = (t_f - t_{in})/m$ by increasing the order of approximation m, the broken line $\mathbf{x}_{W*}^m(t)$ will tend to the trajectory $\mathbf{x}_*(t)$ which supplies the lower bound to the functional (9.86) and, consequently,

$$\lim_{m\to\infty} V_m(\mathbf{x}_0, \mathbf{x}_f) = V(\mathbf{x}_0, \mathbf{x}_f) \tag{9.99}$$

In the simplest case of $m = 1$,

$$\mathbf{x}_W^1(t) = \mathbf{x}_0 + \frac{(\mathbf{x}_f - \mathbf{x}_0)}{T}\, t \tag{9.100}$$

The approximation with $m = 1$ is of course rather rough, however it allows analytical determination of $V_1(\mathbf{x}_0, \mathbf{x}_f)$ and a qualitative estimate of the effects of various factors and system parameters on the probability of loss of control.

The action functional (9.82) takes at $m = 1$ the following form

$$S[\mathbf{x}_W^1(t)] = \frac{1}{2}\int_0^T \left[\frac{\mathbf{x}_f - \mathbf{x}_0}{T} - \mathbf{a}(\mathbf{x}_W^1(t))\right]^{\mathrm{T}} \mathbf{G}^{-1}(\mathbf{x}_W^1(t))$$

$$\times\ [(\mathbf{x}_f - \mathbf{x}_0)/T - \mathbf{a}(\mathbf{x}_W^1(t))]\,\mathrm{d}t \qquad (9.101)$$

To determine $V_1(\mathbf{x}_0, \mathbf{x}_f)$ we need to minimize (9.101) with respect to T at fixed $\mathbf{x}_0$ and $\mathbf{x}_f$. We provide an insight into the procedure with two illustrative examples.

Let $G(x) = g_0$ and $a(x) = -a_p|\mathbf{x}|^p \operatorname{sgn} x$, then

$$S[\mathbf{x}_W^1(t)] = \frac{1}{2g_0}\int_0^T \left[\frac{x_f}{T} + a_p\frac{x_f^p t^p}{T^p}\right]^2 \mathrm{d}t$$

$$= \frac{1}{2g_0}\left\{\frac{x_f^2}{T} + 2a_p\frac{x_f^{p+1}}{p+1} + a_p^2\frac{x_f^{2p}}{2p+1}T\right\}$$

A minimum with respect to T is then achieved at

$$-\frac{x_f^2}{T^2} + a_p^2\frac{x_f^2}{2p+1} = 0 \quad \text{and} \quad T = \frac{\sqrt{2p+1}}{a_p x_f^{p-1}}$$

Thus,

$$V_1(0, x_f) = \min_T S[x_W^1(t)]$$

$$= \frac{a_p x_f^{p+1}}{g_0}\left\{\frac{1}{2p+1} + \frac{1}{p+1}\right\} \qquad (9.102)$$

In agreement with (9.94) the quasipotential of x_f at $a(x) = -a_p|x|^p \operatorname{sgn} x$ is

$$V(0, x_f) = \frac{a_p x_f^{p+1}}{g_0}\left\{\frac{2}{p+1}\right\} = V_0\left\{\frac{2}{p+1}\right\} \qquad (9.103)$$

The expressions (9.102) and (9.103) differ solely by a p-dependent numerical factor. The results of calculations by (9.102, 103) performed with various p are summarized in Table 9.1.

Table 9.1

p	0	0.5	1	2	3	4	5	9
$V_1(0, x_f)/V_0$	2.0	1.3737	1.0774	0.7805	0.6279	0.5333	0.4683	0.3292
$V(0, x_f)/V_0$	2.0	1.3333	1.0	0.6666	0.5	0.4	0.3333	0.2
$\dfrac{V_1(0, x_f)}{V(0, x_f)}$	1.0	1.0291	1.0774	1.1708	1.2558	1.3333	1.4051	1.646

To illustrate the convergence of $V_m(x_0, x_f)$ with increasing m, we take up a case of $m > 1$. For $a(x) = -a_p x^p \operatorname{sgn} x$, the action functional is

$$S[x_W^m(t)] = \frac{1}{2g_0}\left\{\sum_{j=0}^{m-1}\left[\frac{a_p^2}{2p+1}\,\frac{x_{j+1}^{2p+1} - x_j^{2p+1}}{x_{j+1} - x_j}\,\frac{T}{m}\right]\right.$$
$$\left. + \frac{2a_p}{p+1}\sum_{j=0}^{m-1}(x_{j+1}^{p+1} - x_j^{p+1}) + \frac{m}{T}\sum_{j=0}^{m-1}(x_{j+1} - x_j)^2\right\}$$

The minimum with respect to T for this expression is attained at

$$T = T_* = \frac{m}{a_p}\left[(2p+1)\frac{\sum_{j=0}^{m-1}(x_{j+1} - x_j)^2}{\sum_{j=0}^{m-1}[(x_{j+1}^{2p+1} - x_j^{2p+1})/(x_{j+1} - x_j)]}\right]^{1/2}$$

Substituting T_* in $S[x_W^m(t)]$ and introducing the dimensionless coordinates $\eta_j = x_j/x_f$, we have by virtue of (9.98)

$$V_m(0, x_f) = a_p|x_f|^{p+1}/g_0$$
$$\times \min_{\eta_1 \ldots \eta_{m-1}}\left\{\sqrt{\left[\frac{\sum_{i=0}^{m-1}(\eta_{i+1} - \eta_i)^2}{(2p+1)}\right]\sum_{j=0}^{m-1}\frac{(\eta_{j+1}^{2p+1} - \eta_j^{2p+1})}{(\eta_{j+1} - \eta_j)}} + \frac{1}{p+1}\right\}$$

where $\eta_0 = 0$, $\eta_m = 1$, and $m \geqslant 2$.

The search for minimum on the right-hand side of this expression was carried out by the method of deformed polygon. The results are summarized in Table 9.2.

Table 9.2

p	$m = 2$	$m = 3$	$m = 4$	$m = 5$	$m = 6$	$m = 7$
1	1.0346	1.0236	1.0135	1.0097	1.0075	1.0061
2	0.7410	0.7201	0.7077	0.6998	0.6903	0.6877
3	0.5941	0.5740	0.5608	0.5515	0.5446	0.5395
4	0.5034	0.4853	0.4727	0.4633	0.4562	0.4507

Figure 9.13 shows η_j values which define the trajectory $x^m_{W^*}(t)$ at $m = 8$. The broken lines labeled *1-4* are constructed respectively for $p = 1$, 2, 3, and 4.

As the second example, we consider a linear system of the type (9.97). For $S[\mathbf{x}^1_W(t)]$ with $\mathbf{G}^{-1}(\mathbf{x}^1_W(t)) = g_0^{-1}\mathbf{I}$ and $\mathbf{a}(\mathbf{x}^1_W(t)) = \mathbf{B}\mathbf{x}^1_W(t)$ we have

$$S[\mathbf{x}^1_W(t)] = \frac{1}{2g_0}\int_0^T \left\{\left[\frac{x_{f1} + t(x_{f1} + x_{f2})}{T}\right]^2 + \left[\frac{x_{f2} + t(x_{f2} - x_{f1})}{T}\right]^2\right\} dt$$

$$= \frac{1}{2g_0}(x_{f1}^2 + x_{f2}^2)(1/T + 1 + 2T/3)$$

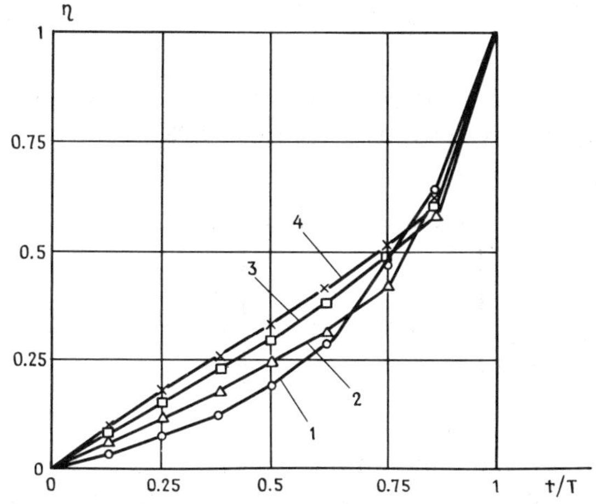

Fig. 9.13

This expression is a minimum at $T = (3/2)^{1/2}$. Hence, for the first order approximation of the quasipotential we get

$$V_1(0, \mathbf{x}_f) = (x_{f1}^2 + x_{f2}^2)(1 + 2\sqrt{2/3})/2g_0$$

It follows that the approximate expression for the quasipotential differs from the exact expression by merely a numerical multiplier of $\cong 1.316$.

For $m > 1$, the expression for the action functional of the system (9.97) has the form

$$S[\mathbf{x}_W^m(t)] = \frac{1}{2g_0}\left\{\left(\frac{m}{T}\right)\sum_{j=0}^{m-1}[(x_{1,j+1} - x_{1,j})^2 + (x_{2,j+1} - x_{2,j})^2]\right.$$
$$+ \left(\frac{T}{m}\right)\sum_{j=0}^{m-1}\left[\frac{2}{3}(x_{1,j+1} - x_{1,j})^2 + \frac{2}{3}(x_{2,j+1} - x_{2,j})^2\right.$$
$$\left. + 2x_{1,j}x_{1,j+1} + 2x_{2,j}x_{2,j+1}\right] + x_{1,f}^2 + x_{2,f}^2$$
$$\left. + \sum_{j=0}^{m-1}[2x_{1,j+1}x_{2,j} - 2x_{2,j+1}x_{1,j}]\right\}$$

This expression is a minimum with respect to T at

$$T = T_* = m\left[\sum_{j=0}^{m-1}(x_{1,j+1} - x_{1,j})^2 + (x_{2,j+1} - x_{2,j})^2\right]^{1/2}$$
$$\times \left[\sum_{j=0}^{m-1}\frac{2}{3}(x_{1,j+1} - x_{1,j})^2 + \frac{2}{3}(x_{2,j+1} - x_{2,j})^2\right.$$
$$\left. + 2x_{1,j}x_{1,j+1} + 2x_{2,j}x_{2,j+1}\right]^{-1/2}$$

The broken lines $\mathbf{x}_{W_*}^m(t)$ computed for $m = 4$ are plotted in Fig. 9.14. Lines labeled *1* and *2* pass, respectively, through points $\alpha_j = x_{1,j}/x_{1f}$ and $\beta_j = x_{2,j}/x_{2f}$ for $j = 0, \ldots, 4$.

The approximate method of evaluation of the quasipotential involving the evaluation of m-th order approximations, $V_m(0, \mathbf{x}_f)$, may be used for analysis of stability and processes leading to loss of control in nonlinear

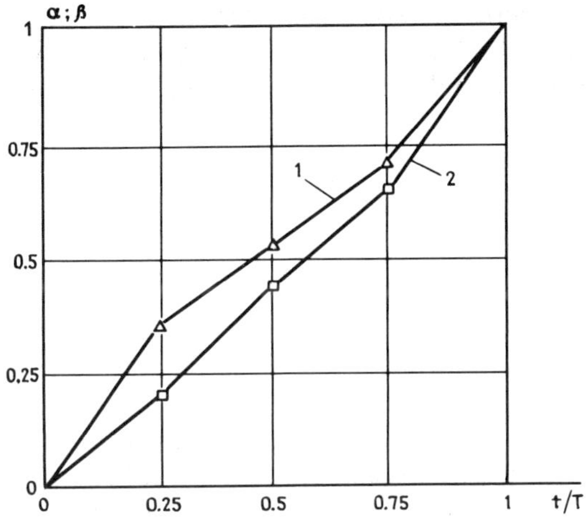

Fig. 9.14

dynamic systems under weak perturbations. By way of example we consider the equations of motion of manipulators given [9.13] in the general form as

$$\mathbf{B}(\mathbf{q})\ddot{\mathbf{q}} = \mathbf{c}(\mathbf{q}, \dot{\mathbf{q}}) + \mathbf{Q}(\mathbf{q}) + \mathbf{w}(t) \tag{9.104}$$

where $\mathbf{B}(\mathbf{q})$ is a symmetric $N \times N$ matrix whose elements $B_{i,j}$ depend on the vector of generalized coordinates of the system $\mathbf{q} = \{q_1, \ldots, q_N\}$; $\mathbf{c}(\mathbf{q}, \dot{\mathbf{q}}) = \{c_1, \ldots, c_N\}$ is the vector of moments of external forces applied to the system; $\mathbf{Q}(\mathbf{q}) = \{Q_1, \ldots, Q_N\}$ is the vector of moments provided by system drive motors; and $\mathbf{w}(t)$ is the vector of perturbing moments expressable through independent Gaussian processes of a white noise.

Multiplying both sides of (9.104) by $\mathbf{B}^{-1}(\mathbf{q})$ and introducing new coordinates $X_i = \dot{q}_i$ and $X_{N+i} = q_i$ $(i = 1, 2, \ldots, N)$, we reduce (9.104) to a canonical form. If in the resultant system of equations the dimensions of the vector $\mathbf{w}(t)$ is N, perturbations excite only N coordinates $X_1, \ldots, X_N$, the other N coordinates $X_{N+1}, \ldots, X_{2N}$ remain unperturbed. The principal diagonal of $\mathbf{B}^{-1}(\mathbf{q})$ is occupied by the inverse effective moments of inertia of the system $J_i^{-1}(\mathbf{q})$. When these parameters are known, some systems are amenable [9.13] to model representations in the form of a set of one degree-of-freedom mechanisms having the equations of motion

$$J_i(q_i)\ddot{q}_i = c_i(\dot{q}_i, q_i) + Q_i(q_i) + w_i(t), \; i = 1, \ldots, N \tag{9.105}$$

subject to constraints imposed on the feasibility domains for q_i.

When the intensity of perturbations is low, a law of control will be governed by a link i with the least quasipotential of escape to the boundary of the region $D_{cr} = \{[q_1^{(-)}, q_1^{(+)}], \ldots, [q_N^{(-)}, q_N^{(+)}]\}$. We assign to this link the number $i = 1$ and assume that its motion is described by the equations

$$\dot{X}_1 = -\frac{B}{m_a}\operatorname{sgn} X_1 - \frac{k_s}{m_a} X_{N+1} + u_1(X_{N+1}) + \frac{\sigma_0}{m_a} v_1(t) \tag{9.106}$$

$$\dot{X}_{N+1} = X_1 \tag{9.107}$$

where $u_1(X_{N+1}) = -\gamma_1 X_{N+1}/m_a$ is the control in the first coordinate.

Suppose that the feasible domain of the coordinate X_{N+1} is given by $|X_{N+1}| \leqslant \Delta$. For $\sigma_0 = 0$, the equilibrium point of the system (9.106, 107) has the coordinates $x_1 = 0$, $x_{N+1} = 0$. For a given $X_1(t)$, the form of $X_{N+1}(t)$ is uniquely determined by equation (9.107). Assuming $X_1(t) = x_W^m(t)$, for the first order approximation of $m = 1$ we have

$$\begin{aligned} x_1(t) &= x_{1f} t/T \\ x_{N+1}(t) &= x_{1f} t^2/2T \end{aligned} \tag{9.108}$$

and the action functional is

$$S[x_W^1(t)] = \frac{1}{2g_0}\int_0^T \left[x_{1f}/T + \frac{B}{m_a}\operatorname{sgn} x_1(t) + \left(\frac{k_s}{m_a} + \frac{\gamma_1}{m_a}\right) x_{N+1}(t)\right]^2 dt \tag{9.109}$$

In considering the escape of the system to the boundary of D_{cr} we assume that $x_{N+1}(T) = \pm\Delta$. Then from (9.108) it follows that $x_{1f} = \pm 2\Delta/T$. For $x_{N+1}(T) = +\Delta$, by virtue of (9.108) we have from (9.107)

$$S[x_W^1(t)] = \frac{1}{2g_0}\left\{\frac{4\Delta^2}{T^3} + \left(\frac{B^2}{m_a^2} + \frac{\alpha^2\Delta^2}{5} + \frac{2\alpha B\Delta}{3m_a}\right) T + (4\alpha\Delta^2/3 + 4\Delta B/m_a)/T\right\} \tag{9.110}$$

where $\alpha = (k_s + \gamma_1)/m_a$ and $g_0 = \sigma_0^2/m_a^2$.

This expression is a minimum provided that

$$T^4 - (a_2/a_1)T^2 - 12\Delta^2/a_1 = 0 \tag{9.111}$$

where

$$a_1 = B^2/m_a^2 + \alpha^2\Delta^2/5 + 2\alpha B\Delta/3m_a$$

$$a_2 = 4\alpha\Delta^2/3 + 4\Delta B/m_a$$

The biquadratic equation (9.111) has one real positive root

$$T_M = \sqrt{a_2/2a_1 + \sqrt{a_2^2/4a_1^2 + 12\Delta^2/a_1}} \tag{9.112}$$

which on substitution into (9.110) yields

$$V_1(0, \Delta) = \frac{1}{2g_0}\left\{\frac{4\Delta^2}{T_M^3} + a_1 T_M + a_2 T_M^{-1}\right\} \tag{9.113}$$

In agreement with (9.88, 99) the logarithm of $\tau_1^{(m)}(0)$ (the m-th order approximation for the expectation of time to loss of control) is given by the approximate value of the quasipotential, namely, $\ln \tau_1^{(m)}(0) = V_m(0, \Delta)$.

Figure 9.15 shows the plots of $\ln \tau_1^{(1)}(0)$ computed as functions of perturbation intensity g_0 for three values of Δ, $\alpha = 1$ and $B/m_a = 0.1$ (in equation (9.110)). The expectation of the time to loss of control is seen to decrease for greater magnitudes of perturbing inputs. The values of $\tau_1^{(1)}$ also reduce where greater accuracy is required, i.e. with decrease in Δ.

The value of $\tau_1^{(1)}(0)$ may be increased by increasing the gain factor γ_1 or the coefficient of friction B. The plots illustrating $\tau_1^{(1)}(0)$ as a function of α and B/m_a computed with $g_0 = 0.01$ are depicted in Figs. 9.16 and 9.17.

This evidence indicates that the procedure outlined above may be used to estimate the effects of various factors on expectation of time to loss of control subject to the given constraints on the feasible region of the vector of state.

Consider the case when the set of equilibrium points has a more complicated form. Assume that in addition to the center of attraction $\mathbf{x}_0$ the system contains a limit cycle L_c associated with the self-oscillation mode $\mathbf{x}_{lc}(t)$. If initially the system is at the state $\mathbf{x}_0$, then the expectation of the time to the self-oscillation mode $\mathbf{x}_{lc}(t)$ is given by (9.88, 87) and the minimum in (9.87) is sought for points $\mathbf{x}_f \in L_c$. It is naturally assumed that the quasipotential of the system's escape from $\mathbf{x}_0$ to the boundary of D_{cr} is higher than the quasipotential of the transition $\mathbf{x}_0 \to L_c$.

Proceeding in a manner similar to (9.86) we define the quasipotential of a point $\mathbf{x}_f$ with respect to the set of points of L_c as follows

$$V(L_c, \mathbf{x}_f) = \inf\{S[\mathbf{x}(t)] : \mathbf{x}(t) \in C_{0T}(R^N);$$

$$\mathbf{x}(0) \in L_c; \quad \mathbf{x}(T) = \mathbf{x}_f \,|\, \mathbf{x}(0)\}$$

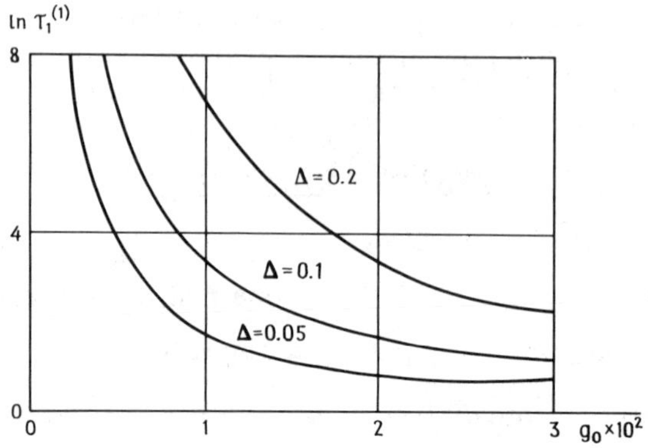

Fig. 9.15

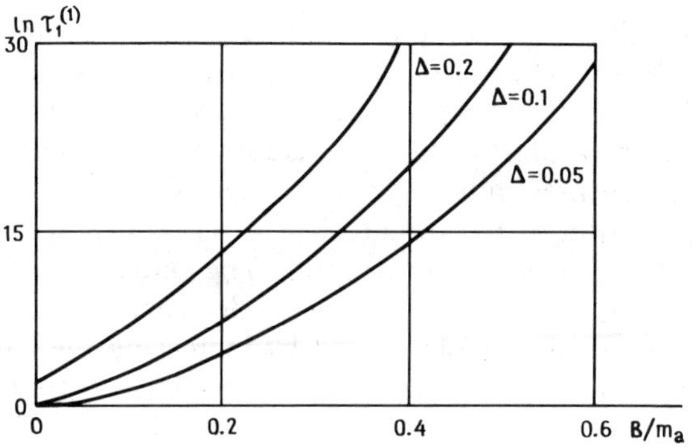

Fig. 9.16

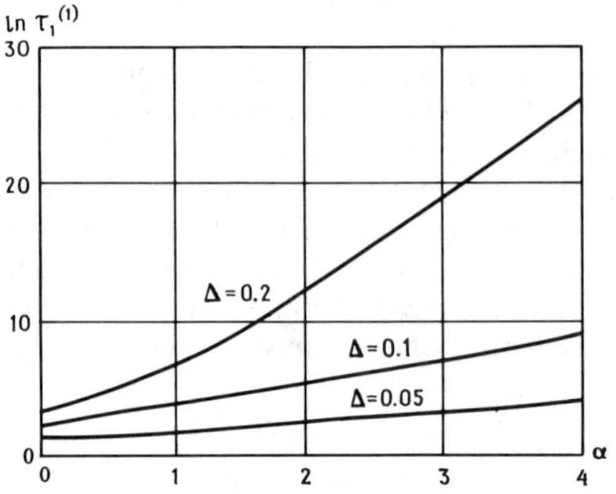

Fig. 9.17

The quasipotential of the system escape from the set L_c to the boundary F of the region D_{cr} is defined by the relation

$$V(L_c, \mathbf{x}_*) = \min_{\mathbf{x}_* \in F} V(L_c, \mathbf{x}_f)$$

This minimum is achieved for some critical trajectory $\mathbf{x}_*(t)$ starting at a point $\mathbf{x}_{L_{c*}} \in L_c$. The time taken by the system to escape from the limit cycle to the boundary F is given by formula similar to (9.88) with one sufficient difference, namely, the exponent in this case should be augmented by a factor of $\tau_{L_c}/\tau_{L_{c*}}$ equal to the inverse of the relative time the system spends in the neighborhood of $\mathbf{x}_{L_{c*}}$. It is worth noting that as the period τ_{L_c} of self-excited oscillations increases, the relative time of system residence in the neighborhood of $\mathbf{x}_{L_{c*}}$ decreases and the characteristic time of the transition $L_c \to F$ increases.

If the set of equilibrium points contains a few limit cycles, say L_{c1} and L_{c2}, then the transitions $L_{c1} \to L_{c2}$ and $L_{c2} \to L_{c1}$ are associated with the change of oscillation modes. The mathematical expectation of the time of these transitions is defined by the relations given above.

In the circumstances with a complex structure of equilibrium points, i.e. in the presence of a few centers of attraction or limit cycles, it would be appropriate to introduce the concept of a *precritical state*, the expectation of excape from which to the boundary F is a minimum.

To prevent the breakdown of control and system's escape beyond the region of feasible values of the state vector additional measures should be taken when the system having one center of attraction moves along a critical trajectory $\mathbf{x}_*(t)$ for which the quasipotential is a minimum (9.87) or when the system having a complicated structure of equilibrium points goes toward a precritical state.

9.5 Constraints on System Coordinates and Speed. The Uncertainty Relation for Diffusion Markov Processes

Some problems of process control need that the system vector of state be monitored subject to the given constraints on system rate. In this case the system operates normally subject to

$$\mathbf{x}(t) \subset D_{cr}$$
$$\frac{d\mathbf{x}(t)}{dt} \subset \Omega_{cr} \tag{9.114}$$

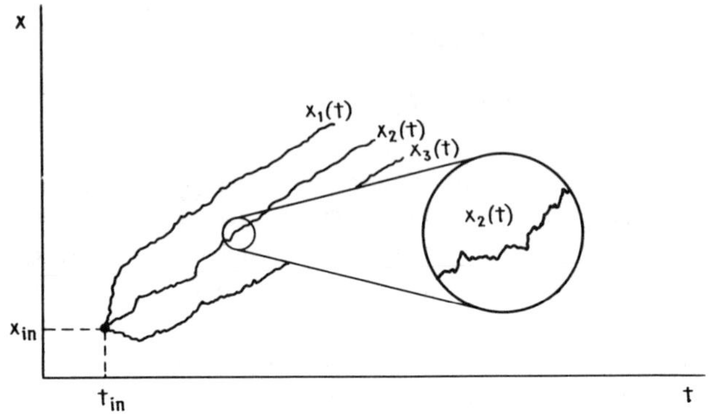

Fig. 9.18

An example may be the control system of a nuclear reactor. In this system, constraints are imposed on admissible deviation of power, n, from the given level, n_g, and on the asymptotic period of the reactor, i.e. the time span over which reactor power varies e times. Practically the monitoring of reactor period boils down to the limitation on the power variation rate $\mathrm{d}n/\mathrm{d}t \leqslant \omega_{cr}$. If at least one of the two conditions

$$(n - n_g)/n_g > \Delta \quad \text{or} \quad \mathrm{d}n/\mathrm{d}t > \omega_{cr}$$

is satisfied, the emergency protection system of the reactor operates and the loss of control, i.e. loss of capacity, occurs. The situation is aggravated by the fact that perturbations of white Gaussian noise type render the trajectory of a random process $x(t)$ is a continuous but not smooth curve with infinitely many break points (Fig. 9.18). Strictly speaking, the trajectory of such a process is not differentiable, with unit probability, at any of its points. Therefore, practical differentiation of such signals is undertaken by a real (time-lag differentiator [9.14]). The equation of motion for this element is

$$\tau_d \frac{\mathrm{d}X_{out}}{\mathrm{d}t} + X_{out} = k_d \frac{\mathrm{d}X_{in}}{\mathrm{d}t} \tag{9.115}$$

where τ_d is the time constant of the differentiating element, k_d is the equivalent gain, and X_{in} and X_{out} are the input and output coordinates of the differentiator.

In operator form, the equations of the time-lag differentiator become

$$\tilde{X}_{out}(s) = \frac{k_d s}{\tau_d s + 1} \tilde{X}_{in}(s)$$

In (9.115) the output signal X_{out} provides a smoothed waveform of the derivative of the input signal. From the operator expression for $\tilde{X}_{\text{out}}(s)$ it follows that the quantity $\omega_{\text{p}} = 1/\tau_{\text{d}}$ defines the bandwidth of the differentiator. For input frequencies within this passband, $\omega < \omega_{\text{p}}$, the output signal coincides, accurate to within a factor k_{d}, with the value of the derivative of the input signal. The choice of τ_{d} is governed by the intensity of perturbations applied to the system, namely, the lower the magnitude of interference, the smaller τ_{d} may be given in (9.115).

We wish to demonstrate that the simultaneous monitoring of the coordinates and speed of the system (9.5) is subject to the inequality

$$\sqrt{\overline{\Delta\mathbf{x}^2}\,\overline{\Delta\mathbf{v}^2}} \gtrsim \text{Tr}\,\{\mathbf{G}(\mathbf{x},\ t)\} \tag{9.116}$$

being the uncertainty relation for diffusion Markov processes. Here, $\sqrt{(\overline{\Delta\mathbf{x}^2})}$ and $\sqrt{(\overline{\Delta\mathbf{v}^2})}$ are the mean square deviations in the coordinates and velocity of the system, and $\text{Tr}\,\{\mathbf{G}(\mathbf{x},\ t)\}$ is the trace of the intensity matrix, i.e. the sum of the diagonal elements of $\mathbf{G}(\mathbf{x},\ t)$.

The physical meaning of (9.116) is as follows. Assume that for the purpose of monitoring of system coordinates and speed we measure the state vector $\mathbf{X}(t_0 + \Delta t)$ at a time $t_0 + \Delta t$ given that the initial state $\mathbf{X}(t_0) = \mathbf{x}_0$ is fixed. The outcome of the measurement at $t_0 + \Delta t$ assumes random values and can be characterized by the distribution of probability density $\varphi(\mathbf{x},\ t_0 + \Delta t)$ which at small Δt is close to a normal distribution law. For $t_0 < t < t_0 + \Delta t$ the state vector of the system $\mathbf{x}(t)$ is not known, i.e. between the measurements there is an uncertainty in the values of system coordinates. This uncertainty will be the smaller, the narrower the width of the distribution $\varphi(\mathbf{x},\ t_0 + \Delta t)$: the uncertainty in the coordinates reduces for smaller Δt. Suppose that the system speed $\mathbf{v}(t_0 + t)$ varies as

$$\mathbf{v}(t_0 + \Delta t) = [\mathbf{X}(t_0 + \Delta t) - \mathbf{X}(t_0)]/\Delta t$$

The values of system velocity $\mathbf{v}(t_0 + \Delta t)$ obtained as the result of measuring $\mathbf{X}(t_0 + \Delta t)$ are also random in character. However, the width of the distribution $\varphi(\mathbf{v},\ t_0 + \Delta t)$ increases for smaller Δt.

Proof of (9.116). If in the stochastic differential equation (9.5) $\sigma_{ik}(\mathbf{x},\ t)$ are bounded continuous functions and $b_i(\mathbf{x},\ t)$ contain a finite number of ordinary discontinuities, then the Markov process $\mathbf{X}(t)$ has continuous trajectories with unit probability and relates to the class of diffusion Markov processes. Such processes are completely described by their local characteristics, namely, by the vector of drag coefficients $\mathbf{a}(\mathbf{x},\ t)$ and the matrix of diffusion coefficients $\mathbf{G}(\mathbf{x},\ t)$, viz.,

$$\mathrm{E}[X_i(t+\Delta t) - X_i(t) \mid \mathbf{X}(t) = \mathbf{x}] = a_i(\mathbf{x}, t)\Delta t + o_1(\Delta t)$$
$$\mathrm{E}[(X_i(t+\Delta t) - X_i(t))(X_j(t+\Delta t) - X_j(t)) \mid \mathbf{X}(t) = \mathbf{x}]$$
$$= G_{ij}(\mathbf{x}, t)\Delta t + o_2(\Delta t) \qquad (9.117)$$

whence

$$\sqrt{(\overline{\Delta \mathbf{x}^2})} \equiv \{\mathrm{E}[\overline{\Delta \mathbf{x}^2}]\}^{1/2} \equiv \left\{\mathrm{E}\left[\sum_{i=1}^{n} (X_i(t+\Delta t) - X_i(t))^2 \mid \mathbf{X}(t) = \mathbf{x}\right]\right\}^{1/2}$$
$$= [\Delta t \,\mathrm{Tr}\,\{\mathbf{G}(\mathbf{x}, t)\}]^{1/2} + o_3(\Delta t) \qquad (9.118)$$

The mean square error arising in velocity measurements is

$$\sqrt{(\overline{\Delta \mathbf{v}^2})} = \sqrt{(\overline{\Delta \mathbf{x}^2})}/\Delta t$$
$$= [\mathrm{Tr}\,\{\mathbf{G}(\mathbf{x}, t)\}/\Delta t]^{1/2} + o_3(\Delta t)/\Delta t \qquad (9.119)$$

Relations (9.118, 119) lead to the inequality (9.116).

We would like to note similarity of the form of (9.116) with the Heizenberg uncertainty relation in quantum mechanics

$$\sqrt{(\overline{\Delta \mathbf{x}_\mathrm{p}^2 \Delta \mathbf{v}_\mathrm{p}^2})} \geqslant \hbar/2m_\mathrm{p} \qquad (9.120)$$

Here, $\sqrt{(\overline{\Delta \mathbf{x}_\mathrm{p}^2})}$ and $\sqrt{(\overline{\Delta \mathbf{v}_\mathrm{p}^2})}$ are uncertainties in the coordinate and velocity of the quantum-mechanical particle, m_p is the particle mass, and $\hbar$ is Planck's constant.

The fact that the expressions (9.116) and (9.120) coincide is explained by the analogy between the equations of Focker-Planck and Schrödinger. We illustrate the existence of such analogy for $N = 1$.

The Focker-Planck equation describes how the probability density $\varphi(\mathbf{x}, t)$ varies in time for a given initial distribution $\varphi(\mathbf{x}_\mathrm{in}, t_\mathrm{in})$. This equation can be derived from the Kolmogorov equation with the aid of the relation

$$\varphi(\mathbf{x}, t) = \int_{-\infty}^{+\infty} f(\mathbf{x}, t; \mathbf{x}_\mathrm{in}, t_\mathrm{in})\varphi(\mathbf{x}_\mathrm{in}, t_\mathrm{in})\,\mathrm{d}\mathbf{x}_\mathrm{in} \qquad (9.121)$$

Multiplying both sides of the Kolmogorov equation (9.12) by $\varphi(x_\mathrm{in}, t_\mathrm{in})$ and integrating with respect to x_in from $-\infty$ to $+\infty$ at $N = 1$, disregarding the time depndence of the coefficients of $a(x, t)$ and $G(x, t)$, we get

$$\frac{\partial \varphi(x, t)}{\partial t} = -\frac{\partial}{\partial x} a(x)\varphi(x, t) + \frac{1}{2}\frac{\partial^2}{\partial x^2} G(x)\varphi(x, t) \qquad (9.122)$$

The Schrödinger equation for a particle with charge e_p and mass m_p moving in a magnetic field of vector potential $B(x)$ and possessing a potential energy $U_p(x)$ will be written as

$$-\frac{\hbar}{i}\frac{\partial\psi}{\partial\tau} = -\frac{1}{2m_{\rm p}}\hbar^2\frac{\partial^2\psi}{\partial x^2} - \frac{e_{\rm p}}{m_{\rm p}c}\frac{\hbar}{i}B(x)\frac{\partial}{\partial x}\psi(x,\tau)$$
$$+\left(\frac{1}{2m_{\rm p}}\frac{e_{\rm p}^2}{c^2}B^2(x) - \frac{1}{2m_{\rm p}}\frac{\hbar}{i}\frac{e_{\rm p}}{c}\frac{\partial B(x)}{\partial x} + U_p(x)\right)\psi(x,\tau) \quad (9.123)$$

where $\psi(x, \tau)$ is the wave function of the particle, c the velocity of light, and $i = \sqrt{-1}$.

Letting in (9.123)

$$U_p(x) = \frac{m_{\rm p}}{2}a^2(x) + \frac{1}{2}\hbar\frac{\partial}{\partial x}a(x)$$

$$B(x) = \frac{m_{\rm p}c}{ie_{\rm p}}a(x), \quad i\tau = t, \quad \hbar/m = g_0$$

we arrive at the Focker-Planck equation (9.122) with $G(x) = g_0$ for the function $\psi(x, t)$.

The expression (9.116) may be used for estimation of the time constant $\tau_{\rm d}$ of the time-lag differentiating unit. Assuming that the admissible uncertainty in velocity values is $\eta\omega_{\rm cr}$ yields

$$\tau_{\rm d} \geqslant {\rm Tr}\,\{\mathbf{G}(\mathbf{x}, t)\}/(\eta\omega_{\rm cr})^2$$

where $\omega_{\rm cr}$ is the characteristic dimension of the region $\Omega_{\rm cr}$, and η is dimensionless parameter less than unity.

To close this topic we emphasize that the form of (9.116) suggests that it is impossible to simultaneously and accurately monitor the values of the coordinates and velocity of a system subjected to perturbations of a white Gaussian noise type. The degree of uncertainty is given as the trace of the intensity matrix of perturbations acting on the system.

References

9.1 I. G. Malkin: *Teoriya ustoichivosti dvizheniya* (Theory of motion stability). Nauka, Moscow 1966.

9.2 Yu. I. Gertner: *Teor. Veroyatn. i eyo Primenen.* **21**(1), 95-106 (1976).

9.3 I. E. Kazakov and S. V. Malchikov: *Analiz stokhasticheskikh sistem v prostranstve sostoyanii* (Analysis of stochastic systems in state space). Nauka, Moscow 1983.

9.4 N. V. Nikitin: *Zhurn. Vychisl. Mat. Mat. Fiz.* **26**(5), 703-710 (1986).

9.5 V. I. Tikhonov and M. A. Mironov: *Markovskie protsessy* (Markov processes). Sovetskoe Radio, Moscow 1977.

9.6 Yu. I. Astapov and M. A. Medvedev: *Staticheskaya teoriya sistem avtomaticheskogo regulirovaniya i upravleniya* (Statistical theory of automatic regulation and control systems). Nauka, Moscow 1982.

9.7 A. D. Egorov, P. I. Sobolevskii, and L. A. Yanovich: *Priblizhennye metody vychisleniya kontinualnykh integralov* (Approximate evaluation of path integrals). Nauka i Tekhnika, Minsk 1986.

9.8 N. V. Nikitin: in *Metody proyectirovaniya slozhnykh sistem* (Methods of complex system design). Energoatomizdat, Moscow 1985, pp. 65-71.

9.9 S. V. Fomin: *Nauchn. Dokl. Vyssh. Shkoly* (*Fiz. Mat.*) No 2, 83-85 (1958).

9.10 A. D. Wentzel and M. I. Freidlin: *Fluktuatsii v dinamicheskikh sistemakh pod deistviyem malykh sluchainykh vozmushchenii* (Fluctuations in dynamic systems under small random disturbances). Nauka, Moscow 1979.

9.11 Yu. I. Topcheyev and N. V. Nikitin: in *Metody rascheta gibkikh avtomatizirovannykh proizvodstv* (Design of flexible automated production systems). VINITI, Moscow 1984. pp. 295-314.

9.12 S. I. Blinnikov and N. V. Nikitin: *Approximatsii kontinualnykh integralov i spectry kvantovykh sistem* (Approximations of path integrals and spectra of quantum systems), Preprint No 111. ITEF, Moscow 1982.

9.13 N. V. Nikitin: *Preobrazovateli i istochniki tepla v avtonomnykh energeticheskikh ustanovkakh* (Heat sources and converters in self-supported power units). MIFI Press, Moscow 1984.

9.14 E. P. Popov, A. F. Vereshchagin and S. A. Zenkevich: *Manipulyatsionnye roboty. Dinamika i algoritmy* (Manipulator robots: dynamics and control algorithms). Nauka, Moscow 1978.

9.15 I. Ya. Emelyanov, A. E. Efanov, and L. V. Konstantinov: *Nauchno-tekhnicheskiye osnovy upravleniya yadernymi reaktorami* (Theoretical and engineering fundamentals of nuclear reactor control). Energoizdat, Moscow 1981.

10 Analysis and Synthesis of Discrete Control Systems Based on Multidimensional z-Transforms

A. D. Modyaev and *A. D. Averina*

The mathematical formalism of z-transforms is widely exploited in design of linear pulse and discrete-continuous systems of automatic control. This method is helpful in deriving transfer functions, calculating transition processes, evaluating performance indices, investigating stability conditions, synthesizing compensatory elements, and performing mathematical and environmental simulation experiments. The principal advantage of the z-transform method is that the designer deals with difference rather than differential equations. Therefore, the analysis and design of linear sampled data systems can be automated to a considerable degree and relegated to a computer.

Extention of z-transforms to nonlinear systems is governed by the possibility of their representation in the form of functional Volterra series [10.1, 2]. If such a representation exists, the nonlinear system is completely described by the set of equivalent transfer functions formed as the result of representing the terms of functional expansion in the multidimensional z-domain. This lends a possibility to determine the response of open-loop and closed-loop nonlinear systems to arbitrary inputs.

The mathematical formalism of multidimensional z-transforms allows extension of the operational methods to nonlinear discrete-continuous systems and evaluatation of the structure and parameters of the compensatory element. This approach allows the designer to compensate for the negative effect of the nonlinear characteristics of the invariable system element and meet the performance specifications.

10.1 System Analysis by Multidimensional z-Transforms

Consider a continuous nonlinear dynamic systems depicted in Fig. 10.1. In the general case, for every time instant t the output can be related to the input by a functional

$$x(t) = W[t;\ u(t_1),\ t_1 \leqslant t] \tag{10.1}$$

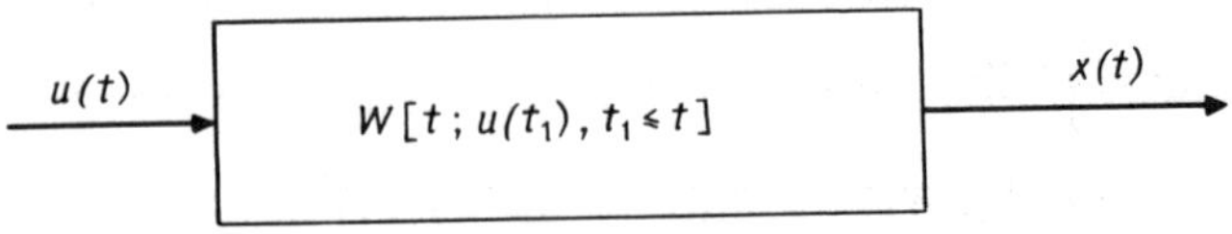

Fig. 10.1

If W is a continuous functional, it can be expanded into a series

$$x(t) = \sum_{n=1}^{\infty} \int_0^{\infty} \dots \int_0^{\infty} w_n(\tau_1, \dots, \tau_n) \prod_{i=1}^{n} u(t - \tau_i)\, d\tau_i \tag{10.2}$$

where in order that the system be physically realizable

$$w_n(\tau_1, \dots, \tau_n) = 0 \tag{10.3}$$

for $\tau_i < 0, \quad i = 1, 2, \dots, n$.

Expansion (10.2) is called a *functional Volterra series*, and the functions $w_n(\tau_1, \dots, \tau_n)$ are the Volterra kernels of the n-th order. If we represent (10.2) in the form

$$x(t) = \sum_{n=1}^{\infty} x_n(t) \tag{10.4}$$

then subject to (10.3)

$$x_n(t) = \int_0^{t} \dots \int_0^{t} w_n(\tau_1, \dots, \tau_n) \prod_{i=1}^{n} u(t - \tau_i)\, d\tau_i \tag{10.5}$$

From (10.4, 5) it follows that in the general case the response of a nonlinear dynamic system contains a linear component $x_1(t)$ and a variety of nonlinear components $x_2(t)$, $x_3(t)$, etc. The linear component $x_1(t)$ results as the convolution of the input signal $u(t)$ and the first order kernel $w_1(\tau)$. The nonlinear components are the results of the n-dimensional convolutions of the input signal and the respective kernels of the n-th order, which may be viewed as multidimensional weight functions of the nonlinear system.

Because the expansion (10.2) can be obtained only for bounded input signals, we introduce the norms of $u(t)$ and $x(t)$ as

$$\begin{aligned} \|u\| &= U = \sup |u(t)|, \quad -\infty < t < +\infty \\ \|x\| &= X = \sup |x(t)|, \quad -\infty < t < +\infty \end{aligned} \tag{10.6}$$

If for all kernels of order $n = 1, 2,$ etc.

$$\|w_n\| = \int_0^{\infty} \dots \int_0^{\infty} |w_n(\tau_1, \dots, \tau_n)| \prod_{i=1}^{n} d\tau_i \tag{10.7}$$

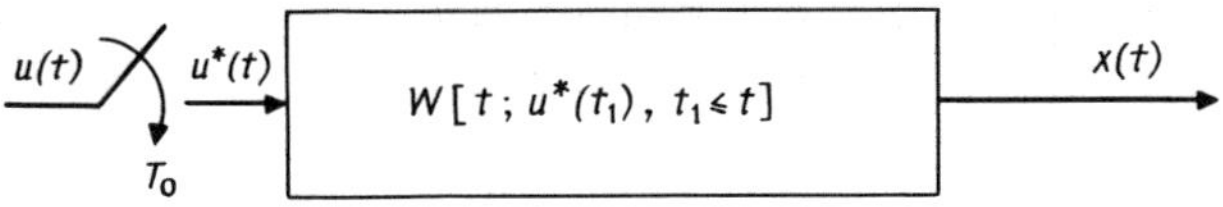

Fig. 10.2

then from (10.2) it follows that

$$X \leqslant \sum_{n=1}^{\infty} \|w_n\| \, U^n \tag{10.8}$$

In this case the right-hand side of (10.8) is a converging power series. The systems for which the condition (10.7) holds true for any n are stable and the respective functional expansion (10.2) possesses a nonzero convergence radius r_0. The value of r_0 is estimated by the radius of convergence of the series in (10.8). A necessary and sufficient condition that a nonzero radius of convergence exists is the stability of the linearized dynamic system, i.e. when $\|w_1\| < \infty$.

Therefore, with a bounded input

$$\|u\| \leqslant r_0 \tag{10.9}$$

the system response can be described to a given accuracy by a section of the functional series of Volterra.

Consider a nonlinear system operating in the sampled-data mode (Fig. 10.2). The sampling switch with the sampling period T_0 converts the continuous input signal $u(t)$ into the discrete quantity $u^*(t)$ so that

$$\begin{aligned} u(kT_0) &= \int_0^{\infty} u(t)\,\delta(t - kT_0) \\ u^*(kT_0) &= \sum_{k=0}^{\infty} u(kT_0)\,\delta(t - kT_0) \end{aligned} \tag{10.10}$$

Substituting $u^*(t)$ from (10.10) for $u(t)$ in (10.2) yields

$$\begin{aligned} x(t) = &\sum_{k=0}^{[t/T_0]} w_1(t - kT_0)u(kT_0) \\ &+ \sum_{k_1=0}^{[t/T_0]} \sum_{k_2=0}^{[t/T_0]} w_2(t - k_1T_0,\ t - k_2T_0)u(k_1T_0)u(k_2T_0) \\ &+ \sum_{k_1=0}^{[t/T_0]} \sum_{k_n=0}^{[t/T_0]} w_n(t - k_1T_0,\ t - k_nT_0) \prod_{i=1}^{n} u(k_iT_0) + \dots \end{aligned} \tag{10.11}$$

Assume that a synchronous sampler is inserted at the output of the nonlinear system. Then the continuous output $x(t)$ is converted into a discrete train

$$x(rT_0) = \sum_{k=0}^{r} w_1(rT_0 - kT_0)u(kT_0)$$
$$+ \sum_{k_1=0}^{r}\sum_{k_2=0}^{r} w_2(rT_0 - k_1T_0,\ rT_0 - k_2T_0)u(k_1T_0)u(k_2T_0) + \dots$$
$$+ \sum_{k_1=0}^{r}\dots\sum_{k_n=0}^{r} w_n(rT_0 - k_1T_0,\ \dots,\ rT_0 - k_nT_0)\prod_{i=0}^{n} u(k_iT_0) + \dots \quad (10.12)$$

where $r = 0, 1$, etc. Rearranging the expressions under the sum symbol, we get

$$x(rT_0) = \sum_{n=1}^{\infty} x_n(rT_0) = \sum_{k=0}^{r} w_1(kT_0)u(rT_0 - kT_0)$$
$$+ \sum_{k_1=0}^{r}\sum_{k_2=0}^{r} w_2(k_1T_0, k_2T_0)u(rT_0 - k_1T_0)u(rT_0 - k_2T_0) + \dots$$
$$+ \sum_{k_1=0}^{r}\dots\sum_{k_n=0}^{r} w_n(k_1T_0,\ \dots,\ k_nT_0)\prod_{i=1}^{n}(rT_0 - k_iT_0) + \dots \quad (10.13)$$

This expression is a discrete analog of the Volterra series (10.2).

Following Alper [10.3], the n-dimensional z-transform of the discrete function of n variables $g(k_1T_0, \dots, k_nT_0)$ is found by the formula

$$G(z_1, \dots, z_n) = Z_n[q(k_1T_0, \dots, k_nT_0)]$$
$$= \sum_{k_1=0}^{\infty}\dots\sum_{k_n=0}^{\infty} g(k_1T_0, \dots, k_nT_0)\prod_{i=1}^{n} z_i^{-k_i} \quad (10.14)$$

where $g(k_1T_0, \dots, k_nT_0) = 0$ for $k_i < 0$, $i = 1, 2, \dots, n$.

The respective inverse n-dimensional transformation is given by

$$g(k_1T_0, \dots, k_nT_0) = Z_n^{-1}[G(z_1, \dots, z_n)]$$
$$= \left(\frac{1}{2\pi_j}\right)^n \oint_{\Gamma_n}\dots\oint_{\Gamma_1} G(z_1, \dots, z_n)\prod_{i=0}^{n} z_i^{k_i-1}\,dz_i \quad (10.15)$$

where $\Gamma_i (i = 1, 2, \dots, n)$ are the integration contours in the domain where the integrand is analytic.

Observing that in (10.13) for every n there holds the relation

$$x_n(rT_0) = \sum_{k_1=0}^{r} \dots \sum_{k_n=0}^{r} w_n(k_1 T_0, \dots, k_n T_0) \sum_{i=1}^{n} u(rT_0 - k_i T_0) \tag{10.16}$$

we introduce the following functions of n variables:

$$x_n(r_1 T_0, \dots, r_n T_0) = \sum_{k_1=0}^{r_1} \dots \sum_{k_n=0}^{r_n} w_n(k_1 T_0, \dots, k_n T_0) \sum_{i=1}^{n} u(r_i T_0 - k_i T_0) \tag{10.17}$$

where $n = 1, 2$, etc.

The transfer from (10.17) to (10.16) can be effected by the substitution $r_i = r(i = 1, 2, \dots, n)$. Obtain the multidimensional z-transform of (10.17) with the aid of (10.15), then

$$x_n(z_1, \dots, z_n) = W_n(z_1, \dots, z_n) \prod_{i=1}^{n} U(z_i) \tag{10.18}$$

where

$$W_n(z_1, \dots, z_n) = \sum_{k_1=0}^{\infty} \dots \sum_{k_n=0}^{\infty} w_n(k_1 T_0, \dots, k_n T_0) \prod_{i=1}^{n} z_i^{-k_i} \tag{10.19}$$

The expressions (10.18, 19) indicate that the multidimensional z-transforms $W_n(z_1, \dots, z_n)$ of discrete kernels may be viewed as equivalent multidimensional transfer functions.

The response $x_n(rT_0)$ is obtained with the help of the inverse multidimensional transformation, namely,

$$\begin{aligned} x_n(rT_0) &= \lim_{r_1, \dots, r_n \to r} x_n(r_1 T_0, \dots, r_n T_0) \\ &= \lim_{r_1, \dots, r_n \to r} Z^{-1}[X_n(z_1, \dots, z_n)] \end{aligned} \tag{10.20}$$

An alternative way is to reduce the multidimensional transform $X_n(z_1, \dots, z_n)$ to the one-dimensional and obtain the inverse transform.

The discrete response of a nonlinear dynamic system

$$x(rT_0) = \sum_{n=1}^{\infty} x_n(rT_0)$$

to an arbitrary impulse input may be obtained by the expressions (10.18-20) or with the relation

$$X(z) = W_1(z)U(z) + F_2[W_2(z_1, z_2)U(z_1)U(z_2)]$$
$$+ \dots + F_n\left[W_n(z_1, \dots, z_n)\prod_{i=1}^{n} U(z_i)\right] + \dots$$

where the operator F_n reduces the n-dimensional z-transform to one-dimensional followed by the inverse transformation.

To find multidimensional equivalent transfer functions it is convenient to resort to the relation

$$W_n(z_1, \dots, z_n) = \left(\frac{1}{2\pi j}\right)^n \int_{c_n - j\infty}^{c_n + j\infty} \dots \int_{c_1 - j\infty}^{c_1 + j\infty} W_n(s_1, \dots, s_n)\prod_{i=1}^{n} \frac{ds_i}{1 - e^{s_i T_0} z_i^{-1}} \tag{10.21}$$

where the integrals are computed by taking residues at all poles of the multidimensional function.

Example 10.1. Evaluate the overall transfer function $W_2(z_1, z_2)$ for the system depicted in Fig. 10.3.

By (10.21) we have

$$W_2(z_1, z_2) = \left(\frac{1}{2\pi j}\right)^2 \int_{c_2 - j\infty}^{c_2 + j\infty} \int_{c_1 - j\infty}^{c_1 + j\infty} W_2(s_1, s_2)$$
$$\times \frac{1}{1 - e^{s_1 T_0} z_1^{-1}} \frac{1}{1 - e^{s_2 T_0} z_2^{-1}} ds_1 s_2$$

where

$$W_2(s_1, s_2) = \frac{k_1}{T_1 s_1 + 1} \frac{k_1}{T_1 s_2 + 1} \frac{k_2}{T_2(s_1 + s_2) + 1}$$

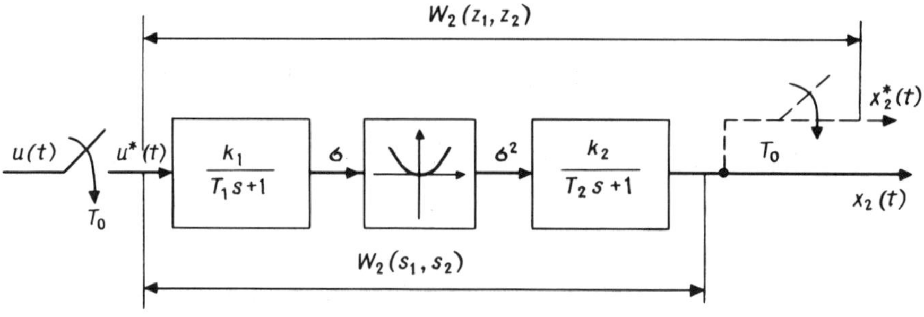

Fig. 10.3

First we compute the internal integral as

$$W_2(z_1, s_2) = \frac{1}{2\pi j}\int_{c_1-j\infty}^{c_1+j\infty} \frac{k_1}{T_1 s_1 + 1}\,\frac{k_1}{T_1 s_2 + 1}\,\frac{k_2}{T_2(s_1 + s_2) + 1}\,\frac{1}{1 - e^{s_1 T_0} z_1^{-1}}\,ds_1$$

$$= \frac{k_1^2 k_2}{T_1^2 T_2 (s_2 + 1/T_1)} \sum_{\substack{s_{1,1} = -1/T_1 \\ s_{1,2} = -(s_2 + 1/T_1)}} \operatorname{Res} \frac{1}{s_1 + 1/T_1}$$

$$\times \frac{1}{s_1 + (s_2 + 1/T_2)}\,\frac{1}{1 - e^{s_1 T_0} z_1^{-1}}$$

$$= \frac{k_1^2 k_2}{T_1^2 T_2 (s_2 + 1/T_1)(s_2 + 1/T_2 - 1/T_1)}$$

$$\times \left[\frac{1}{1 - e^{-T_0/T_1} z_1^{-1}} - \frac{1}{1 - e^{-(s_2 + 1/T_2)T_0} z_1^{-1}}\right]$$

$$= \{k_1^2 k_2 e^{-T_0/T_1} z_1^{-1}[1 - e^{T_0(T_2 - T_1)/T_1 T_2} e^{-s_2 T_0}]\}$$

$$+ \{T_1^2 T_2 (1 - e^{-T_0/T_1} z_1)[1 - e^{-T_0/T_2} z_1^{-1} e^{-s_2 T_0}]$$

$$\times (s_2 + 1/T_1)(s_2 + 1/T_2 - 1/T_1)\}$$

The outer integral is evaluated in a similar manner, viz.,

$$W_2(z_1, z_2) = \frac{1}{2\pi j}\int_{c_2-j\infty}^{c_2+j\infty} W_2(z_1, s_2) \frac{1}{1 - e^{s_2 T_0} z_2^{-1}}\,ds_2$$

$$= \sum_{\substack{s_{2,1} = -1/T_2 \\ s_{2,2} = -1/T_2 + 1/T_1}} \operatorname{Res} \frac{W_2(z_1, s_2)}{1 - e^{s_2 T_0} z_2^{-1}}$$

$$= \frac{k_1^2 k_2 e^{-T_0/T_1} z_1^{-1}}{T_1(T_1 - 2T_2)(1 - e^{-T_0/T_1} z_1^{-1})}\,\frac{1 - e^{T_0(T_2 - T_1)} z_2^{-1}}{1 - e^{-T_0/T_2} z_1^{-1} z_2^{-1}}$$

$$\times \left[\frac{1}{1 - e^{-T_0/T_1} z_2^{-1}} - \frac{1}{1 - e^{T_0(T_2 - T_1)/T_1 T_2} z_2^{-1}}\right]$$

Now, the expression for the two-dimensional overall transfer function takes the form

$$W_2(z_1, z_2) = \frac{k_1^2 k_2 (e^{-2T_0/T_1} - e^{-T_0/T_2})}{T_1(T_1 - 2T_2)}\,\frac{z_1}{z_1 - e^{-T_0/T_1}}$$

$$\times \frac{z_2}{z_2 - e^{-T_0/T_1}}\,\frac{1}{z_1 z_2 - e^{-T_0/T_2}}$$

For the multidimensional z-transform method related to nonlinear sampled systems, there exists a number of important relations similar to the one-dimensional z-transform of linear systems. These are the relations for the response of nonlinear system $x(t)$ to an input $W(t)$ as $t \to 0$, and relations for evaluation of the steady-state overshoot of the stable nonlinear system. The multidimensional z-transform technique also possesses the property of linearity with respect to a combination of multidimensional weight functions, viz.,

$$Z_n\left[\sum_{i=1}^{L} w_{n,i}(k_1 T_0, \ldots, k_n T_0)\right] = \sum_{i=1}^{L} Z_n\left[w_{n,i}(k_1 T_0, \ldots, k_n T_0)\right] \tag{10.22}$$

A time translation, both lead and lag, of the multidimensional weight function by $k_i T_0$ results in multiplying the multidimensional z-transform by $Z_i^{\pm k_i}$. When a multidimensional weight function is multiplied by an exponential in the time domain, then the z-domain acquires an exponential scale factor.

Like in the one-dimensional case, the product of multidimensional z-transforms $G(z_1, \ldots, z_n) W_n(z_1, \ldots, z_n)$ corresponds to the convolution of the discrete weight functions $g_n * w_n$ given as follows:

$$g_n * w_n = \sum_{r_1=0}^{k_1} \cdots \sum_{r_n=0}^{k_n} g_n(r_1 T_0, \ldots, r_n T_0) \times w_n(k_1 T_0 - r_1 T_0, \ldots, k_n T_0 - r_n T_0) \tag{10.23}$$

Finally, the multiplication of a multidimensional discrete weight function $w_n(k_1 T_0, \ldots, k_n T_0)$ by $k_1^{p_1}, \ldots, k_n^{p_n}$ is associated with the q-fold differentiation of the multidimensional z-transform $W_n(z_1, \ldots, z_n)$ with respect to $z_1, \ldots, z_n$.

These properties of multidimensional z-transforms will be used in the evaluation of transfer functions for nonlinear systems.

To determine the response of a nonlinear system in time instants other than sampling points we will use a multidimensional modified z-transform method defined as follows. Represent the response of a nonlinear system operating in the sampled mode as

$$\begin{aligned} x(t) = {} & \sum_{k=0}^{t/T_0} w_1(t - kT_0) u(kT_0) \\ & + \sum_{k_1=0}^{[t/T_0]} \sum_{k_2=0}^{[t/T_0]} w_2(t - k_1 T_0, t - k_2 T_0) u(k_1 T_0) u(k_2 T_0) + \ldots \\ & + \sum_{k_1=0}^{[t/T_0]} \sum_{k_n=0}^{[t/T_0]} w_n(t - k_1 T_0, \ldots, t - k_n T_0) u(k_1 T_0) \ldots u(k_n T_0) + \ldots \end{aligned} \tag{10.24}$$

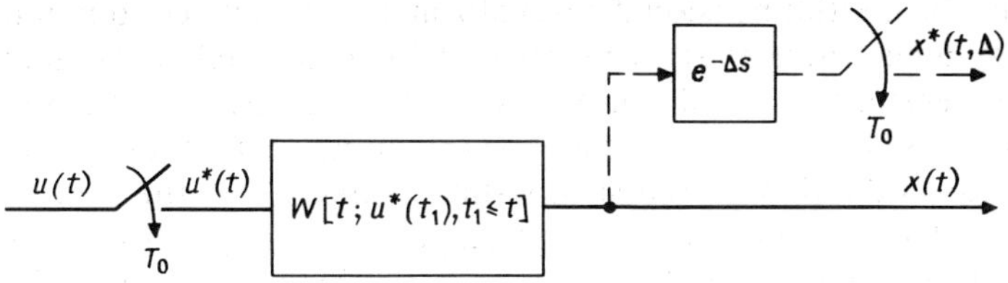

Fig. 10.4

Incorporating a fictitious sampling switch, operating with a delay $\Delta = \theta T_0$, $0 < \theta < 1$, with respect to the main sampler at the system input, at the output of the nonlinear system as shown in Fig. 10.4 allows one to obtain all values of the output in the interval between two adjacent sampling points as follows:

$$x(rT_0 - \Delta) = \sum_{k=0}^{r} w_1(rT_0 - \Delta - kT_0)u(kT_0)$$

$$+ \sum_{k_1=0}^{r} \sum_{k_2=0}^{r} w_2(rT_0 - \Delta - k_1 T_0,\ rT_0 - \Delta - k_2 T_0)u(k_1 T_0)u(k_2 T_0) + \dots$$

$$+ \sum_{k_1=0}^{r} \dots \sum_{k_n=0}^{r} w_n(rT_0 - \Delta - k_1 T_0,\ \dots,\ rT_0 - \Delta - k_n T_0)$$

$$\times\ u(k_1 T_0) \dots u(k_n T_0) + \dots \tag{10.25}$$

On changing the variable $\theta = 1 - m$, $m \in [0, 1]$ in this expression and taking the multidimensional z-transform, we get

$$X_n(z_1, m_1; \dots; z_n, m_n) = W_n(z_1, m_1; \dots; z_n, m_n) \prod_{i=1}^{n} U(z_i) \tag{10.26}$$

where

$$W_n(z_1, m_1; \dots; z_n, m_n) = Z_n^m[w_n((k_1 + m_1 - 1)T_0, \dots, (k_n + m_n - 1)T_0)]$$

$$= \sum_{k_1=0}^{\infty} \dots \sum_{k_n=0}^{\infty} w_n((k_1 + m_1 - 1)T_0, \dots, (k_n + m_n - 1)T_0 z_1^{-k_1} \dots z_n^{-k_n}) \tag{10.27}$$

The equivalent transfer function $W_n(z_1, m_1; \dots; z_n, m_n)$ is a multidimensional modified z-transform of the respective multidimensional weight function. For systems with continuous multidimensional weight functions $w_n(\tau_1 \dots, \tau_n)$ operating in the sampled-data mode, we have $m_1 = m_2 = \dots$

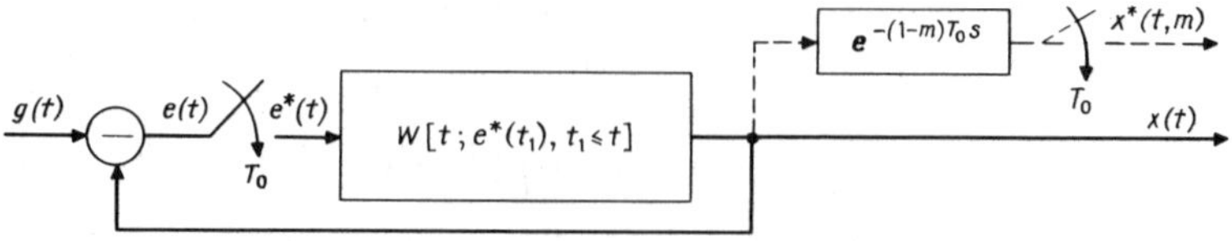

Fig. 10.5

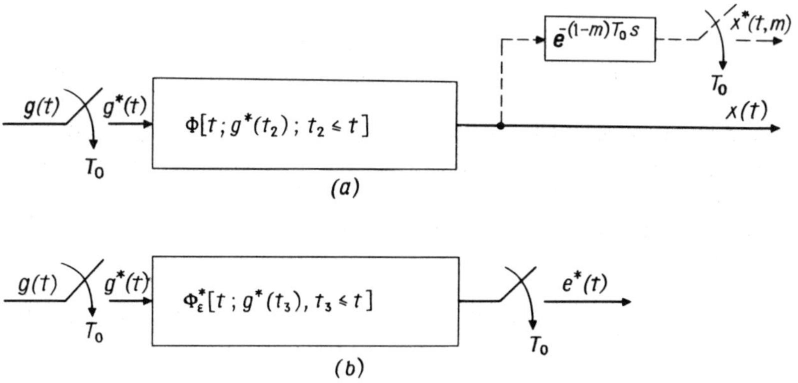

Fig. 10.6

$= m_n = m$, and the respective multidimensional modified z-transform takes the form

$$Z_n^m[w_n((k_1 + m - 1)T_o, \ldots, (k_n + m - 1)T_o)] = W_n(z_1, \ldots, z_n; m) \quad (10.28)$$

The aforementioned methods of evaluation of multidimensional z-transforms may be used to determine equivalent transfer functions of nonlinear systems with the aid of multidimensional modified z-transforms.

Now we turn to the analysis of closed-loop nonlinear discrete systems. Consider the unity feedback error sampled system depicted in Fig. 10.5. Form the equivalent block diagrams (Fig. 10.6) of the system with respect to the input signal $g^*(t)$ for (a) the output $x(t)$ and (b) the error $e^*(t)$. These lead to the functional relationships

$$x(t) = \Phi[t;\ g^*(t_2)], \quad t_2 \leqslant t \quad (10.29)$$

$$e^*(t) = \Phi_\varepsilon^*[t;\ g^*(t_3)], \quad t_3 \leqslant t \quad (10.30)$$

Let for the functional relation of the feedforward loop

$$x(t) = W[t;\ e^*(t_1),\ t_1 \leqslant t] \quad (10.31)$$

there exists an expansion in the Volterra series

$$x(t) = \sum_{n=1}^{\infty} \int_0^{\infty} \ldots \int_0^{\infty} w_n(\tau_1, \ldots, \tau_n) \prod_{i=1}^{n} e^*(t - \tau_i)\, d\tau_i \quad (10.32)$$

We shall seek the respective functionals (10.29, 30) for the closed loop system also in the form of Volterra series

$$x(t) = \sum_{q=1}^{\infty} \int_0^{\infty} \dots \int_0^{\infty} \varphi_q(\tau_1, \dots, \tau_q) \prod_{i=1}^{q} g^*(t - \tau_i)\, \mathrm{d}\tau_i \tag{10.33}$$

$$e^*(t) = \sum_{j=1}^{\infty} \int_0^{\infty} \dots \int_0^{\infty} \varphi_{\varepsilon j}^*(\tau_1, \dots, \tau_j) \prod_{i=1}^{j} g^*(t - \tau_i)\, \mathrm{d}\tau_i \tag{10.34}$$

where $\varphi_q(\tau_1, \dots, \tau_q)$ is the multidimensional weight function of the closed loop system, and $\varphi_{\varepsilon j}^*(\tau_1, \dots, \tau_j)$ is the multidimensional discrete weight function of the closed-loop system error.

Recast (10.34) to the form

$$\begin{aligned} e(rT_0) &= \sum_{j=1}^{\infty} e_j(rT_0) \\ &= \sum_{j=1}^{\infty} \left[\sum_{k_1=0}^{\infty} \dots \sum_{k_j=0}^{\infty} \varphi_{\varepsilon j}(rT_0 - k_1 T_0, \dots, rT_0 - k_j T_0) \prod_{i=1}^{j} e(k_i T_0) \right] \end{aligned} \tag{10.35}$$

Recast in the discrete form (10.33) becomes

$$\begin{aligned} x(rT_0, m) = \sum_{q=1}^{\infty} \Bigg\{ \sum_{d_1=0}^{\infty} \dots \sum_{d_q=0}^{\infty} \varphi_q[&rT_0 - (1 - m)T_0 - d_1 T_0, \dots, rT_0 \\ &- (1 - m)T_0 - d_q T_0] \prod_{i=1}^{q} g(d_i T_0) \Bigg\} \end{aligned} \tag{10.36}$$

Similarly from (10.32) we have

$$\begin{aligned} x(rT_0, m) = \sum_{n=1}^{\infty} \Bigg\{ \sum_{c_1=0}^{\infty} \dots \sum_{c_n=0}^{\infty} w_n[&rT_0 - (1 - m)T_0 - c_1 T_0, \dots, rT_0 \\ &- (1 - m)T_0 - c_n T_0] \prod_{i=1}^{n} e(c_i T_0) \Bigg\} \end{aligned} \tag{10.37}$$

Rewritten in operator form the expressions (10.32—34) become

$$x(t) = W[e^*(t)] \tag{10.38}$$

$$x(t) = \Phi[g^*(t)] \tag{10.39}$$

$$e^*(t) = \Phi_\varepsilon^*[g^*(t)] \tag{10.40}$$

Observing the feedback equation

$$e(t) = g(t) - x(t) \tag{10.41}$$

from (10.38) and (10.40) we obtain

$$[\Phi_\varepsilon^*[g^*(t)]]^* = [g(t) - W[\Phi_\varepsilon^*[g^*(t)]]]^*$$

whence

$$\Phi_\varepsilon^*[g^*(t)] = g^*(t) - W^*[\Phi_\varepsilon^*[g^*(t)]] \tag{10.42}$$

For the stable system (10.32) and stable closed loop system (10.34), the right-hand side of (10.42) is a Volterra series possessing a nonzero radius of convergence. By equating the like terms on the left and right sides of (10.42) and taking z-transforms of each term, we get

for $n = 1$

$$\Phi_{\varepsilon,1}(z) = \frac{1}{1 + W_1(z)} \tag{10.43}$$

for $n = 2$

$$\Phi_{\varepsilon,2}(z_1, z_2) = \left[-W_2(z_1, z_2) \prod_{i=1}^{2} \Phi_{\varepsilon,1}(z_i) \right] \Big/ [1 + W_1(z_1 z_2)] \tag{10.44}$$

for $n = 3$

$$\Phi_{\varepsilon,3}(z_1, z_2, z_3) = \left[-W_3(z_1, z_2, z_3) \prod_{i=1}^{3} \Phi_{\varepsilon,1}(z_i) + 2W_2(z_1, z_2, z_3) \times \Phi_{\varepsilon,1}(z_1)\Phi_{\varepsilon,2}(z_2, z_3) \right] \Big/ [1 + W_1(z_1 z_2 z_3)] \tag{10.45}$$

for $n = 4$

$$\begin{aligned} \Phi_{\varepsilon,4}(z_1, z_2, z_3, z_4) = {} & [1 + W_1(z_1 z_2 z_3 z_4]^{-1}(-1) \\ & \times \left[W_4(z_1, z_2, z_3, z_4) \prod_{i=1}^{4} \Phi_{\varepsilon,1}(z_i) \right. \\ & + 3W_3(z_1, z_2, z_3 z_4)\Phi_{\varepsilon,1}(z_1)\Phi_{\varepsilon,1}(z_2)\Phi_{\varepsilon,2}(z_3, z_4) \\ & + 2W_2(z_1 z_2, z_3 z_4)\Phi_{\varepsilon,2}(z_1, z_2)\Phi_{\varepsilon,2}(z_3, z_4) \\ & + 2W_2(z_1, z_2 z_3 z_4)\Phi_{\varepsilon,1}(z_1)\Phi_{\varepsilon,3}(z_2, z_3, z_4) \end{aligned} \tag{10.46}$$

and so forth.

As a result, the image of the error of the closed-loop nonlinear discrete system in the z-domain will be

$$\begin{aligned} E(z) = {} & \Phi_{\varepsilon,1}(z)G(z) + F_2[\Phi_{\varepsilon,2}(z_1, z_2)G(z_1)G(z_2)] \\ & + F_n\left[\Phi_{\varepsilon,n}(z_1, \ldots, z_n) \prod_{i=1}^{n} G(z_i) \right] \end{aligned} \tag{10.47}$$

where $E(z) = Z[e(kT_0)]$ and $G(z) = Z[g(kT_0)]$.

Substituting (10.40) in (10.33) yields

$$x(t) = W[\Phi_\varepsilon^*[g^*(t)]] \tag{10.48}$$

Then by virtue of (10.38) we find

$$\Phi[g^*(t)] = W[\Phi_\varepsilon^*[g^*(t)]] \tag{10.49}$$

or in discrete form

$$[\Phi[g^*(t)]]_m^* = [W[\Phi_\varepsilon^*[g^*(t)]]]_m^*$$

from which the functional equation results as

$$\Phi_m^*[g^*(t)] = W_m^*[\Phi_\varepsilon^*[g^*(t)]] \tag{10.50}$$

For the stable forward loop (10.32) and stable closed-loop system (10.34), the right-hand side of (10.50) is a Volterra series with a nonzero radius of convergence. Then the solution of this equation can be found by equating the respective terms of like order on either side. In this case taking the multidimensional modified transform of (10.50) results in the following relations:

for $n = 1$

$$\Phi_1(z; m) = W_1(z, m)\Phi_{\varepsilon,1}(z) \tag{10.51}$$

for $n = 2$

$$\Phi_2(z_1, z_2; m) = W_2(z_1, z_2; m)\prod_{i=1}^{2} \Phi_{\varepsilon,1}(z_i) + W_1(z_1z_2; m)\Phi_{\varepsilon,2}(z_1, z_2) \tag{10.52}$$

for $n = 3$

$$\begin{aligned}\Phi_3(z_1, z_2, z_3; m) = {} & W_3(z_1, z_2, z_3; m)\prod_{i=1}^{3} \Phi_{\varepsilon,1}(z_i) \\ & + 2W_2(z_1, z_2z_3; m)\Phi_{\varepsilon,1}(z_1)\Phi_{\varepsilon,2}(z_2, z_3) \\ & + W_1(z_1z_2z_3; m)\Phi_{\varepsilon,3}(z_1, z_2, z_3)\end{aligned} \tag{10.53}$$

for $n = 4$

$$\begin{aligned}\Phi_4(z_1, z_2, z_3, z_4; m) = {} & W_4(z_1, z_2, z_3, z_4; m)\prod_{i=1}^{4} \Phi_{\varepsilon,1}(z_i) \\ & + 3W_3(z_1, z_2, z_3z_4; m)\Phi_{\varepsilon,1}(z_1)\Phi_{\varepsilon,1}(z_2)\Phi_{\varepsilon,2}(z_3, z_4) \\ & + 2W_2(z_1z_2, z_3z_4; m)\Phi_{\varepsilon,2}(z_1, z_2)\Phi_{\varepsilon,2}(z_3, z_4) \\ & + 2W_2(z_1, z_2z_3z_4; m)\Phi_{\varepsilon,1}(z_1)\Phi_{\varepsilon,3}(z_2, z_3, z_4) \\ & + W_1(z_1z_2z_3z_4; m)\Phi_{\varepsilon,4}(z_1, z_2, z_3, z_4)\end{aligned} \tag{10.54}$$

and so on. Now, the response of the closed-loop nonlinear discrete system can be found in the form

$$X(z;\ m) = \Phi_1(z;\ m)G(z) + F_2[\Phi_2(z_1,\ z_2;\ m)G(z_1)G(z_2)] + \ldots$$
$$+ F_n\left[\Phi_n(z_1,\ \ldots,\ z_n;\ m)\prod_{i=1}^{n} G(z_i)\right] + \ldots \qquad (10.55)$$

where $X(z; m) = Z[x((kT_0) - (1 - m)T_0)]$.

Comparison of (10.41) and (10.50) at $m = 1$ indicates that for $n \geqslant 2$ we have

$$\Phi_n(z_1,\ \ldots,\ z_n) = -\Phi_{\varepsilon,n}(z_1,\ \ldots,\ z_n) \qquad (10.56)$$

Example 10.2. Determine the transient behavior of the clossed-loop sampled error system shown in Fig. 10.7*a* at the following parameter values: $T_0 = 0.2$ s, $k_1 = 20$, and $T_1 = 0.2$ s.

Mapped in the z-domain, the block diagram of the system assumes the form shown in Fig. 10.7*b*. Since its forward loop is described by solely two Volterra kernels of the first and third order, we have

$$W_1(z) = (1 - z^{-1})Z[k_1 d_1 k_2/s^2(T_1 s + 1)]$$
$$= (0.294z + 0.211)/(z - 1)(z - 0.368) \qquad (10.57)$$

and

$$W_3(z_1,\ z_2,\ z_3) = [a_0 z_1 z_2 z_3 + a_1(z_1 z_2 + z_1 z_3 + z_2 z_3)$$
$$+ a_2(z_1 + z_2 + z_3) + a_3]/[(z_1 - 0.368)$$
$$\times (z_2 - 0.368)(z_3 - 0.368)(z_1 z_2 z_3 - 1)]$$

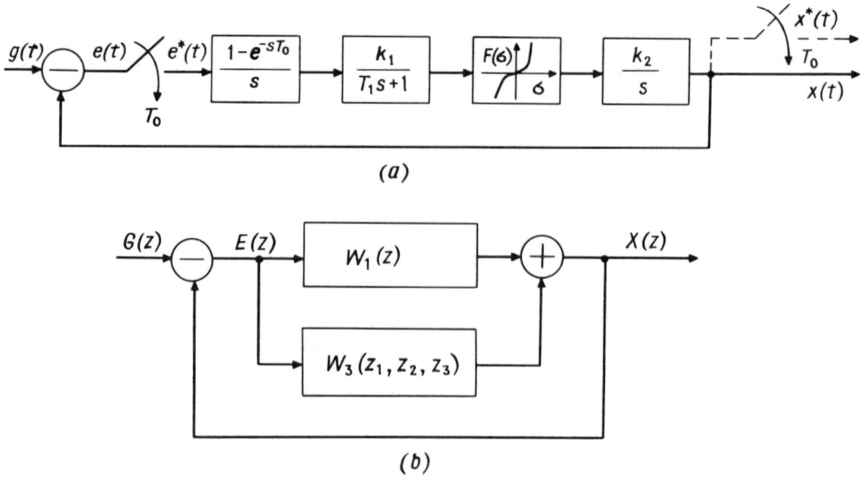

Fig. 10.7

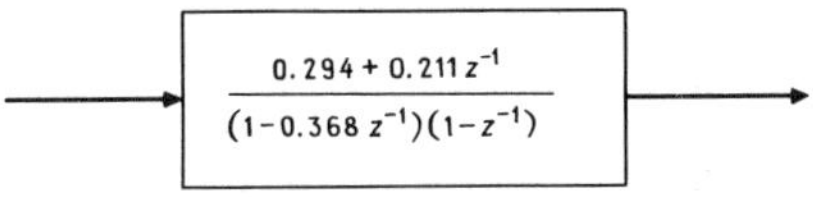

(a)

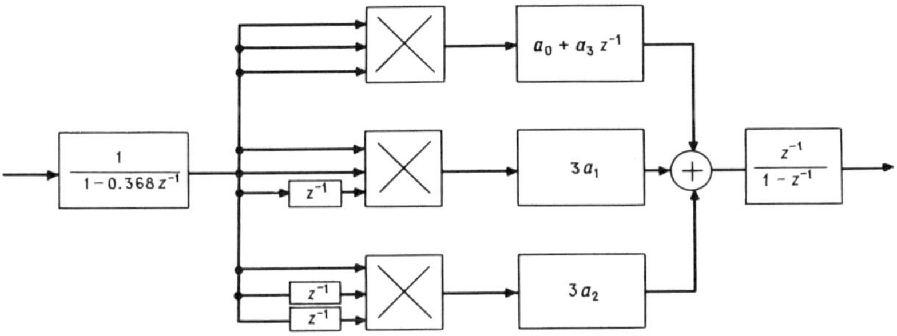

(b)

Fig. 10.8

For the aformentioned parameter values, the coefficients in the numerator are as follows: $a_0 = 0.134$, $a_1 = 0.0358$, $a_2 = 0.0294$, and $a_3 = 0.0743$.

According to (10.57, 58) the block diagrams of the transfer functions $W_1(z)$ and $W_3(z_1, z_2, z_3)$ take the form shown in Fig. 10.8*a* and *b*, respectively. Letting $m = 1$, we obtain the equivalent transfer functions for the closed-loop system with the help of the relations (10.51-54), observing (10.56) and (10.42-45), viz.,

$$\Phi_1(z) = \frac{W_1(z)}{1 + W_1(z)}$$

$$\Phi_3(z_1, z_2, z_3) = \frac{W_3(z_1, z_2, z_3)}{1 + W_1(z_1 z_2 z_3)} \prod_{i=1}^{3} \Phi_{\varepsilon,1}(z_i)$$

where

$$\Phi_{\varepsilon,1}(z) = \frac{1}{1 + W_1(z)}$$

$$\Phi_5(z_1, z_2, z_3, z_4, z_5) = \left[3 W_3(z_1, z_2, z_3 z_4 z_5) \prod_{i=1}^{2} \Phi_{\varepsilon,1}(z_i) \right.$$

$$\left. \times \Phi_3(z_3, z_4, z_5) \right] \Big/ [1 + W_1(z_1 z_2 z_3 z_4 z_5)]$$

and so on, with

$$\Phi_{2k}(z_1, \ldots, z_{2k}) = 0, \quad k = 1, 2, \text{ etc.}$$

Thus, the multidimensional z-transforms of the components of the nonlinear system response take the form

$$\begin{aligned}
X_1(z) &= \Phi_1(z)G(z) \\
X_3(z_1, z_2, z_3) &= \Phi_3(z_1, z_2, z_3)\prod_{i=1}^{3} G(z_i) \\
X_5(z_1, z_2, z_3, z_4, z_5) &= \Phi_5(z_1, \ldots, z_5)\prod_{i=1}^{5} G(z_i) \\
&\cdots\cdots\cdots \\
X_{2k-1}(z_1, \ldots, z_{2k-1}) &= \Phi_{2k-1}(z_1, \ldots, z_{2k-1})\prod_{i=1}^{2k-1} G(z_i) \\
&\cdots\cdots\cdots
\end{aligned} \tag{10.58}$$

where $k = 1, 2$, etc.

Since the respective series (10.35) converges, the components of the response $x_{2k-1}(rT_0)$ will reduce in magnitude as k increases.

Now we construct the block diagram for the response of a closed-loop nonlinear system to an arbitrary input $g(t)$ at sampling points. In agreement with (10.58) the block diagram for the first five components of the response has the form shown in Fig. 10.9. The transient behavior of the nonlinear discrete system computed with this block diagram is presented in Fig. 10.10. The plots represent (*1*) the first component $x_1^*(t)$ of the response, (*2*) $x_1^*(t) + x_3^*(t)$, (*3*) $x_1^*(t) + x_3^*(t) + x_5^*(t)$, and (*4*) the exact response of the nonlinear system $x(t)$ to a step input.

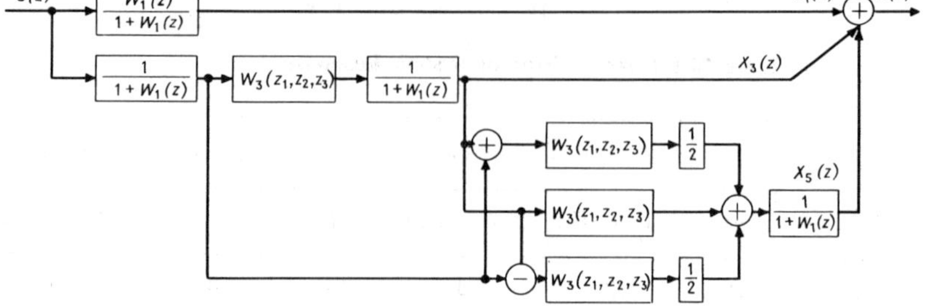

Fig. 10.9

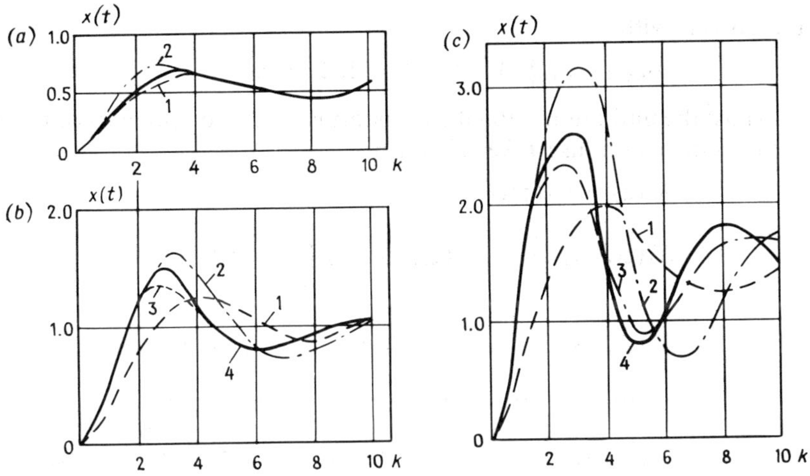

Fig. 10.10

As can be seen from Fig. 10.10*a*, at $g(t) = 0.5$ the system operates practically in a linear mode and the fundamental component $x_1^*(t)$ of the transient process almost coincides with the continuous response $x(t)$. As the input grows in magnitude, the system's nonlinear behavior manifests itself in a greater degree, and the magnitudes of $x_3^*(t)$ and $x_5^*(t)$ become more significant. The diagram at (*c*) exhibits a clear-cut dependence of transient oscillation period on the magnitude of the current system error. Table 10.1 summarizes the percentage accuracy in evaluation of the peak overshoot and the first minimum for three magnitudes of the step input and different number of components involved in the evaluation. The presented evidence suggests that the accuracy in the evaluation of these characteristics is rather high when the third and fifth components of the expansion are incorporated into the picture.

Table 10.1 Percent Error of System Response

Step input $g(t)$	Peak overshoot			Time of first minimum		
	x_1^*	$x_1^* + x_3^*$	$x_1^* + x_3^* + x_5^*$	x_1^*	$x_1^* + x_3^*$	$x_1^* + x_3^* + x_5^*$
0.5	16	22	3	14	8	0
1.0	34	34	8	33	16	8
1.5	52	64	13	60	31	17

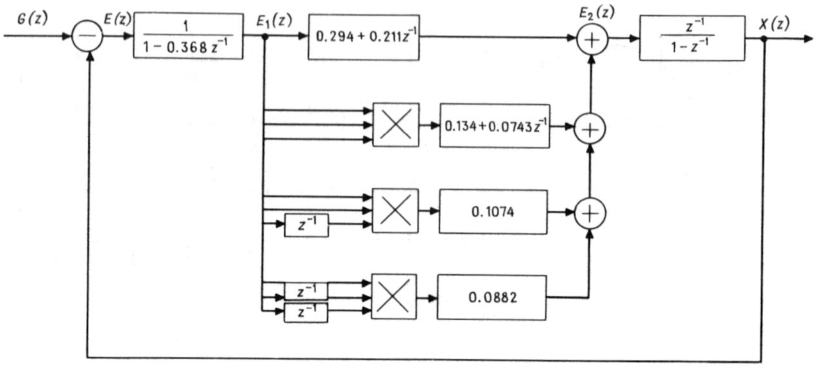

Fig. 10.11

It is worth noting that the exact value of the closed-loop nonlinear system response at sampling periods may be found with the block diagram depicted in Fig. 10.11. By inspection we write the difference equations

$$
\begin{aligned}
e(kT_0) &= g(kT_0) - x(kT_0)\\
e_1(kT_0) &= e(kT_0) + 0.368e(kT_0 - T_0)\\
e_2(kT_0) &= 0.294e_1(kT_0) + 0.211e_1(kT_0 - T_0) + 0.134e_1^3(kT_0)\\
&\quad + 0.0743e_1^3(kT_0 - T_0) + 0.1074e_1^2(kT_0)e_1(kT_0 - T_0)\\
&\quad + 0.0822e_1(kT_0)e_1^2(kT_0 - T_0)\\
x(kT_0 + T_0) &= e_2(kT_0) + x(kT_0 + T_0)
\end{aligned}
$$

When the initial value x_0 is given, the response of the closed-loop nonlinear system to an arbitrary input $g(kT_0)$ is obtained with the equations given above.

10.2 Synthesis of Nonlinear Discrete Compensatory Elements

Consider the closed-loop discrete-continuous system presented in Fig. 10.12, where $g(t)$ is the input, $\varepsilon(t)$ the error, $u^*(t)$ sampled control variable, and $x(t)$ the system response to the input. The continuous part of the system is nonlinear and at each time instant described by the functional relation

$$x(t) = W[t;\ u_t^*(t_1),\ t_1 \leqslant t] \tag{10.59}$$

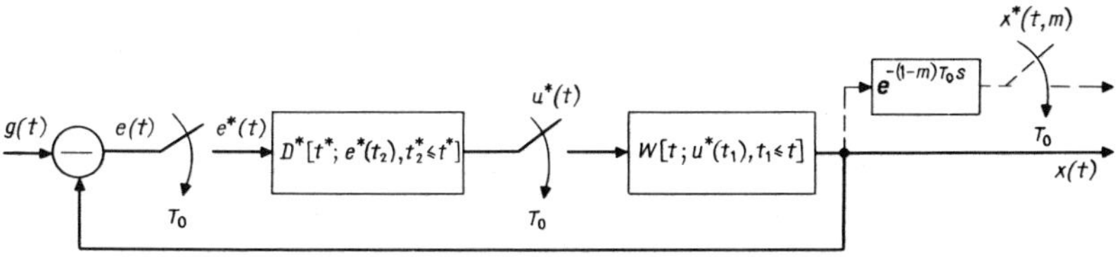

Fig. 10.12

which has the form of truncated Volterra series

$$x(t) = \sum_{n=1}^{N} \int_0^{\infty} \cdots \int_0^{\infty} w_n(\tau_1, \ldots, \tau_n) \prod_{i=1}^{n} u^*(t - \tau_i)\, d\tau_i \tag{10.60}$$

The nonlinear compensation element is to transform the error in agreement with the functional expression

$$u^*(t) = D^*[t^*;\ e^*(t_2),\ t_2^* \leqslant t] \tag{10.61}$$

We shall seek the structure and parameters of the desired compensatory unit in the class of expansion of (10.61) in the Volterra series, viz.,

$$u^*(t) = \sum_{j=1}^{\infty} \int_0^{\infty} \cdots \int_0^{\infty} d_j^*(\tau_1, \ldots, \tau_j) \prod_{i=1}^{j} e^*(t - \tau_i)\, d\tau_i \tag{10.62}$$

Observing that the physically implementable continuous parts of the system possess the property

$$w_n(\tau_1, \ldots, \tau_n) = 0 \tag{10.63}$$

for any $\tau_i < 0$, $i = 1, \ldots, n$, we impose the same constraint on the nonlinear discrete compensating element, viz.,

$$d_j^*(\tau_1, \ldots, \tau_j) = 0 \tag{10.64}$$

for $\tau_i < 0$, $i = 1, \ldots, j$.

In view of (10.64) the expressions (10.60, 62) in discrete form become

$$x(rT_0, m) = \sum_{n=1}^{N} \left[\sum_{k_1=0}^{\infty} \cdots \sum_{k_r=0}^{\infty} w_n(rT_0 - (1 - m)T_0 - k_1 T_0, \ldots, rT_0 - (1 - m)T_0 - k_n T_0) \prod_{i=1}^{n} u(k_i T_0) \right] \tag{10.65}$$

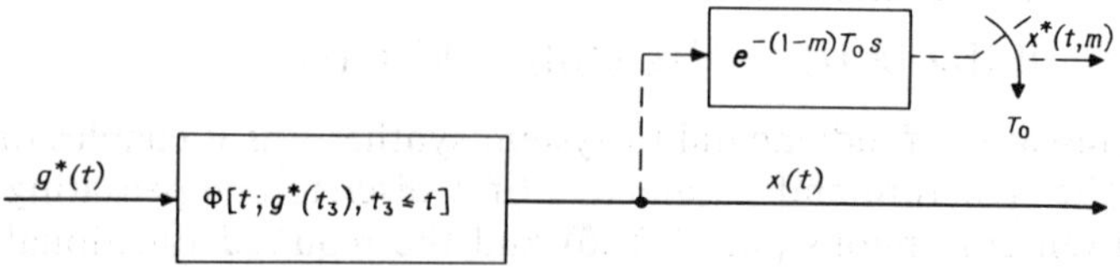

Fig. 10.13

$$u(rT_0) = \sum_{j=1}^{\infty} u_j(rT_0)$$

$$= \sum_{j=1}^{\infty} \left[\sum_{a_1=0}^{r} \dots \sum_{a_j=0}^{r} d_j(rT_0 - a_1 T_0, \ \dots, \ rT_0 - a_j T_0) \prod_{i=1}^{j} e(a_i T_0) \right] \quad (10.66)$$

On replacing the closed-loop nonlinear sampled system concerned with the equivalent one, as shown in Fig. 10.13, we get the functional relation

$$x(t) = \Phi[t; \ e^*(t_3); \ t_3 \geqslant t] \quad (10.67)$$

which can also be represented as the Volterra series

$$x(t) = \sum_{q=1}^{\infty} \int_0^{\infty} \dots \int_0^{\infty} \varphi_q(\tau_1, \ \dots, \ \tau_q) \prod_{i=1}^{q} g^*(t - \tau_i) \, d\tau_i \quad (10.68)$$

where

$$\varphi_q(\tau_1, \ \dots, \ \tau_q) = 0 \quad (10.69)$$

for $\tau_i < 0$, $i = 1, \ \dots, \ q$.

Recast in discrete form, (10.68) becomes

$$x(rT_0, \ m) = \sum_{q=0}^{\infty} \left[\sum_{b_1=0}^{\infty} \dots \sum_{b_q=0}^{\infty} \varphi_q(rT_0 - (1 - m)T_0 \right.$$

$$\left. - b_1 T_0, \ \dots, \ rT_0 - (1 - m)T_0 - b_q T_0) \prod_{i=1}^{q} g(b_i T_0) \right] \quad (10.70)$$

In operator form, (10.65, 66, 70) are as follows:

$$x^*(t, \ m) = W_m^*[u^*(t)]$$
$$u^*(t) = D^*[e^*(t)] \quad (10.71)$$
$$x^*(t, \ m) = \Phi_m^*[g^*(t)]$$

Being augmented by the feedback equation

$$e^*(t) = g^*(t) - x^*(t) \quad (10.72)$$

these operator equations lead to the functional relation

$$W_m^*[D^*[g^*(t) - \Phi^*[g^*(t)]]] = \Phi_m^*[g^*(t)] \tag{10.73}$$

This expression is fundamental to system synthesis as it enables one to determine the structure and parameters of a discrete compensating element by the given continuous part (10.60) and the required functional relation (10.68) of the system output and the input. Selection of a specific functional $\Phi_m^*[g^*(t)]$ and, consequently, the values of the equivalent multidimensional transfer functions $\Phi_n(z_1, ..., z_n; m)$ are decided by the performance criterion of the closed-loop sampled data system.

Consider the synthesis of a nonlinear discrete compensation element for such a performance criterion which ensures system linearity with respect to the system input varying within a certain range. For $q \geqslant 2$

$$\varphi_q(\tau_1, ..., \tau_q) = 0 \tag{10.74}$$

and (10.68, 70) become

$$x^*(t) = \int_0^\infty \varphi_1(\tau) g^*(t - \tau)\,\mathrm{d}\tau \tag{10.75}$$

$$x(rT_0, m) = \sum_{b_1=0}^{\infty} \varphi_1(rT_0 - (1 - m)T_0 - b_1 T_0) g(b_1 T_0) \tag{10.76}$$

Taking the multidimensional modified z-transform of (10.73), by virtue of (10.74) we get for $n \geqslant 2$

$$\Phi_n(z_1, ..., z_n; m) = 0 \tag{10.77}$$

and, consequently, for $n = 1$

$$W_1(z, m) D_1(z)[1 - \Phi_1(z)] = \Phi_1(z, m) \tag{10.78}$$

for $n = 2$

$$W_2(z_1, z_2; m) \prod_{i=1}^{2} D_1(z_i) + W_1(z_1 z_2; m) D_2(z_1, z_2) = 0 \tag{10.79}$$

for $n = 3$

$$W_3(z_1, z_2, z_3; m) \prod_{i=1}^{3} D_1(z_i) + 2W_2(z_1, z_2 z_3; m) D_1(z) D_2(z_2, z_3) + W_1(z_1 z_2 z_3; m) D_3(z_1, z_2, z_3) = 0 \tag{10.80}$$

and for $n = 4$

$$\begin{aligned} W_4(z_1, z_2, z_3, z_4; m) \prod_{i=1}^{4} D_1(z_i) &+ 3W_3(z_1, z_2, z_3 z_4; m) D_1(z_1) D_1(z_2) D_2(z_3, z_4) \\ &+ 2W_2(z_1 z_2, z_3 z_4; m) D_2(z_1, z_2) D_2(z_3, z_4) \\ &+ 2W_2(z_1, z_2 z_3 z_4; m) D_1(z_1) D_3(z_2, z_3, z_4) \\ &+ W_1(z_1 z_2 z_3 z_4; m) D_4(z_1, z_2, z_3, z_4) = 0 \end{aligned} \tag{10.81}$$

From these expressions it follows that at $n = 1$ the equivalent transfer functions of the compensatory element are invariant with respect to the parameter m, and $D_1(z)$ remains constant as m varies from 1 to 0. For $n \geqslant 2$ the structure and parameters of the equivalent transfer functions $D_n(z_1, ..., z_n)$ will already depend on the particular m. Therefore a certain m should be specified in synthesizing a compensatory element. Set $m = 1$, then from (10.78-80) it follows for $n = 1$ that

$$D_1(z) = \frac{\Phi_1(z)}{W_1(z)[1 - \Phi_1(z)]} \tag{10.82}$$

for $n = 2$

$$D_2(z_1, z_2) = -\frac{1}{W_1(z_1 z_2)}[W_2(z_1, z_2)D_1(z_1)D_1(z_2)] \tag{10.83}$$

$$D_3(z_1, z_2, z_3) = -\frac{1}{W_1(z_1 z_2 z_3)}[W_3(z_1, z_2, z_3)\prod_{i=1}^{3} D_1(z_i)$$

$$+ 2W_2(z_1, z_2 z_3)D_1(z_1)D_2(z_2, z_3)] \tag{10.84}$$

for $n = 4$

$$D_4(z_1, z_2, z_3, z_4) = -\frac{1}{W_1(z_1 z_2 z_3 z_4)}[W_4(z_1, z_2, z_3, z_4)\prod_{i=1}^{4} D_1(z_i)$$

$$+ 3W_3(z_1, z_2, z_3 z_4)D_1(z_1)D_1(z_2)D_2(z_3, z_4)$$

$$+ 2W_3(z_1 z_2, z_3 z_4)D_2(z_1, z_2)D_2(z_3, z_4)$$

$$+ 2W_2(z_1, z_2 z_3 z_4)D_1(z_1)D_3(z_2, z_3, z_4)] \tag{10.85}$$

and so forth.

For $n > N$, in these equations one should let

$$W_n(z_1, ..., z_n) = 0 \tag{10.86}$$

It is important to note that for the equivalent transfer functions (10.82-85), the condition (10.74) will be satisfied only at the discrete time instants $t = rT_0$, since for $n \geqslant 2$ and $m \neq 1$ the nonlinear components of system response

$$x_n(rT_0, m) = Z^{-1}\left[\Phi_n(z_1, ..., z_n; m)\prod_{i=1}^{n} G(z_i)\right]$$

$$= \sum_{k_1=0}^{\infty} ... \sum_{k_n=0}^{\infty} \varphi_n(rT_0 - (1 - m)T_0 - k_1 T_0, ..., rT_0$$

$$- (1 - m)T_0 - k_n T_0)\prod_{i=1}^{n} g(rT_0) \tag{10.87}$$

can in this case be other than zero.

The transfer function $D_1(z)$ in (10.82) can be obtained with one of the methods of synthesis of ordinary linear discrete-continuous systems by the known transfer function $W_1(z)$.

The outlined method of synthesis is applicable also when the desired functional $\Phi_m^*[g^*(t)]$ is nonlinear. The synthesis of the suitable compensatory element is then performed with nonzero values of the equivalent discrete transfer functions of the closed loop system, $\Phi_n(z_1, ..., z_n; m)$ for $n \geqslant 2$.

Taking the multidimensional z-transform of the expression (10.66) describing the compensatory elements in the form of Volterra series, we have

$$\begin{aligned} &U_1(z) = D_1(z)E(z) \\ &U_2(z_1, z_2) = D_2(z_1, z_2)\prod_{i=1}^{2} E(z_i) \\ &\cdots\cdots\cdots\cdots\cdots\cdots \\ &U_n(z_1, ..., z_n) = D_n(z_1, ..., z_n)\prod_{i=1}^{n} E(z_i) \end{aligned} \tag{10.88}$$

These expressions suggest that in this case the compensatory element has the network indicated in Fig. 10.14. The transfer functions $D_n(z_1, ..., z_n)$ it involves will be found by the expressions (10.82-85). It follows that each of the equivalent transfer functions $D_n(z_1, ..., z_n)$, will be found as a combination of transfer functions $D_i(z_1, ..., z_i)$, $i = 1, 2, ..., n-1$, of lower order and the equivalent transfer functions of the continuous part of the system $W_j(z_1, ..., z_j)$, $j = 1, 2, ..., n$. For $j > N$, $W_j(z_1, ..., z_j) = 0$.

To construct the associated block diagrams we denote the bracketed expressions in (10.83-85) by $H_n(z_1, ..., z_n)$. Then

$$D_n(z_1, ..., z_n) = C_1(z_1...z_n)H_n(z_1, ..., z_n) \tag{10.89}$$

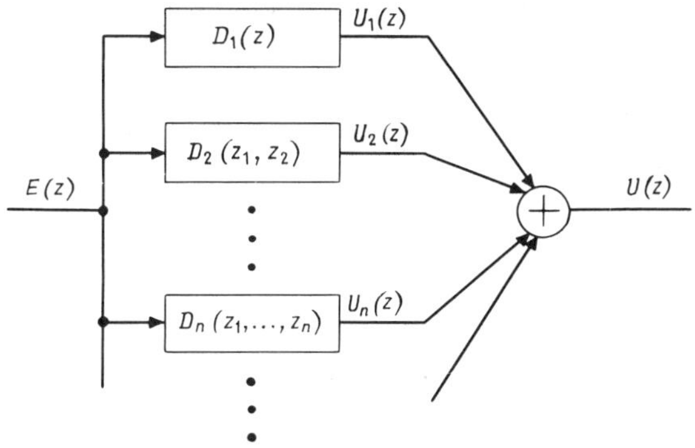

Fig. 10.14

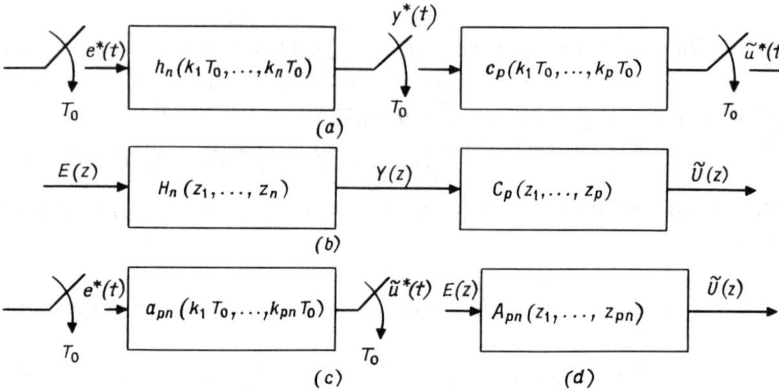

Fig. 10.15

where

$$C_1(z_1 \ldots z_n) = -1/W_1(z_1 \ldots z_n) \tag{10.90}$$

To begin with we set up the block diagram of the equivalent transfer function $D_n(z_1, \ldots, z_n)$ for (10.89). Consider the cascade connection of nonlinear elements indicated in Fig. 10.15*a* and *b*. In the general case these elements have the equivalent transfer functions $H_n(z_1, \ldots, z_n)$ and $C_p(z_1, \ldots, z_p)$ whose weight functions satisfy the conditions

$$\begin{aligned} h_n(k_1 T_0, \ldots, k_n T_0) &= 0 \\ c_p(k_1 T_0, \ldots, k_p T_0) &= 0 \end{aligned} \tag{10.91}$$

for $k_i < 0$, $i = 1, \ldots, \max(p, n)$.

For this type of connection, the nonlinear element response will be represented as

$$y(rT_0) = \sum_{i_1=0}^{\infty} \ldots \sum_{i_n=0}^{\infty} h_n(rT_0 - i_1 T_0, \ldots, rT_0 - i_n T_0) \prod_{q=1}^{n} e(i_q T_0) \tag{10.92}$$

$$\tilde{u}(rT_0) = \sum_{j_1=0}^{\infty} \ldots \sum_{j_p=0}^{\infty} c_p(rT_0 - j_1 T_0, \ldots, rT_0 - j_p T_0) \prod_{q=1}^{p} y(j_q T_0) \tag{10.93}$$

where $r = 0, 1$, etc.

Substituting (10.92) in (10.93) yields

$$D_n(z_1, \ldots, z_n) = -\frac{H_n(z_1, \ldots, z_n)}{W_1(z_1 \ldots z_n)} \tag{10.94}$$

where

$$a_{pn}(rT_0 - i_1T_0, \ldots, rT_0 - i_{pn}T_0) = \sum_{j_1=0}^{\infty} \ldots \sum_{j_p=0}^{\infty} c_p(rT_0 - j_1T_0, \ldots, rT_0 - j_pT_0)$$

$$\times h_n(j_1T_0 - i_1T_0, \ldots, j_1T_0 - i_nT_0) \ldots h_n(j_pT_0 - i_{n(p-1)+1}T_0, \ldots, j_pT_0 - j_{pn}T_0) \tag{10.95}$$

The change of variables $r - i_b = k_b$, $r - j_q = v_q$, $b = 1, \ldots, pn$ in (10.94, 95) gives

$$a_{pn}(k_1T_0, \ldots, k_{pn}T_0) = \sum_{v_1=0}^{\infty} \ldots \sum_{v_p=0}^{\infty} c_p(v_1T_0, \ldots, v_pT_p)h_n(k_1T_0 - v_1T_0, \ldots, k_nT_0 - v_1T_0) \ldots h_n(k_{n(p-1)+1}T_0 - v_pT_0, \ldots, k_{pn}T_0 - v_pT_0) \tag{10.96}$$

Thus, the cascade connection of nonlinear elements (Fig. 10.15*a*) may be replaced with one nonlinear element with the multidimensional weight function (10.96) as shown in Fig. 10.15*c*.

Taking the multidimensional *z*-transform of (10.96) gives

$$A(z_1, \ldots, z_{pn}) = C_p(z_1 \ldots z_n, \ldots, z_{n(p-1)+1} \ldots z_{np}) \times H_n(z_1, \ldots, z_n) \ldots H_n(z_{n(p-1)+1}, \ldots, z_{np}) \tag{10.97}$$

It follows that the cascade connection of the equivalent transfer functions $H_n(z_1, \ldots, z_n)$ and $C_p(z_1, \ldots, z_p)$, as shown in Fig. 10.15*b*, is equal to the equivalent transfer function $A_{pn}(z_1, \ldots, z_{pn})$ defined by (10.97), as shown in Fig. 10.15*d*.

In particular case at $p = 1$,

$$A_n(z_1, \ldots, z_n) = C_1(z_1, \ldots, z_n)H_n(z_1, \ldots, z_n) \tag{10.98}$$

whence

$$D_n(z_1 \ldots z_n) = A_n(z_1, \ldots, z_n) \tag{10.99}$$

In view of (10.90) the block diagram of the equivalent transfer function (10.89) takes the form shown in Fig. 10.16. At $n = 2$, we get

$$D_2(z_1, z_2) = C_1(z_1z_2)H_2(z_1, z_2) \tag{10.100}$$

In view of (10.90) and (10.83) the transfer function $D_2(z_1, z_2)$ is a cascade consisting of a linear element with the transfer function $D_1(z)$, a nonlinear

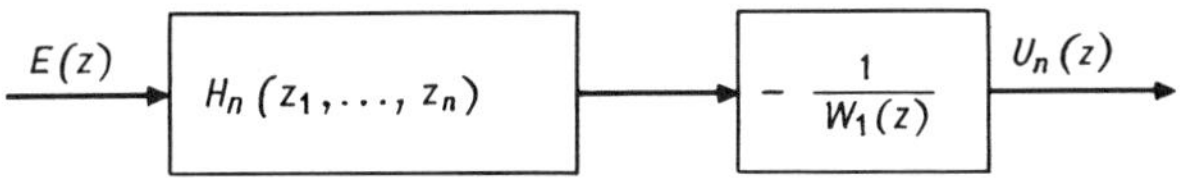

Fig. 10.16

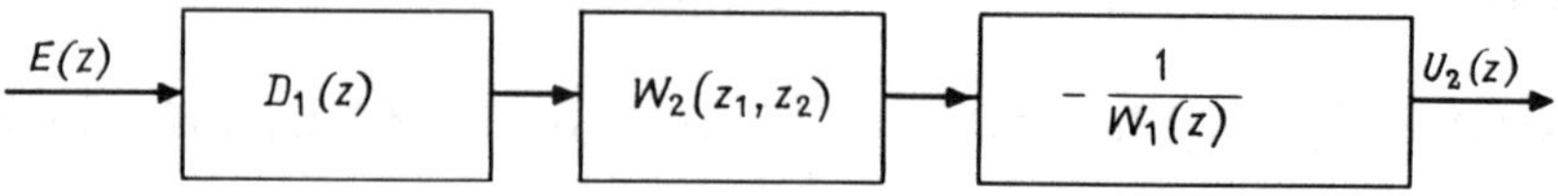

Fig. 10.17

element with the equivalent transfer function $W_2(z_1, z_2)$, and a linear element $-1/W_1(z)$, as shown in Fig. 10.17.

Now we obtain the block diagram for $H_n(z_1, ..., z_n)$ for $n > 2$. In agreement with (10.84, 85) we have

$$H_n(z_1, ..., z_n) = \sum_{\substack{p=1,...,n \\ \alpha_1,...,\alpha_p=1,...,n \\ \alpha_1+...+\alpha_p=n}} \tilde{H}_p(z_1, ..., z_n) \tag{10.101}$$

where

$$\tilde{H}_p(z_1, ..., z_n) = W_p(z_1 ... z_{\alpha_1}, ..., z_\gamma ... z_n) D_{\alpha_1}(z_1, ..., z_{\alpha_1}) ... D_{\alpha_p}(z_\gamma, ..., z_n)$$
$$\gamma = \alpha_1 + ... + \alpha_{p-1} + 1, \; p \leqslant n \tag{10.102}$$

Therefore, the block diagrams of the transfer functions (10.102) must be available to construct the block diagram for $H_n(z_1, ..., z_n)$. Comparison of (10.102) and (10.101) indicates that the equivalent transfer function $\tilde{H}_p(z_1, ..., z_n)$ cannot be implemented as cascaded nonlinear elements $D_i(z_1, ..., z_i)$ and $W_p(z_1, ..., z_p)$ only [10.4]. An appropriate construction procedure is as follows. Develop the block diagram for

$$\tilde{H}_2(z_1, ..., z_n) = W_2(z_1 ... z_{\alpha_1}, z_{\alpha_1+1} ... z_n) D_{\alpha_1}(z_1, ..., z_{\alpha_1}) D_{\alpha_2}(z_{\alpha_1+1}, ..., z_n)$$
$$\alpha_1 + \alpha_2 = n \tag{10.103}$$

The nonlinear element (10.103) performs the nonlinear transformation of $e(rT_0)$ into $\tilde{u}_n(rT_0)$ as follows:

$$\begin{aligned}\tilde{u}_n(rT_0) &= \sum_{f_1=0}^{r} ... \sum_{f_n=0}^{r} \tilde{h}_p(rT_0 - f_1T_0, ..., rT_0 - f_nT_0) \prod_{i=1}^{n} e(f_iT_0) \\ &= \sum_{k_1=0}^{r} \sum_{k_2=0}^{r} w_2(rT_0 - k_1T_0, rT_0 - k_2T_0) \\ &\times \left\{ \left[\sum_{q_1=0}^{k_1} ... \sum_{q_{\alpha_1}=0}^{k_1} d_{\alpha_1}(k_1T_0 - q_1T_0, ..., k_1T_0 - q_{\alpha_1}T_0) \right] \right. \\ &\times \left[\sum_{b_1=0}^{k_2} ... \sum_{b_{\alpha_2}=0}^{k_2} d_{\alpha_2}(k_2T_0 - b_1T_0, ..., k_2T_0 - b_{\alpha_2}T_0) \right] \\ &\times \left. \prod_{i=1}^{\alpha_1} e(q_iT_0) \prod_{j=1}^{\alpha_2} e(b_jT_0) \right\}, \quad r = 0, 1, ..., \infty \end{aligned} \tag{10.104}$$

Introducing the intermediate quantities

$$y_1(rT_0) = \sum_{q_1=0}^{r} \dots \sum_{q_{\alpha_1}=0}^{r} d_{\alpha_1}(rT_0 - q_1 T_0, \dots, rT_0 - q_{\alpha_1} T_0) \prod_{i=1}^{\alpha_1} e(q_i T_0) \tag{10.105}$$

$$y_2(rT_0) = \sum_{b_1=0}^{r} \dots \sum_{b_{\alpha_2}=0}^{r} d_{\alpha_2}(rT_0 - b_1 T_0, \dots, rT_0 - b_{\alpha_2} T_0) \prod_{j=1}^{\alpha_2} e(b_j T_0) \tag{10.106}$$

carries (10.104) to the form

$$\tilde{u}_n(rT_0) = \sum_{k_1=0}^{r} \sum_{k_2=0}^{r} w_2(rT_0 - k_1 T_0, rT_0 - k_2 T_0) y_1(k_1 T_0) y_2(k_2 T_0) \tag{10.107}$$

Now, we evaluate how the nonlinear element with the weight function $w_2(k_1 T_0, k_2 T_0)$ handles the sum $y_1(rT_0) + y_2(rT_0)$ and the difference $y_1(rT_0) - y_2(rT_0)$,

$$\begin{aligned} v_1(rT_0) &= \sum_{k_1=0}^{r} \sum_{k_2=0}^{r} w_2(rT_0 - k_1 T_0, rT_0 - k_2 T_0) \\ &\quad \times [y_1(k_1 T_0) + y_2(k_1 T_0)][y_1(k_2 T_0) + y_2(k_2 T_0)] \\ &= \sum_{k_1=0}^{r} \sum_{k_2=0}^{r} w_2(rT_0 - k_1 T_0, rT_0 - k_2 T_0) y_1(k_1 T_0) y_1(k_2 T_0) \\ &\quad + \sum_{k_1=0}^{r} \sum_{k_2=0}^{r} w_2(rT_0 - k_1 T_0, rT_0 - k_2 T_0) y_2(k_1 T_0) y_2(k_2 T_0) \\ &\quad + 2 \sum_{k_1=0}^{r} \sum_{k_2=0}^{r} w_2(rT_0 - k_1 T_0, rT_0 - k_2 T_0) y_1(k_1 T_0) y_2(k_2 T_0) \end{aligned} \tag{10.108}$$

and

$$\begin{aligned} v_2(rT_0) &= \sum_{k_1=0}^{r} \sum_{k_2=0}^{r} w_2(rT_0 - k_1 T_0, rT_0 - k_2 T_0) \\ &\quad \times [y_1(k_1 T_0) - y_2(k_1 T_0)][y_1(k_2 T_0) - y_2(k_2 T_0)] \\ &= \sum_{k_1=0}^{r} \sum_{k_2=0}^{r} w_2(rT_0 - k_1 T_0, rT_0 - k_2 T_0) y_1(k_1 T_0) y_1(k_2 T_0) \\ &\quad + \sum_{k_1=0}^{r} \sum_{k_2=0}^{r} w_2(rT_0 - k_1 T_0, rT_0 - k_2 T_0) y_2(k_1 T_0) y_2(k_2 T_0) \\ &\quad - 2 \sum_{k_1=0}^{r} \sum_{k_2=0}^{r} w_2(rT_0 - k_1 T_0, rT_0 - k_2 T_0) y_1(k_1 T_0) y_2(k_2 T_0) \end{aligned} \tag{10.109}$$

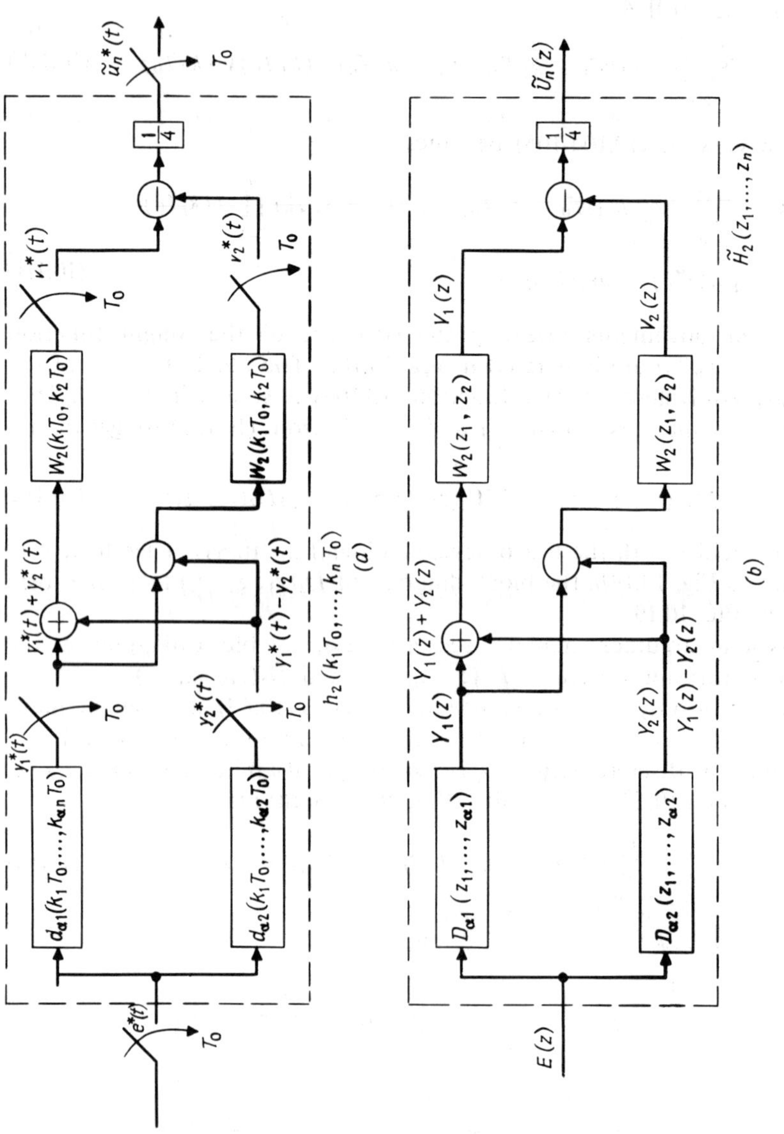

$e^*(t)$
T_0
$d_{\alpha 1}(k_1T_0, \ldots, k_{\alpha n}T_0)$
$d_{\alpha 2}(k_1T_0, \ldots, k_{\alpha 2}T_0)$
$y_1^*(t)$
$y_2^*(t)$
$y_1^*(t) + y_2^*(t)$
$y_1^*(t) - y_2^*(t)$
$W_2(k_1T_0, k_2T_0)$
$W_2(k_1T_0, k_2T_0)$
$r_1^*(t)$
$r_2^*(t)$
$\frac{1}{4}$
$\tilde{u}_n^*(t)$
$h_2(k_1T_0, \ldots, k_nT_0)$
(a)
$E(z)$
$D_{\alpha 1}(z_1, \ldots, z_{\alpha 1})$
$D_{\alpha 2}(z_1, \ldots, z_{\alpha 2})$
$Y_1(z)$
$Y_2(z)$
$Y_1(z) + Y_2(z)$
$Y_1(z) - Y_2(z)$
$W_2(z_1, z_2)$
$W_2(z_1, z_2)$
$V_1(z)$
$V_2(z)$
$\frac{1}{4}$
$\tilde{U}_n(z)$
$\tilde{H}_2(z_1, \ldots, z_n)$
(b)

Fig. 10.18

Subtracting (10.109) from (10.108) and dividing the difference by 4 gives

$$[v_1(rT_0) - v_2(rT_0)]/4$$
$$= \sum_{k_1=0}^{r} \sum_{k_2=0}^{r} w_2(rT_0 - k_1T_0, rT_0 - k_2T_0)y_1(k_1T_0)y_2(k_2T_0) \qquad (10.110)$$

which in view of (10.104-105) becomes

$$\tilde{u}_n(rT_0) = \sum_{k_1=0}^{r} \dots \sum_{k_n=0}^{r} \tilde{h}_p(rT_0 - k_1T_0, \dots, rT_0 - k_nT_0) \prod_{i=1}^{n} e(k_iT_0)$$
$$= [v_1(rT_0) - v_2(rT_0)]/4 \qquad (10.111)$$

The rearrangements revealing the structure of the weight function $\tilde{h}_p(k_1T_0, \dots, k_nT_0)$ are illustrated in Fig. 10.18*a*. The block diagram of the respective equivalent transfer function (10.103) is shown in Fig. 10.18*b*.

In a particular case of $\alpha_1 = 1$ and $\alpha_2 = 2$, from (10.103) we get the relation

$$2\tilde{H}_2(z_1, z_2, z_3) = 2W_2(z_1, z_2z_3)D_1(z_1)D_2(z_2, z_3) \qquad (10.112)$$

which coincides with the last bracketed addend in (10.84). Therefore, with reference to Fig. 10.18*b*, the block diagram of $D_3(z_1, z_2, z_3)$ takes the form shown in Fig. 10.19.

This same argument may be used to develop the block diagrams of the equivalent transfer functions $D_4(z_1, \dots, z_4)$ and $D_5(z_1, \dots, z_5)$.

In applications, the convergence of the series (10.62) controls the number of terms required for satisfactory approximation. It was mentioned earlier that the discrete Volterra series converges if all its kernels are stable. For the series (10.62) this implies that the conditions

$$\sum_{k_1=0}^{\infty} \dots \sum_{k_n=0}^{\infty} |d_n(k_1T_0, \dots, k_nT_0)| = \|d_n\| < \infty \qquad (10.113)$$

must be satisfied for all $n = 1, 2$, etc.

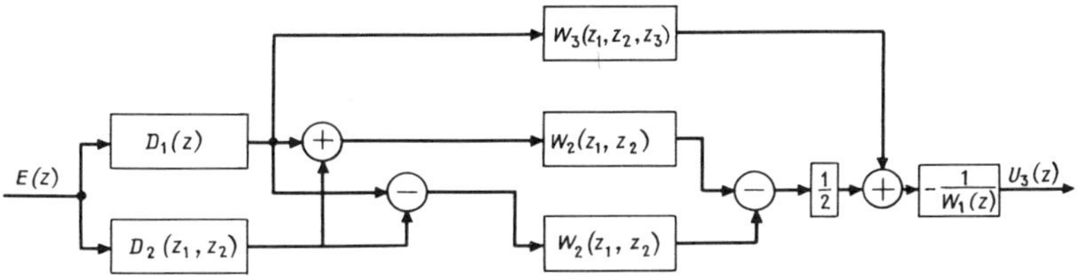

Fig. 10.19

Defining the norms of $e^*(t)$ and $u^*(t)$ as

$$\|e\| = E = \sup |e(rT_0)|, \ 0 \leqslant r < +\infty \tag{10.114}$$

and

$$\|u\| = U = \sup |u(rT_0)|, \ 0 \leqslant r < +\infty \tag{10.115}$$

from (10.66) it follows

$$U \leqslant \sum_{n=1}^{\infty} \|d_n\| E^n \tag{10.116}$$

The right-hand side of this inequality is a power series whose radius of convergence is also the radius of convergence of the discrete Volterra series (10.66). A nonzero radius of convergence is ensured by the stability of the transfer function of discrete compensatory element $D_1(z)$, which corresponds to the requirement

$$\sum_{k=0}^{\infty} |d_1(kT_0)| = \|d_1\| < \infty \tag{10.117}$$

In some cases the radius of convergence derived for (10.66) can be so small that the compensatory element has to operate in the linear mode only, thus spoiling the system dynamic behavior in handling sizeable inputs.

The idea to synthesize a compensating element in a closed loop form embodies the evaluation of its structure and parameters over the class of nonlinear elements with feedback so that the relation (10.74) be satisfied in the closed-loop system. If so, the analysis of convergence radius is not needed any longer.

Using the operator algebra we set up the expression for the nonlinear operator D^* of a discrete compensatory unit

$$D^* = (W_m^*)^{-1}[(\Phi_m^*)^{-1} - I^*]^{-1} \tag{10.118}$$

where I^* is the discrete unit operator. The nonlinear operator D^* is seen to depend on the parameter m. Therefore the condition (10.74) is satisfied for one value of m only. Let $m = 1$, then (10.118) becomes

$$D^* = (W^*)^{-1}[(\Phi^*)^{-1} - I^*]^{-1} \tag{10.119}$$

According to (10.77) the second factor in (10.119) is a linear operator,

$$C_1^* = [(\Phi_1^*)^{-1} - I^*]^{-1} \tag{10.120}$$

We represent the linear operator W^* as a sum of two operators,

$$W^* = W_1^* + W_{2,N}^* \tag{10.121}$$

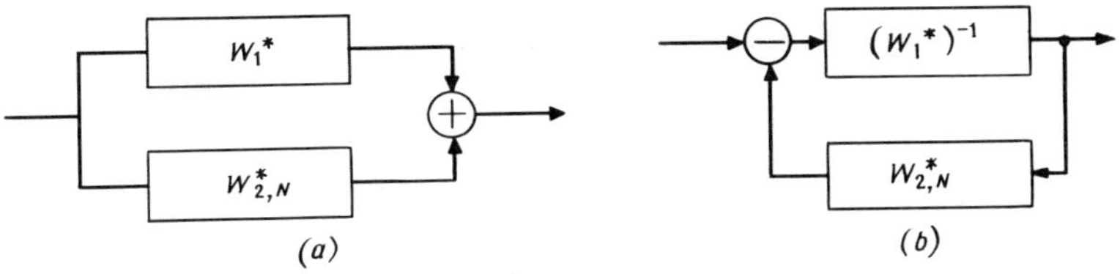

Fig. 10.20

where W_1^* is linear and $W_{2,N}^*$ is nonlinear,

$$W_{2,N}^* = \sum_{n=2}^{N} W_n^* \tag{10.122}$$

The operator W_1^* is associated with the transfer function $W_1(z)$, whereas $W_{2,n}^*$ with a set of equivalent transfer functions $W_n(z_1, \ldots, z_n)$, $n = 2, \ldots, N$. Assuming that the block diagram of W^* is as shown in Fig. 10.20*a*, the inverse operator $(W^*)^{-1}$ may be set in accordance with the diagram at (*b*) [10.5].

In view of (10.120) the expression (10.119) becomes

$$D^* = (W^*)^{-1} C_1^* \tag{10.123}$$

or, observing that the operator of the linear part of the compensatory element is

$$D_1^* = (W_1^*)^{-1} C_1^* \tag{10.124}$$

we finally get

$$D^* = (W^*)^{-1} W_1^* D_1^* \tag{10.125}$$

The block diagram of the nonlinear discrete compensatory element constructed in the closed loop form by the above expression taking into account the network of $(W_1^*)^{-1}$ is presented in Fig. 10.21. In terms of multidimensional z-transforms, it takes the form shown in Fig. 10.22.

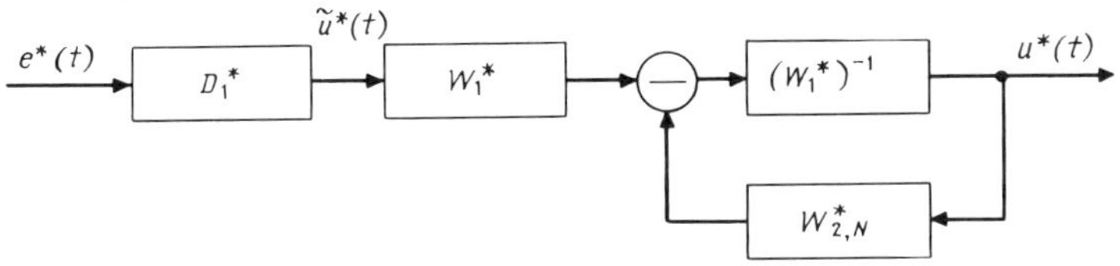

Fig. 10.21

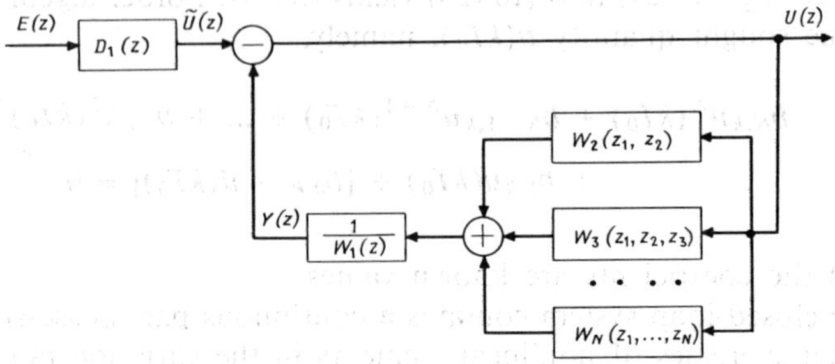

Fig. 10.22

In view of (10.120) the expression (10.124) is associated with the transfer function

$$D_1(z) = \frac{\Phi_1(z)}{W_1(z)[1 - \Phi_1(z)]} \tag{10.126}$$

The outlined structure of the compensatory element is convenient in that it allows the synthesis by a known transfer function $W_1(z)$ followed by forming a loop compensating for the nonlinear properties of the continuous part of the system. This nonlinear compensation element does not produce an explicit dependence of the control action $u^*(t)$ on the error $e^*(t)$. To evaluate this dependence we have to reveal the relation between the intermediate quantities $y^*(t)$, $\tilde{u}^*(t)$ and the control $u^*(t)$.

By inspecting the block diagram in Fig. 10.22 we have

$$U(z) = \tilde{U}(z) - Y(z) \tag{10.127}$$

therefore

$$u(kT_0) = \tilde{u}(kT_0) - y(kT_0), \; k = 0, \; 1, \; \text{etc.} \tag{10.128}$$

where the current value of $\tilde{u}(kT_0)$ is known.

Because the block diagrams of the equivalent transfer functions $W_n(z_1, \; ..., \; z_n)$, $n = 2, \; ..., \; N$, are known, the quantity $y(kT_0)$ may be represented as a polynomial of degree N, viz.,

$$\begin{aligned} y(kT_0) &= b_{N,k}u^N(kT_0) \\ &= b_{N-1,k}u^{N-1}(kT_0) + \; ... \; + b_{1,k}u(kT_0) + b_{0,k} \end{aligned} \tag{10.129}$$

where the coefficients $b_{i,k}$, $i = 0, \; 1, \; ..., \; N$, are also known and depend on k in the general case.

Substituting (10.129) into (10.128) yields the N-th order algebraic equation in the sought quantity $u(kT_0)$, namely,

$$b_{N,k}u^N(kT_0) + b_{N-1,k}u^{N-1}(kT_0) + \dots + b_{2,k}u^2(kT_0) + b_{1,k}u(kT_0) + [b_{0,k} - \tilde{u}(kT_0)] = 0 \qquad (10.130)$$

where all the coefficients are known values.

If the closed-loop system contains a continuous part possessing single values characteristics of nonlinear elements in the variation range of the control action

$$u(rT_0) \in (-U_1,\ U_2), \qquad r = 0,\ 1,\ \text{etc.} \qquad (10.131)$$

then equation (10.130) has in this range a single real root which is the sought value of control at time $t = kT_0$.

Evaluation of the real root of (10.130) in the given interval (10.131) is carried out by simple iteration techniques and can be implemented in real time even for large N.

Example 10.3. It is desired to synthesize a nonlinear discrete compensatory unit for the automatic control system shown in Fig. 10.23. The system parameters are $T_0 = T_1 = 0.2$ s, $k_1 = 20$, and $k_2 = 1$.

The nonlinear element is described by the function

$$\begin{aligned} F(\gamma) &= a_1\gamma + a_3\gamma^3, \quad &\text{for } \gamma \leqslant \gamma_1 \\ &= c \operatorname{sgn} \gamma, \quad &\text{for } \gamma > \gamma_1 \end{aligned} \qquad (10.132)$$

where $a_1 = 0.2$, $a_3 = 0.001$, $\gamma_1 = 16.2$, and $c = 10$.

The nonlinear switching function of the element is shown in Fig. 10.24 from which we infer that subject to condition

$$\sup_{0<r\leqslant+\infty} |u(rT_0)| \leqslant \gamma_1/k_1 = 0.81 \qquad (10.133)$$

the continuous part of the system may always be represented as a section

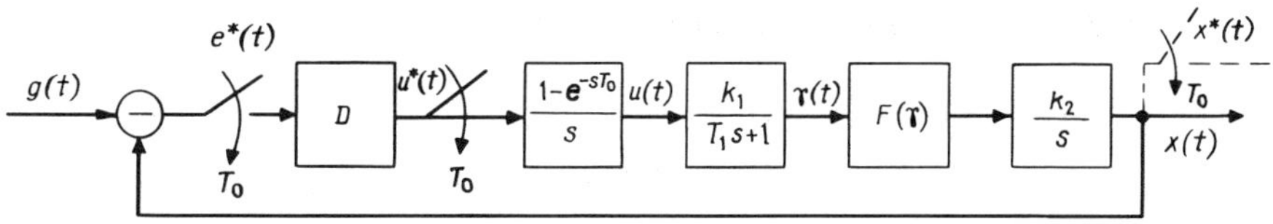

Fig. 10.23

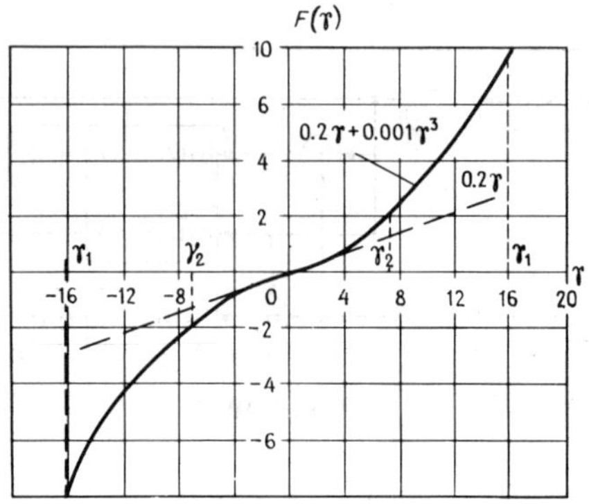

Fig. 10.24

of Volterra series with $N = 3$. Then, in agreement with Example 10.2, the equivalent transfer functions have the form

$$W_1(z) = (0.294z + 0.211)/(z - 1)(z - 0.368) \tag{10.134}$$

$$W_2(z_1, z_2) = 0 \tag{10.135}$$

$$\begin{aligned} W_3(z_1, z_2, z_3) = {} & [0.134z_1z_2z_3 + 0.0358(z_1z_2 + z_1z_3 + z_2z_3) \\ & + 0.0294(z_1 + z_2 + z_3) + 0.0743]/[(z_1 - 0.368) \\ & \times (z_2 - 0.368)(z_3 - 0.368)(z_1z_2z_3 - 1)] \end{aligned} \tag{10.136}$$

Consider first the synthesis of a nonlinear compensatory element in the form of a discrete series of Volterra. We confine ourselves to the first five terms of the series (10.66). To begin with we note that the continuous part of the system satisfies the condition

$$W[u^*(t)] = -W[-u^*(t)] \tag{10.137}$$

because of which its equivalent transfer functions of even order are identically zero and, consequently,

$$D_2(z_1, z_2) = D_4(z_1, z_2, z_3, z_4) = 0 \tag{10.138}$$

Thus, in agreement with (10.134) and (10.138), the block diagram of the compensatory unit takes the form shown in Fig. 10.25.

Let us require that, given the condition (10.33) is satisfied in the compensated system, the transients have the least and finite duration. Then the transfer function found, say, by the polynomial equation technique [10.6]

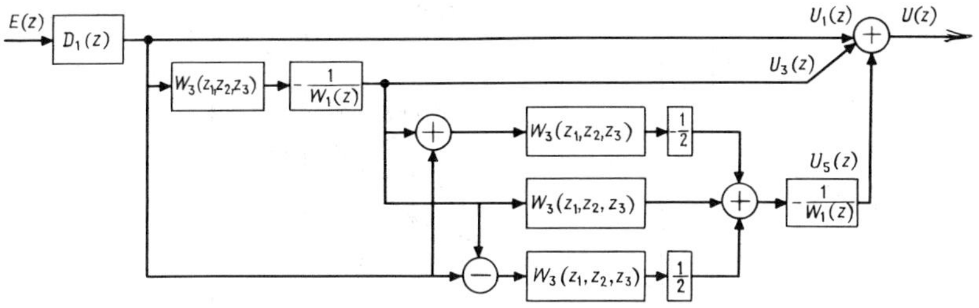

Fig. 10.25

will be

$$D_1(z) = \frac{1.98z - 0.727}{z + 0.418} \tag{10.139}$$

In this case, using (10.134, 136) the control law formed by the nonlinear compensatory element may be written as

$$u_1(kT_0) = 1.98e(kT_0) - 0.727e(kT_0 - T_0) - 0.418u_1(kT_0 - T_0) \tag{10.40}$$

$$\left.\begin{aligned} e_1(kT_0) &= e(kT_0) + 0.368e(kT_0 - T_0) \\ e_2(kT_0) &= 0.134e_1^3(kT_0) + 0.0743e_1^3(kT_0 - T_0) \\ &\quad + 0.1074e_1^2(kT_0)e_1(kT_0 - T_0) + 0.0822e_1(kT_0)e_1^2(kT_0 - T_0) \\ u_3(kT_0) &= -3.39e_2(kT_0) + 1.25e_2(kT_0 - T_0) - 0.718u_3(kT_0 - T_0) \end{aligned}\right\} \tag{10.41}$$

$$\left.\begin{aligned} e_4(kT_0) &= u_1(kT_0) + u_3(kT_0) \\ e_5(kT_0) &= \frac{1}{2}\Big[0.134e_4^3(kT_0) + 0.0743e_4^3(kT_0 - T_0) \\ &\quad + 0.1074e_4^2(kT_0)e_4(kT_0 - T_0) + 0.0822e_4(kT_0)e_4^2(kT_0 - T_0)\Big] \\ e_6(kT_0) &= u_1(kT_0) - u_3(kT_0) \\ e_7(kT_0) &= \frac{1}{2}\Big[0.134e_6^3(kT_0) + 0.0743e_6^3(kT_0 - T_0) \\ &\quad + 0.1074e_6^2(kT_0)e_6(kT_0 - T_0) + 0.0822e_6(kT_0)e_6^2(kT_0 - T_0)\Big] \\ e_8(kT_0) &= 0.134u_3^3(kT_0) = 0.0743u_3^3(kT_0 - T_0) \\ &\quad + 0.1074u_3^2(kT_0)u_3(kT_0 - T_0) + 0.0822u_3(kT_0)u_3^2(kT_0 - T_0) \\ &\quad - e_5(kT_0) + e_7(kT_0) \\ u_5(kT_0) &= -3.39e_8(kT_0) + 1.25e_8(kT_0 - T_0) - 0.718u_5(kT_0 - T_0) \end{aligned}\right\} \tag{10.142}$$

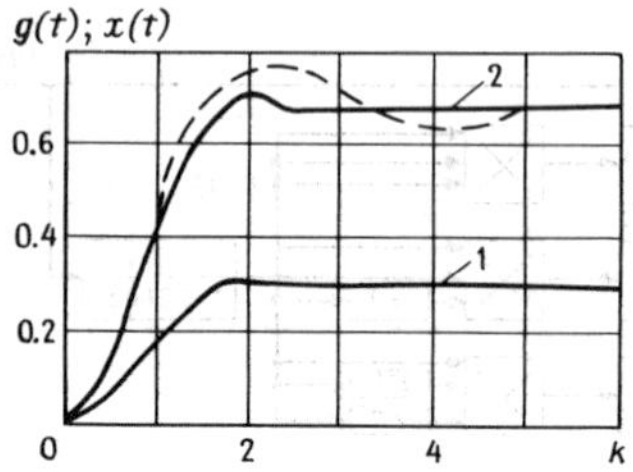

Fig. 10.26

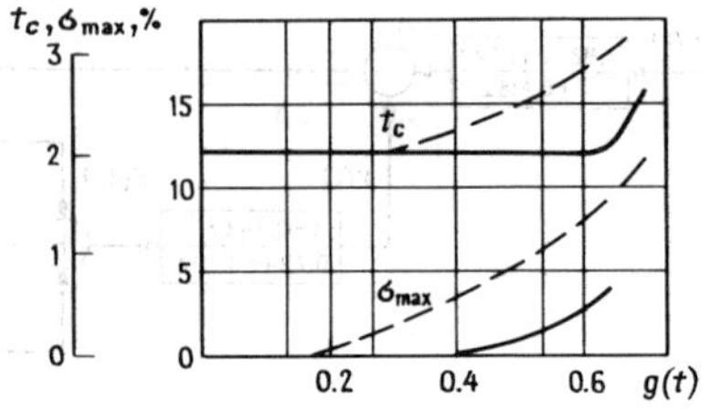

Fig. 10.27

and

$$u(kT_0) = u_1(kT_0) + u_3(kT_0) + u_5(kT_0), \quad k = 0,\ 1,\ \text{etc.} \tag{10.143}$$

Simulation of the nonlinear system on a computer resulted in the transients both with the linear control law (10.140), corresponding to a compensatory element with the transfer function (10.139), and with a nonlinear compensatory element having the control law (10.140-143). The simulated response curves obtained with two step inputs are shown in Fig. 10.26. Curve *1* represents the system response to $g(t) = 0.3$. With this excitation the system exhibits a practically linear behavior. The nonlinear properties become significant for stronger stimuli. Curve *2* represents the system response to $g(t) = 0.7$. The broken line shows a transient in the system with a linear compensatory element. The settling time $t = 3T_0$ and the overshoot $\sigma_{max} = 12\%$. The system with a nonlinear compensatory unit exhibits a superior transient behavior with $t = 2.2T_0$ and $\sigma_{max} = 4\%$.

Figure 10.27 compares the transient characteristics t_c and σ_{max} for the systems with a linear (dashed lines) and nonlinear (solid lines) compensation element in handling various step inputs. The system with a nonlinear compensatory element is seen to exhibit a superior performance. However, its error variation range is not very large as the radius of convergence r_M of the respective series (10.66) is about 0.75. Therefore, for normal operation when the input signals are such that the error exceeds the radius of convergence, the input of the compensation unit, implemented as a trancated Volterra series, should be equipped with a bilateral limiter with a level of $r_M = 0.75$. Simulation experiments indicated that in all dynamic modes, the quantity $\gamma(t)$ at the input of the nonlinear element of the continuous system part is within $|\gamma(t)| \leqslant \gamma_2 = 7$. This circumstance constrains the power potentialities of the actuator of the control system in handling errors

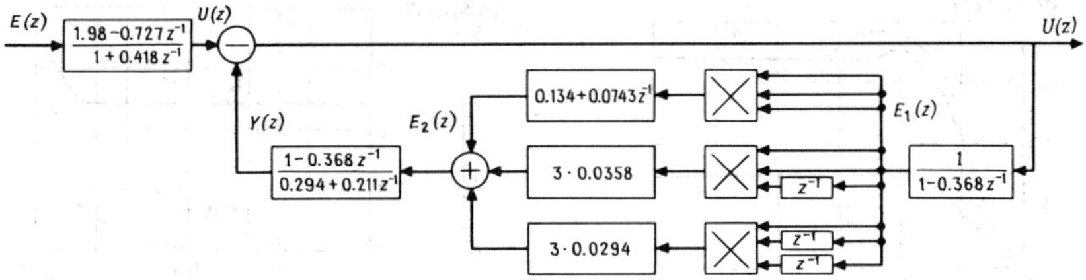

Fig. 10.28

exceeding the radius of convergence and considerably penalizes the dynamic behavior of the system.

This limitation is eliminated in the closed loop nonlinear compensatory unit shown in Fig. 10.28. This element is associated with the difference equations

$$\tilde{u}(kT_0) = 1.98e(kT_0) - 0.727e(kT_0 - T_0) - 0.418\tilde{u}(kT_0 - T_0) \qquad (10.144)$$

$$\left.\begin{aligned} e_1(kT_0) &= u(kT_0) + 0.368e_1(kT_0 - T_0) \\ e_2(kT_0) &= 0.134e_1^3(kT_0) + 0.0743e_1^3(kT_0 - T_0) \\ &+ 0.1074e_1^2(kT_0)e_1(kT_0 - T_0) + 0.0822e_1(kT_0)e_1^2(kT_0 - T_0) \\ y(kT_0) &= 3.39e_2(kT_0) - 1.25e_2(kT_0 - T_0) - 0.718y(kT_0 - T_0) \\ u(kT_0) &= \tilde{u}(kT_0) - y(kT_0) \end{aligned}\right\} \qquad (10.145)$$

In (10.144, 145) we know the current value of system error $e(kT_0)$ and seek for $u(kT_0)$. Using (10.145), we set up the cubic equation in the sought quantity

$$b_{3,k}u^3(kT_0) + b_{2,k}u^2(kT_0) + b_{1,k}u(kT_0) + [b_{0,k} - \tilde{u}(kT_0)] = 0 \qquad (10.146)$$

where

$$\left.\begin{aligned} b_{3,k} &= 0.454 \\ b_{2,k} &= 0.863e_1(kT_0 - T_0) \\ b_{1,k} &= 1 + 4.45e_1^2(kT_0 - T_0) \\ b_{0,k} &= 1.07e_1^3(kT_0 - T_0) - 1.25e_2(kT_0 - T_0) \\ &\quad - 0.718y(kT_0 - T_0) \end{aligned}\right\} \qquad (10.147)$$

To compute the current values of the coefficients $b_{i,k}$, $i = 0, 1, 2$, in (10.146) we employ the relation derived with (10.145), namely,

$$\left.\begin{aligned} e_1(kT_0) &= u(kT_0) + 0.368e_1(kT_0 - T_0) \\ e_2(kT_0) &= 0.134u^3(kT_0) + 0.255e_1(kT_0 - T_0)u^2(kT_0) \\ &\quad + 1.312e_1^2(kT_0 - T_0)u(kT_0) + 0.316e_1^2(kT_0 - T_0) \\ y(kT_0) &= 3.39e_2(kT_0) - 1.25e_2(kT_0 - T_0) - 0.718y(kT_0 - T_0) \end{aligned}\right\} \quad (10.148)$$

In solving (10.146) by iterations it would be convenient to take zero value as the initial approximation of the sought control action, i.e. $u_0(kT_0) = 0$. In view of single-valuedness of the nonlinear characteristic $\tilde{F}(\gamma) = a_1\gamma + a_3\gamma^3$, for any γ cubic equation (10.146) has one real root at all values of $\tilde{u}(kT_0)$. Therefore, the variation range for $u(kT_0)$ in which the root is sought may be taken arbitrary wide, say, $u(kT_0) \in (-5, 5)$. If so, then in handling large error values a situation may occur with $\gamma > \gamma_1$, and the actuator enters the saturation mode.

Figure 10.29 shows the transients obtained by computer simulation of the system with a nonlinear closed loop compensatory element. Curve *1* shows the system response to the step input $g(t) = 0.6$, curve *2* is for $g(t) = 1.2$, and the dashed lines represent the respective transients in the system with a linear compensation unit.

With the nonlinear compensatory unit, the condition $\gamma \leqslant \gamma_1$ was satisfied up to $g(t) = 1.8$. For $g(t) = 1.2$, the overshoot decreased from 42% to 3%, and the settling time reduced to less then one third of the previous value. The presence of the small overshoot in the system with the nonlinear closed loop compensatory unit is associated with the occurrence of damped transient oscillations when the input is considerable because zero error is guaranteed in sampling instants only.

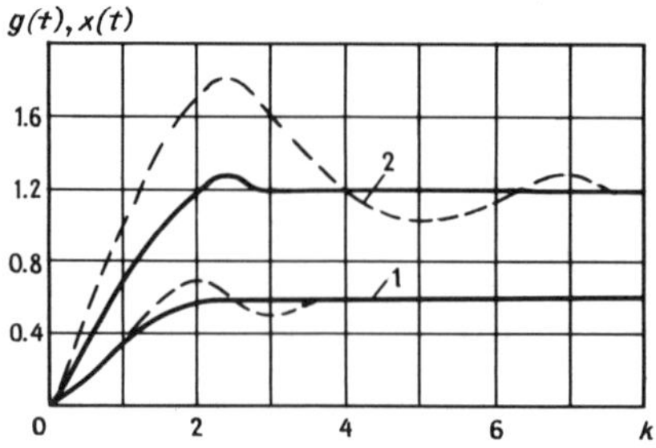

Fig. 10.29

References

10.1 Ya.Z. Tsypkin and Yu.S. Popkov: *Teoriya nelineinykh impulsnykh system* (Theory of nonlinear sampled data systems). Nauka, Moscow 1973.

10.2 H.A. Barker and S. Ambati: *Proc. IEE* **119**(9), 1407-1413 (1972).

10.3 P. Alper: *Revue A* **6**(4), 199-212 (1964).

10.4 M. Kielkiewicz: *Int. J. Control* **11**(6), 1005-1010 (1970).

10.5 V.V. Solodovnikov and V.L. Lenskii: in *Sovremennye metody proyektirovaniya sistem avtomaticheskogo upravleniya* (Modern methods of control system design). Mashinostroyeniye, Moscow 1967. pp. 176-200.

10.6 A.I. Shipilov, A.D. Modyayev, and V.G. Ivanenko: in *Nelineinye korrektiruyushchiye ustroistva v sistemakh avtomaticheskogo upravleniya* (Nonlinear compensatory elements in automatic control systems). Mashinostroyeniye, Moscow 1971, pp. 335-363.

11 Sequential Decision Making in Multicriterial Optimization of Nonlinear Systems

B. V. Sazykin

In the design of any engineering system there is a need for formulating and choosing the most efficient alternative among a variety of feasible versions satisfying the performance specification. The choice is effected on the basis of a preference structure devised by the decision maker (DM). If this structure allows linear preferential ordering of alternatives, then the choice may be reduced to a scalar optimization problem which is solvable in a rather straightforward manner. In multiattribute optimization of nonlinear systems, however, preference ordering of objective functions $g(\cdot)$ of alternative solutions $\mathbf{a}_i$, $i = 1, ..., n$, can no longer be constructed as a linear scheme $g(\mathbf{a}_1) > g(\mathbf{a}_2) > ... > g(\mathbf{a}_n)$ — as a rule it is induced by a partial ordering of points on vector objective space. A decision as to the more suitable alternative is then made on the basis of a Pareto-optimum model.

One should be careful, however, in using Paretian models. An example may be a selection from a random set of an independent random sample of vector quantities with independent continuously distributed coordinates. Dubov *et al.* [11.1] have indicated that with large samples the Paretian set, on average, receives a number of points equal to about the logarithm of the sample size. In other words, to obtain 14 Pareto optimal points one has to exercise a sample consisting of one million of alternatives, which is too large to be practical. Therefore in order to solve these problems it is desirable to decompose the partial ordering into a few quasi-orderings and after that apply suitable optimization procedures in sequence.

Several techniques are available for solving the decomposition problems, for example, covering of a partial order graph by the minimum number of trials [11.2] and Paretian stratification of elements of a random sample [11.2]. In the method of sequential decision making, the formulation and choice of alternatives are combined into a single problem. The ensuing chapter addresses the problem of multicriterial optimization with the nonlinear preference functions known accurate to the vector of parameters. A problem solver developed by the authors [11.3] for this situation relies on a multiple step procedure of formulation and choice of a solution. It eliminates in succession the areas of parameter space that do not contain an optimal solution. A sequential decomposition of the parameter space area into feasible and unfeasible subregions is performed at each step of the iteration procedure with the aid of pattern recognition theory [11.4] and ordered statistics [11.5].

11.1 Multiobjective Optimization

Consider a multiobjective optimization problem in the following statement. Let $\mathbf{f} = [f_1(\mathbf{x}), f_2(\mathbf{x}), ..., f_p(\mathbf{x})]$ be a vector of nonlinear objective functions in a p-dimensional Euclidean space $\mathbf{E}^p$ where p is the number of objectives thought of as all performance criteria and pertinent constraints, $\mathbf{x} \in \mathbf{E}^k$ is a parameter vector to be determined, also $\mathbf{x} \in X$ where X is a parallepiped of the parameters $x_j^- \leqslant x_j \leqslant x_j^+$; $j = 1, ..., k$.

Assume that the preference functions μ_i scaled between 0 and 1 are given on the sets $\{f_i\} \subset \mathbf{E}^1$, i.e.,

$$\mu_i(f_i \,|\, \mathbf{v}_i)\colon\ \mathbf{E}^1 \to [0,\ 1];\ \boldsymbol{\mu} = (\mu_1,\ ...,\ \mu_p) \in Y \subset \mathbf{E}^p$$

In the general case the functions are nonlinear and given accurate to the parameter vector $\mathbf{v}_i$ which can be changed by the DM in the optimization process. Some versions of giving the preference function are shown in Fig. 11.1. It is required to find

$$\max_{x \in X} \hat{U}(\mu_1[f_1(\mathbf{x})],\ \mu_2[f_2(\mathbf{x})],\ ...,\ \mu_p[f_p(\mathbf{x})];\ \mathbf{v}) \tag{11.1}$$

where $\hat{U}$ is the multiattribute (global) utility function (MUF) given in an implicit form, convex on the set Y and twice differentiable; and $\mathbf{v} =$ $= (\mathbf{v}_1, ..., \mathbf{v}_p)$ is the vector of the parameters of the preference functions.

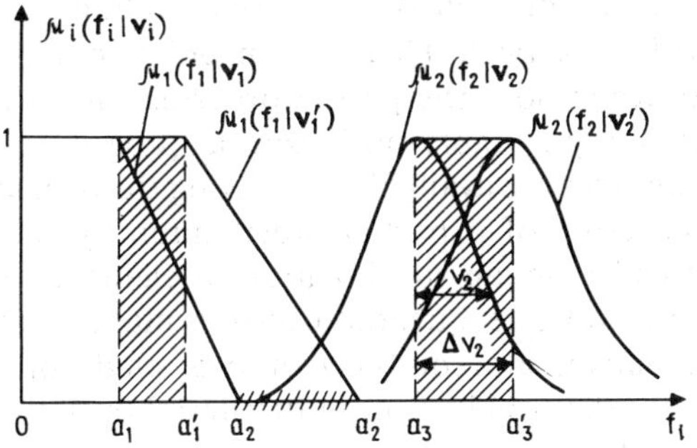

Fig. 11.1

Because $\hat{U}$ is not given explicitly we avoid application of the methods based on approximation of $\hat{U}$ [11.6] as being rather laborious and hard to implement. Instead we convert the continuous problem (11.1) to a discrete problem of independent choice [11.7]. For this purpose, as alternatives, we shall view the set $F = \{f^1, ..., f^n\}$ of the elements $f^i \in \mathbf{E}^p$ satisfying the condition $f^i \in \Omega^i$, $i = 1, ..., n$, where $\Omega^1, ..., \Omega^n$ is a sequence of bounded sets contained in $\mathbf{E}^p$.

Consider the problem $\langle \Omega, R_\psi \rangle$, where

$$\Omega = \prod_{i=1}^{n} \Omega^i$$

R_ψ is the binary relation defined by the formula

$$FR_\psi G \Leftrightarrow \psi(F) > \psi(G), \qquad F, G \subset \Omega \tag{11.2}$$

and $\psi(F) = \sum_{i=1}^{n} \hat{U}(\mu[f^i])$. We reduce the problem $\langle \Omega, R_\psi \rangle$ to a single objective problem $\langle \tilde{\Omega}, > \rangle$ with quasilinearly ordered classes, where

$$\tilde{\Omega} = \prod_{i=1}^{n} \tilde{\Omega}^i$$

and a class $\tilde{\Omega}^i$ is the image of Ω^i in the mapping g: $\mathbf{E}^p \to \mathbf{E}^1$ given by the expression

$$g(\mathbf{f}^i \mid \mathbf{v}^i) = \sum_{j=1}^{p} \lambda_j(\mathbf{v}^i)\mu_j(\mathbf{f}_j^i \mid \mathbf{v}_j^i) \tag{11.3}$$

where $\lambda_j(\mathbf{v}^i)$ is the weight factor reflecting the preference structure at point $\mathbf{f}^i$ of the i-th step, and $\mathbf{v}$ is a parameter varying the preference structure. For $\lambda_j(\mathbf{v}^i)$, the following relations hold

$$\sum_{j=1}^{p} \lambda_j(\mathbf{v}^i) = 1 \quad \text{and} \quad \lambda_j(\mathbf{v}^i) > 0 \tag{11.4}$$

A proof of the problem $\langle \Omega, R_\psi \rangle$ reducibility to the problem $\langle \tilde{\Omega}, > \rangle$ for a convex, bounded and closed Ω and λ-separable relations can be found in the book by Makarov *et al.* [11.7]. For a nonconvex problem, the mapping g has the form

$$g(\mathbf{f}^i \mid \mathbf{v}^i) = \min_{1 \leqslant j \leqslant p} \lambda_j(\mathbf{v}^i)\mu_j(f_j^i \mid \mathbf{v}_j^i) \tag{11.5}$$

The solution to the problem $\langle \tilde{\Omega}, > \rangle$ is a set of points $\{z_*^1, ..., z_*^n\}$, where

$$z_*^i = \max g(\mathbf{f}^i \mid \mathbf{v}^i), \qquad z_*^i \in \tilde{\Omega}^i, \qquad i = 1, ..., n \tag{11.6}$$

The solution of the problem $\langle \tilde{\Omega}, > \rangle$ in the form (11.6) has been obtained on the space of objective functions Y. We shall seek a solution lying in the parallelepiped of parameters X. It can be found by the transformations

$$\Psi_i: z_*^i \to \mathbf{x}_*^i \in \pi X^i \subset X \tag{11.7}$$

given with the aid of multivalued relations [11.8].

In (11.7), πX^i is a Pareto class of quasiordered alternatives for which $\mathbf{x} > \mathbf{x}'$ if $\mathbf{x} \in \pi X^i$, $\mathbf{x}' \in \pi X^k$, and $k < i$. To construct the sets $\pi X^i \subset X$, $i = 1, \ldots, n$, and the sequence $\{x_*^i[\lambda(\mathbf{v}^i)]\}$ that converges to the optimal solution

$$\mathbf{x}_0[\lambda(\mathbf{v}_0)] = \lim_{n \to \infty} \mathbf{x}_*^n[\mathbf{v}^n] \tag{11.8}$$

an appropriate procedure may be as follows.

Let F be interpreted as a set containing a system of subsets $\Omega_j^i \subset \Omega^i$, $j = 1, 2$; $i = 1, \ldots, n$, with the same collection of properties within a class j. The methods of pattern recognition may be used to evaluate a required Ω_j^i. We construct a step-by-step procedure where at each i-th step the DM varies M times the parameter v^i in an appropriate range, and the m-th variation is followed by computing the quantity

$$z_m^i = g(\mathbf{f}^i \,|\, \mathbf{v}_m^i), \qquad m = 1, \ldots, M \tag{11.9}$$

at points $\mathbf{f}^i \in \Omega^i$. Then at step i one can formulate two conditional classes of images, Ω_1^i and Ω_2^i: $(\mathbf{f}_1^i, \ldots, \mathbf{f}_m^i, \ldots, \mathbf{f}_M^i) \subset \Omega_1^i$, for which $z_m^i \in [\mathbf{z}^i, z_*^i]$, where $\mathbf{z}^i$ is some level of significance established by the DM; and $(f_1^i, \ldots, f_l^i, \ldots$ $\ldots, f_L^i) \subset \Omega_2^i$, for which $z_m^i \in B_k^m$, where B_k^m are k-dimensional blocks of equal significance to be discussed in Sect. 11.3.

Figure 11.2 illustrates the formation of conditional classes of images for the case of a two-criterial problem ($p = 2$, $M = L = 2$, $m = 5$). Since $f^i = f^i(x^i)$, the classes Ω_1^i and Ω_2^i in the space of criteria are associated with the classes X_1^i and X_2^i in the space of parameters. Consequently, with the aid of the transformation Ψ_i, a sample $(f_1^i, \ldots, f_M^i) \subset \Omega_1^i$ may be associated with the sample

$$(\mathbf{x}_1^i, \ldots, \mathbf{x}_m^i, \ldots, \mathbf{x}_M^i) \subset X_1^i \tag{11.10}$$

and a sample $(f_1^i, \ldots, f_l^i, \ldots, f_L^i) \subset \Omega_2^i$ may be associated with

$$(\mathbf{x}_1^i, \ldots, \mathbf{x}_l^i, \ldots, \mathbf{x}_L^i) \subset X_2^i \tag{11.11}$$

We shall refer to the samples (11.10, 11) as *teaching samples* for classes X_1^i and X_2^i respectively. These teaching samples will be used in separating the initial set of parameters $X^i = X_1^i \cup X_2^i$ into a "good" set X_1^i, i.e. the one containing the optimal solutions πX^i and a "bad" set X_2^i which does not. We shall assume that the separating frontier $d_i(\mathbf{x})$ can be constructed at

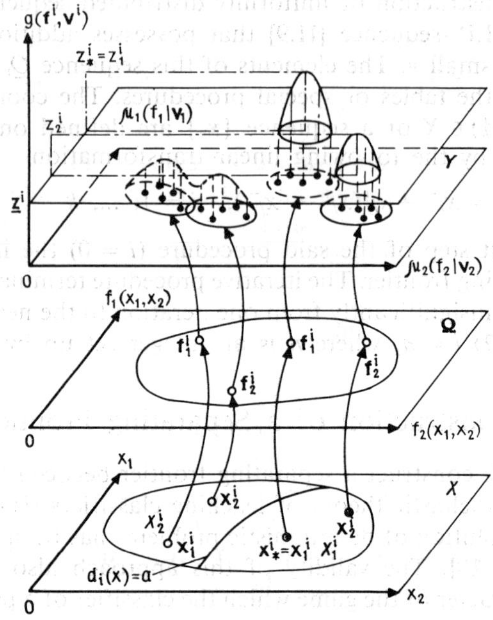

Fig. 11.2

each i-th step and the decision rule assigning any vector $\mathbf{x} \in X$ to the class X_1^i has the form

$$d_i(\mathbf{x}) \leqslant a, \quad i = 1, \ldots, n \tag{11.12}$$

where a is a real number. Suppose now that we have a generator of sequences $\{\mathbf{x}_\nu\}$, $\mathbf{x}_\nu \in X^i$, uniformly distributed in X^i. By screening these sequences with the help of the decision rule (11.12) we obtain at each $(i + 1)$th step of the optimization a set of improved feasible solutions

$$\tilde{X}^{i+1} = \{\mathbf{x}_\nu^{i+1} \,|\, d_i(\mathbf{x}_\nu) \leqslant a, \quad \nu = 1, \ldots, N_{i+1}\} \tag{11.13}$$

where N_{i+1} is the total number of points $\mathbf{x}_\nu^{i+1}$ which will be called the points of exploration of the set X^{i+1} at step $i + 1$.

Based on (11.6, 7) we find

$$\mathbf{x}_*^{i+1}[\lambda(\mathbf{v}^{i+1})] \in \pi X^{i+1} \tag{11.14}$$

for which $z_m^{i+1} = z_*^{i+1}$. Repeating this procedure n times, we arrive at the sequence $\{\mathbf{x}_*^i[\lambda(\mathbf{v}^i)],\ i = 1, \ldots, n\}$. This sequence converges as n tends to infinity because it is an infinite subset of the bounded set $X \subset \mathbf{E}^k$ and consequently has at least one limiting point.

For the construction of uniformly distributed sequences $\{\mathbf{x}_\nu\}$ we may resort to the LP_τ-sequence [11.9] that possesses additional properties of uniformity at small ν. The elements of this sequence $Q_\nu = (q_1^\nu, ..., q_k^\nu)$ are computed by the tables or special procedures. The coordinates of points $\mathbf{x}_\nu = (x_1^\nu, ..., x_k^\nu) \in X$ of a sequence $\{\mathbf{x}_\nu\}$ are defined on the basis of the LP_τ-sequence by the following linear transformation

$$x_j^\nu = x_j^- + q_j^\nu(x_j^+ - x_j^-), \quad j = 1, ..., k \tag{11.15}$$

At the first step of the said procedure ($i = 0$) the initial set $X^0 = X$ has no separating frontier. The iterative procedure terminates if (1) solutions $\mathbf{x}_*^i[\lambda(\mathbf{v}^i)]$ vary insignificantly from one iteration to the next and they satisfy the DM, or (2) $i = n$, where n is an integer set up by the DM.

11.2 Construction of a Separating Frontier

We construct a separating frontier between the classes X_1^i and X_2^i using the stochastic theory of teaching classifiers [11.4]. The stochastic approach to solution of deterministic problems has been justified by many authors [11.4, 10]. The validity of this approach also follows from the statistical character of the game which the classifier of a pattern recognition system plays with the nature in the process of decision making. For two classes, the separating frontier $d_i(\mathbf{x})$ is defined [11.10] by the equation

$$d_i(\mathbf{x}) = 2P(X_1^i | \mathbf{x}) - 1, \qquad i = 1, ..., n$$

where $P(X_1^i | \mathbf{x})$ is the conditional probability density for the class X_1^i. In this case the classification rule is as follows:

$$\begin{aligned} \mathbf{x} \in X_1^i, &\quad \text{if} \quad P(X_1^i | \mathbf{x}) > 1/2 \\ \mathbf{x} \in X_2^i, &\quad \text{if} \quad P(X_1^i | \mathbf{x}) < 1/2 \end{aligned} \tag{11.16}$$

With a complicated image taxonomy, the separating frontier is as a rule nonlinear, therefore we represent the decision function in the form of a decomposition over the basis of nonlinear functions $\varphi_j(\mathbf{x})$ constituted by multidimensional orthonormal Hermite polynomials in the orthogonality interval $-\infty < x < \infty$. Then

$$P(X_1^i | \mathbf{x}) \simeq \sum_{j=1}^{J} w_j^i \varphi_j(\mathbf{x}) = \mathbf{W}^i \mathbf{\Phi}(\mathbf{x}) \tag{11.17}$$

where $\mathbf{W}^i = (w_1^i, ..., w_J^i)$ is the row vector of weights for the class X_1^i, which is to be determined in the teaching process, and $\mathbf{\Phi}(\mathbf{x}) = [\varphi_1(\mathbf{x}), ..., \varphi_J(\mathbf{x})]^T$ is the column vector of basic functions.

To construct the teaching algorithm and evaluate the vector of weights $\mathbf{W}^i$ we resort to the least square approximation. Consider the function of the least square criterion

$$D(\mathbf{W}^i, \mathbf{x}) = \frac{1}{2} E\{[r(\mathbf{x}) - \mathbf{W}^i \mathbf{\Phi}(\mathbf{x})]^2\}$$

where E designates mathematical expectation, and

$$r(\mathbf{x}) = 1, \quad \text{if } \mathbf{x} \in X_1^i$$
$$= 0, \quad \text{otherwise}$$

The minimum of the function D can be evaluated by taking the partial derivative with respect to $\mathbf{W}^i$ and setting it equal to zero

$$\frac{\partial D(\mathbf{W}^i, \mathbf{x})}{\partial \mathbf{W}^i} = E\{-\mathbf{\Phi}(\mathbf{x})[r(\mathbf{x}) - \mathbf{W}^i \mathbf{\Phi}(\mathbf{x})]\} = 0 \tag{11.18}$$

To estimate the root of this equation we use the Robbins-Monro algorithm

$$\mathbf{W}^i(k+1) = \mathbf{W}^i(k) - \alpha_k \left[\frac{\partial D(\mathbf{W}^i, \mathbf{x})}{\partial \mathbf{W}^i} \right]_{\mathbf{W}^i = \mathbf{W}^i(k)} \tag{11.19}$$

Here, the weight $\mathbf{W}^i(1)$ is selected arbitrarily, and the coefficients α_k are defined by the expression $\alpha_k = 1/k$, $k := k + 1$. In view of (11.18, 19) we find

$$\mathbf{W}^i(k+1) = \mathbf{W}^i(k) + \alpha_k \mathbf{\Phi}(\mathbf{x}_k)[r(\mathbf{x}_k) - \mathbf{W}^i(k)\mathbf{\Phi}(\mathbf{x}_k)] \tag{11.20}$$

The iteration procedure given by (11.20) terminates when all the images of the teaching samples (11.10 and 11), belonging to the classes X_1^i and X_2^i, are classified correctly, i.e. obey the rule

$$\begin{aligned} \mathbf{W}^i \mathbf{\Phi}(\mathbf{x}) > 1/2 \quad &\text{for} \quad \mathbf{x} \in X_1^i \\ \mathbf{W}^i \mathbf{\Phi}(\mathbf{x}) < 1/2 \quad &\text{for} \quad \mathbf{x} \in X_2^i \end{aligned} \tag{11.21}$$

Once the vector of weights $\mathbf{W}^i$ has been found and the coefficients of the decision function $d_i(\mathbf{x})$ determined, the teaching process terminates. A non-linear separating frontier between the classes X_1^i and X_2^i is defined by the equation

$$\mathbf{W}^i \mathbf{\Phi}(\mathbf{x}) - 1/2 = 0$$

To classify the unknown vector of parameters X^{i+1} in forming a set of exploration points X^{i+1} at the $(i + 1)$th step, one is to use the decision rule (11.21) with the weight vector found as the result of teaching at step i.

11.3 Generation of a Teaching Sample

In procedures delineated in the preceding section, the number of elements in the teaching sample for class X_2 should be equal to the number of elements in the teaching sample for class X_1 (we have dropped the superscript for simplicity). In practice, however, the number of elements for X_1, i.e. M, is far less than the number L for X_2 due to the limited possiblility of variation of the parameters υ interactively performed by the DM on a computer. There is a need, therefore, for generating the teaching sample for X_2 so that sample size should be diminished down to $M(L = M)$ while retaining its informational significance. To solve this problem we resort to the sample theory for order statistics and construct some analogy of this method for this case.

As will be recalled from mathematical statistics, the expectation of the probability to occur in the tolerance interval equals, accurate to a factor, to the number of samples appearing in this interval. The proportion of a tolerance interval with respect to some measure corresponding to a sample $\mathbf{x}_1, ..., \mathbf{x}_L$ is the so-called probability mass in this interval [11.5]. With reference to our case, an analog of tolerance interval may be constructed (called T-region) each proportion of which will characterize an equal significance of the points appearing in it (called "significant mass" in the region). Suppose that $\{\mathbf{x}_l; l = 1, ..., L\}$ is a sample of a uniformly distributed sequence of points belonging to X_2 and given by the LP_τ-sequence generator. We define the "significance" of a point $\mathbf{x} \in X_2$ by the function

$$g_N(\mathbf{x}) = A_N \sum_{j=1}^{p} \lambda_j \mu_j(\mathbf{x})$$

where A_N is a normalizing factor such that $0 \leqslant g_N(x) \leqslant 1$.

Introduce the ordering nonlinear function

$$d(\mathbf{x}) = \mathbf{W\Phi}(\mathbf{x}) \tag{11.22}$$

where $\mathbf{W}$ is the weight row vector obtained for the class X_2 in the preceding $(i - 1)$th iteration, and $\mathbf{\Phi}(\mathbf{x})$ is the column vector of basis functions. Then, the quantities $b_\xi = d(\mathbf{x}_\xi)$, $\xi = 1, ..., L$, can be ordered as $b_{(1)} < b_{(2)} < ... < b_{(L)}$ and arranged into k-dimensional blocks $B_k^{(1)}, ..., B_k^{(L+1)}$ in which the region X_2 is divided by the hypersurfaces $d(\mathbf{x}) = b_\xi$, $\xi = 1, ..., L$. For $k = 2$ this case is illustrated in Fig. 11.3. The parts of the T-region will be defined by the relations

$$\hat{u}_{(\xi)} = \int_{B_k^{(\xi)}} dg_N(\mathbf{x}), \qquad \xi = 1, ..., L \tag{11.23}$$

To construct the blocks $B_k^{(1)}, ..., B_k^{(L+1)}$ corresponding to the parts $\hat{u}_{(\xi)}$ and possessing "equal significance" an appropriate procedure would be as

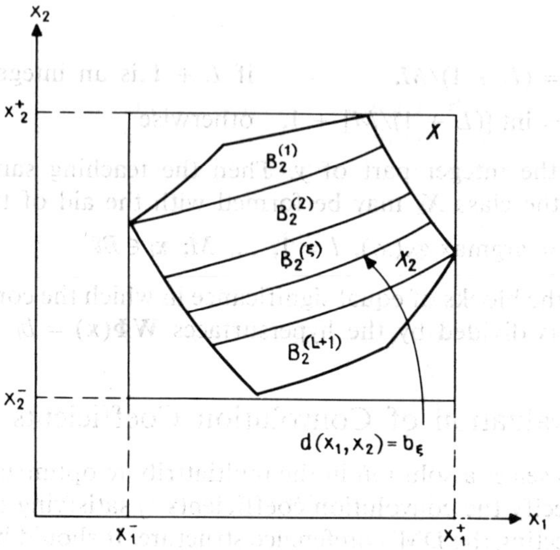

Fig. 11.3

follows. Select the maximum value of the L values $g_N(x_1)$, ..., $g_N(x_L)$ and divide the region X_2 into two subregions

$$B_k^{(L+1)} = \{\mathbf{x}\colon\ g_N(\mathbf{x}) > \max g_N(\mathbf{x}_i), \quad i = 1, \ldots, L\}$$

and

$$C_k^{(L+1)} = \{\mathbf{x}\colon\ g_N(\mathbf{x}) < \max g_N(\mathbf{x}_i), \quad i = 1, \ldots, L\}$$

Divide the subregion $C_k^{(L+1)}$ into

$$B_k^{(L)} = C_k^{(L+1)} \cap \{\mathbf{x}\colon\ g_N(\mathbf{x}) > \max g_N(\mathbf{x}_i), \quad \mathbf{x}_i \in C_k^{(L+1)}\}$$

and

$$C_k^{(L)} = C_k^{(L+1)} \cap \{\mathbf{x}\colon\ g_N(\mathbf{x}) \max g_N(\mathbf{x}_i), \quad \mathbf{x}_i \in C_k^{(L+1)}\}$$

The process continues until blocks $B_k^{(L+1)}$, ..., $B_k^{(1)}$ are formed. The separation of X_2 into a given number of blocks M possessing "equal significance" is effected by the union operations

$$\begin{aligned}
\tilde{B}_k^{(M)} &= \bigcup_{i=(L+1)-(p-1)}^{L+1} B_k^{(i)} \\
\tilde{B}_k^{(M-1)} &= \bigcup_{i=L-(p-1)}^{L} B_k^{(i)} \qquad (11.24) \\
&\ldots\ldots\ldots\ldots\ldots\ldots\ldots\ldots \\
B_k^{(1)} &= \bigcup_{i=1}^{S} B_k^{(i)}
\end{aligned}$$

where

$$S = (L + 1)/M, \qquad \text{if } L + 1 \text{ is an integer}$$
$$= \text{int}\,[(L + 1)/M] + 1, \quad \text{otherwise}$$

and int $[y]$ is the integer part of y. Then the teaching sample $\{\mathbf{x}_l, l = 1, ..., M\}$ for the class X_2 may be formed with the aid of the expression

$$\mathbf{x}_l = \text{argmax}\, g_N(x_i), \; l = 1, ..., M; \; \mathbf{x}_i \in B_k^{(l)} \tag{11.25}$$

where $B_k^{(l)}$ are the blocks of equal significance in which the confined bounded region X_2 is divided by the hypersurfaces $\mathbf{W\Phi(x)} = b_l$.

11.4 Evaluation of Convolution Coefficients

To select a solution in the multiattribute optimization problem we need to specify the convolution coefficients λ_j satisfying the conditions (11.4) and reflecting the DM's preference structure. It should be emphasized that in contrast to the traditional methods of multicriterial problem solving the approach developed in this chapter allows a rather simple way to do this because the problem statement (11.1) assumes that the preference functions $\mu_i(f_i | \mathbf{v}_i)$ are known, accurate to parameters $\mathbf{v}_i$. The suggested computational procedure of search for the coefficients λ_j relies on a theorem proving [11.11] that the efficient alternative $\mathbf{x}_*$ is the solution of the system of inequalities

$$\lambda_j u_j^0(\mathbf{x}_*) \leqslant C_0, \quad j \in J = \{1, ..., p\} \tag{11.26}$$

for the minimum parameter C_0^* which renders this system consistent. Here, u_j^0 is some monotonous transformation $u_j^0[f_j(\mathbf{x})]$ reducing $f_j(\mathbf{x})$ to a dimensionless form. We select u_j^0 in the form

$$u_j^0[f_j(\mathbf{x})] = \frac{f_j(\mathbf{x}) - f_j^m}{f_j^M - f_j^m}$$

where $f_j^M = \max |f_j|$, $f_j \in \Omega$, and $f_j^m = \min |f_j|$, $f_j \in \Omega$, and Ω is the set of exploration points on the space of criteria at step i of the optimization procedure for the j-th performance criterion.

In veiw of the said theorem and the expressions of (11.4) we may write

$$u_k^*/u_j^* = \lambda_j/\lambda_k, \quad \sum_{k \in J} \lambda_k = 1; \quad k, j \in J \tag{11.27}$$

where $u_j^* = [f_j^*(\mathbf{v}_j) - f_j^m]/[(f_j^M - f_j^m)]$, and the values $f_j^*(\mathbf{v}_j)$, for example for the preference functions μ_1 (see Fig. 11.1), may be defined as follows:

$$f_1^*(\mathbf{v}_1) = \sup \text{argmax}\, \mu_1(f_1 | \mathbf{v}_1) = a_1$$

and

$$f_1^*(\mathbf{v}_1') = \sup \operatorname{argmax} \mu_1(f_1 \mid \mathbf{v}_1') = a_1'$$

subject to $f_1 \in F_1$, where F_1 is the domain of μ_1.

For the preference functions μ_2 (see Fig. 11.1), the respective values $f_j^*(\mathbf{v}_j)$, $j = 2$, are defined by the expressions

$$f_2^*(\mathbf{v}_2) = \operatorname{argmax} \mu_2(f_2 \mid \mathbf{v}_2) = a_3$$

and

$$f_2^*(\mathbf{v}_2') = \operatorname{argmax} \mu_2(f_2 \mid \mathbf{v}_2') = a_3'$$

subject to $f_2 \in F_2$.

Resolving the system (11.27) for λ_j, we obtain

$$\lambda_j(\mathbf{v}_i) = \prod_{k \in J; k \neq j} u_k^* \Big/ \prod_{\substack{q \in J; k \in J \\ k \neq q}} u_k^* \tag{11.28}$$

The vector of weights $\boldsymbol{\lambda} = (\lambda_1, \ldots, \lambda_p)$, satisfying (11.28) for $j = 1, \ldots, p$, characterizes the relative importance of the objective functions at the i-th step of the optimization procedure.

11.5 Sensitivity Analysis

In this section we shall follow the conceptual framework devised by Kolmogorov and Fomin [11.12] and Tanino and Sawaragi [11.13]. Let, as before, $X \subset \mathbf{E}^k$ be the space of parameters, and the vector function μ: $X \to Y \subset \mathbf{E}^p$ indicate the worth of each solution $\mathbf{x} \in X$ in the p-criterial space. Consider in $\mathbf{E}^p$ a λ-separable relationship R induced by a cone $\mathbf{K}$,

$$\mathbf{y}_\alpha R \mathbf{y}_\beta \Leftrightarrow \mathbf{y}_\alpha - \mathbf{y}_\beta \in \mathbf{K}, \qquad \forall \mathbf{y}_\alpha,\ \mathbf{y}_\beta \in \mathbf{E}^p$$

The set $\{\mathbf{K}[\mathbf{y}] \mid \mathbf{y} \in \mathbf{Y}\}$ is referred to as a *preference structure* henceforth denoted by $\mathbf{K}$. Figure 11.4 shows the upper and lower sections of R at a

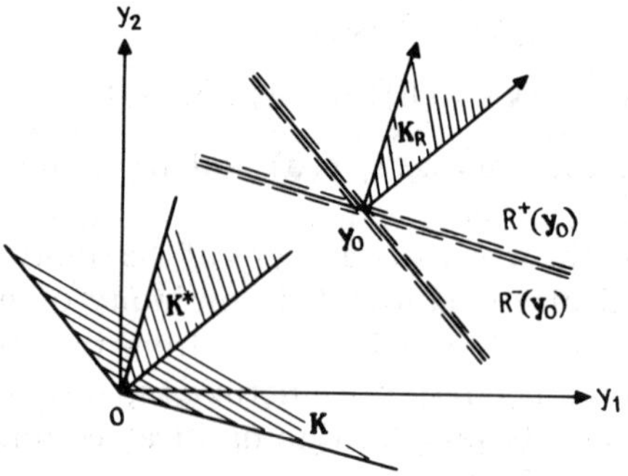

Fig. 11.4

point $\mathbf{y}_0 \in \mathbf{E}^2$. The set of vectors $\lambda \in \mathbf{E}^p$ for which the relationship R is λ-separable represents a cone denoted by $\mathbf{K}_R$. The figure also shows a cone $\mathbf{K}^*$ dual to the cone $\mathbf{K}$.

In the outlined method of multicriterial optimization, the set X and the dominance structure $\mathbf{K}$ are not constant, rather they vary depending on some parameters. By construction the closed bounded set X is independent of the number of exploration points N and is defined by the boundary $d(\mathbf{W}, \mathbf{x})$, while the dominance structure $\mathbf{K}$ is defined by a selection of $\lambda(\mathbf{v}) \in \mathbf{E}^p$ depending on $\mathbf{v}$. Therefore, the sensitivity analysis of the method should be carried out allowing for the dependence of the solution on two types of parameters. One of the parameters $\mathbf{u}(N, \mathbf{W}) \in U$ defines the set of feasible solutions and the objective function, the other $\mathbf{v} \in V$ defines the preference structure in the space of objectives. Both U and V are subsets of linear topological spaces where convergence is defined in terms of sequences. Then the set of feasible solutions in objective space may be defined as

$$Y(\mathbf{u}) = \{\mathbf{y} \in \mathbf{E}^p \mid \mathbf{y} = \mu(\mathbf{x} \mid \mathbf{u}),\ \mathbf{x} \in X(\mathbf{u})\}$$

The optimization problem will be solved on constructing the set $S(Y(\mathbf{u}), \mathrm{K}(\mathbf{v}))$ and the set $\mathrm{P}(\mathbf{u}, \mathbf{v}) = \{\mathbf{x} \in X(\mathbf{u}) \mid \mu(\mathbf{x} \mid \mathbf{u}) \in S(Y(\mathbf{u}), \mathrm{K}(\mathbf{v}))\}$.

Consider two cases of the stability of solutions to variation of the parameters.

Case 1. Suppose that the parameter $\mathbf{v} = \hat{\mathbf{v}}$ is fixed, then the multivalued mapping $S_{\hat{\mathbf{v}}}$ from U into $\mathbf{E}^p$ is defined as

$$S_{\hat{\mathbf{v}}}(\mathbf{u}) = S(Y(\mathbf{u}), \mathrm{K}(\hat{\mathbf{v}}))$$

The multivalued mapping $P_{\hat{\mathbf{v}}}$ from U into $\mathbf{E}^k$ is defined as

$$P_{\hat{\mathbf{v}}}(\mathbf{u}) = \{\mathbf{x} \in X(\mathbf{u}) \mid \mu(\mathbf{x} \mid \mathbf{u}) \in S_{\hat{\mathbf{v}}}(\mathbf{u})\}$$

Case 2. Parameter $\mathbf{u} = \hat{\mathbf{u}}$ is fixed. The respective multivalued mappings are defined in a similar manner

$$S_{\hat{\mathbf{u}}}(\mathbf{v}) = S(Y(\hat{\mathbf{u}}), \mathrm{K}(\mathbf{v}))$$

$$P_{\hat{\mathbf{u}}}(\mathbf{v}) = \{\mathbf{x} \in X(\hat{\mathbf{u}}) \mid \mu(\mathbf{x} \mid \hat{\mathbf{u}}) \in S_{\hat{\mathbf{u}}}(\mathbf{v})\}$$

We shall denote the fixed sets $Y(\hat{\mathbf{u}})$ and $\mathrm{K}(\hat{\mathbf{v}})$ simply by Y and K, and $\mu(\mathbf{x} \mid \hat{\mathbf{u}})$ by $\mu(\mathbf{x})$.

In this section the sensitivity analysis of the solution to the multicriterial problem is based on the concept of continuity of multivalued mapping. Sufficient conditions of stability follow from the conditions of semicontinuity from below. The respective definitions of continuity and semicontinuity from below (above) related to multivalued mappings may be found for instance in Refs. [11.12, 13]. The proof of the method stability will be preceded by the following lemma.

Lemma. Let Y be a nonempty compact set in $\mathbf{E}^p$, and K be a preference structure well defined on Y. If $K[\mathbf{y}]$ is a multivalued mapping from Y on $\mathbf{E}^p$ semicontinuous from above, then for any $\mathbf{y}_0 \in Y$ there exists $\bar{\mathbf{y}} \in S(Y, K)$ such that $\mathbf{y}_0 \in \bar{\mathbf{y}} + K[\mathbf{y}]$. The lemma is proved in Ref. [11.13]. The following two theorems formulate sufficient conditions of the method's stability.

Theorem 11.1. A mapping $P_{\hat{\mathbf{v}}}$ is semicontinuous from below at point $\hat{\mathbf{u}} \in U$ if

(1) the mapping of X is continuous in $\hat{\mathbf{u}}$ and uniformly compact in the neighborhood of $\hat{\mathbf{u}}$,

(2) $X(\mathbf{u})$ is a nonempty compact set for any $\mathbf{u} \in \mathbf{U}$ in the vicinity of $\hat{\mathbf{u}}$,

(3) $\mu(\mathbf{x}|\mathbf{u})$ is continuous and single valued on $UX(\mathbf{u})T$, where T is a compact neighborhood of $\hat{\mathbf{u}}$ in U,

(4) the mapping of K is semicontinuous from above on $Y(\mathbf{u})$ for any $\mathbf{u} \in U$ in the neighborhood of $\hat{\mathbf{u}}$ and semicontinuous from above on $Y(\hat{\mathbf{u}})$,

(5) K is a well defined structure on $Y(\mathbf{u})$ for all $\mathbf{u} \in U$ in the vicinity of $\hat{\mathbf{u}}$.

Proof. Let $\{\mathbf{u}_i\} \subset U$, $\mathbf{u}_j \to \hat{\mathbf{u}}$, and $\mathbf{x} \in P_{\hat{\mathbf{v}}}(\hat{\mathbf{u}})$. It is required to demonstrate that there exists a sequence $\{\bar{\mathbf{x}}_j\}$ with $\bar{\mathbf{x}}_j \in P_{\hat{\mathbf{v}}}(\mathbf{u}_j)$ converging to $\hat{\mathbf{x}}$ and this point is unique. From the continuity of X in $\hat{\mathbf{u}}$ there follows the semicontinuity from below for X in $\hat{\mathbf{u}}$. This implies that there is a number m' and a sequence $\{\mathbf{x}_j\}$ such that $\mathbf{x}_j \in X(\mathbf{u})$ for $j \geqslant m'$ and $\mathbf{x}_j \to \hat{\mathbf{x}}$. Suppose $\mathbf{y}_j = \mu(\mathbf{x}_j|\mathbf{u}_j)$, then for $j \geqslant m'$, $\mathbf{y}_j \in Y(\mathbf{u})$ and $\mu(\mathbf{x}_j|\mathbf{u}_j) \to \mu(\hat{\mathbf{x}}|\hat{\mathbf{u}}) = \hat{\mathbf{y}}$ in view of the continuity of $\mu(\mathbf{x}|\mathbf{u})$.

From the condition (2) and semicontinuity from above for K on $Y(\mathbf{u})$ it follows that there exists a number m'' such that $Y(\mathbf{u}_j)$ is a nonempty compact set and the constriction of the map of K on $Y(\mathbf{u}_j)$ is semicontinuous from above on $Y(\mathbf{u}_j)$ for all $j \geqslant m''$. Choose $m = \max(m', m'')$. In accordance with the lemma there exists $\bar{\mathbf{y}}_j \in S_{\hat{\mathbf{v}}}(\mathbf{u}_j)$ for which $\mathbf{y}_j \in \bar{\mathbf{y}}_j + K[\bar{\mathbf{y}}_j]$, $j \geqslant m$ or

$$\mu(\mathbf{x}_j|\mathbf{u}_j) \in \mu(\bar{\mathbf{x}}_j|\mathbf{u}_j) + K[\mu(\bar{\mathbf{x}}_j|\mathbf{u}_j)], \quad j \geqslant m \tag{11.29}$$

From the continuity of $\mu(\mathbf{x}, \mathbf{u})$ and (11.29) it follows that there is $\bar{\mathbf{x}}_j$ such that

$$\bar{\mathbf{x}}_j \in P_{\hat{\mathbf{v}}}(\mathbf{u}_j) \tag{11.30}$$

Let $\bar{\mathbf{x}}_j$ be an arbitrary vector in $\mathbf{E}^k$ for $1 \leqslant j < m$. Since X is uniformly compact in the vicinity of $\hat{\mathbf{u}}$, the sequence $\{\bar{\mathbf{x}}_j\}$ converges to a limit point denoted by $\bar{\mathbf{x}}$. This point is contained in $X(\hat{\mathbf{u}})$ because X is semicontinuous from above in $\hat{\mathbf{u}}$. On the other hand, from the continuity of $\mu(\mathbf{x}|\mathbf{u})$ it follows that the sequence $\bar{\mathbf{y}}_j = \mu(\mathbf{x}_j|\mathbf{u}_j)$ also converges to a limit point. We denote this point by $\bar{\mathbf{y}} = \mu(\bar{\mathbf{x}}|\hat{\mathbf{u}})$. Then the sequence $\{\mu(\mathbf{x}_j|\mathbf{u}_j) - \mu(\mathbf{x}_j|\mathbf{u}_j))\}$ converges to the limiting point $\hat{\mathbf{y}} - \bar{\mathbf{y}}$. From the semicontinuity from above of K on $Y(\hat{\mathbf{u}})$ it follows that $\hat{\mathbf{y}} - \bar{\mathbf{y}} \in \mathrm{K}[\bar{\mathbf{y}}]$ or

$$\mu(\hat{\mathbf{x}} \mid \hat{\mathbf{u}}) \in \mu(\chi \mid \hat{\mathbf{u}}) + \mathrm{K}[\mu(\chi \mid \hat{\mathbf{u}})] \tag{11.31}$$

According to the definition of $P_{\hat{\mathbf{v}}}(\mathbf{u})$,

$$\mu(\hat{\mathbf{x}} \mid \hat{\mathbf{u}}) \in S_{\hat{\mathbf{v}}}(\hat{\mathbf{u}}) \tag{11.32}$$

From (11.31, 32) it follows that

$$\mu(\hat{\mathbf{x}} \mid \hat{\mathbf{u}}) = \mu(\bar{\mathbf{x}} \mid \hat{\mathbf{u}}) \tag{11.33}$$

for otherwise $\mu(\bar{\mathbf{x}} \mid \hat{\mathbf{u}})$ would not belong to the set of nondominated alternatives. From (11.33) and single valuedness of μ on X it follows that $\hat{\mathbf{x}} = \bar{\mathbf{x}}$, i.e. that the point $\hat{\mathbf{x}}$ is unique.

Thus, for any sequence $\{\bar{\mathbf{u}}_j\} \subset U$, $\mathbf{u}_j \to \hat{\mathbf{u}}$, $\hat{\mathbf{x}} \in P_{\hat{\mathbf{v}}}(\hat{\mathbf{u}})$ we are able to point out m such that for all $j \geqslant m$ the sequence $\{\bar{\mathbf{x}}_j\}$ with $\bar{\mathbf{x}}_j \in P_{\hat{\mathbf{v}}}(\mathbf{u}_j)$ converges to $\hat{\mathbf{x}}$ and, consequently, $P_{\hat{\mathbf{v}}}$ is semicontinuous from below in $\hat{\mathbf{u}}$.

Theorem 11.2. The map $P_{\hat{\mathbf{u}}}$ is semicontinuous from below in $\hat{\mathbf{v}} \in V$ if

(1) X is a compact set,

(2) $\mu(\mathbf{x})$ is continuous and single valued on X,

(3) the map $\mathrm{K}(\mathbf{v})[y]$ is semicontinuous from above on Y for any $\mathbf{v} \in V$ in the neighborhood of $\hat{\mathbf{v}}$ and K is semicontinuous from above on $\mathbf{v} \times Y$,

(4) $\mathrm{K}(\mathbf{v})$ is a well defined structure on Y for any $\mathbf{v} \in V$ in the vicinity of $\hat{\mathbf{v}}$.

Proof. Let $\{\mathbf{v}_j\} \subset V$, $\mathbf{v}_j \to \hat{\mathbf{v}}$, $\hat{\mathbf{x}} \in P_{\hat{\mathbf{u}}}(\hat{\mathbf{v}})$. By the conditions (3) and (4) the map $\mathrm{K}(\mathbf{v}_j)\,[\cdot]$ is semicontinuous from above on Y and $\mathrm{K}(\mathbf{v}_j)$ has a well defined structure on Y for all $j \geqslant m$, where m is some number. By the lemma proved earlier a point $\mathbf{y}_j \in S_{\hat{\mathbf{u}}}(\mathbf{v}_j)$ exists such that

$$\hat{\mathbf{y}} \in \mathbf{y}_j + \mathrm{K}(\mathbf{v}_j)[\mathbf{y}_j], \quad j \geqslant m \tag{11.34}$$

$\mathbf{y}_j$ is some point from Y for $1 \leqslant j \leqslant m$. From functional analysis it is known that the continuous mapping of a metric compact X into a metric space is uniformly continuous and by Arcell's theorem the set $Y = \mu(X)$ is also compact. Therefore, the sequence $\{\mathbf{y}_j\}$ converges to a limit point in Y, denoted by $\bar{\mathbf{y}}$. Then the sequence $\{\hat{\mathbf{y}} - \mathbf{y}_j\}$ converges to the limit point $\hat{\mathbf{y}} - \bar{\mathbf{y}}$. Since $\hat{\mathbf{y}} - \mathbf{y}_j \in \mathrm{K}(\mathbf{v}_j)[\mathbf{y}_j]$, $j \geqslant m$, which follows from (11.34), and K is semicontinuous from above in $\hat{\mathbf{v}} \times Y$, then

$$\hat{\mathbf{y}} - \bar{\mathbf{y}} \in \mathrm{K}(\hat{\mathbf{v}})[\bar{\mathbf{y}}] \tag{11.35}$$

From the definition of $P_{\hat{\mathbf{u}}}(\hat{\mathbf{v}})$ we have

$$\hat{\mathbf{y}} = \mu(\hat{\mathbf{x}}) \in S_{\hat{\mathbf{u}}}(\hat{\mathbf{v}}) \tag{11.36}$$

From (11.35, 36) it follows that

$$\bar{\mathbf{y}} = \hat{\mathbf{y}} \tag{11.37}$$

that is, $\hat{\mathbf{y}}$ is the unique limit point to which the sequence $\{\mathbf{y}_j\}$ converges.

Thus, the sequence $\{\mathbf{y}_j\}$ satisfies the conditions of $\mathbf{y}_j \in S_{\hat{\mathbf{u}}}(\mathbf{v}_j)$ for $j \geqslant m$ and $\mathbf{y}_j \to \hat{\mathbf{y}}$, that is, $S_{\hat{\mathbf{u}}}(\hat{\mathbf{v}})$ is semicontinuous from below in $\hat{\mathbf{v}}$. Since $S_{\hat{\mathbf{u}}}$ is semicontinuous from below in $\hat{\mathbf{v}}$ and $S_{\hat{\mathbf{u}}}(\mathbf{v}) \subset Y(\mathbf{v})$, there exists a number m' and a sequence $\{\mathbf{x}_j\}$ such that for $j \geqslant m'$ and $\mu(\mathbf{x}_j) \to \mu(\hat{\mathbf{x}})$, $\mathbf{x}_j \in P_{\hat{\mathbf{u}}}(\mathbf{v}_j)$. From the compactness of the set X it follows that the sequence $\{\mathbf{x}_j\}$ has a point of convergence denoted by $\bar{\mathbf{x}}$. From (11.37) we have $\mu(\bar{\mathbf{x}}) = \mu(\hat{\mathbf{x}})$, and from single valuedness of μ on X it follows that $\bar{\mathbf{x}} = \hat{\mathbf{x}}$.

Thus, for any sequence $\{\mathbf{v}_j\} \subset V$, $\mathbf{v}_j \to \hat{\mathbf{v}}$, $\hat{\mathbf{x}} \in P_{\hat{\mathbf{u}}}(\hat{\mathbf{v}})$ there exists m' such that $\{\mathbf{x}_j\} \subset X$, where $\mathbf{x} \in P_{\hat{\mathbf{u}}}(\mathbf{v}_j)$ for $j \geqslant m'$ and $\{\mathbf{x}_j\} \to \bar{\mathbf{x}}$ $(j \to \infty)$, that is, $P_{\hat{\mathbf{u}}}$ is semicontinuous from below in $\hat{\mathbf{v}}$.

11.6 Interactive Computational Procedure

The algorithm of sequential decision making in nonlinear system optimization is based on the construction of a region πX of optimal alternatives by screening the elements $\mathbf{x} \in X$, generated by the LP_τ-sequence, through the filtering boundaries $d_i(\mathbf{x} \mid \mathbf{W}^i)$ obtained at each i-th step of the interactive procedure. The boundaries $d_i(\mathbf{x} \mid \mathbf{W}^i)$ are built with the aid of teaching procedures outlined in Sects. 11.2 and 11.3 by some additional information received from the decision maker. The flow diagram of the algorithm is shown in Fig. 11.5. Below we consider the operational framework of this algorithm.

The construction begins by the preparation and loading of the initial data including parameter values $\mathbf{v}_j$, preference functions $\mu_j(f_j \mid \mathbf{v}_j)$ and possible variations $\Delta\mathbf{v}_j$, a given number of iterations n, dimensionality of the space of parameters N, the boundaries (x_+^j, x_-^j) of the parallelepiped of parameters X, and so on.

Step 1. Explore the parameter region X^i.

At the first step, $i = 1$, X^1coincides with X. The LP_τ-sequence generator produces a given number of points K in the N-dimensional space with the coordinates $q_k^{ij} \in [0, 1]$. The transfer to the real coordinates is effected by the formula

$$x_k^{ij} = x_+^{ij} + q_k^{ij}(x_+^{ij} - x_-^{ij}); \quad k = 1, \ldots, K, \; j = 1, \ldots, N$$

Step 2. Compute the convolution coefficients $\lambda_i(\mathbf{v}_i)$. The variation of the parameters $\mathbf{v}^i = \{\mathbf{v}_j^i\}$ is achieved by the scheme

$$v_j^i := v_j^i - \Delta v_j^i + 2(j - 1)\Delta v_j^i/(m_1 - 1)$$

where $j = 1, \ldots, m_1$, and m_1 is the number of points in the sample specified by the DM.

Step 3. Compute the functions $g(f_k^i \mid \mathbf{v}^i)$ at the exploration points $\mathbf{x}_k^i$, $k = 1, \ldots, K$.

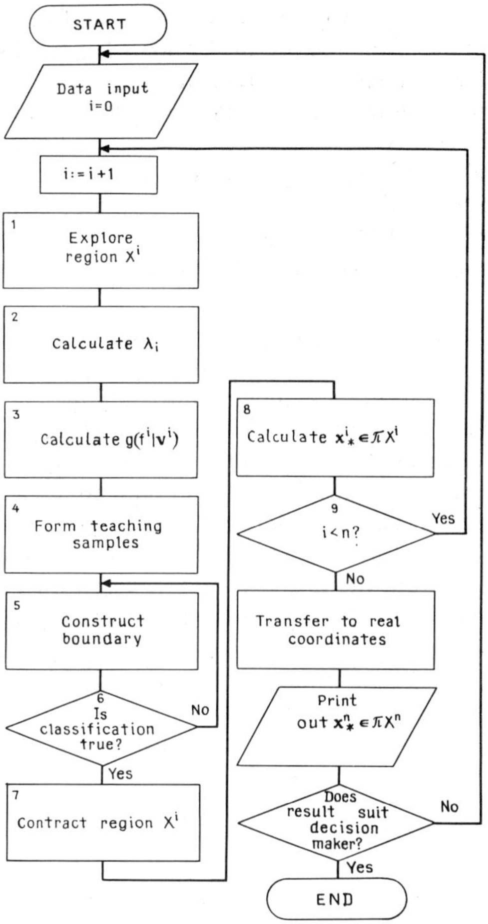

Fig. 11.5

Step 4. Form in the dialog mode the teaching samples

$$(\mathbf{x}_1^i, \ldots, \mathbf{x}_L^i) \subset X_1^i \quad \text{and} \quad (\mathbf{x}_1^i, \ldots, \mathbf{x}_L^i) \subset X_2^i$$

Step 5. Construct the separating frontier $d_i(\mathbf{x} \mid \mathbf{W}^i)$ defined by the equation $\mathbf{W}^i\mathbf{\Phi}(\mathbf{x}) - 1/2 = 0$, and compute vector $\mathbf{W}^i$ on the base of the teaching samples.

Step 6. Check whether the classification is correct. The frontier is deemed to be constructed correctly if

$$d_i(\mathbf{x} \mid \mathbf{W}^i) > 0 \quad \text{for} \quad \mathbf{x} \in X_1^i$$

and

$$d_i(\mathbf{x}\,|\,\mathbf{W}^i) < 0 \quad \text{for} \quad \mathbf{x} \in X_2^i$$

If this decision rule is not met, return to Step 5.

Step 7. Contract the region X^i and construct the bounds of the new region

$$\tilde{X}^i = \{\mathbf{x}\,|\,\mathbf{x}_{\min}^i - 2\mathbf{z}^i \leqslant \mathbf{x} \leqslant \mathbf{x}_{\max}^i + 2\mathbf{z}^i;$$
$$d_m(\mathbf{x}\,|\,\mathbf{W}^m) > 0; \quad m = 1, \ldots, i\}$$

where the coordinates of the vector $\mathbf{z}^i$ are evaluated as

$$z^{ij} = (x_{\max}^{ij} - x_{\min}^{ij})/(l - 1)$$

l is the number of points appearing in X^i, j indicates the j-th coordinate of the vector, $x_{\max}^{ij} = \max\{x_k^{ij}\}$, $x_{\min}^{ij} = \min\{x_k^{ij}\}$, $k = 1, \ldots, K$, and $\{x_k^{ij}\}$ is the set of exploration points appearing in X_1^i. The bounds of X^i should coincide with the bounds of the region X^{i+1} at step $i + 1$.

Step 8. Compute $\mathbf{x}_*^i = \operatorname{argmax}\{z_m(\mathbf{x}_k^i) : \mathbf{x}_k^i \in \tilde{X}^i\}$.

Step 9. Check whether the condition $i < n$ is met. If so, then move to Step 1. If this conditon is not satisfied, then the transformation from the dimensionless coordinates to real coordinates is executed and the finite set of values $\mathbf{x}_*^n[\lambda(\mathbf{v}^n)] \in \pi X^n$ is presented to the DM for subsequent analysis. If this result does not suit the DM, the procedure is restarted after altering the initial data in the preference functions. The algorithm terminates if the result satisfies the decision maker.

Let us analyze the operation of this algorithm with an illustrative example. The problem at hand will be to maximize

$$z = -(x - 3)^2 - (y - 2)^2 + 3$$

subject to $X = \{(x, y)\,|\,0 \leqslant x \leqslant 9,\ 0 \leqslant y \leqslant 9\}$.

The solution of this problem is obviously $z_0 = 3$, $x_0 = 3$, and $y_0 = 2$. Upon the normalization for $0 \leqslant x \leqslant 1$ we have

$$x_0 \simeq 0.333 \quad \text{and} \quad y_0 \simeq 0.222$$

Suppose that the selected preference function is μ_2 (see Fig. 11.1) with the parameters $a_3 = 4$, $v_2 = 7$, $\Delta v_2 = 1$, and the number of iterations $n = 10$. The optimization process will be checked at step $i = 4$.

Figure 11.6 shows the results of algorithm operation with the contraction of the parameter parallelepiped. The separation frontiers of unfeasible regions at various steps i of the optimization procedure are shown by solid lines. The contracted regions of the parameter space at steps $i = 4$ and $i = n = 10$ are indicated by hatching. The circles represent the exact solu-

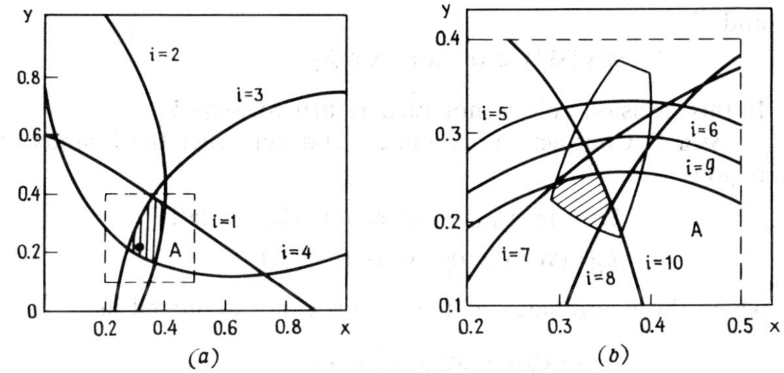

Fig. 11.6

tion, the asterisk represents the result of the algorithm at $n = 10$: $x_* = 0.328$ and $y_* = 0.217$.

The accuracy of the solution depends on the number of exploration points K. Ovbiously, the accuracy increases with K, although the computing time increases proportionally. It has been established [11.3] that the sample size should not be smaller than $2(2N + 1)$, where N is the dimensionality of the parameter space. For our illustrative example, $N = 2$ and $K = 2(2N + 1) = 10$. In the general case, the DM specifies K to suit the problem at hand.

Consider now the application of the delineated procedure to solving the problem of optimal design of large systems (LS) under uncertainty and incomplete information. This problem is discussed in depth in Ref. [11.14, 15]. The structural model of a particular large system is a hexagonal truss with a tetrahedral base. The optimization model requires to minimize the relative mass of the structure subject a number of constraints on the system geometry due to the load capacity and length of the cargo bay of the transporter craft carrying the structural elements of the system to the site of assembly, on the resonance phenomena in the platform and its elements, and on the strength and stability of the system elements.

We formulate the evaluation of system's design parameters as a constrained optimization problem with $p = 8$ objective functions summarized in Table 11.1. Five of these objective functions $(f_1, f_3, f_5, f_7, f_8)$ are nonlinear with respect to the parameters, and the remaining three (f_2, f_4, f_6) are linear. The sought vector of parameters has the form

$$x = \{h, l_f, \varphi, b_1, b_2)$$

where h is the height of the base tetrahedron, l_f is the length of outer rods, φ is a coefficient limiting the level of stress in the rods losing stability in

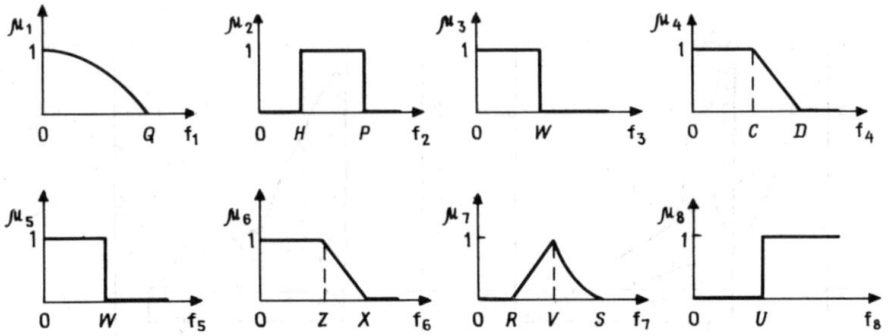

Fig. 11.7

the Euler scheme, and b_1 and b_2 are the diameter of the cargo bay and the load capacity of the carrier.

The coordinates of the vectors of maximum and minimum parameters were fixed and remain unvariable during the optimization:

$$x_- = (3,\ 5,\ 0,\ 0,\ 1.0 \times 10^4)$$

$$x_+ = (10,\ 18,\ 1,\ 7,\ 4.0 \times 10^4)$$

The form of preference functions is shown in Fig. 11.7, and the respective parameter values in Table 11.1. The optimization procedure of sequen-

Table 11.1 Objective Functions and Preference Function Parameters

No	Objective function	Designation	Parameters
1	Relative mass criterion	$f_1 = \mathcal{M}(h, \varphi, l_f)$	$Q = 0.6$
2	Constraint on cargo bay length	$f_2 = l_f$	$H = 5,\ P = 12$
3	Constraint on package area in transport position	$f_3 = \psi_1(b_1, l_f)$	$W = 0$
4	Constraint on bay diameter	$f_4 = b_1$	$C = 4.2,\ D = 6.6$
5	Constraint on load capacity	$f_5 = \psi_2(l_f, b_2)$	$W = 0$
6	Constraint on load capacity parameter	$f_6 = b_2$	$Z = 2.5 \times 10^4$, $X = 3.7 \times 10^4$
7	Constraint on system natural frequency	$f_7 = \omega_1(h, \varphi, l_f)$	$R = 0.01,\ V = 0.1$, $S = 0.2$
8	Constraint on relation of natural frequencies of system and individual rod	$f_8 = \omega_s(h, \varphi, l_f)$	$U = 0$

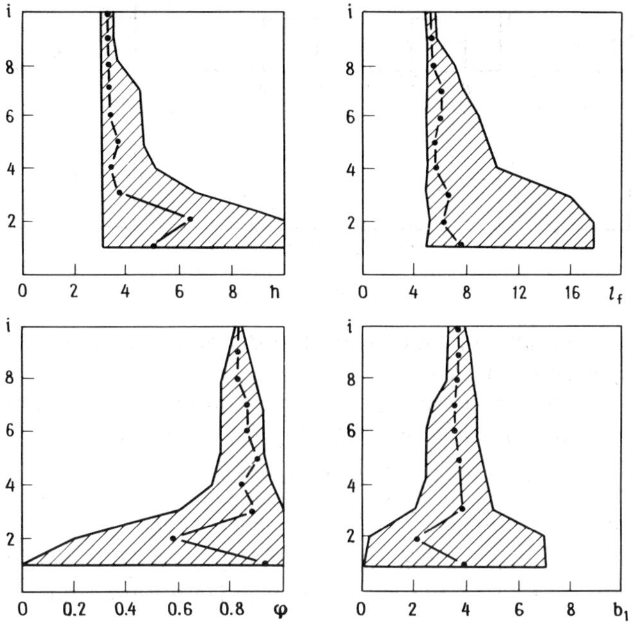

Fig. 11.8

tial decision making produced the following vector

$$x_0 = (3.31,\ 5.01,\ 0.84,\ 3.99,\ 1.7 \times 10^4)$$

Figure 11.8 shows how the parameters vary at various steps of the optimization procedure. The hatched areas correspond to the boundaries of contraction of the parameter parallelepiped. The points connected by straight lines reflect variation of the trajectory $\{\mathbf{x}_k^i\}$, $\mathbf{x}_k^i \in \pi X^{\mathbf{i}}$. The optimization was terminated at $i = 10$ because beginning from this step the parameters varied only insignificantly and the result suited the DM.

The algorithm was implemented as a software written in PL1. For the above example the optimization run on a ES-1033 minicomputer took about 25 minutes.

To close this topic we note that the method (LLP) of sequential decision making suggested for multicriterial problems may be used in any field. Like for the Zadeh fuzzy set technique, the presence of a large number of linear and nonlinear functional constraints is immaterial for this method as they can be readily taken into account in generating the vector of objective functions. The method is noncritical to the form of objective functions — analytical expressions are not mandatory. The numerical information on the objective functions derived, for example, from the available computational programs suffices for the method. The constraints of the method are dictated by the sufficient condition of stability formulated in Theorems 11.1 and 11.2.

References

11.1 Yu.A. Dubov, S.I. Travkin, and V.N. Yakimets: *Mnogokriterialnye modeli formirovaniya i vybora variantov system* (Multicriterial models of alternative generation and selection). Nauka, Moscow 1986.

11.2 V.A. Gorbatov: *Teoriya chastichnogo uporyadocheniya sistem.* (Theory of system partial ordering). Sovetskoe Radio, Moscow 1976.

11.3 B.V. Sazykin and Yu.I. Topcheev: *Posledovatelnoe prinyatiye reshenii v zadache optimizatsii s neskolkimi kriteriyami* (Multiattribute sequential decision making. The LLP method). Preprint MIFI, MIFI Press, Moscow 1985, pp. 36-85.

11.4 J. Tou and B. Gonzalez; *Pattern Recognition Principles.* Reading, Massachusetts 1974.

11.5 S.S. Wilks: *Mathematical Statistics.* John Wiley & Sons, New York 1963.

11.6 L.E. Bell: *Manag. Sci.* **25**(8), 744-753 (1979).

11.7 I.M. Makarov, T.M. Vinogradskaya, A.A. Rubchinsky, and V.B. Sokolov: *Teoriya vybora i prinyatiya reshenii* (Decision making theory). Nauka, Moscow 1982.

11.8 L.B. Abraitis, R.I. Sheinauskas, and V.A. Zhilevichus: *Avtomatizatsiya proektirovaniya EVM* (Automation of computer design). Sovetskoe Radio, Moscow 1978.

11.9 I.M. Sobol and R.B. Statnikov: *Nailuchshie resheniya — gde ikh iskat* (Multicriterial optimization theory). Znaniye, Moscow 1982.

11.10 B.V. Sazykin: *Atomnaya Energiya* **55**(7), 46-48 (1983).

11.11 V.S. Mikhalevich and V.L. Volkovich: *Vychislitelnye metody issledovaniya i proyektirovaniya slozhnykh sistem* (Computational methods of analysis and design of large systems). Nauka, Moscow 1982.

11.12 A.N. Kolmogorov and S.V. Fomin: *Elementy teorii funktsii i funktsionalnogo analiza* (Topics of theory of functions and functional analysis). Nauka, Moscow 1981.

11.13 T. Tanino and Y. Sawaragi: *J. Opt. Theor.* Appl. **30**(2), 229-253 (1980).

11.14 B.V. Sazykin: *Kosmicheskie Issledovaniya* **23**(1), 81-91 (1985).

11.15 V.Yu. Malov, Yu.M. Pochtman, B.V. Sazykin, and M.M. Yakushin: *Optimalnoye proyektirovaniye krupnogabaritnykh konstruktsii* (Multicritrial optimization of large elastic structures under uncertainty). Preprint MIFI, MIFI Press, Moscow 1986, pp. 38-85.

Index